Water, Food and Poverty in River Basins

Conventional wisdom says that the world is heading for a major water crisis. By 2050, the global population will increase from 7 billion to a staggering 9.5 billion and the demands this will place on food and water systems will inevitably push river basins over the edge.

The findings from this book present a different picture. While it is convenient to visualize an inevitable global water and food crisis in which increasing demands result in increasing poverty, food insecurity and conflict, the reality is far more nuanced and revolves around the politics of equitable and sustainable development of resources.

The first part of this book provides detailed insight into conditions of water flows within nine river basins. In the second part, authors summarize and re-analyze the outcome of the nine basins, providing a coherent global picture of water, water productivity and development. They assess the impacts of variations of these attributes on development and approaches for poverty alleviation, and explore the institutional factors that support or obstruct change.

How people will manage river systems while protecting vital ecosystem functions will make the difference between catastrophe and survival. As Professor Asit Biswas points out, "... the world is facing a water crisis not because of physical scarcity of water but because of poor management practices in nearly all countries of the world."

The book is based on the four years (2006–2010) of extensive research into the state of ten of the world's major river basins carried out under the CGIAR Challenge Program for Water and Food's Basin Focal Project.

This book was published as two special issues of *Water International*.

Myles Fisher is an Emeritus Scientist at the Centro Internacional de Agricultura Tropical CIAT in Cali, Colombia. He worked in the agronomy and physiology of tropical pastures in the Northern Territory of Australia and southeast Queensland with CSIRO and in Colombia at CIAT. He was Lead Scientist for the CGIAR InterCenter Working Group on Climate Change. He consults with CIAT on simulation modeling, soil carbon dynamics and climate change, and to the UNDP on the preservation of Tajikistan's (Central Asia) agrobiodiversity in the face of climate change.

Simon Cook is Director of the CGIAR's Research Program 5, Water, Land and Ecosystems, based in Colombo, Sri Lanka. He coordinated the Basin Focal Projects of the Challenge Program for Water and Food on which this work is based. Associated with the Centre for Water Research at the University of Western Australia, Perth, he led the working group on Global Drivers for the Challenge Program. Prior to that, he was theme leader within the Challenge Program, and project leader with CIAT.

Water, Food and Poverty in River Basins

Defining the Limits

Edited by

Myles Fisher and Simon Cook

LONDON AND NEW YORK

First published 2012
by Routledge
2 Park Square, Milton Park, Abingdon, Oxfordshire OX14 4RN

Simultaneously published in the USA and Canada
by Routledge
711 Third Avenue, New York, NY 10017

Routledge is an imprint of the Taylor & Francis Group, an informa business

First issued in paperback 2012

This book is based on a reproduction of two special issues of *Water International*, vol. 35, issue 5 and vol. 36, issue 1. The Publisher requests to those authors who may be citing this book to state, also, the bibliographical details of the special issue on which the book was based.

British Library Cataloguing in Publication Data
A catalogue record for this book is available from the British Library

ISBN13: 978-0-415-59207-9 (hbk)
ISBN13: 978-0-415-53891-6 (pbk)

Typeset in Times New Roman
by Taylor & Francis Books

Disclaimer
The publisher would like to make readers aware that the chapters in this book are referred to as articles as they had been in the special issue. The publisher accepts responsibility for any inconsistencies that may have arisen in the course of preparing this volume for print.

Contents

Acknowledgements

Phase I of the Challenge Program on Water and Food of the Consultative Group for International Agricultural Research was funded by the UK Department for International Development, the World Bank, the Netherlands, the European Commission, Switzerland, France, Germany, Denmark, Norway, Sweden, the International Fund for Agricultural Development, and the United States Agency for International Development. We gratefully acknowledge that support. Laure Collet drew the relief maps and Jenny Correa Gutiérrez provided secretarial assistance to MJF.

Preface

Harold Macmillan, the eminent former British Prime Minister in his final speech to the House of Lords said: "After a long life I have come to the conclusion that when all the establishment is united, it is always wrong." The prevailing water establishment strongly believes that the world is running out of water. The global population is likely to increase to over 9 billion by 2050. Water requirements to produce food, energy and to satisfy other human requirements will increase as well. The world has only limited quantity of water, and many countries, according to several major international institutions, are already facing serious water stress. Thus, the thinking goes, the world will face an unprecedented water crisis within the next two to three decades, which no other earlier generation had to face.

Writings about water crisis and water wars because of physical shortages of water have been a growth industry for many years. Dozens of books have been published on this issue in recent years. If one puts water crisis in Google, over 49 million items are identified, mostly in the English language alone.

In my view, Prime Minister Macmillan was correct. The prevailing consensus view of the water establishment that world is facing a water crisis because of physical scarcity of this resource is wrong. Such a pessimistic view is based on the implicit assumption that water is like oil or coal, which once used cannot be used again. In reality, water is a renewable resource, and with good governance practices as well as advances in technology and management, each drop of water can be used numerous times. Based on the research carried out at our Third World Centre for Water Management, it can be categorically said that the world is facing a water crisis not because of physical scarcity of water but because of poor management practices in nearly all countries of the world. Water is being used very inefficiently in most countries of the world and for nearly all purposes.

Unlike the simplistic and linear thinking of the vast majority of the people in the water profession, the book takes a refreshing approach and comes out with a different view. The book is the result of a group of specialists from over 30 national and international research institutions, representing several disciplines like water sciences, geography, agriculture, economics, sociology and political sciences. They studied nine river basins from different parts of the world. These included Andes, Indus-Ganges, Karkheh, Limpopo, Mekong, Nile, Volta and Yellow River. The scale and nature of the basins studied were diverse. For example, they varied from the highly fragmented basins of the Andes to mega-basins like the Indus and the Nile. It included essentially closed basins like the Karkheh and the Yellow to the Ganges and the Indus basins in which vast flows and frequent inundations form integral and important components of the river systems. The vast majority of the basins studied were transboundary in nature, which contributed to an additional layer of complexity.

Most of the basins studied had a strongly agricultural base. However, in recent years urbanization and industrialization have been important. The level and extent of this transition differs from one basin to another. Thus, the book analyses critically and objectively different types of problems prevalent in the nine river basins, and also their different magnitudes and complexities. Rainfed agriculture remains an important driver of water use in all of these basins. Irrigation is important to most of the Asian basins. For the people living in the Mekong, fishing is an important activity.

The book provides a wealth of valuable information from different basins, including water availability and distribution, water productivity, and efficacy of the institutions that plan and manage water. An important component of the book is poverty analysis and its linkage to water. In spite of the belief of many people in the water profession that the linkages between water and poverty are direct and significant, they are often weak. For example, more water does not automatically translate to less poverty, or vice versa.

The picture that emerges from these studies is that these river basins provide a range of ecosystem services that are used to varying degrees of effectiveness by a host of different groups with differing results. These include provisioning of services, including water for urban, rural and industrial consumption, agricultural production (both irrigated and rainfed) hydropower generation, and a variety of services on which survival and integrity of ecosystems depends. In the final analysis, it is the human organizations through institutions at several levels of governments, private sector and the civil society that manage or mismanage water. These institutions, not surprisingly, respond to political and/or social pressures. To a significant extent, success or failure of managing water, food and poverty nexus is determined by the efficiency and functioning of the institutions.

The book is a refreshing one. Unlike numerous water- and food-related books, it does not provide solutions that could be common to all the nine basins. It eschews what I call solution-in-search-of-a-problem approach, which is very common in the development sector, especially as history shows that such approaches often fail. Instead it tries to analyse and explain the complex interrelationships that invariably exist in the water-food-poverty nexus. Most importantly, it does not provide a general prescriptive solution that could be valid for the nine river basins. Rather, it outlines a roadmap as to how individual problems or issues in specific areas could be understood and approached in the various basins under very different physical, social, economic, political and institutional conditions.

I have no doubt that this publication will prove to be very useful and valuable to scholars and policy-makers who are engaged in finding solutions to the complex problems of water-food-poverty nexus in the various river basins of the world.

Asit K. Biswas
President, Third World Centre for Water Management, Mexico
and
Distinguished Visiting Professor
Lee Kuan Yew School of Public Policy

Introduction: water, food and poverty in river basins

Myles Fisher and Simon Cook

The world faces emerging crises of water and food. Its population already nears 7 billion and is forecast to be at least 9 billion by 2050. The increased population will need 70 percent more food than it does today (Bruinsma 2009), which has major implications for the global environment that supports the food system. That goal will be a lot harder to achieve than the green revolution last century.

Where will the water come from to produce 70% more food when agriculture already uses 70% of the world's freshwater resources? In the case of the Indus, the Yellow, and the Nile, the basins are essentially closed, that is, all the water is used. In some cases, the environmental flows essential to maintain ecosystem function towards the river mouths have ceased or are threatened. The only solution in closed basins is to increase water productivity (WP) of crops by using the water more efficiently, what former UN Secretary General Kofi Annan called, "More crop per drop" (Annan 2000).

The CGIAR took up this challenge in its Challenge Program on Water and Food (CPWF). One of the CPWF's approaches was to examine in detail the issues of development, poverty, and water productivity in ten river basins worldwide in the Basin Focal Projects. The first part of this book, which was published as a Special Issue of *Water International* in late 2009, has papers from nine of these basins: the Andes system of basins in South America; the Limpopo, the Niger, the Nile, and the Volta in Africa; the Karkheh in Iran; and the Ganges, the Indus, the Mekong, and the Yellow in Asia. The nine basin papers are followed by analyses of processes across basins, which were published as a second Special Issue of *Water International* in early 2011.

The fascinating outcome is that all the basins are different. Many of them have high levels of poverty and there are some similarities, but each of them presents different underlying problems, which must be addressed if the goals of increasing food production and overcoming poverty are to be met.

Water scarcity

Population growth has reduced available water in some basins below 1700m^3/capita/yr, the level considered secure (Falkenmark 1989). Absolute water scarcity worsens when the growing population depends on unsustainable irrigation as in the Yellow, Indus, Karkheh, and upstream Limpopo basins.

Water productivity

Apart from the need to increase WP in closed basins, populations in the sub-Sahel are doubling every 30 years, with every indication that they will continue to do so. Food production has kept pace with the increase over the last 20 years, largely by increasing the cropped area, which can continue in the short term. For the longer term, however, it is necessary to address the cause of low WP of rainfed agriculture in the sub-Sahel (the Volta and the Niger Basins), which Cai *et al.* identify as "lack of inputs, and poor water and crop management". WP could be increased with appropriate agronomy (high-yielding varieties and fertilizer) as demonstrated by the Millennium Villages project (Sachs 2007). As pressure on the available land increases, however, higher WP is the only solution to providing the food that will be needed with the water that is available.

Water quality

In some basins, water quality for the rural poor is a more important than quantity. Indeed water quality is a universal issue for the rural poor as reported in the Nile, the Indus-Ganges, and the Volta papers, but also in the relatively developed basins of the Andes, where mining and other uses threaten water quality. Moreover, it is difficult to provide safe water to the invariably dispersed populations of the rural poor. The success of "Thai jars" (small, artisanal, ferro-concrete water tanks) in Nepal, however, suggests that there are feasible solutions, which could be applied more widely. Rainwater harvesting for domestic water receives little attention, but it is viable even in semi-arid countries. Rainfall collected from dwelling and other roofs was the source of domestic water for much of rural Australia during its pioneering phase, and still is in many places.

Water-related hazards

Water-related hazards of drought, flood and water-borne diseases have major impacts on development. The hazards cause more hardship where countries have little capacity to manage them, such as in the Niger, the Volta, or the Nile Basins, or where the events can be extreme as in the Limpopo, and the 2011 floods in the Indus and the Niger.

Fish and the commons

Fish in general are a common resource and at least in the case of maritime fisheries have been plundered to the point of collapse as fishing became industrialized last century. Will the Mekong suffer a similar fate? There is evidence that the total catch has remained static for the last ten years (see Kirby et al.'s Mekong paper), so that per capita consumption has fallen as the population has increased. The productivity of the Tonle Sap fishery in Cambodia, which provides livelihoods for over one million people, depends on the seasonal ebb and flow of the Mekong. Will hydropower dams reduce this ebb and flow, reduce the fish catch, and cause wrenching social change? Will economic development based on hydropower provide compensation for the population that now depends on fishing? If there is a parallel between possible loss of the commons of the fish in the Mekong and the misery and migration caused by the enclosure of the commons in the UK 250 years ago, the effects may be severe and long lasting.

Legal duality

Legal duality of institutions leads to the inability of herders, migrants and fishers to get access to land and water resources in West Africa (see the Niger and the Volta papers). Central governments have been unable to insist that rights to land and water should be by means of formal land title. The breakdown of traditional cattle herders' access to forage and water is having a profound effect on their livelihoods in West Africa. Fisher *et al.* show that a more comprehensive, integrated approach to institutions and organizations could make them more relevant to basin-wide needs. The mismatch between the needs of development and providing ecosystem services is a key issue in many basins and impacts food, poverty, livelihoods, and sustainable ecosystems.

Some thoughts on poverty

It is easy to say that water quality is an indicator of poverty. Yes, the poor often have bad water. But is this a cause or an effect? Certainly, poor-quality water brings with it problems like water-borne diseases and infant mortality. But if they had good water would they still be poor? Probably, but their quality of life would be improved. Moreover, Sachs (2007) argues that lower infant mortality will cause an immediate, voluntary reduction in fertility rates and thus reduce the rate of population increase. Access to water is influenced by interactions between local, regional and national institutions and organizations (see Kemp-Benedict *et al.*'s paper). Their influence on livelihood strategies decreases at higher levels of economic development.

With education too, are people poor because they are illiterate? Almost certainly, yes they are. And are they illiterate because they are poor? Again it is plausible that they are, but the remarkable success of the *Bolsa Familia* component of the *Fome Zero* (Zero Hunger) programme in Brazil, in which the subsidy to the poor is contingent upon the household children remaining in school, suggests that there are effective solutions

Transboundary issues

Transboundary organizational weakness is a common theme, identified in all but the Yellow and the Karkheh, which are entirely within one country. Boundaries do not have to be international to be problematic; provinces in China and states in federal systems such as India are quite proprietary over the waters within their borders.

One reviewer of the Limpopo paper commented that the river is notable by not having any large dam that would encourage transnational cooperation. There is no large dam on the lower Limpopo because there is no suitable dam site. But it begs the question of whether large dams do indeed encourage transnational cooperation. Giordano *et al.* (2005) show that despite tensions, transboundary rivers are more a subject of agreement than conflict. The papers in this book support that conclusion.

Salman (2010) describes how upstream riparian countries can be "harmed by downstream [riparian countries] through foreclosure of their future uses [of water]". He concludes that cooperation amongst riparian countries is the cardinal principle of the law of international waters, and that the interests and concerns of both upstream and downstream riparian countries need to be considered by all parties. It is hard to argue against that conclusion, but implementing it requires good will on all sides, which is difficult to achieve if all parties continue to pursue their own narrow interests, as they often seem to do.

Transboundary organizations

Most of the transnational rivers do have a statutory organization, nominally with a coordinating role, but the participating countries in general have not ceded any useful authority to the institutions they have created. They remain bodies that support dissemination of research, and convene conferences and meetings, but they do little to influence political outcomes, which can only be arrived at by consensus of the constituent countries. The River Nile Commission is dominated by the downstream countries, Egypt and Sudan, who insist on adherence to the arrangements made in colonial times, which did not consider upstream countries. Indeed, Egypt threatens to go to war with any country that presumes to reduce downstream flows of the Nile.

The Volta River Commission does achieve some useful collaboration between Ghana and Burkina Faso, which together occupy 84% of the Basin. In contrast, each of the members of the Mekong River Commission (MRC), Cambodia, the Lao PDR, Thailand, and Vietnam insist on their right to do whatever is in their own best interests. China's participation in the MRC is limited to observer status, and although it appears to be increasingly willing to cooperate, there is little reason to expect that it will be any less protective of its interests than other Mekong basin states. Even though the number of nations involved is fewer, conflict in the Ganges is more intense. The Farakka Barrage in India, 10 km upstream from the border with Bangladesh, controls the Ganges by diverting it to the Hooghly River from its course through Bangladesh. India closes it during the dry season, but opens it when the Ganges floods so that Bangladesh gets no Ganges water in the dry season, but is inundated when the river floods. Repeated efforts to resolve the issue have not been successful.

Climate change

The threat of climate change hangs over all. Climate change will have both positive and negative impacts, both between and within basins (see Mulligan *et al.*'s paper). The common feature is that the poor are almost always the most vulnerable. The global circulation models forecast that temperatures will rise by 2-3°C by 2050, which will increase water losses to evaporation. The effect of the higher temperatures per se on crop yield is harder to predict, but there are some indications with maize and rice that higher temperatures reduce yields.

Precipitation is not so clear-cut, but most basins are likely to decrease somewhat, which coupled with higher temperatures, will cause more water stress in crops. Moreover, with less snow and ice to spread river flows, timing of flow peaks will change and there will be more floods. In some places, there will be plant breeding solutions, such as crops that flower earlier in the day to avoid the heat, but these are possibilities rather than off-the-shelf solutions. There are also agronomic solutions, such as later planting to avoid high rates of evaporation during the very hot weather that precedes the monsoon to reduce the demand on groundwater in the Indian Punjab (see the Indus-Ganges paper).

Basin summaries

We summarize briefly our opinion of the outstanding features of each basin:

Andes

The Andes are a complex system of independent basins in which biophysical and developmental diversity are confronting change. The economies of the Andean countries are developing, although there are still large populations who do not share the benefits. The pressing issue the countries confront is how to share the benefits of development more equitably.

Ganges

The basin is under extreme population pressure. Low WP downstream contrasts with high WP upstream but unsustainable groundwater use. There were great benefits from the green revolution in the western states, but much less in the eastern states. The Farakka Barrage is a transnational issue, which forced Bangladeshi farmers to adapt to less water by changing from flooded, dry-season rice to other crops and irrigation by groundwater.

Indus

The Indus is a closed basin that is under extreme population pressure, with aging, unreliable water infrastructure, and increasing, unsustainable use of groundwater. The challenge is to upgrade the infrastructure to reduce dependence on groundwater, and to manage use of groundwater to maintain the resource.

Karkheh

The basin is under pressure to meet the political demand for Iran's food self-sufficiency. In general, the rural population of the basin is not the poorest in Iran. Water for the downstream Hoor-al-Azim wetlands on the border with Iraq is not a priority.

Limpopo

The riparian countries have vulnerable populations, unreliable water and low WP, in addition to dealing with their colonial heritage. Upstream is a juxtaposition of productive commercial agriculture and unproductive subsistence farming. Downstream is a poor population vulnerable to the basin's damaging floods and droughts.

Mekong

The Mekong is a diverse basin facing the tensions of development. The commons of the fishery resource on which many depend for their livelihoods is vulnerable to changed hydrology by hydropower dams. The countries as a whole may benefit, but those whose livelihoods depend on fishing likely will not. China's role remains an enigma.

Niger

Water and actual poverty in the Niger are caused by illiteracy, poor-quality water, and dysfunctional institutions. Planned dams upstream of the Inland Delta threaten its annual flood on which much of its productivity and the livelihoods of a million people depend.

Nile

The basin is characterized by downstream-upstream conflict and unmet agricultural potential in the upstream countries. Eighty percent of the water that arrives at the Aswan Dam comes from Ethiopia, which wants to develop some of its irrigation potential. Egypt and Sudan want to maintain the flows agreed in colonial times.

Volta

Ghana is regarded as a model in West Africa being further along the development pathway than Burkina Faso or any of the Niger countries except oil-rich Nigeria. Ghana's "rural households accounted for a large share of a steep decline in poverty induced in part by agricultural growth" (World Bank 2007), and the fertility rate is falling as a consequence. Upstream small dams will have little effect on hydropower at Akasombo.

Yellow

China's burgeoning economy puts increasing pressure on agricultural water, and in the case of the Yellow, has caused extreme basin closure and increasing water scarcity. The Yellow shows that a centrally-directed economy can facilitate dramatic shifts in water allocations in the absence of firm and litigable water-use rights, but it is not without cost. After not reaching the sea for a number of years in the late 1990s, there is now a minimum year-round flow, but achieving it caused hardship to upstream water users.

Resilience

Resilience to changes in the quantity of water that is available is critical (see Cumming's paper). When water becomes limiting, the social-ecological system must adapt rapidly to avoid damage to communities and biodiversity. When the quantity of water becomes limiting, it "alters political and institutional links between actors". Resilience can be managed, but it requires participatory processes that are politically acceptable and that take account of social, ecological and economic issues in defining the problems and their solutions.

Development

Ogilvie *et al* conclude in the Niger paper that "improved agriculture and water management require technical, sociological, and regulatory changes to address the wider causes of poverty". This could be said of all basins. The tough question is how to make these changes happen. The short answer is economic development. But how can that be achieved? According to the World Bank (2007), the solution is through support of agriculture.

"Agriculture has served as a basis for growth and reduced poverty in many countries, but more countries could benefit if governments and donors were to reverse years of policy neglect and remedy their underinvestment and misinvestment in agriculture" (World Bank 2007). The *World Development Report* (World Bank 2007) goes on to argue that agriculture was heavily taxed to support industrialization, which, coupled with continued anemic investment in agriculture, reflects a political economy in which urban interests dominate policy that "proved lethal in Africa" (Byerlee *et al.* 2009).

In the 21st century, agriculture continues to be a fundamental instrument for sustainable development and poverty reduction, even while countries move beyond agriculture to more industrial economies; "The global development agenda will not be possible without explicitly focusing on the role of agriculture for development" (Byerlee et al. 2009).

Using agriculture as the basis for economic growth in the agriculture-based countries requires a productivity revolution in smallholder farming, which papers presented here show to have low WP. To pursue agriculture-for-development agendas, local, national, and global governance for agriculture need to be improved. Growth in GDP from agriculture is at least twice as effective in reducing poverty as growth in GDP in sectors outside agriculture. In the case of China, growth in agriculture reduced poverty 3.5 times more than growth outside agriculture, while in Latin America it was 2.7 times more (World Bank 2007). Development is successful when increased productivity satisfies increased demand, but fails when development of land and water resources is prevented for whatever reason (see Cook *et al.*'s paper.)

Agriculture is therefore the basis for economic growth, even though development moves economies beyond it. But increased agricultural activity has major impacts on the river basin systems that support it. Furthermore, as development moves beyond agriculture, demand increases from other sectors and from the populations they sustain. Achieving processes that support balanced development of water and food systems requires detailed insight of conditions as they occur in basins, together with analysis of processes that cause them.

Myles Fisher
Simon Cook
Editors

References

Annan, K., 2000. *We the Peoples: The Role of the United Nations in the 21st Century*. New York: United Nations Department of Public Information.

Bruinsma, J., 2009. The resource outlook to 2050: By how much do land, water and crop yields need to increase by 2050? Paper presented to *FAO expert meeting, on how to feed the world in 2050*, 24-26 June 2009. Rome, Italy: Food and Agriculture Organization of the United Nations. Available from: ftp://ftp.fao.org/docrep/fao/012/ak971e/ak971e00.pdf [Accessed 22 August, 2010].

Byerlee, D., de Janvry, A., and Sadoulet, E., 2009. Agriculture for development: Toward a new paradigm. Annual Review of Resource Economics, 1 (1), 15-31.

Falkenmark, M., 1989. The massive water scarcity now threatening Africa: Why isn't it being addressed? Ambio, 18 (2), 112-118.

Giordano, M.F., Giordano, M.A., and Wolf, A.T., 2005. International resource conflict and mitigation. Journal of Peace Research, 42 (1), 47-65.

Sachs, J., 2007. *Bursting at the seams: The BBC Reith lectures, 2007*. London: Radio 4, British Broadcasting Corporation. Available from: http://www.bbc.co.uk/radio4/reith2007/ [Accessed 22 August, 2010].

Salman, S.M.A., 2010. Upstream riparians can also be harmed by downstream riparians: The concept of foreclosure of future uses. Water International, 35 (4), 350-364.

World Bank, 2007. World development report 2008: Agriculture for development. Washington DC: World Bank.

The Andes basins: biophysical and developmental diversity in a climate of change

Mark Mulligan, Jorge Rubiano, Glenn Hyman, Douglas White, James Garcia, Miguel Saravia, Juan Gabriel Leon, John J. Selvaraj, Tatiana Guttierez and Luis Leonardo Saenz-Cruz

To understand how agriculture and poverty interact, we analysed water availability, productivity and institutions for the Andes basins. Water limits agricultural productivity in the southern basins but is plentiful in the northern basins where steep slopes or poor land and water management limit productivity. The dominance of small, steep basins results in important upstream–downstream linkages. The greatest challenge to improving the productivity of water in the Andes basins is to regulate water quality better for multiple uses and to negotiate fair and transparent compensation for upstream providers of water-based ecosystem services for the benefits that they provide to downstream users.

Introduction

In this paper we review the status of water availability, access and productivity in the Andes basins. We focus on the relationship between water and poverty with a view to identifying the role that water and its use currently plays in supporting or constraining development, rural livelihoods and environmental sustainability within the region. Further, we examine the likely water future of the basins and evaluate the development implications of hydrological change over the coming decades. Finally, we evaluate the potential for different intervention and adaptation strategies to lessen the water-related constraints to sustainable development in different parts of the basins and the role that institutions must play in facilitating this adaptation.

Characteristics of the Andes basins

We include in the Andes basins all river basins in the Andes from 500 masl elevation and higher (Figure 1). The basins run from Venezuela, Colombia, through Ecuador and Peru to Bolivia, Chile and Argentina. We include Venezuela, Chile and Argentina in the biophysical analysis but exclude them from the socio-economic analysis on the basis of their more advanced level of development compared with the other countries. In addition

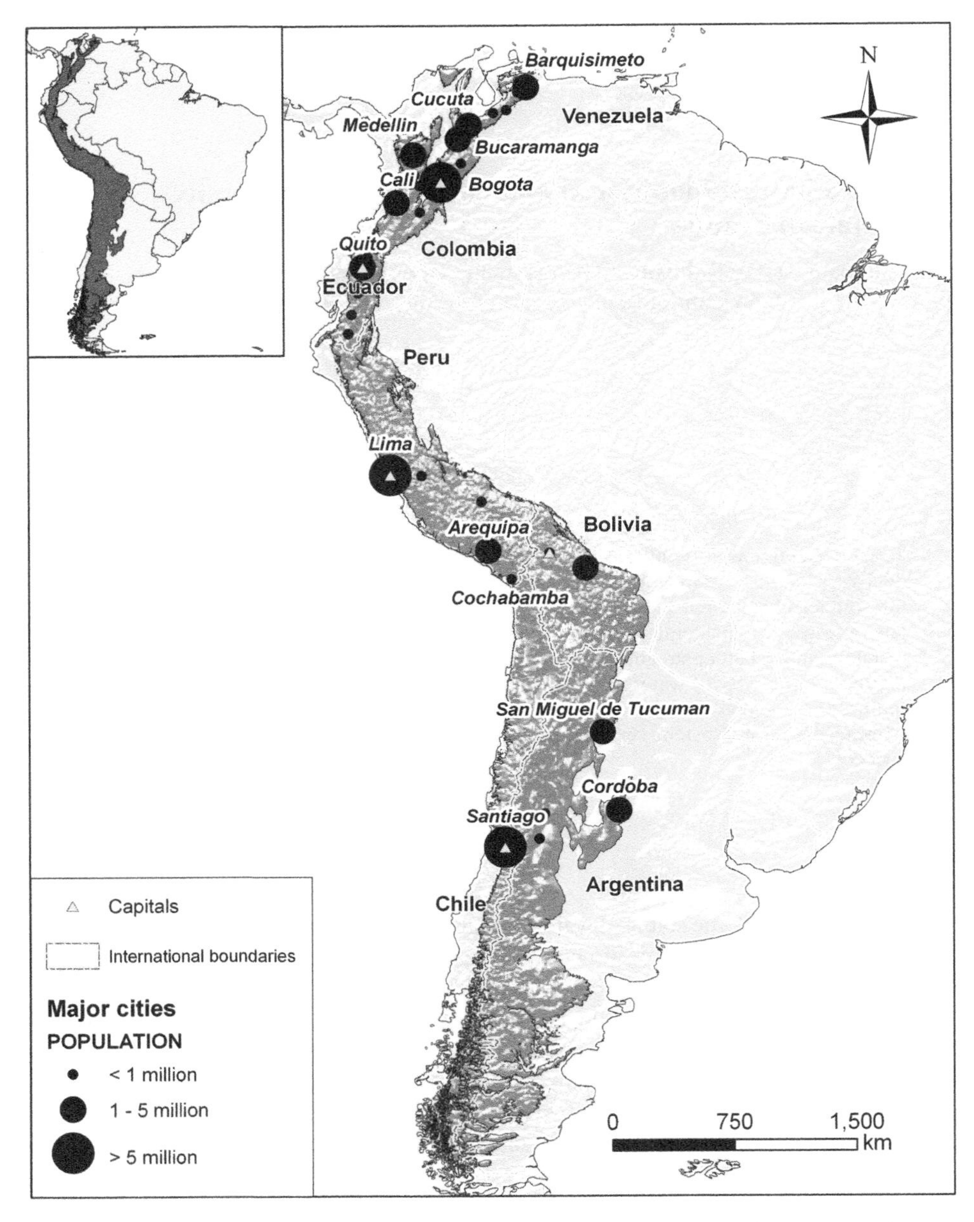

Figure 1. Map of the Andes basins. We define the basins as all catchments above 500 masl.

to an analysis of the Andes basins as a whole, we carried out detailed local studies in three sub-basins located in Colombia, Ecuador and Peru. In biophysical terms, the Andes basins are characterized by being 83% mountainous (Mulligan *et al.* 2011) with rainfall extremes from zero to over 10,000 mm annually, and spatial heterogeneity in climate, geology and soils. The Andes basins consist of a series of small eastwards- or westwards-draining basins with some connecting valleys draining south or north. This culminates in a large number of separate basin outlets, in contrast to the very large, dendritic and single outlet basins that are common elsewhere. This proliferation of small basins crossing large altitudinal gradients creates important upstream–downstream linkages and inequalities over short distances.

Colombia, Ecuador and Peru are classified as urbanized and industrialized (Byerlee *et al.* 2009) on the basis of proportion of rural population, poverty and the relatively low contribution of agriculture to gross domestic product (GDP). Agriculture in the basins ranges from purely subsistence, through mixed subsistence and cash cropping for local markets, to intensive commercial agriculture that takes place in some of the inter-mountain valleys. Profitability and demand for domestically grown basic staple crops has decreased markedly in Andean upland communities along with most other global regions. The basic crops of smallholder agriculture, such as maize and potatoes, have changed from generating cash income to become subsistence crops (Katz 2003). About half (47 million) of the population in Colombia, Ecuador and Peru, and almost two-thirds in Bolivia are classified as poor according to national poverty criteria. Of these 47 million poor people, 26 million depend on rural livelihoods (Mulligan *et al.* 2009)

The socio-political environment of the Andes

Political and economic insecurity in Andean countries have led to migration to urban centres and a depopulation of rural areas, particularly in the upland regions (Chomitz *et al.* 2005). Even where the rural population continues to grow, its growth rate is slower than in the nearby urban populations (Brea 2003, Cohen 2003, CEPAL 2005). The main driving factors for this trend have been the fall in profitability of smallholder agriculture, which has led to structural changes away from smallholdings towards larger, more commercial agriculture (Rodríguez and Busso 2009). The downside is that it has been the difficult for the cities to cope with the social and economic problems caused by rapid growth and increasing rural inequality. A further problem of rapid urbanization is the water pollution generated by poor management of wastewater and effluents from urban and industrial centres.

The increase in urban population means higher demand for water, food and other goods and services provided by rural areas. This demand provides economic benefit to the rural areas that are able to meet the demands, but sometimes to the detriment of the environment upon which rural populations depend (Valdés and López 1999, Yancari 2009). Rural areas are often neglected in the investment policies of national governments, even though there is growing recognition of the importance of agriculture in stimulating economic development (WDR 2007, Byerlee *et al.* 2009).

Eggertsson (1996) and Saleth and Dinar (2004) document the formal regulations and informal practices of water resource management in the Andes basins. In recent decades, basin countries have enacted legislation incorporating the concepts of integrated water resource management (IWRM). The new laws are administered by new national agencies responsible for water management and water availability. IWRM seeks integration across agricultural, industrial and domestic sectors, taking account of the need for environmental flows,[1] but are often disputed by traditional water users and have been difficult to implement in the Andes.

The spatially and culturally diverse context of the Andes warrants that national priorities take local conditions into account carefully. Although there are formal mechanisms for public participation in matters of resource management, marginal sectors of society are often excluded. Water user associations are growing and gaining more influence, however, leading to conflict over the management of water resources in places, even where they may be plentiful.

The biophysical environment of the Andes basins

The dominant characteristic of the basins' landscape is heterogeneity. Climate, geology, soils and vegetation are spatially variable, while temporal variability is pronounced due to the seasonal passage of the inter-tropical convergence zone and inter-annual changes related to the El Niño-southern oscillation (ENSO) phenomenon. Rainfall is spatially, seasonally and inter-annually variable, ranging from hyper-humid (>10,000 mm/yr) in parts of Colombia, to hyper-arid (a few hundred mm/yr or less) in Bolivia and the Atacama desert of Chile.

Rainfall variability in mountain environments is complex because of orographic and wind-driven effects. Moreover, gauge networks are often sparse, especially at high elevations, so that there is no definitive measure of precipitation distribution at the basin scale. This makes hydrological and water resource analysis highly uncertain. The Tropical Rainfall Monitoring Mission (TRMM) (Mulligan 2006a) and WorldClim (Hijmans *et al.* 2005) are the most detailed data of rainfall distribution available for the Andes (Figure 2) and show that precipitation is greatest in the Chocó region of Colombia, the Caribbean lowlands and the eastern slopes of the Andes from Colombia through Ecuador to Peru and Bolivia. The driest areas include the Bolivian Altiplano, Chile and northwest Argentina.

While the patterns of rainfall in the two datasets are qualitatively similar, estimates for any one point can differ by up to 2000mm (Figure 2, scatter diagram) (Mulligan *et al.* 2009). It is likely that the real incident rainfall is somewhere between the remotely sensed but spatially detailed TRMM estimate and the ground-measured but highly interpolated WorldClim estimate. There are 2157 rainfall stations in the Andes basin, but even if each

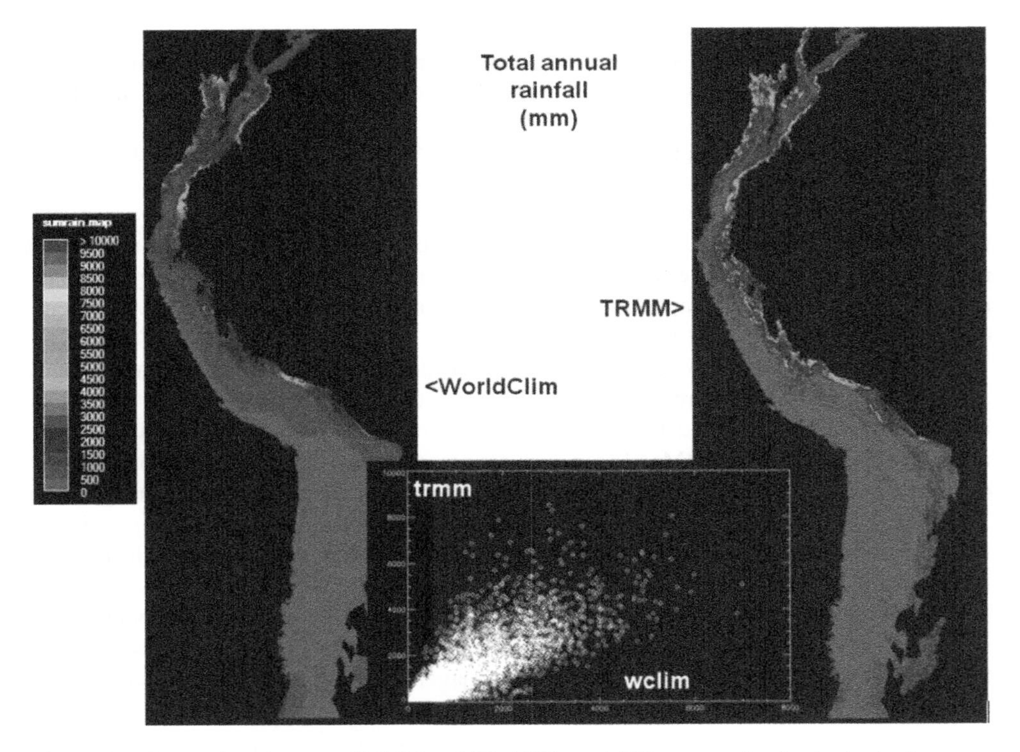

Figure 2. Comparison between TRMM and WorldClim rainfall patterns for the Andes basins.

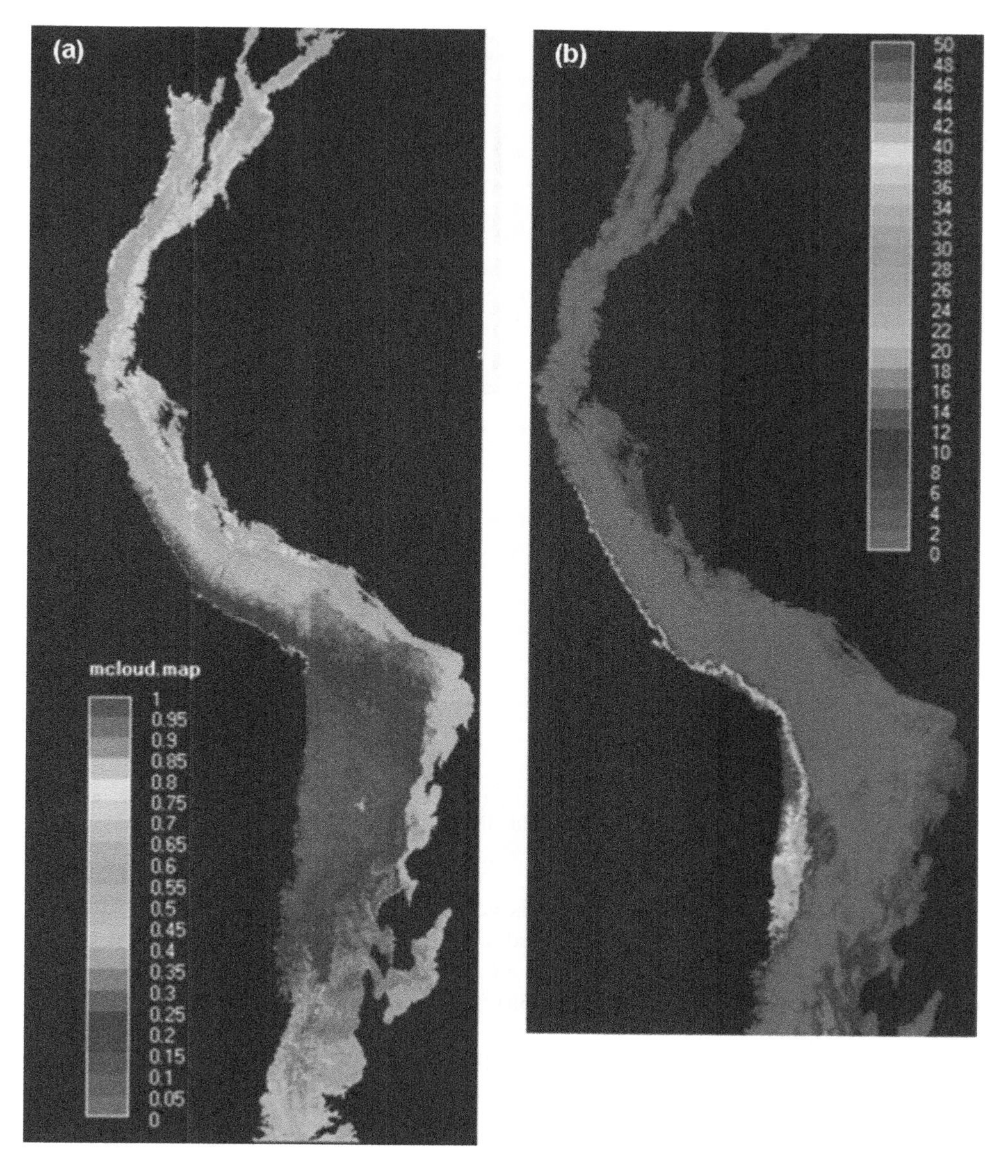

Figure 3. Cloud and fog in the Andes basins: (a) cloud frequency; and (b) fog inputs as a percentage of river discharge.

one represented 1 km^2 surrounding the station, 2157 km^2 (for which a rainfall measurement is available) is less than 0.01% of the area of the Andes basins. On the other hand, the TRMM data are more spatially detailed, but lack temporal detail as the satellite only passes over each 10 days. The uncertainty of these basic inputs means estimates of water input for water accounts at the scale of the basins are potentially subject to error so that it is difficult to make recommendations for policy formulation.

The basins have mountains to over 6000 masl, and they include desert, savannah, rainforest, cloud forest and alpine biomes. Cloud frequency is greatest in the Amazonian and

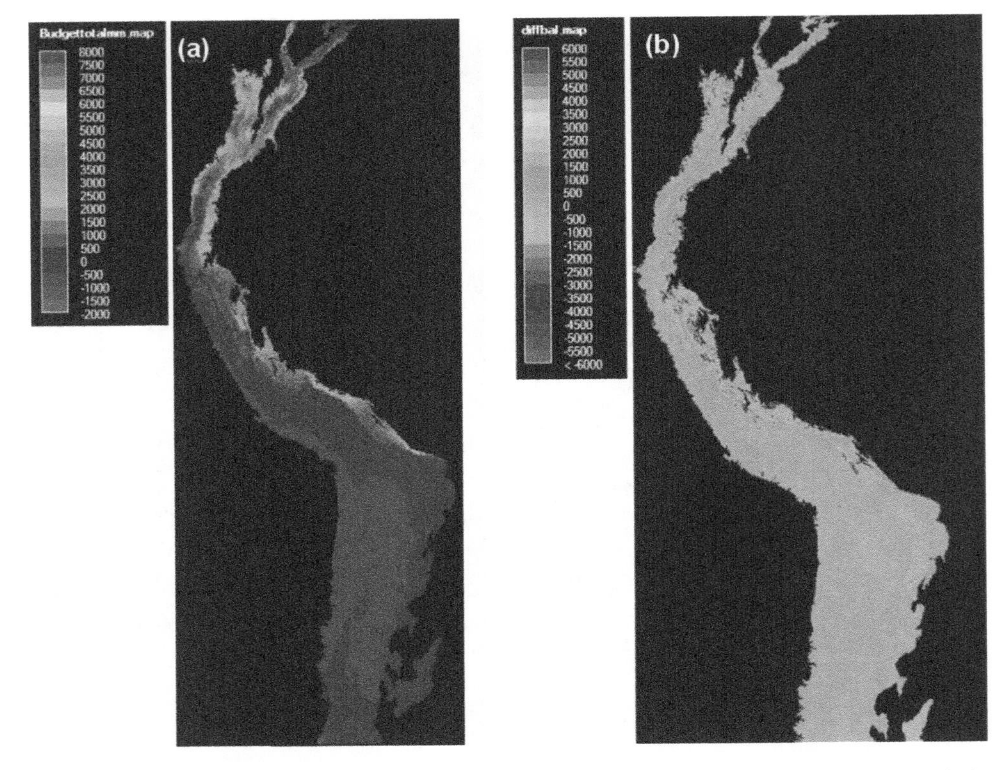

Figure 4. Water balance for the Andes basins: (a) Mean of WorldClim and TRMM; and (b) difference between water balance based on WorldClim and TRMM.

Pacific flanks of the northern Andes and is very rare in parts of the Bolivian Altiplano (Mulligan 2006b). Cloudiness is associated with high rainfall but also reduced evapotranspiration because solar radiation is reduced. Parts of the basins have persistent cloud cover, often to ground level as fog (Figure 3a). These cloud forests have a unique and little understood hydrology (Bruijnzeel 2005). It is known, however, that the volume of water intercepted from fog can be important, especially in the dry season. Fog inputs can be up to 10% of annual flows for some Andean slopes and >50% in the dry coastal parts of Chile and southern Peru (Mulligan and Burke 2005).

The Fog Interception for the Enhancement of Streamflow in Tropical Areas (FIESTA) model (Mulligan and Burke 2005) parameterized with both WorldClim or TRMM rainfall data gave positive water balances of 4500 mm/yr on average, indicating large discharges, for much of the northern Andean slopes throughout Colombia and in the eastern Andes of Ecuador, Peru and Bolivia (Figure 4a). Water balances are only 500–1000 mm/yr throughout the rest of the basins, and there are some very dry areas in the high southern Andes (especially in the southwest). Given the very low local rainfall, these areas are totally dependent on inflows to maintain evapotranspiration. Water balance based on the two rainfall estimates (TRMM and WorldClim) differed somewhat (Figure 4b), but a few areas exhibited large differences.

The Andes rest on three active tectonic plates so they are prone to natural hazards such as earthquakes, landslides and volcanic eruptions. These can be locally damaging. When Nevado del Ruiz in Colombia erupted in 1985, the subsequent lahar (volcanic mudflow) destroyed the town of Armero with a death toll of 23,000 (McDowell 1986). Thouret *et al.* (1990) estimate that 100,000 people are still at high risk in the Ruiz-Tolima area from lahars, which are also a constant threat at locations elsewhere in the basins. The cost of natural disasters can be high. The Ruiz lahars reportedly cost Colombia US$7.7 billion, about one fifth of its GDP for the year. Damage from recent earthquakes in Peru and Chile has also had an impact on national economies, with inevitable consequences for the funds available to address issues such as agricultural development and poverty alleviation.

The Andes basins are naturally a site of extreme climates, which take the form of successive and lengthy droughts, floods and strong winds (Comunidad Andina 2004). The most important water-related natural hazards that affect livelihoods in the basins are droughts and floods related to the ENSO phenomena. Ongoing climate change is projected to increase the magnitude and frequency of these climatic extremes (Wetherald and Manabe 2002). Communities in the basins are highly vulnerable to disasters for economic and geographic reasons such as the occupation of unsafe sites (Armero was re-built on the site of two earlier lahar events since first colonization) and environmentally inappropriate or non-resilient livelihood options.

Agriculture is the dominant human use of water resources in the Andes basins. Mean actual evapotranspiration (ET_a) for the basins is 643 mm/yr. Mulligan *et al.* (2009) calculate that the actual evapotranspiration of the crop and pasture land defined by Ramankutty *et al.* (2008) accounts for 28% of the ET of the basins. The area in Latin America dedicated to agriculture was about 770 Mha in 2001, almost twice that at the end of the 1950s (Flavio Dias Avila *et al.* 2010), with 34%, 38%, 27% and 17% (mean 1975–2007), of the total land area under agriculture in Bolivia, Colombia, Ecuador and Peru, respectively. For Colombia, Ecuador and Peru, the agricultural area has remained stable since the mid 1970s while fertilizer consumption has increased several-fold, indicating intensification of agriculture rather than extensification. In contrast, fertilizer consumption in Bolivia remained static but the agricultural area expanded by about 7 Mha (Mulligan *et al.* 2009), reflecting conversion of the eastern lowlands to commercial cropping, rather than changes in the basins, which have become more urban but not more agricultural.

There are three broad categories of agricultural production in the Andes basins. In the highlands, agriculture is characterized by smallholder farms, which have been the main regional and local food supply. They produce low yields of maize, potatoes, beans, wheat, and pasture for livestock (Mulligan *et al.* 2009). In the more remote areas these are essentially for subsistence, but in areas with access to markets they are a mix of subsistence and cash crops. These smallholder systems impose low demands on natural and external resources and in doing so provide ecosystem services. They have become increasingly eco-inefficient, however, due to distorted markets and economic policies that do not internalize the environmental externalities associated with agriculture. In this regard, eco-efficiency is about achieving a greater quantity and quality of agricultural output for lower inputs of land, water, nutrients and energy (Keating *et al.* 2010), that is, more efficient use of all inputs so that more are available for users elsewhere.

In contrast, agricultural production in the lowlands and inter-Andean valleys is dominated by large-scale commercial agriculture producing sugarcane, oil palm, soybean and rice. They make an important contribution to rural development since they are labour intensive and are currently expanding in response to demand (Mulligan *et al.* 2009). Commercial agriculture is a heavy user of environmental resources and can be environmentally degrading, especially through soil erosion and pollution.

The third category is smallholder farming in specific environments supplying high value products, like coffee, to global markets (Weinberger and Lumpkin 2005). These systems generally have a low resource demand and are relatively eco-efficient, especially when they supply an assured market. In the Andes basins, dairying to supply local cities is an important example (Bernet *et al.* 2001), although cut flowers, aquaculture and wool/textiles are also important and provide considerable employment.

Mining has been important in the Andes basins since pre-colonial times, and there has been an increasing number of mine concessions in the last two decades. For Colombia, the state granted 7343 concessions by 2008, collectively covering 3.5% of the national territory. The number of applications for new mining concessions grew from 900 in the year 2000 to over 4800 in 2007 (Revista Latinomineria 2009). Some of these mines are very large: the Escondida copper mine at 3000 m altitude in northern Chile is the world's largest producing copper mine. All countries have regulations requiring that effluent from mines is treated to minimize negative impacts on downstream populations, their health, and their livelihoods as well as environmental flows, but communities often argue that they are not enforced adequately or consistently.

Water supply is critical for the large urban centres in the Andes basins, especially those at high altitude like Bogotá and La Paz, which have limited areas at higher altitude from which water can be supplied cheaply by gravity. Protection of these water-producing zones, *páramos* and cloud forests, is key for the sustained supply of urban water. Water as an environmental flow is essential for the functioning of other ecosystems such as wetlands, which then help maintain water quality and seasonal regulation of flows. Other services such fisheries, aquatic ecosystems and irrigation schemes are important downstream users with legitimate needs for these environmental flows.

Water-productivity–poverty interactions at the scale of the Andes basins

Water availability and accessibility across the Andes

Water in the Andes basins comes mainly from rainfall, snowmelt and fog as ground-level occult precipitation. Snow and ice supplies up to 58% downstream of the large ice-fields in Peru (Mark *et al.* 2005). Snow and ice are still precipitation, but seasonally redistributed from temporary storage. Permanent snow and ice cover of the Andes is only 0.5% (GlobCover 2008), so it is not an important store at the basin scale although seasonally it is important in the extra-tropical southern part of the basins. Groundwater plays only a minor role in the Andes, except in large inter-Andean valleys, due to steep slopes and deep, inaccessible aquifers. Lima, on the coastal plain, depends on groundwater but is below the Andes basins as we define them. About 11% of the Andean surface area drains into dams, which regulate seasonal water availability, mainly for irrigation, urban supply and hydropower generation. Fog contributes only 10% overall, but can be up to 50% on the Chilean coast where there is very low rainfall and high cloud frequency.

There are two broad trends in water availability in the Andes: in the north water is abundant and in the south it is more limited. In the north, per capita water resources are surplus except in areas with high population densities such Bogota, Quito, and other large urban centres (Figure 5), where they are somewhat lower. In the southern Andes the lack of water supply relative to demand is apparent in the driest areas, so that per capita water balances are very low, with water deficits in the south and west of Peru above the coastal strip, and in the clear, high-altitude southwest of the Bolivian Altiplano, Chile, and northwest Argentina. The Falkenmark scale defined 1700 m^3/person as the level below which water scarcity arises at increasing levels of severity (Falkenmark 1989). Where these areas with water deficit are not supplied by incoming flows or groundwater, Falkenmark's criteria

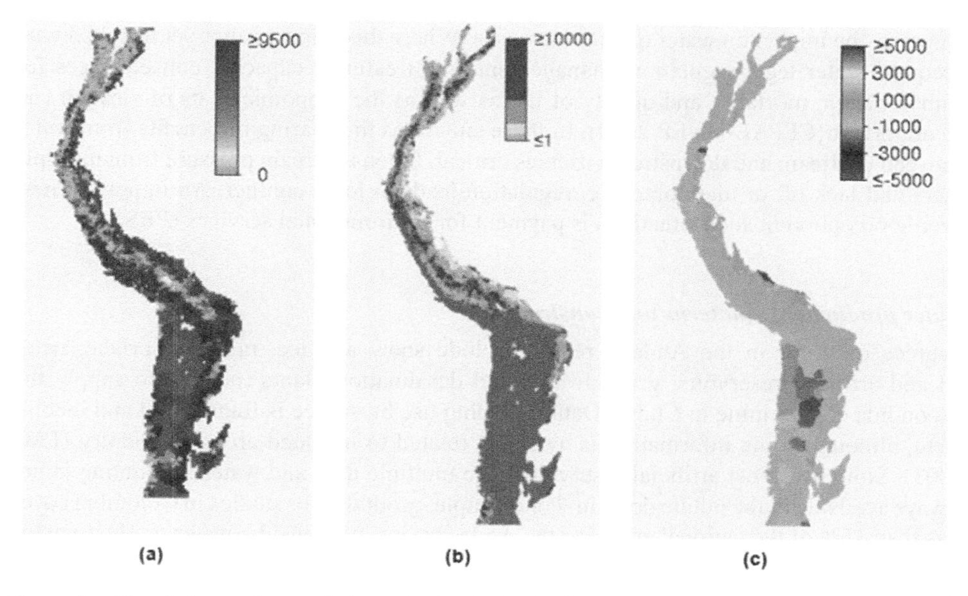

Figure 5. Population and water balance in the Andes basins: (a) population density (persons/km^2, 2005); (b) mean per capita water balance (mm/km^2/person or 1000s m^3/person); and (c) mean per capita water balance on a pixel basis (mm/km^2/person or 1000s m^3/person) by municipality (right). Log scale from green to red. Pink and purple colours are below the minimum shown on the legend.

for water scarcity, which are really focused on societies with a high level of development where water availability is equivalent to water access, will certainly be met. The reality for much of the rural Andes is that even if water is available, the level of infrastructural development may not be sufficient to make that water accessible or usable. The Department with the very least per capita water balance in the Andes derived from this analysis is found in Antofagasta de la Sierra in the Department of Catamarca, Argentina. Although water demands can be greater in the urban areas because of population concentration, urban areas have better developed supply infrastructures based on dams, aqueducts or groundwater reserves. Access to enough water, in quality as well as quantity, can be more limiting in rural areas, which often have no water treatment or aqueducts, and competing water uses such as irrigation, which can limit infrastructural or economic access to water by the poor even where availability is high. That is why, despite high availability through much of the basins, a quarter of the rural population in the Andes basins does not have adequate water access. Moreover, more than half of the population have insufficient sanitation facilities (WHO 2008), which threatens access to quality water downstream, especially for the poor.

Water quality in the Andes basin is affected mainly by contamination from point sources such as domestic, industrial or mining wastewater. Non-point sources also contribute, for example from livestock and agriculture. Especially in the densely populated northern Andes, this human footprint on water quality is often more important in determining water poverty than is the quantity of water available. Protection of cloud forests and *páramos* in national parks and nature reserves serves to dilute the human footprint on water quality and thus confers benefits downstream. In poorly developed countries, population densities and industry are usually low and water quality is not usually the most important issue for human health and development. As population growth and economic development progress, the impact on water quality increases. Where this impact is not yet matched with adequate water legal regulation, management, and treatment capacity, consequences for human health, mortality and quality of life as well as the economic costs of cleanup can be important (CEPAL-UNEP 2002). In these situations the sharing of benefits from water between upstream and downstream users is critical. Often upstream pressure from multiple users and lack of, or unenforceable, regulation leads to local conflict. An important new livelihood option in such situations is payment for environmental services (PES).

Water productivity: patterns and constraints

Sources of water in the Andean region include snow and ice, rainfall, surface, natural and artificial reservoirs, groundwater, and desalination plants (notably to supply the Escondida copper mine in Chile). Data regarding use by source is fragmented and incomplete, although some information is available related to irrigated areas by country (FAO 2003). Similarly, most artificial reservoirs have multiple uses and water accounting is not always available in the public domain. For example, groundwater studies in Colombia cover less than 15% of the national area. For the Andes basins, we consider water productivity as more than agricultural productivity, but also to include hydropower and water for domestic and industrial use.

Agriculture

We calculate agricultural water productivity (WP) as the mean crop production per unit of mean rainfall (WP_{rain}) from data of dry matter production every ten days over the last ten years using 1 km resolution SPOT vegetation data, excluding areas of tree cover (Mulligan *et al.* 2009). Again we used TRMM and WorldClim data to indicate uncertainty in the rainfall estimates. The highest crop WP_{rain} is in those parts of Peru, Ecuador, and Chile where rainfall is very low (Mulligan *et al.* 2009). The patterns are broadly similar between the two rainfall data sets, although the absolute values sometimes differ. Averaging across elevation and catchment indicates that the lowest elevations and small, lowland-dominated Pacific and eastern foothill catchments have the highest crop WP_{rain}.

WP_{rain} for agriculture (and separately measured aquaculture productivity) is lower in the Andean basins compared with the lowlands of the same countries. Colombia and Ecuador in the north and parts of Argentina have the highest WP_{rain} averaged over all the agricultural areas delineated by Ramankutty *et al.* (2008) and Siebert *et al.* (2007). WP_{rain} is a coarse measure of productive efficiency since in wetter areas it includes runoff water that has little impact on crop yield since it does not necessarily contribute to transpiration. Better measures include $WP_{evaporation}$ (a measure of the efficiency of the land-use system's water use) and $WP_{transpiration}$ (a measure of the efficiency of the crop's water use) (Mulligan *et al.* 2009, Xueliang *et al.* 2011). WP_{rain}, $WP_{evaporation}$ (WP per unit of modelled actual evaporation [AET]) and $WP_{transpiration}$ for the Andes are estimated as 0.44 kg/m^3 of rain, 1.43 kg/m^3 of AET and 1.56 kg/m^3 of transpiration (Mulligan and Burke 2010).

Hydropower

Hydropower makes important contributions to national energy supplies in most countries in the Andes basins, especially in Peru and Ecuador. There are some 174 dams in the Andes basins, which capture water from 11% of the surface area and 20% of the streamflow. The total installed hydroelectric capacity in the seven countries that have areas in the Andean basins is more than 46,000 MW. All countries have reliable data at the country level, but the data do not differentiate between dams in the Andes basins and those elsewhere. For the Andes basins themselves, including Chile and Argentina, Mulligan *et al.* (2009) estimated installed capacity at 20,000 MW. Production for Colombia, Peru, Ecuador and Bolivia is over 60,000 GWh/yr, representing a strong contribution to the energy sector. The Spanish Endesa Energy Group, with about five million customers in the region, controls the majority of energy generation and commercialization in Colombia, Peru, Chile and Argentina (Endesa 2009). For the seven Andean countries, hydropower produces some US$20 billion though it is not known how much of this is generated in the Andes basins and how much in the lowlands.

Aquaculture

Of total food production, capture fisheries and aquaculture contribute 1.3% in Colombia, 3.6% in Ecuador and 0.2% in Bolivia, with per capita consumption of 4 and 6 kg/yr in Colombia and Ecuador respectively (FAO 2004), although most of this is likely to be marine capture fisheries. There are no aquaculture-only figures available for the Andes basins. Aquaculture production in the Andes basins (excluding marine fisheries and marine aquaculture and with the exception of Peru) is mostly for local consumption, not for export, and total production of fish is low compared with production of livestock and crops (FAOSTAT 2008). However, with appropriate interventions and technical support to build infrastructure and access to markets as well as efficient marketing, aquaculture is a promising livelihood opportunity in the Andes basins.

Threats to water productivity

WP in the Andes is threatened by human interventions, in particular from continued deforestation, leading to land-use change in key water-producing zones, but also from growth in industry, mining and hydropower, and the increasing demands, influence of and effluent from urban centres. Deforestation for agriculture can increase water yields, because of reduced infiltration and evaporation, but at the cost of increased flooding, soil erosion, sediment yield, and deterioration of water quality, which all have negative effects on downstream populations and water uses. The impact of deforestation will depend on the local biophysical conditions but also the nature and management of the land use that replaces forest. Good land management can reduce negative impacts.

Water, poverty and development within the Andes basins

Poverty cannot be reduced by maximizing water productivity alone. There are two main influences of water on development: firstly, biophysical attributes that can limit water productivity and access and secondly, markets and policies that fail to share the benefits of water productivity with those in need of them.

Although biophysical attributes can limit water productivity and development, water scarcity is not a widespread problem in the region, especially for the northern basins, although it can be important in the drier southern basins. More influential are water-related hazards of droughts and floods related to the ENSO phenomenon. Major constraints arise in steep terrain through its impact on soil erosion, soil water storage, and soil quality. Soil erosion and sedimentation are key water-related problems in the basins influencing both agricultural productivity on steep lands and downstream water quality (Mulligan *et al.* 2009). These water-quality issues are mainly the consequence of steep slopes and marginal agriculture carried out with little attention to erosion control, coupled with intense rainfall, which leads to soil erosion and landslides. In the highly urbanized countries of the northern basins there are also important inputs from urban wastewater, industrial and mining sources. Biophysical problems are compounded by a lack of institutional capacity to cope with the challenges that the natural environment presents.

Secondly, markets and policies are both important and related influences on rural development in the Andes basins. Global market forces, supported by favourable national economic policies and free-trade agreements, have encouraged the spread of commercial and high-value agriculture, which, in combination with a lack of investment in rural areas and political unrest and violence (for example in Colombia), has led to increased marginalization and isolation of smallholder agriculture. This is true particularly in the uplands of the basins and has led to continued rural–urban migration. These drivers negatively affect rural development, and have led to a demographic shift towards increasing urbanization. As a consequence, natural-resource degradation is becoming a critical issue for sustainability, affecting the most vulnerable (the urban and the rural poor) most strongly although all four water-use sectors (rural users, lowland commercial agriculture, urban areas and industry) are impacted to some degree.

Poverty persists in the Andean region despite the generally high level of economic development. Forty-seven million people in the four study countries of Colombia, Ecuador, Bolovia and Peru are considered below national poverty lines (Mulligan *et al.* 2009). The poor include about half of the population in Colombia, Ecuador and Peru, and almost two-thirds in Bolivia. Of these total poor, 26 million are rural. Falling productivity and changing market patterns put pressure on smallholder agriculture and aggregation of holdings is rising (USAID and ARD 2008) leading to patterns of land tenure that cause social unrest and civil conflict (Ibañez and Querubin 2004). Land-use changes also change local power relationships and affect the way water is distributed and used.

In Ecuador and Peru, poverty is higher in the highlands compared to the coastal regions. This pattern is reversed in Colombia, where poverty is more severe in hyper-humid coastal areas. Poverty levels are higher in the dry Bolivian and Peruvian Andes than in the humid Ecuadorian and Colombian Andes. It is noteworthy that national poverty lines are substantially lower than the international levels used by the World Bank (2009) (Table 1). Inequity, as measured by the Gini coefficient, in the four countries reached 55 in 2009, comparable with Mexico, Namibia, Botswana and South Africa (Table 1). Understanding interactions between water, agriculture and development is needed to support policies and interventions that could improve living standards in the region.

Studies in Bolivia and Peru have implicated factors such as lack of health care and education, increasing pressure on land, and land degradation (Swinton and Quiroz 2003, Zimmerer 1993) as pathways into (and consequences) of poverty. Conversely, in Colombia

Table 1. Poverty and inequality of four countries of the Andes basins.

		National			International		Gini
Country	Year	Rural %	Urban %	National %	<US$1/d %	<US$2/d %	Ranking *of worst*
Bolivia	1999	81.7	50.6	62.7	23.2	42.2	7
Colombia	1999	79.0	55.0	64.0	7.0	17.8	14
Ecuador	1998	69.0	30.0	46.0	17.7	40.8	38
Peru	2004	72.1	42.9	53.1	10.5	30.6	19

Source: World Bank (2009).

and elsewhere, dairying (Holman *et al.* 2003), growing high-value crops such as flowers and coffee (Rushton and Viscarra 2006), rural non-farm income and rural non-farm employment (Berdegué *et al.* 2001, Deininger and Olinto 2001, Escobal 2001, Dirven 2004) in many cases have provided pathways out of poverty. Urban development seems to be influential and is notable in Bolivia, Ecuador and Colombia, but less so in Peru. In many cases, labour migration (particularly males) from upstream communities causes labour scarcity upstream, which can lead to consequent problems of disinvestment in agriculture (Jokisch 2002).

Detailed studies within sub-basins in Colombia, Ecuador and Peru have identified the key factors associated with poverty (Farrow *et al.* 2005, Gomez *et al.* 2005, Chapalbay *et al.* 2007, Johnson *et al.* 2009). Several common factors occur: rural non-farm income and employment, diversified economic activities, ownership of livestock, and market-oriented agriculture all tended to be associated with higher levels of wellbeing. A key poverty-reducing activity in these basins is dairying, which is increasing throughout the basins. Other influential activities include manufacturing of artisanal goods and textiles and production of high-value crops. Direct links between poverty and water are relatively weak, although in dry or seasonally-dry areas, households with access to irrigation tended to fare better than those without.

All three sub-basins have strong upstream-downstream interactions, with land management decisions upstream affecting water quantity and quality downstream. The interaction between upstream and downstream areas is a consequence of the strong altitudinal contrast within small basins, which leads to corresponding land-use and livelihood contrasts. In the Jequetepeque Basin in Peru, poverty levels, measured by indicators of unmet basic needs, are twice as high in the upstream parts of watersheds as they are in the downstream areas. Access to irrigation was significantly correlated with lower levels of poverty, supported by Gomez *et al.* (2005). Downstream irrigation obviously depends on water services supplied by the upper basin so that the less-poor lower basin benefits from water services supplied by the poorer upper basin. Households in the upper basin, however, receive no compensation for adopting prudent land-use practices that supply water resources, nor do they suffer the downstream consequences of not adopting such practices. The difference in livelihood conditions is a clear reflection of the unequal distribution of the benefits of water. Work is underway by the Worldwide Fund for Nature (WWF)/CARE, amongst others, to develop payments for environmental services schemes in this basin to share the benefits of water more equitably.

Bayesian analysis of the Ambato Basin in Ecuador showed a significant relationship between poverty and irrigation (Mulligan *et al.* 2009), but other factors related to education were more important. Education is a strong predictor of poverty outcomes in both Ambato and Jequetepeque, while irrigation is more important in certain agricultural settings, such as the arid lower-altitude Jequetepeque Basin. Those areas most affected by poverty are sometimes the ones that are only poorly suited for large-scale irrigation development, thus small-scale irrigation schemes may be a suitable alternative for poverty alleviation through agricultural improvement.

We identified few general factors influencing poverty levels at a broad scale for the Andes basins. The key ones were related to education, including literacy, educational infrastructure, attendance and completion rates at school. An additional factor relates to internal conflicts and the displacement of rural population. The statistics are controversial, but according to estimates during 1980s and 1990s, between 600,000 and one million people were displaced in Peru (Hurtado 2003, p. 68). In Colombia, independent and official agencies provide different numbers due to their varying interpretations of the problem and their differing study periods. Values range from 1.9 to 4.6 million displaced, mostly from rural areas (IDMC 2009).

Better water management is one of many factors that contribute to development and poverty alleviation. It is helpful to distinguish between interventions intended to support development broadly (education, health) and water-specific interventions. Education and health requirements tend to be general, occurring throughout the region, whereas water-related problems tend to be locationally specific. Hydropower is an exception since it has both general effects on economic development by providing energy, but can have local specific effects as well, while water treatment plants have direct benefits on local health and wellbeing. Water management interventions are often viewed as the solution to poverty problems.

Our review of the literature and data analysis for the Andes basins suggests that water-related interventions may have only limited effects on poverty in much of the basins (Mulligan *et al.* 2009). A better approach may be to focus on water-related problems in their own right and solve these for the direct impact on agricultural and economic development and health that they may have. Due to the lower living standards in the upstream portion of Andean basins, payment for environmental services (PES) schemes may have potential for improving water (and other natural resources) management issues in those areas through an improved sharing of the benefits accruing from good land and water management. Such schemes are being developed around water funds for the major cities (Bogotá, Cali, Quito, Medellin) as collaborations between water companies, regional government and large conservation non-governmental organizations (NGOs) (see for example TNC 2010). Such schemes are on the increase globally with Landell-Mills (2002) reporting 287 PES schemes for forests in their global review. Some 101 different programmes were recently reported for Latin America covering 2.3 M ha and involving transactions up to US$31 million in 2008 alone (Stanton *et al.* 2010).

The water future of the Andes basins

Population change and water demand

The average population growth rate for Andean countries in the last 25 five years (1980–2005) was around 2%/yr, ranging from 1.9%/yr in Colombia and Peru to 2.3%/yr in Bolivia (United Nations [UN] 2007). Continued growth will be concentrated in urban areas with negative growth in some rural areas as rural–urban migration continues. Pressures will continue to be greatest in the large urban centres of Bogotá, Cali, Medellin, Quito, Lima, La Paz and Santiago, which will continue to demand more water from the surrounding catchments.

Mining

The greatest threat to equitable sharing of flows in the Andes basins may be continued growth in the hydropower, irrigation and mining sectors as well as population growth and urban demands. To understand better the potential quality of water and impacts upon flows, we looked at the origins of water arriving at major cities. There are 24,454 known mines and quarries in the Andes basins (Hearn *et al.* 2003). By analysing the runoff generated in the areas of these mines and its contribution to the total runoff of the basins we can better understand the likely impact of these mines on water quality. If we assume that the runoff water from mines has lower quality than that from other areas, then any presence of a runoff signal from mines that represents more than, say, 0.0001% of the runoff (equivalent one part per million, a representative threshold for toxin concentrations) could be considered an impact. Results from analysing, at the scale of the basins, the mean proportion of runoff in rivers used for urban supply derived from mining, showed that between 0 and 10% of the runoff that reaches most basin cities comes from areas with upstream mines (for example, 13% for Quito, 5% for Bogota, 8% for Lima). For particular contaminants, specific threshold concentrations will need to incorporate results of further investigations into the consequences of this contamination and strategies for reducing it.

The impacts of land use change and nature conservation

We assessed the change in water balance resulting from land-use change from pre-agricultural times to current patterns of land use using the FIESTA model and scenarios representing pre-agricultural forest cover and current forest cover determined from satellite data. Deforestation has likely produced minor increases in water balance (up to 12.5 mm/yr and runoff <1%), caused by decreased evapotranspiration in places where forest has been replaced by other vegetation covers. At the scale of the Andes basins, a small runoff increase is the dominant effect of this deforestation. Whilst more runoff can mean more available water, in the absence of forest cover on steep slopes this runoff may generate more soil erosion, carry more sediment and be more seasonal with lower flows in the dry season.

The spatial coincidence of pasture, cropland, urban areas and protected areas in the Andes basins indicates a highly impacted system in terms of human land use but also one in which protected areas play a key part in the maintenance of ecosystem functions and the supply of clean water to end users. On the basis of our analyses, the continued conversion of the remaining natural landscape, especially the Andean forests, to pasture and cropland will lead to continued increases in water yield with concomitant soil erosion, sediment yield, and deterioration of water quality, which will impact upon on the increasingly urbanized populations downstream.

Responses to change: development of payment for ecosystem services

Compensation schemes for the provision of ecosystem services can provide incentives for better agriculture and land-use practices. The Andes basins have much to offer in terms of environmental services and there are many pioneering schemes for payments for PES. The schemes compensate land stewards for environmental services and functions such as carbon sequestration, biodiversity or watershed services. The latter, also called payment for watershed services (PWS), are especially relevant in the context of the Andes basins since they provide incentives for upstream land managers to manage land prudently to provide positive impacts on downstream water quantity, quality and regulation. They promise to reduce negative environmental impacts of unsustainable land-management practices, while at the same time contributing to poverty alleviation because poverty in the Andes basins is usually greater in upstream watersheds than downstream. PWS schemes can only be developed, however, where upstream services can be reliably identified and where providers can agree their value with downstream consumers (Jack *et al.* 2008).

If PWS (and other PES) schemes are to achieve both positive environmental impacts and poverty alleviation, hydrological analyses must be incorporated to support negotiation and policy-making processes. This includes a more systemic analysis of water resources at the local level so that all existing water interventions, such as water infrastructure and sources of pollution, are accounted for in the analysis and the downstream impacts of anticipated changes in land management are clear. To legitimize these types of schemes and achieve sustainability, it is necessary to formalize property rights, reduce potentially excessive transaction costs, and increase political will (Jack *et al.* 2008). PWS schemes have great potential in tropical mountains that enjoy large extents of high-quality, water-producing cloud forest and *páramo* ecosystems and populous areas downstream, such as in the Andes basins. They are particularly promising in watersheds where there are dams, since agreements between local authorities and dam projects can use the institutions established for the dam projects, such as water utilities and energy companies, to collect and make payments (Mulligan *et al.* 2009).

Climate change impacts

We used the FIESTA delivery model to understand better the threats of climate change to current provision of water resources in the Andes basins (Mulligan *et al.* 2009). We expect climate change to increase global temperatures and rainfall, but the impact at the regional scale on the basins varies between various global climate models (GCMs). We therefore analysed the projections for 2050 of 17 GCMs for the Intergovernmental Panel on Climate Change (IPCC 2007) fourth assessment business as usual scenario (IPCC AR4 A2a) downscaled to 1 km resolution by Ramirez and Jarvis (2008). Over the whole of the basins, temperatures may increase by 2.9°C and precipitation by 68 mm/yr. For this scenario FIESTA generates a mean increase in water balance of 75 mm/yr for the basins compared with the WorldClim (Hijmans *et al.* 2005) baseline. This mean value disguises important within-basin variations, with many areas showing small increases or decreases, while in a few areas changes are much larger (Figure 6).

Climate change is likely to increase water availability and runoff in parts of the Andes basins, and will affect agriculture as well as urban centres. Threats to water productivity

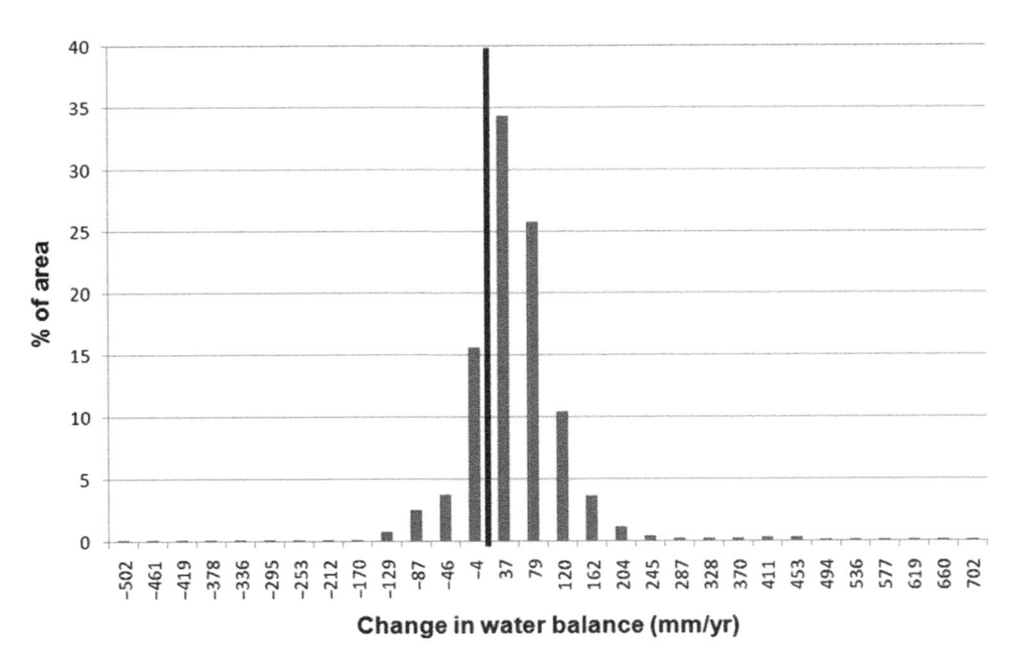

Figure 6. Impact of a mean climate change scenario on water balance for the Andes for business as usual (SRES AR4 A2a) calculated as mean of 17 GCMs.

will be greatest in regions that are already most exposed to land-use change and thus more sensitive to further impacts from climate change, such as parts of the western and central Andes at higher elevation. The dry slopes of coastal Peru and Chile may experience significant increases in total annual runoff.

Conclusions: the Andes basins in a climate of change

The combined Andes basins above 500 m altitude occupy an area of 2.89 million km^2, which would rank them fifth after the Amazon, Congo, Mississippi and Parana basins. The region is made up of a series of neighbouring basins flowing independently to the Pacific, Amazon/Atlantic and Caribbean. In biophysical terms it is mountainous and diverse, has great extremes of rainfall, and great spatial heterogeneity in climate, geology and soils. The Andes basins have greater social inequality between the upper reaches of watersheds and the lower valleys compared to basins with less relief and larger sub-basins in other parts of the world.

In contrast to economies in African and Asian basins described elsewhere in this series, the Andes countries are classified as industrial on account of relatively low overall levels of rural poverty and the low contribution of agriculture to GDP. Crop production is mainly for income generation (cash cropping), and subsistence agriculture is less important than elsewhere, although there remain areas that have poor access to markets or foci of poverty where agriculture is essentially for subsistence. Poverty remains an issue. About half of the population of Colombia, Ecuador, and Peru and almost two thirds in Bolivia are considered poor based on national poverty lines. Of these 47 million poor people, 26 million depend on rural livelihoods. Poverty in the region is largely the result of unequal distribution of the benefits of natural resources. Broad-scale interventions in the education and health sector could reduce poverty and increase opportunities, especially for the rural poor. The heterogeneity of the Andes basins calls for site-specific interventions in the agricultural and hydrological domains to ensure local applicability.

We found no broad causal relationships at either the local or national level between water availability or water management as drivers of poverty, although poverty has implications for the availability of good-quality water at the household level in some cases. Investments in water infrastructure at the regional scale (dams for hydropower and irrigation) support regional development and thus increase the diversity of livelihood options. Specific local interventions such as water treatment are useful in solving immediate problems, but do not necessarily provide routes out of poverty in the Andes basins since access to good-quality water is only one of many factors that defines poverty in the basins.

We found water availability not to be an issue in the northern basins, in contrast to the drier, southern basins. Long-term drought in the Andes is not severe, though seasonal droughts and floods related to the ENSO phenomenon do affect livelihoods. Upstream contamination from agriculture, mining and urban areas constrain access to water since most access to water is from reservoirs or rivers with very little use of groundwater. Access is also constrained by poverty given the high cost of water infrastructure and its paucity for rural populations. Agricultural water productivity is generally high, especially in Colombia and Ecuador but is higher in the lowlands of these countries than in the Andes basins, for a variety of reasons, including the steep land and lower temperature and solar radiation at altitude.

Changes in water demand due to increasing urbanization and industrialization present major uncertainties to future management of water resources. Increased demand from urban centres, mining and irrigation combine with uncertainties of climatic change to increase the potential for conflict and competition amongst uses. Where this takes place within the context of continuously deteriorating natural environments, the potential for a worsening of the water environment in the Andes basins increases. This necessitates careful management of downstream impacts of development and an increased emphasis on more equitable sharing of benefits from land and water use. Better decision making in such a complex environment requires support of detailed and locally relevant analyses of the likely downstream impacts of any new upstream intervention.

As elsewhere, the process requires transparent involvement of, and communication with, stakeholders to legitimize schemes for negotiating and sharing the benefits of water. Negotiations around benefit sharing need to be supported by locally relevant and readily-applicable tools for spatial planning to understand the water-resource environment and the likely impacts of scenarios for climate and land-use or management interventions upon it.[2]

Notes

1. Environmental flow is a "flow regime capable of sustaining a complex set of aquatic habitats and ecosystem processes" (Smakhtin and Anputhas 2006).
2. Such a tool has been developed for the Andes and is available as a web based policy support system at http://www.policysupport.org/links/aguaandes for application throughout the basins.

References

Berdegué, J.A., *et al.*, 2001. Rural non-farm employment and incomes in Chile. *World Development*, 29 (3), 411–425.

Bernet, T., Staal, S., and Walker, T., 2001. Changing milk production trends in Peru: small-scale highland farming versus coastal agro-business. *Mountain Research and Development*, 21 (3), 268–275.

Brea, J.A., 2003. Population dynamics in Latin America. *Population Bulletin*, 58 (1). Washington, D.C.: Population Reference Bureau.

Bruijnzeel, L.A., 2005. Tropical montane cloud forests: a unique hydrological case. *In*: M. Bonell and L.A. Bruijnzeel, eds. *Forests, water and people in the humid tropics*. Cambridge: Cambridge University Press, 462–483.

Byerlee, D., de Janvry, A., and Sadoulet, E., 2009. Agriculture for development: toward a new paradigm. *Annual Review of Resource Economics*, 1, 15–31.

CEPAL, 2005. *América Latina: proyecciones de población urbana y rural 1970–2025*. Boletín Demográfico No. 76. Santiago, Chile: Comisión Económica para América Latina y el Caribe (CEPAL).

CEPAL-UNEP, 2002. *The sustainability of development in Latin America and the Caribbean: challenges and opportunities*. Libros de la CEPAL No. 68. Santiago de Chile: Comisión Económica para América Latina (CEPAL) and United Nations Environment Programme (UNEP).

Chapalbay, W., *et al.*, 2007. *Pobreza, agua y tierra en Ambato, Ecuador: perfil de pobreza y el acceso y manejo de agua y tierra en la cuenca de Ambato, Ecuador*. Copenhagen: Danish Institute for International Studies.

Chomitz, K.M., Buys, P., and Thomas, T.S., 2005. *Quantifying the rural-urban gradient in Latin America and the Caribbean*. Policy Research Working Paper 3634. Washington D.C.: World Bank.

Cohen, B., 2003. Urban growth in developing countries: a review of current trends and a caution regarding existing forecasts. *World Development*, 32 (1), 23–51.

Comunidad Andina, 2004. Andean strategy for disaster prevention and relief. Quito, Ecuador: Andean Council of Ministers of Foreign Affairs.

Deininger, K. and Olinto, P., 2001. Rural non-farm employment and income diversification in Colombia. *World Development*, 29 (3), 455–465.

Dirven, M., 2004. Rural non-farm employment and rural diversity in Latin America. *CEPAL Review*, 83, 47–66.

Eggertsson, T., 1996. A note on the economics of institutions. *In*: L.J. Alston, T. Eggertsson, and D.C. North, eds. *Empirical studies in institutional change*. New York: Cambridge University Press, 6–21.

Endesa, 2009. Electricidad en Latinoamérica. Empresa Nacional de Electricidad S.A. Available from: http://www.endesa.es/Portal/es/negocios/electricidad/latinoamerica/default.htm [Accessed November 2009].

Escobal, J., 2001. The determinants of non-farm income diversification in rural Peru. *World Development*, 29 (3), 497–508.

Falkenmark, M., 1989. The massive water scarcity now threatening Africa. Why isn't it being addressed? *Ambio*, 18 (2), 112–118.

Farrow, A., *et al.*, 2005. Exploring the spatial variation of food poverty in Ecuador. *Food Policy*, 30 (5–6), 510–531.

Flavio Dias Avila, A., Romano, L., and Garagorry, F., 2010. Agricultural productivity in Latin America and the Caribbean and sources of growth. *In*: P. Pingali and R. Evenson, eds. *Handbook of agricultural economics, Volume 4*. Amsterdam: Elsevier, 3713–3768.

FAO, 2003. *Aquaculture: key terms and concepts* [online]. Rome: Food and Agriculture Organization of the United Nations. Available from: www.fao.org/english/newsroom/focus/2003/aquaculture-defs.htm [Accessed 7 August 2010].

FAO, 2004. *State of world fisheries and aquaculture.* Rome: Food and Agricultural Organization of the United Nations. Available from: http://www.fao.org/DOCREP/007/y5600e/y5600e00.htm [Accessed 6 October 2010].

FAOSTAT 2008. *Food balance sheets*. Rome: Food and Agriculture Organization of the United Nations. Available from: http://faostat.fao.org/ [Accessed 6 October 2010].

GlobCover, 2008. *Land Cover v2 database. European Space Agency European Space Agency GlobCover Project* [online]. Available from: http://ionia1.esrin.esa.int/index.asp [Accessed 7 August 2010].

Gomez, L.I., *et al.,* 2005. Pobreza, agua y tierra en Jequetepeque, Peru. DIIS working paper no. 2005/14. Copenhagen: Danish Institute for International Studies.

Hearn, P.P., *et al.*, 2003. Global GIS - Global Coverage DVD. 1st edition. Alexandria, VA: American Geological Institute.

Hijmans, R.J., *et al.*, 2005. Very high resolution interpolated climate surfaces for global land areas. *International Journal of Climatology*, 25, 1965–1978.

Holman, F., *et al.*, 2003. Evolution of milk production systems in tropical Latin America and its interrelationship with markets: an analysis of the Colombian case. *Livestock Research for Rural Development,* 15 (9). Available from: www.lrrd.org/lrrd15/9/holm159.htm [Accessed 7 August 2010].

Hurtado, A.D., 2003. *Los desplazados en el Perú* [online]. Lima, Peru: Comité Internacional de la Cruz Roja and Programa de Apoyo al Repoblamiento y Desarrollo de Zonas de Emergencia. Available from: www.internal-displacement.org/8025708F004CE90B/(httpDocuments)/59E7F182DE8D651A802570B700599C67/$file/ICRC+PAR+Los+desplazados+en+el+Perú.pdf [Accessed 13 August 2010].

Ibañez, A.M. and Querubin, P., 2004. *Desplazamiento forzoso, conflicto armado, acceso a tierras, deseo de retorno y migración*. Documento CEDE 2004-23. Bogotá, Colombia: Universidad de Los Andes.

IPCC 2007. *Climate change 2007: synthesis report. Contribution of working groups I, II and III to the fourth assessment report of the Intergovernmental Panel on Climate Change.* R.K. Pachauri and A. Reisinger, eds. Geneva: IPCC.

IDMC, 2009. *Continúan nuevos desplazamientos, respuesta aún ineficaz* [online]. Geneva: Internal Displacement Monitoring Center. Available from: www.internal-displacement.org [Accessed 11 August 2010].

Jack, B.K., Kousky, C., and Sims, K.R.E., 2008. Designing payments for ecosystem services: lessons from previous experience with incentive-based mechanisms. *Proceedings of the National Academy of Sciences of the United States of America*, 105 (28), 9465–9470.

Johnson, N., *et al.*, 2009. Water and poverty in two Colombian watersheds. *Water Alternatives*, 2 (1), 34–52.

Jokisch, B., 2002. Migration and agricultural change: the case of smallholder agriculture in highland Ecuador. *Human Ecology*, 30 (4), 523–550.

Katz, E., 2003. Is there a feminization of agriculture? *In:* B. Davis, ed. *Current and emerging issues for economic analysis and policy research. Vol. 1: Latin America and the Caribbean.* Rome: FAO, 31–66. Available from: ftp.fao.org/docrep/fao/006/y4940e/y4940e00.pdf [Accessed 7 August 2010].

Keating, P.S., *et al.*, 2010. Eco-efficient agriculture: concepts, challenges and opportunities. *Crop Science*, 50 (2), S-109–S-119.

Landell-Mills, N., 2002. Developing markets for forest environmental services: an opportunity for promoting equity while securing efficiency? *Philosophical Transactions of the Royal Society of London A*, 360 (1797), 1817–1825.

McDowell, B., 1986. Eruption in Colombia: 23,000 villagers perish in volcanic mudflows. *National Geographic*, 169, 640–653.

Mark, B., McKenzie, J.M., and Gomez, J., 2005. Hydrochemical evaluation of changing glacier meltwater contribution to stream discharge: Callejon de Huaylas, Peru. *Hydrological Sciences Journal*, 50 (6), 975–987.

Mulligan, M., 2006a. Global gridded 1 km TRMM rainfall climatology and derivatives. Version 1.0. Database 2006a [online]. Available from: www.ambiotek.com/1kmrainfall

Mulligan, M., 2006b. MODIS MOD35 pan-tropical cloud climatology. Version 1 [online] September 2006. Available from: www.ambiotek.com/clouds

Mulligan, M. and Burke, S.M., 2005. *FIESTA: Fog interception for the enhancement of streamflow in tropical areas* [online]. Available from: http://www.ambiotek.com/fiesta

Mulligan, M. and Burke, S.M., 2010. *Spatial analysis of productivity in the CGIAR Challenge Programme on Water and Food basins* [online]. Colombo: Challenge Programme on Water and Food. Available from: http://www.ambiotek.com/cpwfproductivity [Accessed 7 August 2010].

Mulligan, M., *et al.*, 2009. *The Andes basin focal project. Final report to the CGIAR Challenge Program on Water and Food* [online]. Available from: http://www.bfpandes.org [Accessed 7 August 2010].

Mulligan *et al.,* 2011. Water availability and use in ten major river basins. *Water International*, forthcoming.

Ramankutty, N.A., *et al.*, 2008. Farming the planet: 1. Geographic distribution of global agricultural lands in the year 2000. *Global Biogeochemical Cycles*, 22, 1–19.

Ramirez, J. and Jarvis, A. 2008. *High resolution statistically downscaled future climate surfaces* [online]. Cali, Colombia: International Centre for Tropical Agriculture, CIAT. Available from: http://gisweb.ciat.cgiar.org/GCMPage [Accessed 7 August 2010].

Revista Latinomineria, 2009. La nueva frontera de la mineria Colombiana. *Latinomineria*, 67, 8–15. Available from: www.latinomineria.com/revistas/PDF/LM%2067%20CHILE.pdf [Accessed 12 August 2010].

Rodríguez, J. and Busso, G., 2009. *Migración interna y desarrollo en América Latina entre 1980 y 2005. Un estudio comparativo con perspectiva regional basado en siete países* [online]. Santiago, Chile: Comisión Económica para América Latina y el Caribe (CEPAL). Available from: www.eclac.org/cgi-bin/getProd.asp?xml=/publicaciones/xml/6/36526/P36526.xml&xsl=/celade/tpl/p9f.xsl&base=/celade/tpl/top-bottom_minterna.xslt [Accessed 7 August 2010].

Rushton, J. and Viscarra, R., 2006. *Productive strategies for poor rural households to participate successfully in global economic processes: country report for Bolivia, central Andes.* London: Overseas Development Institute (ODI).

Saleth, R.M. and Dinar, A. 2004. *The institutional economics of water: a cross-country analysis of institutions and performance.* Washington D.C. and Cheltenham: World Bank and Edward Elgar.

Siebert, S., *et al.*, 2007. Global map of irrigation areas version 4.0.1. Frankfurt and Rome: University of Frankfurt (Main) and FAO.

Smakhtin, V. and Anputhas, M., 2006. *An assessment of environmental flow requirements of Indian river basins* [online]. Research report No. 107. Colombo: International Water Management Institute. Available from: www.iwmi.cgiar.org/Publications/IWMI_Research_Reports/PDF/PUB107/RR107.pdf [Accessed 13 August 2010].

Stanton, T., *et al.*, 2010. *State of watershed payments: an emerging marketplace* [online]. Washington D.C.: Ecosystem Marketplace, Forest Trends Association. Available online: www.foresttrends.org/documents/files/doc_2438.pdf [Accessed 11 August 2010].

Swinton, S. and Quiroz, R., 2003. Is poverty to blame for soil, pasture and forest degradation in Peru's altiplano? *World Development*, 31 (11), 903–1919.

TNC, 2010. *The Nature Conservancy in South America: Creating water funds for people and nature* [online]. Available from: www.nature.org/wherewework/southamerica/misc/art26470.html [Accessed 12 August 2010].

Thouret, J.-C., *et al.*, 1990. Stratigraphy and quaternary eruptive history of the Ruiz-Tolima volcanic massif, Colombia. Implications for assessement of volcanic hazards [online]. *Symposium international Géodynamique andine, 15–17 May, 1990 Grenoble, France: résumés des communications*. Paris: ORSTOM, 391–393. Available from: horizon.documentation.ird.fr/exl-doc /pleins_textes/pleins_textes_4/colloques/31077.pdf [Accessed 7 August 2010].

UN, 2007. *Panorama de la población mundial, revisión 2006. Edición en CD-ROM - Base de datos en formatos Excel y ASCII*. Publicación ST/ESA/SER.A/266. New York: UN.

USAID and ARD, Inc. 2008. Global land tenure master database. 2007. Unpublished data. Original graphics published in USAID and ARD, Inc. 2007. *Land tenure and property rights tools and regional reports*. Washington, DC: United States Agency for International Development, EGAT/Natural Resources Management/Land Resources Management Team and Burlington, Vermont: ARD, Inc.

Valdés, A. and López, R. 1999. Fighting rural poverty in Latin America: new evidence of the effects of education, demographics, and access to land. *Economic Development and Cultural Change*, 49 (1), 197–211.

Weinberger, K. and Lumpkin, T.A. 2005 Horticulture for poverty alleviation: the unfunded revolution. Working Paper No 15. Shanhua, Taiwan: AVRDC - The World Vegetable Center.

Wetherald, R.T. and Manabe, S., 2002. Simulation of hydrologic changes associated with global warming. *Journal of Geophysical Research*, 107 (D19), 4379–4393.

World Bank, 2009. *World development indicators*. Washington D.C.: World Bank.

WHO, 2008. *Progress in drinking water and sanitation: special focus on sanitation*. World Health Organization and United Nations Children's Fund Joint Monitoring Programme for Water Supply and Sanitation (WHO/UNICEF-JMP). New York and Geneva: UNICEF and WHO.

Xueliang *et al.*, 2011. Producing more food with less water in a changing world: assessment of water productivity in ten major river basins. *Water International*, forthcoming.

Yancari, J., 2009. *Crisis y pobreza rural en América Latina: el caso de Perú*. Documento de Trabajo No. 41. Programa Dinámicas Territoriales Rurales. Santiago, Chile: Rimisp-Centro Latinoamericano para el Desarrollo Rural.

Zimmerer, K., 1993. Soil erosion and labor shortages in the Andes with special reference to Bolivia 1953–91: implications for conservation-with-development. *World Development*, 21 (10), 1659–1675.

The Indus and the Ganges: river basins under extreme pressure

Bharat Sharma, Upali Amarasinghe, Cai Xueliang, Devaraj de Condappa, Tushaar Shah, Aditi Mukherji, Luna Bharati, G. Ambili, Asad Qureshi, Dhruba Pant, Stefanos Xenarios, R. Singh and Vladimir Smakhtin

The basins of the Indus and Ganges rivers cover 2.20 million km and are inhabited by more than a billion people. The region is under extreme pressures of population and poverty, unregulated utilization of the resources and low levels of productivity. The needs are: (1) development policies that are regionally differentiated to ensure resource sustainability and high productivity; (2) immediate development and implementation of policies for sound groundwater management and energy use; (3) improvement of the fragile food security and to broaden its base; and (4) policy changes to address land fragmentation and improved infrastructure. Meeting these needs will help to improve productivity, reduce rural poverty and improve overall human development.

Introduction

Although the basins of the Indus and Ganges rivers are separate entities, they are contiguous. We shall therefore consider them together as the IGBs to enable comparisons and contrasts. The IGBs cover 2.2 million km^2 and have a population of about one billion, one of the world's densest. The IGBs' diverse agro-climatic, social and economic conditions in four countries, India, Pakistan, Nepal and Bangladesh, make them very complex. The IGBs have been cultivated and irrigated since 5000 B.C. Cultivation in the Indus Basin has intensified with massive resource development, especially recently. The Indus Basin irrigation system (IBIS) in Pakistan and the Bhakra irrigation system in India are the world's largest, generating surplus food that provides regional food security outside the IGBs (Amarasinghe *et al.* 2005). The IGBs were the seat of green revolution in Asia.

The performance of the large publicly funded irrigation infrastructure of the Indus has started to decline, however. Anarchic development of groundwater currently drives the rural economy (Mukherji *et al.* 2009a, Shah 2009) and meets the large domestic and industrial water needs. The groundwater is being exploited unsustainably, but is still expanding, creating large regional negative water balances. Falling water tables (Rodell *et al.* 2009) have also created an unviable nexus between water and energy (Scott and Sharma 2009). The Indus Basin is now practically closed with near zero environmental flows in most years.

The Ganges Basin has large supplies of both surface and groundwater but faces wide economic water scarcity (Molden *et al.* 2003), especially in Nepal, Bangladesh and eastern India. Land and water productivity for most crops and fisheries are low and the population dependent on agriculture (>85%) is very poor. There is prolific shallow groundwater, but it is poorly developed due to low rural electrification, high prices for diesel fuel, and tiny and scattered land holdings (Shah *et al.* 2009). The Ganges often floods during the monsoon season, and coastal Bangladesh is subject to cyclones, but there are dry spells and even droughts. Groundwater downstream is widely contaminated with arsenic (Chakraborti 2004).

There are complex transnational boundary issues in the IGBs. These, coupled with the weak water institutions, governance and policies at the local, national and regional level, seriously constrain implementation of basin-wide strategies and policies. Furthermore, climate change will alter the patterns of snow and glacier melt, and increase the frequency and intensity of floods and droughts. Its potential impact on agricultural yields, water resources, production systems and livelihoods in the IGBs calls for large-scale, multi-sectoral adaptation (Sharma 2009, Sharma and McCornick 2006, Hosterman *et al.* 2009). Fortunately, there are vast areas under rainfed mono-crops and more than 45% of the available water resources are un-appropriated, which presents opportunities to intensify farming systems and alleviate poverty.

This paper summarizes two years' research (2007–2009) supported by the Basin Focal Projects of the CGIAR Challenge Program on Water and Food. We shall review the challenges and opportunities in the IGBs in terms of hydrology, water resources and the variations in water and land productivity at the regional and local levels. We report on the prevalence of poverty at national, sub-national and household level, and its nexus with poverty for land and water and the opportunities to alleviate it. We further examine the state and effectiveness of existing water laws, policies and institutions in the region. Finally, we recommend interventions to improve productivity, reduce poverty and ensure sustainable use of resources.

Regional perspective

The IGBs (Figure 1) have a mosaic of interactions of poverty and prosperity, and problems and possibilities. Population growth reduced cultivable land per capita from 0.35 ha in 1880 to 0.19 ha in 1970. It halved to again to 0.10 ha in 2000. Agricultural water use increased rapidly, so access to water is central for livelihoods of the rural poor.

The Indus Basin covers an area of about 1.10 million km^2 distributed between Pakistan (63%), India (29%), and China and Afghanistan (8%) (Jain *et al.* 2009). The Indus has two main tributaries, Panjnad from the east formed by the five rivers Jhelum, Chenab, Ravi, Beas and Sutlej, and the Kabul River on the west. The Indus Water Treaty (1960) shares the Indus' waters between Pakistan and India.

The Ganges rises at Gangotri glacier in India. Its basin is about 1.09 million km^2, distributed between India (79%), Nepal (13%), Bangladesh (4%), and the rest in Tibet (China). The major tributary of Ganges is the Yamuna, which joins it at Allahabad. The Ganges then flows east joined by the further tributaries Ramganga, Gomti, Ghaghra, Gandak, Bagmati, Kosi, Sone and Damodar.

The upper reaches of both the Indus and the Ganges are in the high Karakoram and Himalayan mountains with many peaks over 7000 masl. The intensively cultivated area

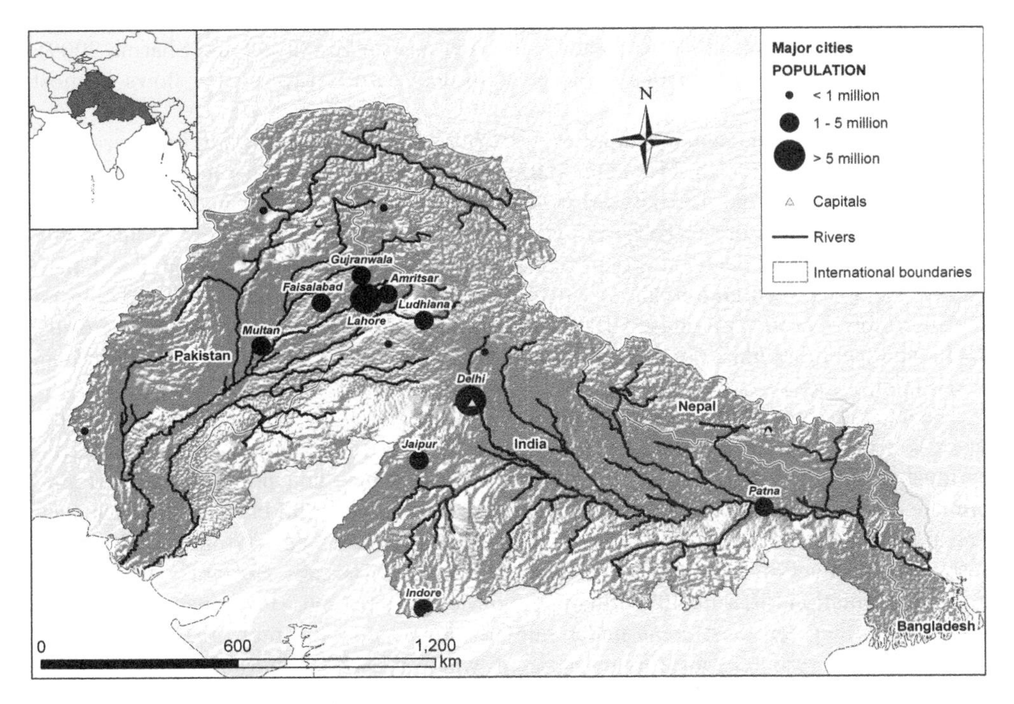

Figure 1. Map of the basins of the Indus and Ganges rivers.

Table 1. Socio-economic indicators of the countries of the Indus and Ganges basins.

Parameters	Bangladesh	India	Nepal	Pakistan
Access to improved water resources, %	74	86	90	91
Access to improved sanitation, %	39	33	35	59
Per capita electricity consumption, kWh	145	594	91	493
Population below national poverty line, %	49.8	28.6	30.9	32.6
Agriculture, % of gross domestic product (GDP)	20.1	18.3	38.2	21.6
Per capita GDP (USD)	406	640	252	632
IRWR (m^3/capita/yr)	688	1149	7539	325

Note: IRWR = Internally renewable water resources.
Source: Adapted from Babel and Wahid (2008).

of the contiguous Indus-Gangetic plains (IGP) is on fertile soils formed from alluvium. The Ganges delta lies downstream of Farakka barrage on the India–Bangladesh border.

The IGBs, with about one seventh of the world's population, are characterized by social and economic heterogeneity. In Nepal, 86% of the total population is rural, in Bangladesh 80%, in India 75%, and in Pakistan 68%. Poverty amongst the rural population is about 31% (2000) or more than 220 million, over half the total rural poor of South Asia. Table 1 shows indicators of the socio-economic conditions of the IGBs.

The net cropped area of the IGP is 1.14 million km^2, mainly to wheat, rice, cotton and sugarcane. The IGP produces 93% of the wheat and 58% of the rice produced in the IGBs' countries. Inland fisheries are important producers of food in the lower parts of the Ganges in India and Bangladesh. The economy and income levels are rising in the IGBs as a whole and in India in particular, which has implications for future water and food requirements (Amarasinghe *et al.* 2008). The eastern part of the Ganges Basin, where population densities are highest, was bypassed by the green revolution. It is still weak in rural infrastructure, market development, institutions, energy, credit for agricultural operations, location-specific technologies, surface-storage irrigation systems and development of groundwater resources. In contrast, the western region was the seat of green revolution and has high productivity, good irrigation (now dominated by groundwater), rural infrastructure and markets.

Water resources: use, abuse and non-use

Overall, the IGP has passed from common welfare irrigation systems in the ancient past, to imperialism and strict governance during the British colonialism, to the recent private-venture groundwater irrigation. There are important regional differences from the dynamic and progressive Indus Basin compared to the under-developed, introspective eastern part of the Ganges Basin. In the west, the overarching issue is sustainable management of water and land resources; in the east, it is access to resources and institutions, livelihoods, poverty and food security.

Annual flow in the Indus is 120–230 km^3 (1957–97) with only about 10% net discharge, with glacier melt providing stream flow in the upper basin. Average annual flow in the Ganges varies from 5.9 km^3 upstream at Tons River to 459 km^3 at Farakka barrage. Northern tributaries, especially in Nepal, contribute substantially (Jain *et al.* 2007). The mean annual input from rainfall to the Ganges Basin is 1170 km^3. Net discharge of the Ganges is about 429 km^3 (37% of the input). It is the largest water use, followed by rainfed agriculture. Groundwater is not uniformly available across the IGP; recharge in some parts of the Indus is < 2 mm/yr, while in Nepal and Bangladesh groundwater recharge is medium to high. In addition to the dynamic groundwater resource, which is commonly pumped, there is a massive static (unused) groundwater resource of about 9170 km^3, 85% of which is in the Ganges Basin.

Water use in the IGBs is dominated by agriculture, which uses 91% followed by 7.8% for domestic use (Eastham *et al.* 2010). Irrigated agriculture uses about 286 km^3, 23% of the total water use. IBIS in Pakistan is the world's largest irrigation network, with an area of 17 Mha and an intake of almost 75% of the water in the Indus River. A large proportion of the groundwater in the Indus Basin is pumped by irrigators; in contrast, they pump only 54% of the groundwater available in the Ganges Basin. Demographic pressures and industrial development influence the distribution of water between sectors (Table 2). Amarasinghe *et al.* (2007) forecast demand for domestic and industrial water to increase in the near future.

Floods and droughts kill people and seriously damage crops, livestock and infrastructure. With climate change they are predicted to increase in frequency and severity

Table 2. Water demands in the Indus (down to Kotri), and the Ganges (down to Farakka).

	Demands			
Basin	Irrigation	Domestic	Industries	Total
	km³/yr (%)	*km³/yr (%)*	*km³/yr (%)*	*km³/yr*
Ganges	93.9 (82.1)	10.7 (9.4)	9.8 (8.6)	114.4
Indus	168.7 (95.5)	4.4 (2.5)	3.4 (2.0)	176.5

Source: Adapted from Amarasinghe *et al.* (2007), Habib (2004), and the Bhakra Beas Management Board (http://bbmb.gov.in/english).

(Aggarwal *et al.* 2004). The Indus Basin ranked in the top ten of the world's most vulnerable with inflows predicted to fall by 27% by 2050 (IPCC 2001). Glacier melt is the major water source for the IGBs and hence of water for irrigation so that reduced glacier melt will therefore impact food production in the IGP. There was an overall deglaciation of 21% since 1962 in 466 glaciers in the Chenab, Parbati and Bapsa basins. The Gangotri glacier retreated an alarming 100 m between 1994 and 1998 (Tangri 2000).

The combination of glacial retreat, decreasing ice mass, early snowmelt and increased winter stream flow suggest that climate change is already affecting the Himalayan cryosphere (Kulkarni *et al.* 2007). Reduced surface runoff will reduce groundwater recharge and affect the groundwater dynamics in the region, which will be critical in the western region, where most irrigation is from groundwater. Strength-weakness-opportunity-threat (SWOT) analysis showed an overall reduction in runoff in both rivers under all greenhouse gas emission scenarios (Gosain and Rao 2007).

Floods are common in the Ganges Basin, with the Ganges itself and its tributaries Yamuna, Sone, Ghagra, Gandak, Koshi and Mahananda all flooding annually. Bangladesh is flooded almost every year by the Ganges, Brahmaputra and Meghna rivers. These are transnational rivers, so it is difficult for the individual government to implement appropriate and timely measures for forecasting and moderating floods.

Water control and land use

We did Soil Water Assessment Tool (SWAT) modelling of three sub-basins, Gorai (Bangladesh), Upper Ganga (India) and Koshi (Nepal). Completion of the upstream Farakka barrage in 1974 changed the inflow to Gorai, which is critical for growing rice in the dry season. The 1965–75 inflows in February and March averaged 17 MCM, but only averaged 2 MCM in 1990–9. The annual downstream runoff from Gorai averaged 188 mm in 1990–9, 26% less than in 1965–75. Additionally, abstraction between Farakka and Gorai further lessens inflow. Farmers in Gorai had to abandon rice for less demanding crops and develop groundwater to make up the shortfall.

Dams, barrages, and canals regulate flow in the upper Ganges River (Figure 2). Rainfall and evapotranspiration are less and water yield is higher in the upper catchments, especially where structures control runoff. Evapotranspiration was highest for the forested areas followed by irrigated areas.

The southern part of the Koshi sub-basin is wetter than the trans-Himalayan northern part. Evapotranspiration is higher in the south than in the north but runoff exceeds evapotranspiration in the south because of higher rainfall. Optimal resource planning at the regional level requires detailed modelling of the existing and future scenarios at the sub-basin level.

Water productivity: assessment and improvement

Crop water productivity (WP) indicates food security at a basin scale and can guide policies to adapt to climate change. We mapped WP in the whole IGBs by integrating crop census data with remote sensing and GIS. First we mapped dominant crops using census data and existing maps of land use and land cover. We calculated yields from national statistics and interpolated them to pixel level (250 m × 250 m) using the normalized difference vegetation index (NDVI) satellite data. We mapped crop evapotranspiration (ET_a) using a simplified surface energy balance (SEBAL) model based on satellite data of land surface temperature and data from 56 weather stations. We mapped WP for rice and wheat

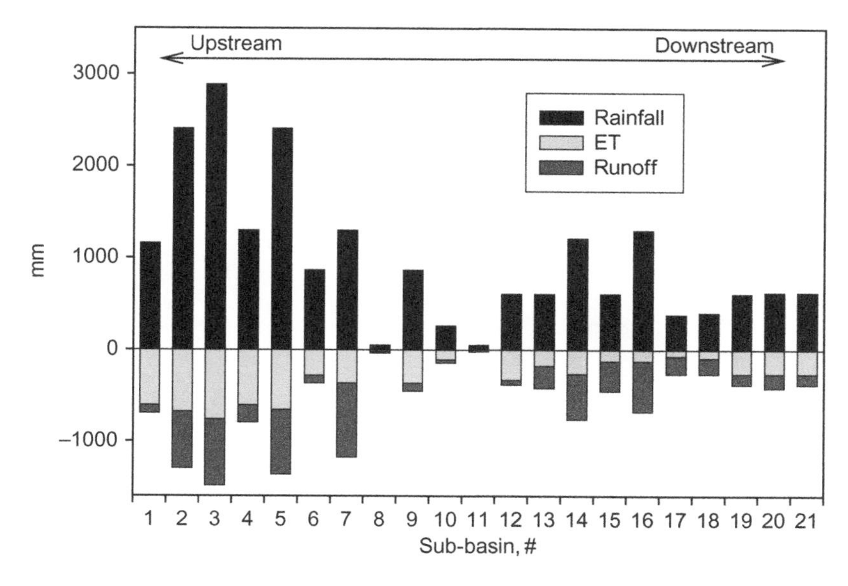

Figure 2. Impact of water control and land uses on the water balance components of rainfall, evapotranspiration (ET), and runoff from sub-basins within the catchment of the upper Ganga (Ganges) River in Uttar Pradesh.

and total agricultural yield by dividing crop yield by ET_a for each pixel (Ahmad *et al.* 2009, Cai and Sharma 2010).

The WP map displays the spatial variation in great detail. We can identify well-performing bright spots and low performing hot spots regardless of administrative boundaries. Linking them to rainfall distribution, topography, groundwater level and other spatial information can indicate causal relationships, which is useful to provide information for improved intervention planning.

The IGBs are intensively cropped with more than 50% of the total area under cultivation. The dominant system is a rice–wheat rotation, mixed with cotton, sugarcane, pulses, oilseeds, millet and jute. Rice (32.3% of the total area) and wheat (27.5%) occupy more than half the area under cultivation. Both crops are irrigated, but need better water management. We look at the WP of the rice–wheat system in more detail below.

Physical water productivity of rice

The average rice yields for the Pakistan, India, Nepal and Bangladesh parts of the IGBs (Figure 3) are 2.60, 2.53, 3.54 and 2.75 t/ha, respectively, but differ widely across the region.

The Indian Punjab state, with some adjacent areas of Haryana and Rajasthan states (dark patch in Figure 3[c]) has an average yield of 6.18 t/ha, significantly higher than most other parts of the Indus Basin. The rice yields of the Indian states of Madhya Pradesh (1.18 t/ha), Rajasthan (1.49 t/ha), and Bihar (2.04 t/ha) in the Ganges Basin are low, which accounts for India's low average yield. The map of rice yield at pixel resolution shows yield variations at the local scale. For example, about 1% of the area (adjacent to the foothills) around the high-yielding Punjab yields less than 3 t/ha. In contrast, although the mean yield of Bihar is only 2.04 t/ha, there is a 37 km circular area southwest of the Bhojpur district that yields 4 t/ha. This area in Bihar is well served by the conjunctive use of the

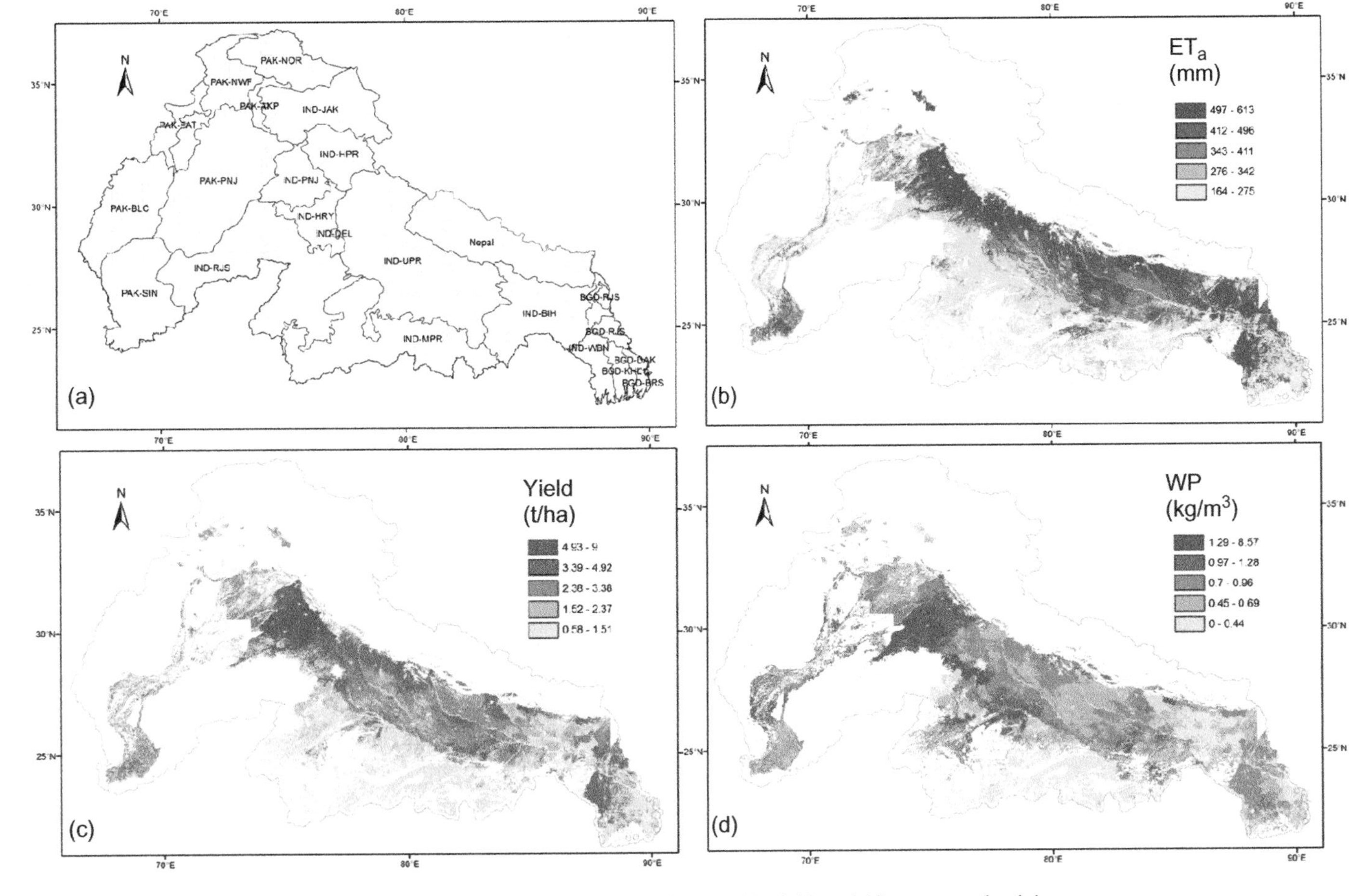

Figure 3. Rice in the IGBs for the year 2005: (a) administrative boundaries; (b) ET_a; (c) yield; and (d) water productivity.

Sone canal system and tubewells, as well as widespread adoption of improved agronomic practices.

ET_a is a measure of how much water a particular ground cover uses. A crop's potential ET_a is its evapotranspiration without water or nutrient limitation. ET_a data indicate the water-use efficiency of irrigation, and low ET_a values indicate stressed rice or mixed cropping.

The ET_a of paddy rice in the IGBs for the growing season 10 June–15 October 2005 averaged 416 mm (range 167–608 mm, standard deviation (SD) 104.6 mm), which is significantly less than the rice potential evapotranspiration (ETp, 610 mm). There is significant variation of ET_a across the region (Figure 3[b]). The adjoining areas of the Indus and Ganges basins in Punjab, Haryana and west Uttar Pradesh, covering 8% of the total rice area, have ET_as averaging 551 mm. ET_a is also high in north West Bengal (528 mm), and in a belt from the Khulna division of Bangladesh to the Indian states of West Bengal, northern Bihar, central Uttar Pradesh, Haryana and Punjab (Figure 3[b]). ET_a is low in Madhya Pradesh and Rajasthan states in India, which are distant from the Ganges River, and in south Punjab and north Sind provinces in Pakistan, where there is more mixed cropping. Overall, ET_a follows yield, although in Bihar ET_a is high but yields are low, indicating low irrigation efficiency.

Average WP of rice in the IGBs is 0.74 kg/m^3 (0.18–1.8 kg/m^3, SD 0.329), with the pattern of WP generally following that of yield. WP of the Indian Punjab and adjoining areas that cover 6% of the IGBs' rice area averages 1.32 kg/m^3. About 23% of the IGBs' rice areas, mainly in Madhya Pradesh and Bihar in India and the Dhaka division of Bangladesh, all in the Ganges Basin, have WP less than 0.5 kg/m^3. These areas are not adequately served by surface or groundwater resources and could benefit from improved access to irrigation.

Physical water productivity of wheat

Wheat yield in the IGBs averaged 2.65 t/ha (SD 1.0 t/ha.). Yield distribution is similar to rice (Figure 4[a]). Bright spots are in the Indian states of Punjab and Haryana where yields average 4.4 t/ha and somewhat lower in the upstream Indus in Pakistan, in west Uttar Pradesh, and northeast Rajasthan in India. Hot spots, with yield levels of only 0.70–1.58 t/ha, are in the Ganges Basin in Bihar and West Bengal states in India and in Bangladesh. The yield differences in irrigated wheat are likely caused by soil, fertilizer and crop variety rather than water.

Wheat is grown in the cooler season, so that evapotranspiration 24 November 2005–14 April 2006 averaged 299 mm (SD 61 mm). Wheat ET_a follows the same pattern as yields (Figure 4[b]) with the bright spots in the areas where yields were high. Elsewhere, ET_a is relatively low and uniform. WP of wheat averages 0.94 kg/m^3 (SD 0.66 kg/m^3). Despite low yields in Rajasthan and Madhya Pradesh, WP is higher because of the low ET_a and short growing season. These states grow low-yielding traditional varieties with high cooking quality, which fetch premium prices. The areas with the highest yields had high WP, but not the highest. Bihar in India has large areas of low WP, indicating scope for improvement.

Annual economic water productivity of rice–wheat cropping system

WP of rice in 2005 at local market prices was US$0.121/m^3, and for wheat was US$0.148/m^3, reflecting the differing physical WPs of the crops and the higher price

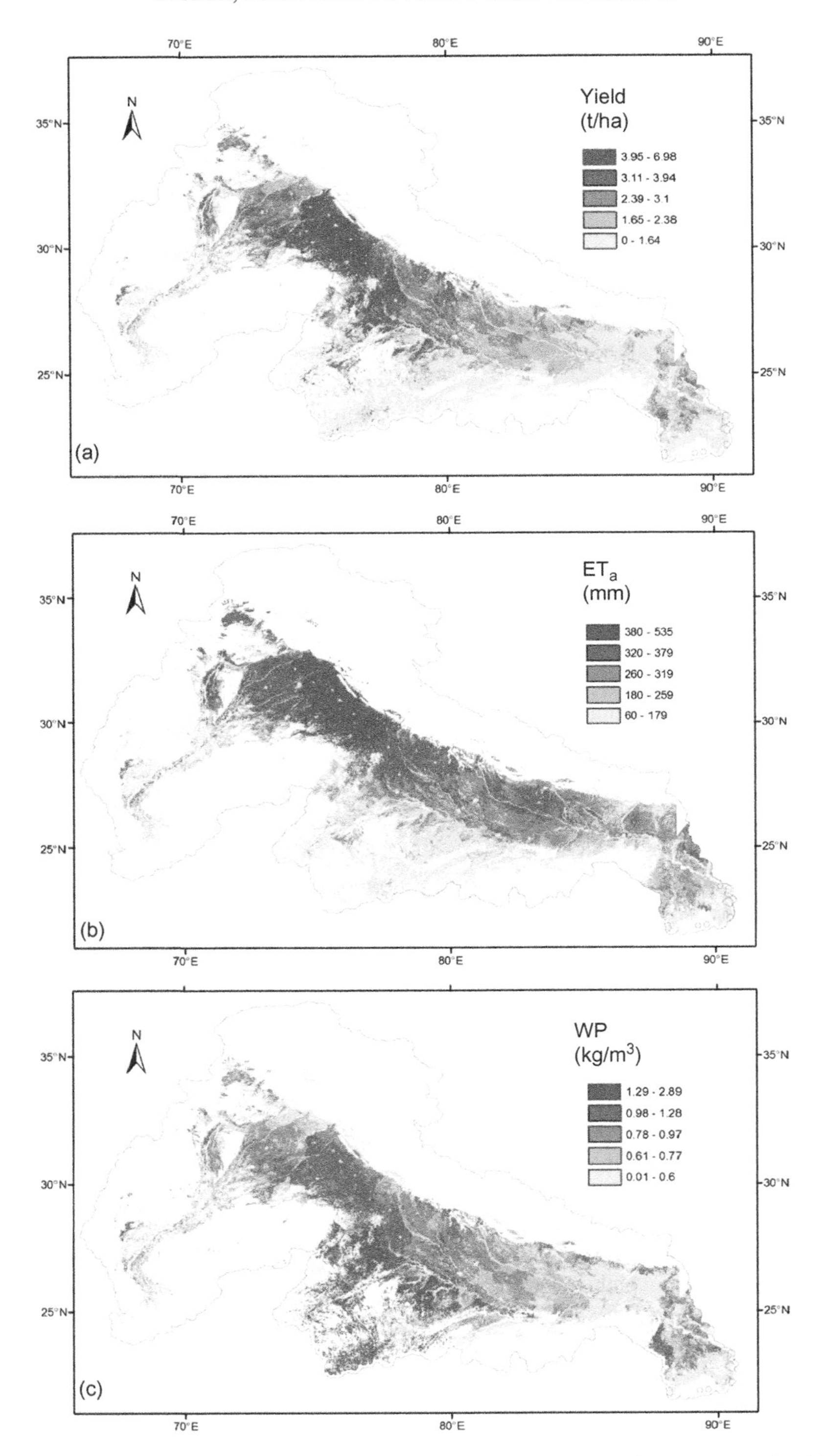

Figure 4. Wheat in the IGBs for the year 2005/6: (a) yield; (b) ET_a; and (c) water productivity.

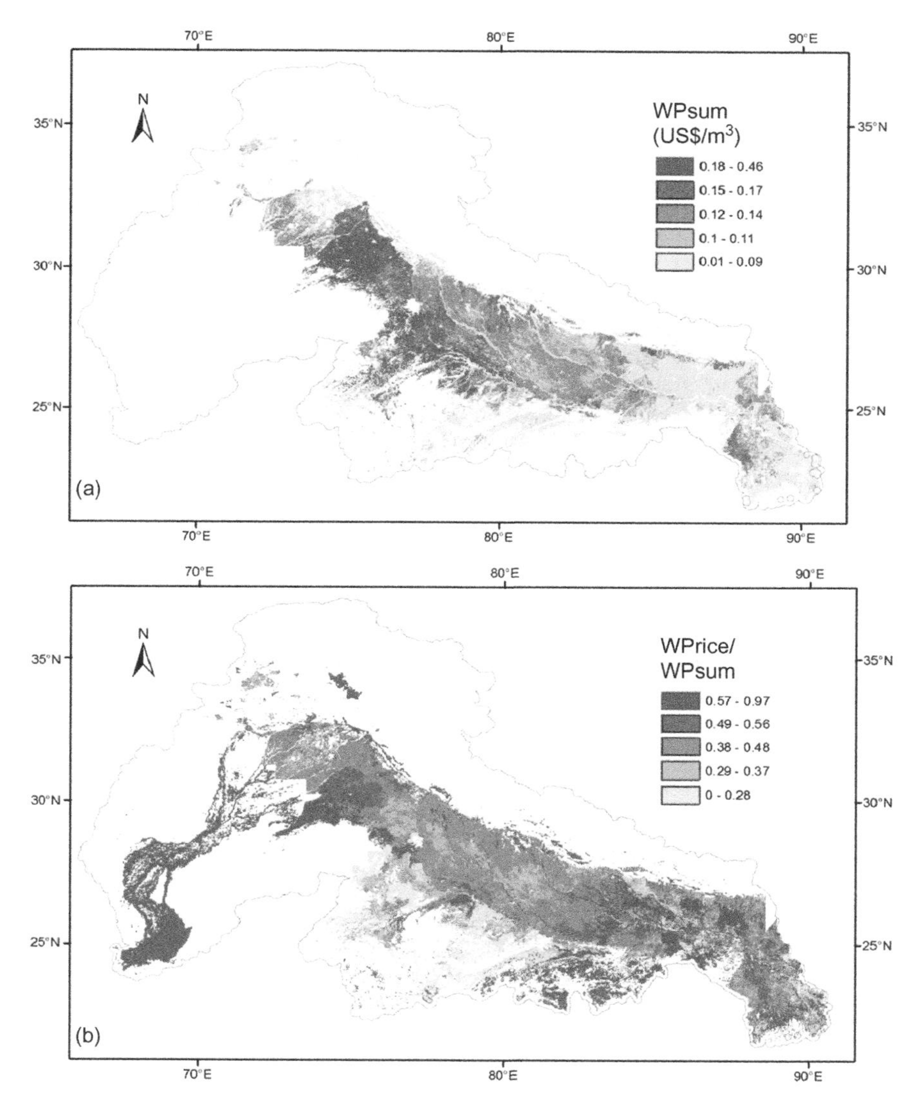

Figure 5. The rice–wheat rotation in the IGBs for the year 2005/6: (a) economic WP (WPsum); and (b) the contribution of rice (WPrice/WPsum).

for rice. WP for the rice–wheat rotation is US$0.131/m^3 (Figure 5). Economic WP for the rice–wheat system is different from either crop alone. The border areas of Rajasthan, Madhya Pradesh and Uttar Pradesh together with the Indian Punjab perform best with the combined system. Rice contributed 50.7% to the combined system despite lower WP because the cropped area is larger.

Factors affecting variation in water productivity

Rainfall in the IGBs is much higher during rice growing season (*kharif*) than that of wheat (*rabi*). Moreover, rainfall in the Indus Basin is much lower than in the Ganges Basin during *kharif* but is reversed in *rabi*. Temperatures of the land surface are similar across basins and seasons. Temperatures are hottest in the deserts of the downstream Indus Basin, and coldest in the northern mountains. Topographically the IGP is flat although the southern Ganges Basin is higher.

Potential evapotranspiration (ET_P) is lower in the Ganges Basin than the Indus, reflecting the differences in rainfall distribution. Crop stress (ET_a/ET_P, the lower the ratio the greater the stress) is widespread in the Indus Basin and in the south Ganges Basin. As a closed basin, the Indus faces severe water scarcity. In contrast, the well-performing Indian Punjab shows little water stress in spite of the deep groundwater table and low rainfall. High water availability in the eastern Ganges does not give higher yields or WP.

Rainfed crops are more vulnerable, so that rainfed paddy may have more standing water at times but can still suffer from water stress at critical growth stages, such as grain filling, which can reduce grain yield. Rice is also vulnerable to excess water particularly soon after establishment. Adequate irrigation and drainage systems and appropriate management can maximize utilization of rainfall and river flows to achieve high yield and WP. Other land and crop interventions, such as laser land levelling, furrow-irrigated raised beds (FIRBS), control of insects and diseases, optimum fertilizer and suitable varieties, are part of the agronomic package.

Yields, ET_a and WP of wheat are more consistent across the IGBs. High ET_a, yields and WP occur together. Irrigation is the main control of ET_a in the dry *rabi* season, so that wheat yields are more related to irrigation volume. The high wheat yields in the Haryana, and the Indian and Pakistani Punjab come from the extensive irrigation from groundwater, which is over-exploited and unsustainable.

Yield and WP vary consistently where water availability is the major constraining factor, such as the downstream Indus Basin and downstream areas in the southern Ganges Basin. Elsewhere in the Ganges, yields and WP are not constrained by water availability, or climate, but the irrigation infrastructure, access to irrigation water and crop management.

The bright spot for both rice and wheat in the Indian Punjab and adjacent areas, with only 5% of the IGBs' area cropped to rice and wheat, has the high WP of US\$0.190/m^3. If the basin average of US\$0.131/m^3 could be increased to the same as in bright spots, the basin could use 31% less water for the same production, or increase production by 31% with the same amount of water. There are many constraints to achieving this theoretical optimum, but there are clearly possibilities to increase WP, which would be important for regional food security.

Potential interventions for improving water productivity

Cropping patterns and sources of irrigation water have changed greatly in the IGBs. The area under paddy has grown exponentially, while water tables have dropped due to pumping of groundwater in the Indian and Pakistani Punjab, Haryana and western Uttar Pradesh. Intensive production of irrigated rice and wheat with limited water resources is hydrologically unsustainable. In contrast, the abundant water available in the eastern Ganges Basin is underdeveloped and underutilized, so that land and water productivity are low. A variety of physical, crop-related, and policy interventions have been tried in the IGBs but with varying degrees of success. We analyse some of the successful ones below.

Potential interventions for selected crops of the Indus and Ganges basin

We used the analytic hierarchy process (AHP) (Saaty 1980) to rank potential interventions. We selected eight crops from the major cropping systems in the IGBs (Biswas *et al.* 2006),

Table 3. Priority ranking of three most important interventions for productivity improvement of the selected crop in the Indus and Ganges basins.

Crops	Interventions	Rank
Rice	Crop diversification with legume-chickpea with furrow-irrigated raised bed planting technique in sequence and intercropping.	1
	Transplanted rice on raised bed (bed-planted system).	2
	Use of suitable cultivars for direct seeded rice.	3
Wheat	Use of 100% recommended dose of nitrogen and phosphorus + FYM/Gypsum/ under no till	1
	Timeliness in sowing operation: sowing by third week of November	2
	Proper irrigation scheduling and maintenance of moisture regime	3
Maize	Use of hybrid variety seeds	1
	Cultivation on permanent raised-bed planting system	2
	Proper irrigation scheduling and maintenance of moisture regime	3
Sugarcane	Proper irrigation scheduling and alternate or skip furrow method	1
	Timeliness in planting operation	2
	Furrow planting	3
Pulses	Crop diversification: short-duration mung bean as summer crop	1
	Multi-crop zero-till-drill *cum* bed planting	2
	Crop diversification: chickpea and lentil in rotation with wheat	3
Oilseeds	Selection of suitable cultivars and hybrid seeds	1
	Laser land levelling	2
	Crop diversification: intercropping with Indian mustard and sugarcane	3
Potato	Planting with quality seed	1
	Proper irrigation scheduling and moisture regime	2
	Basal application of farmyard manure	3
Tomato	Integrated pest management	1
	Use of quality seed and seedlings	2
	Integrated nutrient management	3

and identified multiple, crop-specific interventions to improve WP. We sorted the interventions for each crop based on values in the literature of their impact on water productivity. We made up questionnaires that included the interventions for each crop and sought the expert opinions of 216 experts/stakeholders covering the whole IGBs, including 88 scientists, 18 government officials, 106 farmers and four non-governmental organizations (NGOs).

We analysed the responses and selected the three top-ranked interventions for each of the selected crops (Table 3). Each of these interventions has a large potential to improve land and water productivity of the specific crops. The benefits of each of the interventions need to be explained to farmers, and market and infrastructure linkages need to be established for adequate and timely supply of the required inputs.

The study showed that the potential interventions for improving WP are crop-specific. Proper irrigation scheduling is important for sugarcane, potato, maize and wheat, while high-quality hybrid seeds of suitable cultivars were important for maize, potato and oilseeds. Suitable cultivars and crop diversification with legumes were important for rice and pulses.

Farmers identified the amount and timing of irrigation scheduling as important. Specifically, farmers need to know the critical growth stages for each crop in the different regions so as to facilitate proper irrigation scheduling. Similarly, farmers need up-to-date knowledge on crop diversification, hybrid seeds, quality seeds, suitable cultivars and integrated nutrient management to enable them to improve WP.

Adoption of resource conservation technologies (RCTs)

Improved physical WP in agriculture would reduce the need for additional water and land, and is a critical response to water scarcity. RCTs can save water and increase productivity, especially in the arid and semi-arid parts in the northwest Indus Basin where water scarcity limits crop yields. RCTs emphasize minimum till and mulch with crop residues and have been adopted in about 2 M ha in the IGBs. Technologies promoted in the IGP include reduced/zero tillage, laser land levelling, raised bed planting, and direct seeding of rice, either alone or combined. Zero till (ZT) in wheat and laser land levelling is widely adopted in the IGBs.

ZT gives 12% more utilization of water (Sikka *et al.* 2003) and reduces the number of irrigations per season. In the rice–wheat system at Rechna Doab in the Indus Basin (Pakistan), farmers used 24% less water with ZT, and saved 52% on fuel implying much less groundwater pumping. The influence of different RCTs in the field on WP is summarized in Table 4. However, factors involved in scaling up to sub-basin or basin scale are not fully understood (Ahmad *et al.* 2007). Higher WP with RCTs may come from yield gains, or saved water, or both. Water savings reduce flows to saline groundwater sinks, as in Rechna Doab where groundwater is more saline, which could have broader implications for water quality.

Water saving through watercourse improvements: case of Indus Basin irrigation system

High conveyance losses contributed to low WP in the IBIS. The national program for the improvement of watercourses (NPIW) in Pakistan expected to improve a total of 28,000 watercourses in IBIS in the Punjab at a cost in 2004 of US$1.1 billion. We found that improved watercourses improved the efficiency of water delivery (Table 5). Before improvement, the average time taken to irrigate one hectare of land, irrespective of farm size, at the head, middle and tail sections of a watercourse was 4.5 h, 6.75 h, and 5.25 h,

Table 4. Water productivity improvements in rice–wheat systems of the IGBs by different resource conservation technologies.

Technology/ Location	Irrigation water productivity	Increase (from conventional)
	kg/m³	%
Zero tillage		
Haryana[a]		
Canal	3.69	33.2
Groundwater	2.23	21.5
Mona Project, Pakistan[b]	1.43	30.0
Sheikhupura district, Pakistan[c]	3.00	19.5
Bed planting		
Sheikhupura district, Pakistan[c]	2.98	18.7
Mona Project, Pakistan[b]	1.81	64.5
Laser land levelling		
Modipuram, Uttar Pradesh, India[d]		
Laser land levelling (wheat)	1.31	59.8
Laser levelling and raised bed (wheat)	1.90	37.7
Laser levelling (rice)	0.91	65.5
Mona project, Pakistan[b]	1.67	51.8

Sources: [a] Erenstein *et al.* (2007), [b] Hobbs and Gupta (2007), [c] Jehangir *et al.* (2007), [d] Jat *et al.* (2005).

Table 5. Improvement in delivery efficiency (%) as a result of watercourse improvement in the Indus Basin irrigation system.

	Delivery efficiency of main water course[1]				Overall delivery efficiency[2]			
Project	Head	Middle	Tail	Average	Head	Middle	Tail	Average
Pre-improvement	76	71	63	70	66	63	54	61
Post-improvement	82	81	74	80	79	73	64	72

[1] Efficiency of the section between watercourse outlet *(mogha)* and farm outlet *(nakka)*
[2] Efficiency of the section between watercourse outlet *(mogha)* and field outlet *(nakka)*

respectively. After improvement, irrigation time was reduced about 22%, 40% and 19%, and cropping intensities increased from 115%, 121% and 115% to 138%, 151% and 132%, respectively.

Policy intervention to save water: Punjab (Indian) Preservation of Subsoil Water Act 2009

The Indus Basin depends heavily on groundwater, which is over-exploited. Current energy subsidies and grain price support encourage farmers to grow flooded rice, even in areas where groundwater is overused. In the Indian Punjab, the *Punjab Preservation of Subsoil Water Act 2009* prohibits transplanting rice before 10 June to avoid water losses from flooded fields in the hot weather that precedes the monsoon. Synchronizing the rice crop with the monsoon decreases the water required by flooded rice by 260–300 mm and saves about 2.10 km^3 over the rice-growing area of the Indian Punjab (Figure 6). The water table will still fall but at 60–65% of its current rate (Sharma and Ambili 2010), and less water pumped will save up to 175 million kW h electricity. Growing short-duration varieties like PR115 will save even more water (Figure 6).

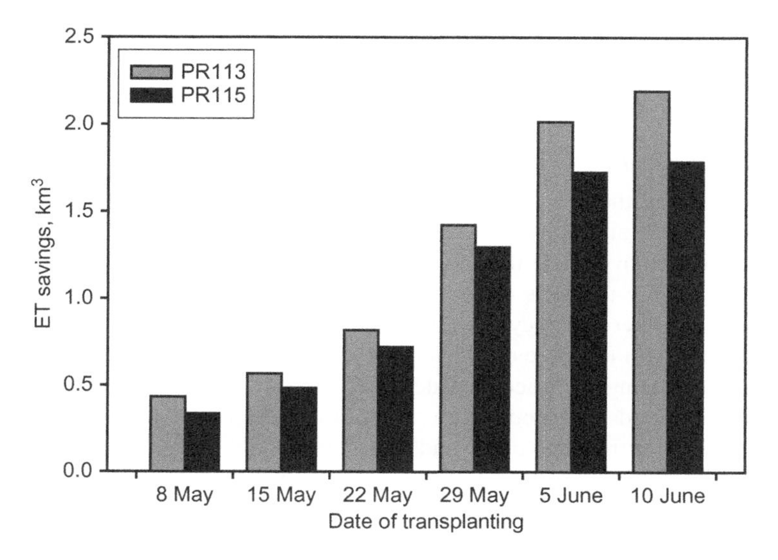

Figure 6. Savings in water by delayed transplanting over the total rice area in Punjab for two rice varieties (PR113 and PR115).

Multiple water-use schemes in hill areas

Short, sloping terraces in the hills of the upper Indus and Ganges Basins are unsuitable for surface and groundwater irrigation. A multiple water-use scheme (MUS) in the Nepalese hills provides a solution by supplying household water and water for homestead gardens (Mikhail and Yoder 2009). Water is piped from upstream springs to a 3000 L ferro-cement water tank (Thai jar), which provides household water (45 L/person/day), the first priority. Overflow is collected in a larger 10 m^3 underground tank, which supplies irrigation water to household vegetable gardens (400–600 L/household) where it is applied by a low-cost drip system (Sharma *et al.* 2010). The average cost is US$100/household in cash and kind. Sale of surplus vegetables (about 90% of production) meets the cost in the first year (Sharma *et al.* 2010) and provides continuing cash income.

MUSs cover 10 to 40, occasionally 80, households depending on the community and the availability of water. They provide potable water to households, reduce the incidence of water-borne diseases, and eliminate the drudgery for women in fetching water. Where conditions are suitable, MUSs are expanding rapidly, supported by partnerships between government agencies, local institutions, NGOs and private parties.

Agrarian change and water institutions

Although the water sector has successfully developed technical solutions to meet the growing water needs of the economy and society in the past, it is getting harder. All sectors demand more water, but development of additional water resources is difficult and expensive, and already there are conflicts over sharing the water that is available. Hence governance plays a critical role in the water sector.

The water sector of the IGBs has many informal institutional arrangements, which co-exist with large formal institutions. We know or understand little about how they operate at the regional level or over different tiers of regional management. In particular, we do not know their relative strengths, weaknesses and efficacy in response to variations in resource availability, economics and politics. In short, there is no comprehensive analysis of water governance in the IGBs. We discuss below some issues of water governance and water laws in the IGBs.

Is irrigation water really free in the Indus-Ganges Region?

It is widely believed that pricing water according to its actual cost will prevent waste, misallocation and scarcity. But in the IGBs it does not appear to be that simple. When the cost of water rises above a low threshold, many small farmers do improve the efficiency of water use by lining channels or using pipes, and by switching to water-saving crops. When the cost of water rises beyond some upper threshold, however, farmers are forced to make drastic cuts in their use of irrigation water, stop growing irrigated crops, or even quit farming. Increasing the price of water does increase the efficiency of water use but threatens the livelihoods and food security of millions of agrarian poor (Shah *et al.* 2009).

Water in public irrigation (canal and tanks) on 20–22 Mha is cheap (US$0.0025–0.02/$m^3$) (Figure 7). The dearest water (US$0.15–0.25/$m^3$) is from hired diesel-powered generators to power submersible electric pumps on deep tubewells, used only in case of emergency. The dearest water costs up to 100 times the cheapest, but about 80% of land in the IGBs is irrigated at non-trivial cost. In sub-economy II (Figure 7), water from electrified tubewells, electricity is the main cost (US$0.01–0.0175/kW h) and volumetric in Punjab,

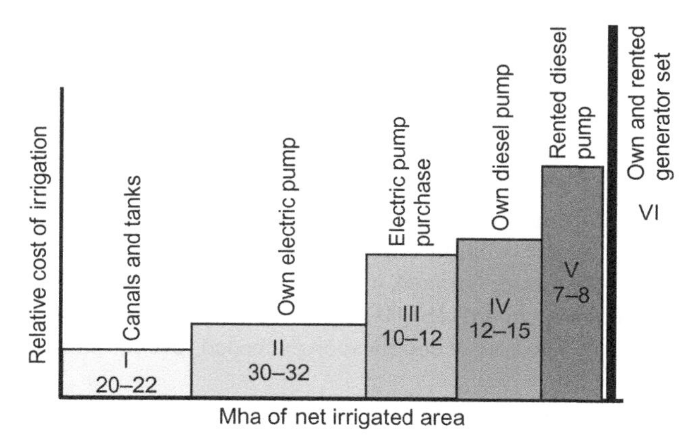

Figure 7. Irrigation sub-economies of South Asia.
Source: Shah *et al.* (2009).

and Sind in Pakistan, in Nepal and in Bangladesh. In India, electric tubewell owners pay a flat tariff (except in Punjab where electricity for pumping is free), but there is stringent rationing of the increasingly unreliable power supply. Farmers in sub-economies I, II and III are increasingly using diesel-powered pumps to cope with the unreliability of surface irrigation (I) and electricity supplies (II and III).

The diesel-fuel price squeeze

Indian Punjab and Haryana have many electric tubewells, but the rest of the IGBs depend on diesel pumps for water because in Bangladesh and Pakistan farm electricity is metered and expensive, or is not available as in Bihar and Nepal Terai. The price of diesel fuel increased by 670% between 1990 and 2006, while the farm-gate price of rice rose only about 60%. The squeeze on small-scale irrigation is therefore approaching crisis in all IGBs' countries but particularly in eastern India and Nepal Terai (Table 6).

Case studies of 19 villages of the Indian Ganges Basin and the Pakistan Indus Basin showed two responses: increased efficiency or distress. Farmers increase efficiency by improving distribution, by lining field channels or using pipes, by just-in-time irrigation, by reducing irrigation frequency, and by switching to crops that need less water. In eastern India, farmers also commonly switch from Indian to cheaper Chinese diesel pumps. Distress responses include quitting farming, mostly by marginal farmers and sharecroppers dependent on expensive purchased water. Other smallholders switch to

Table 6. Farm-gate price relative to the price of diesel fuel in IGBs countries.

	Diesel price: February 2007	Farm-gate rice price: February 2007	Rice needed to buy one litre of diesel
	/litre	*/kg*	*kg*
India (Indian Rs.)	34.0(US$0.85)	6.4	5.7
Pakistan (Pak. Rs.)	37.8(US$0.64)	11.8	3.2
Bangladesh (Taka)	35.0(US$0.50)	9.0	3.9
Nepal terai (Nepal Rs.)	57.0(US$0.84)	10.0	5.7

Source: Data collected from case studies undertaken for this research.

high-risk, high-value crops on a small plot leaving the rest of the farm fallow or under rainfed crop.

Energy and irrigation in the IGBs

Over the last 50 years, groundwater has become the main source of water for irrigation in the IGBs, with the green revolution often titled the tubewell revolution (Repetto 1994). But while the use of groundwater boomed, it did so at the cost of the energy economy. In the late 1970s all state electricity boards (SEBs) in India changed from metered electricity to a subsidized flat-rate tariff. The cost of supply soon exceeded revenue earned and most SEBs made huge losses.

Since the marginal cost of pumping groundwater was near zero, there was over-pumping and in places groundwater markets developed. In arid and semi-arid regions with hard-rock aquifers and low recharge, water tables fell sharply, putting the livelihoods of millions of poor farmers, who depended on groundwater irrigation, in jeopardy (Moench 1996). In contrast, in areas with abundant rainfall and alluvial aquifers with adequate recharge, such as Bihar and West Bengal (Mukherji 2007, 2009b) the groundwater system is more robust and was not over-pumped, due in part to inadequate electricity infrastructure.

There are therefore two energy-irrigation stories in the IGBs. The first is the vicious cycle of low, flat electricity tariffs, leading to over-pumping of groundwater and bankrupt SEBs, which is difficult to break due to the political power of the farmers with entrenched interests. This is true for Punjab, Haryana and western UP in India. The second is less often told. In water-abundant eastern India, Nepal Terai and Bangladesh, farmers mostly use diesel pumps, which have become increasingly costly to run, and with different outcomes. We examine the two scenarios below.

Electricity policy and groundwater use: evidence from Gujarat, West Bengal and Uttarakhand states in India

West Bengal, in eastern India in the lower Ganges Basin, has a groundwater potential of 31 km^3 at shallow depths. Only 42% of the available groundwater resource is used (WIDD 2004). Agriculture uses only 6.1% of total electricity consumption (WBSEB 2006). There is an active market for groundwater. The state of Uttarakhand in the upper Ganges Basin has groundwater resources of 2.1 km^3, of which 66% is currently used (GoI 2005). Agriculture uses only 12% of the total electricity in the state, although 70% of tubewells are electric. Water markets are less developed than in West Bengal because landholdings are larger so that farmers have their own tubewells, and water is also widely available from reliable government tubewells. Gujarat, a western state of India, is outside the IGBs, but because it successfully resolved its agricultural energy problem, we include it. Gujarat has a groundwater potential of 15.8 km^3 of which 76% is withdrawn each year. Groundwater in 61% of the administrative blocks is rated by the Central Ground Water Board as over-exploited, critical or semi-critical. Until recent reforms, Gujarat had one of the highest electricity subsidies in India. The SEB made heavy losses and the quality of power supply deteriorated, especially in rural areas. Gujarat also has an active market for groundwater.

The process of reform in three states. The government of West Bengal installed remotely sensed, tamper-proof meters, which differentiate the cost of electricity during the day. Prices discourage pumping during peak hours and encourage it during the slack hours of the night. The government of Uttarakhand installed electronic meters, but uses conventional meter reading and bills a fixed low tariff of Rs 0.70/kW h, which is lower than the off-peak rate in West Bengal. In 2003, the government of Gujarat pioneered the *Jyotirgram* Scheme (JGS) to separate agricultural and non-agricultural feeders. This involved rewiring of rural Gujarat at cost of US$255 million. The government now provides high-quality, predictable, reliable but *rationed* power supply to farmers who daily get eight hours' power at full voltage on a pre-announced schedule that alternates weekly between day and night.

Impact of electricity reforms. Soon after metering in West Bengal, the pump owners (2% of households) increased their rates by 30–50%. Water buyers have to pay for dearer water and also receive supply at inconvenient hours. In Uttarakhand, tubewell owners will pay less than one third the amount for electricity than under the flat tariff, but we found that it was usually unpaid in any case. As in West Bengal, the Uttarakhand SEB will earn less revenue than before, and the very low, metered tariffs are unlikely to impact the volume of water pumped (Mukherji *et al.* 2006, 2009a).

In contrast, the JGS in Gujarat has improved the quality of village life, spurred non-farm activities, halved the power subsidy to agriculture and reduced groundwater pumping. It offered a mixed bag to medium and large farmers but disadvantaged marginal and landless farmers. Electricity use of tubewells has fallen 37% from over 15.7 billion kW h/yr in 2001 to 9.9 billion kW h in 2006. Groundwater use in Gujarat agriculture has fallen and has reduced the over-exploitation. Furthermore, JGS has created a groundwater economy that is amenable to regulation at different levels. Elsewhere in India and the rest of the world, groundwater management has largely been ineffective and costly. In contrast JGS has shown that effective rationing of the power supply can indeed act as a powerful tool for managing groundwater use for irrigation.

Laws and water resources governance

Water laws provide an effective basis for water governance at national, sub-national and basin levels. An understanding of how past laws worked can provide insights into how the current policy and legal frameworks will evolve in the future, and how well they address current needs in the water sector, which is a critical component of economic and socio-political development. Policy and legal frameworks have progressed from a focus on water development up to the 1970s towards water management in which water governance has become prominent. In India as a whole there was a flurry of legislation in the last decade, in which the water sector also took part. However, water resources are a state subject and any law regarding its management must be passed by the state concerned, for which the federal government provides guidance and model frameworks.

Analysis of laws applicable to groundwater shows greater attention to the topic across the IGBs since the 1990s (Figure 8), and it is a key issue in India at both the federal and state levels. Groundwater featured in 20 of the 25 instruments assessed for the Indian states in 1990–2009, with 15 having it as either a primary or substantial focus. While groundwater received important legislative attention, the relevant instruments adopted by different states were substantially similar in content and language. This was most obvious in three federal draft Groundwater Bills (1992, 1996 and 2005) and the closely similar 2005 bills of the groundwater legislation of West Bengal, Bihar and Himachal Pradesh.

At the federal level in India, two trends emerge. The first is the gradual move towards regulation of groundwater use. The need for limiting extraction to the amount of recharge was recognized by the National Water Policy of 1987 (Section 7.2), which also called

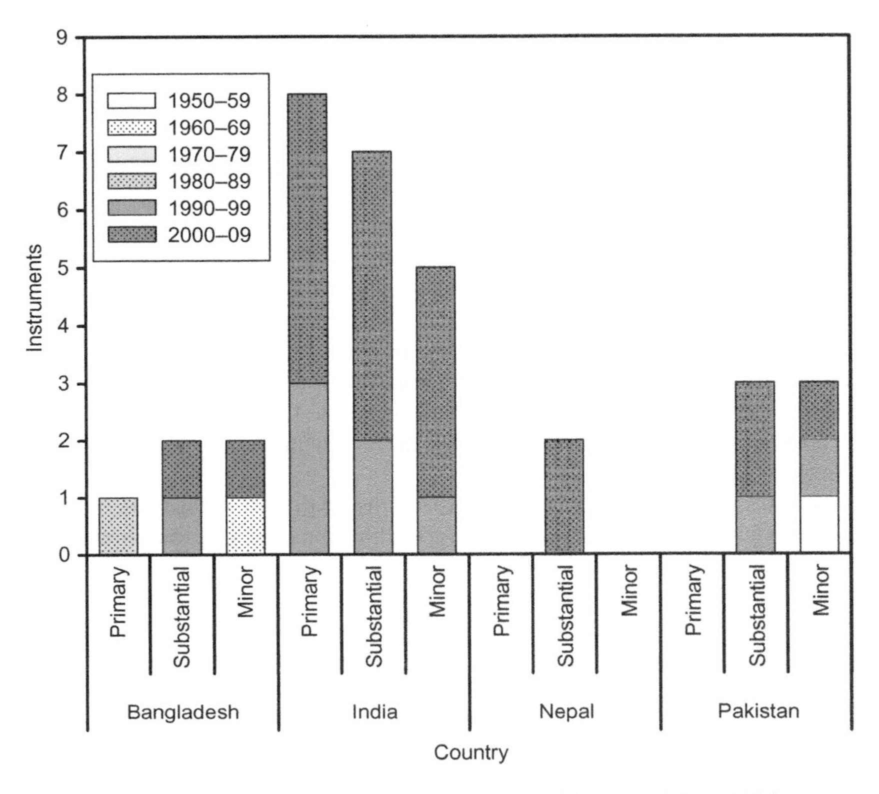

Figure 8. Dominance of groundwater laws in the countries of the IGBs (1950–2009).

for a periodic scientific reassessment of the groundwater potential, taking into account the quality of the water available and economic viability (Section 7.1). This was followed by the proposed introduction of permits for and registration of new and existing wells[1], as well as the regulation of commercial well digging in the *Draft Groundwater Bill of 1992*. The Bill also envisaged the creation of a National Ground Water Authority with the power to advise the state/union territory government to declare any area for the purposes of controlling the extraction or use of groundwater and to provide licences for further development of groundwater.

The *Draft Groundwater Bill of 1996* follows many of the provisions of the 1992 Bill, while introducing additional criteria to be considered when evaluating applications for new wells. The exceptions to the need for permits were also amended to include anyone who proposes to install a well that is to be fitted with a hand-operated manual pump or when water is proposed to be withdrawn by manual devices. The third *Draft Groundwater Bill* developed in 2005 continues the thinking of its predecessors, and introduces further criteria when reviewing applications for permission to construct wells.

The second trend, and an important addition in the 2005 Bill, is the emphasis placed on enhancing the supply side through groundwater recharge systems. The Bill envisages permits for digging new wells to include a mandatory provision requiring artificial recharge structures to be built as part of the well (Article 6.3). The proposed Ground Water Authority would also be charged with identifying areas that need recharge, and issuing guidelines for adoption of rainwater harvesting in these areas (Article 19.1). The Authority may give directions to the concerned departments of the state/union territory government to include "rainwater harvesting" in all developmental schemes falling under notified areas.

In urban areas that fall within notified areas, the Authority may issue directives for constructing appropriate rainwater harvesting structures in all residential, commercial and other premises having an area of 100 m^2 or more. Community participation through watershed management was identified as another means of facilitating ground water recharge in rural areas (Article 19.1). This is to be supported by the promotion of mass awareness and training programs on rainwater harvesting and artificial recharge (Article 19.3).

In Bangladesh, permits for wells were introduced by the *Ground Water Management Ordinance* of 1985 (Article 5.1). The Bangladesh *National Water Policy* of 1999 calls for preserving natural depressions and water bodies, underground aquifers (Section 4.6b) and the prohibition of filling in publicly owned water bodies and depressions in urban areas (Section 4.12e). It also encourages massive afforestation and tree coverage, specifically in areas with declining water tables (Section 4.12h), and proposes the regulation of extraction in identified scarcity zones (Section 4.3c).

The primary concern with groundwater in Nepal has been the dependence of large industries on groundwater extraction through deep tubewells and the need to regulate this use through licensing and effective monitoring (Section 4.2.2.4, *National Water Plan 2002*). The approach to groundwater management in Pakistan appears to be area-specific. The *Draft National Water Policy* promotes groundwater recharge wherever technically and economically feasible (Policy 8.4). It calls for the evaluation of the various technologies being used for undisturbed extraction and skimming of fresh groundwater layers overlying saline water. It also calls for optimal groundwater pumping in waterlogged areas to lower the water table (Policy 8.7), while areas with falling water tables are to be delineated for restricting uncontrolled abstraction (Policy 8.8).

The importance given to groundwater in water legislations shows the importance of this resource in the overall water resources in the IGBs.

Water, land and poverty

Poverty in the countries of the IGBs is intense and multi-faceted. Despite marked progress in the last 50 years, India still has the world's largest number of poor people. Lack of access to water and sanitation services and exposure to extreme events reinforce a cycle of vulnerability. Given that 90% of the poor live in rural areas, poverty is still a rural phenomenon in the IGBs, and agriculture is the main source of livelihood for most of the rural population in the countries of the IGBs. Thus reducing rural poverty through improving agricultural income is a major pathway for poverty alleviation in the IGBs. Adequate access to reliable water and quality land resources are crucial for an increase in agricultural productivity. With increasing pressure on limited land resources, however, enhancing the value of agricultural productivity per unit of water is crucial for the reduction of rural poverty. But the spatial linkages between poverty and WP are not clearly understood. We need this knowledge to design appropriate and geographically targeted interventions for increasing WP and reducing rural poverty.

Spatial and temporal variation of poverty

Low ranking of the human development index (HDI) of the four riparian countries – India (134), Pakistan (141), Nepal (144), Bangladesh (146) (UNDP 2008) – shows the plight of the progress of health, education and economic growth in the IGBs. Unfortunately, these values are far below those of other Asian countries. The headcount ratio (HCR, the fraction of the population below the official poverty line) of the riparian countries of IGBs has lately shown some improvement, however. In India, income of more than half the population was below the poverty line before the mid 1970s. HCR has decreased since then, to about 36% by 1993, and 26% by 2000. About 21% of the Indian population still lived below the poverty line in 2006. In Pakistan, the HCR increased in the latter part of the 1990s, but has since declined. About 22% of the population in Pakistan remained poor in 2005. In Nepal and Bangladesh, about 31% and 40% of the population were poor in 2003 and 2005, respectively. In spite of these gains, the poverty associated with high rural population is a major concern in all four countries. The depth and severity of poverty in all demographic groups in the four countries of the IGBs have declined over the last decade, however. Pakistan still has a high poverty gap index and squared poverty gap index, indicating that the income of a large part of the poor population is still well below the poverty line, which is highly likely to be in chronic poverty in the near term. The severity of poverty in Bangladesh and Nepal is comparatively lower, indicating smaller inequities of the income amongst the poor.

Poverty maps show large variations of poverty across the region. The HCR is relatively lower in north and northwestern parts of the IGBs (Figure 9), which mainly includes the northern part of Pakistan and the Indian Punjab, parts of Rajasthan in the Indus Basin and Haryana and the western parts of Uttar Pradesh of India, and Kathmandu region in Nepal in the Ganges Basin. The low HCR regions in India and Pakistan are known to have high agricultural productivity and growth, and poverty in these regions, especially in rural areas, fell much faster than elsewhere. In Nepal, areas of low poverty are concentrated in urban centres. The HCR is high in the southern and eastern parts of the IGBs, consisting of states of Bihar, Chattisgarh, Jharkand, and the eastern part of Uttar Pradesh, and Madhya Pradesh in India, the southwest divisions of Bangladesh, and all Nepal Living Standard Survey regions except Kathmandu in Nepal in the Ganges Basin, and southern Punjab, North West Frontier (NWFP), Sind, and Baluchistan provinces in Pakistan in the Indus

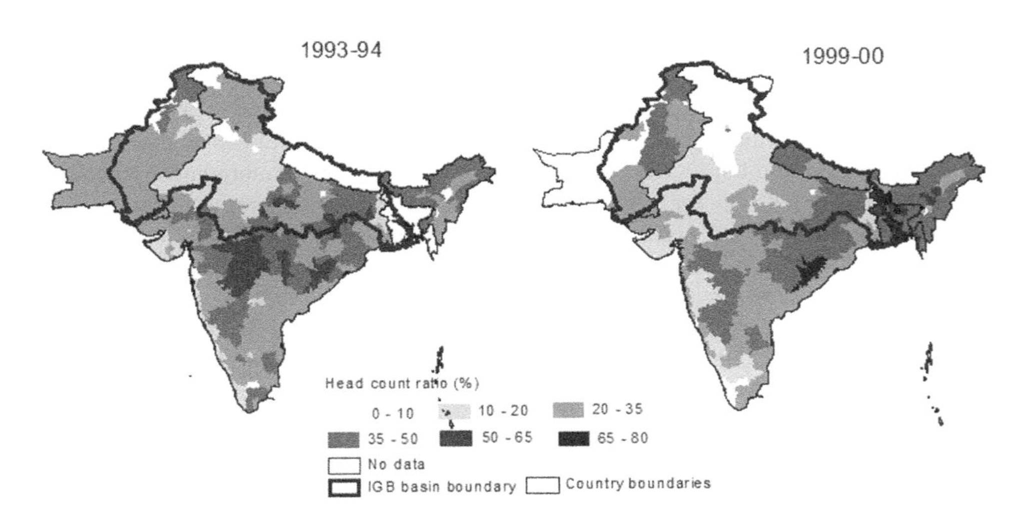

Figure 9. Poverty maps across national sample survey (NSS) regions in India, districts in Pakistan, and *zilas* in Bangladesh showing changes over the last six years of the 1990s.

Basin. Many of the rural poverty hotspots, i.e. those on the third and fourth quartile of the HCR distribution, are clustered in the eastern Ganges Basin.

Water–land–poverty nexus (NWLP)

Poverty is both a cause and an effect involving many factors. The agriculture sector contributes only about 20% of GDP in India, Bangladesh and Pakistan but provides livelihoods to more than 60% of the rural population. Our hypothesis is that growth in agriculture will reduce the incidence of rural poverty in many parts of the IGBs. Among the poorest Indian states in the Ganges Basin, the agriculture sector in Bihar contributes to 39% of the state's GDP, 34% in Uttar Pradesh and 28% in West Bengal. Analysis across Indian states in 1990/2000 shows that, on average, a 1% increase in GDP/person can decrease total HCR by 1.4%. Analysis based on 2004/5 data shows that a 1% increase in overall GDP and in agriculture GDP should contribute to a reduction of 0.96 and a 1.19% reduction of total and of rural poverty. This is further evidence that there is a large scope for reducing rural poverty through growth in agriculture, which in turn can have a large impact on reducing overall poverty in the IGBs.

Linkages between water and poverty

Access to irrigation, especially through groundwater, gives a greater control of water supply. A reliable supply of irrigation water is a key determinant for better use of inputs, such as improved seed varieties, machinery, fertilizer and pesticides. Inputs increase productivity and income, and reduce poverty. This is clearly the case in Punjab in the Indus Basin and Haryana in the Ganges Basin (Figure 10), where irrigation, much of it from groundwater, covers a large part of the cropped land. These two states have some of the lowest rates of rural poverty. There are some exceptions, however. Although not as high as in western parts of the IGBs, access to irrigation in the eastern Ganges Basin (comprised of Bihar, eastern Uttar Pradesh) and in western Bangladesh is substantially high, but poverty is also high. Low productivity due to recurrent floods and inadequate infrastructure, such as roads, markets and electricity, are major constraints. Access to irrigation is low in West Bengal and northern Madhya Pradesh, and a large part of the rural population depends on rainfed agriculture. Recurrent droughts are a major constraint on high productivity.

Linkages between land and poverty

While the incidence of poverty among the marginal to small landholders in central and eastern Ganges Basin states (Bihar, West Bengal, Uttar Pradesh and Madhya Pradesh) is very high, it is not strikingly lower among the larger landholders in these states. In comparison, poverty among the medium to large landholders in the states of western Ganges Basin is almost non-existent. Poverty among the landless in Pakistan and Bangladesh is even worse. In Bangladesh, more than 50% of landless people are poor, and the poverty among small landholders is three times higher than among large landholders. Reduction of poverty among the landless was very slow in Bangladesh, where 61% of the landless rural population was poor in 1988/89 (Ravallion and Sen 1994). This only decreased to 53% by 2000 and to 49% by 2005 (BBS 2006). The incidence of poverty in the near landless (0–0.5 ha) and marginal landholders (0.5–0.7 ha), 48% and 33%, are more than twice the poverty levels of small (0.7–1.1 ha) and medium (1.1–3.0 ha) landholders, respectively. Moreover, rural poverty in Bangladesh is high in both lowland and highland areas (Kam *et al.* 2005). Hence land ownership, holding-size and land-quality aspects are closely associated with rural poverty in Bangladesh.

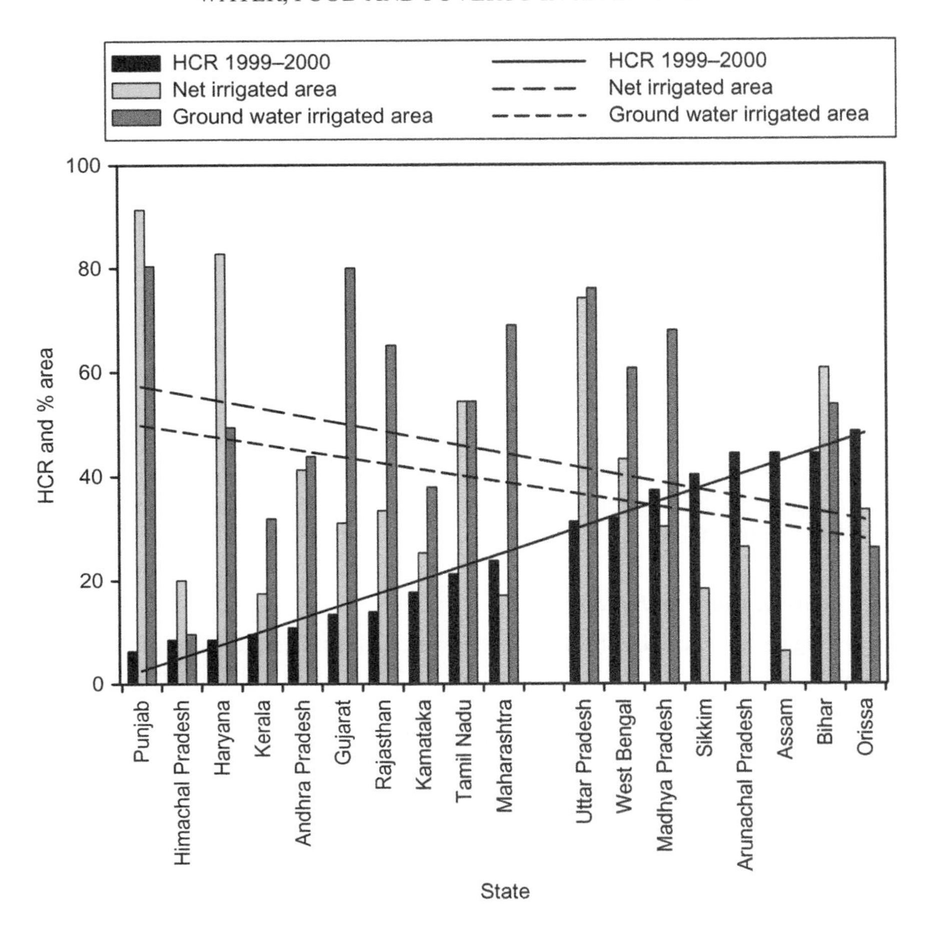

Figure 10. Headcount ratio in rural areas versus net irrigated and groundwater-irrigated area in Indian states.

In Pakistan, more than half the landless population whose livelihoods depend on agriculture, are poor (Anwar *et al.* 2004). Skewed land ownership, where two thirds of the population has no access to land and another 18% possesses small agriculture landholdings (<2.25 ha), is a major determinant of poverty in Pakistan. In Nepal, however, more than half the population classified as landless to small landholders are poor, but poverty among the large landholders is not strikingly lower (Chhetry 2001, Pant and Gautam 2008). Land productivity is a major factor that separates the rural poor from non-poor in Nepal.

Although access to irrigated land is a major determinant of decreasing rural poverty, intensive irrigation could also contribute to land degradation and threaten the very benefits that irrigation has delivered to rural people by lowering productivity and profitability. Degraded lands in turn contribute to low productivity and profitability. There is also high frequency of poverty among those who use wasteland for agriculture. Furthermore, lack of access to safe drinking water and sanitation, another dimension of human deprivation, is related to poverty (Abeywardane and Hussain 2002). Over 299 million people or 31% of the rural population in the countries of the IGBs lack access to safe drinking water (UNDP 2008), and over 796 million or 84% of the rural population lack access to safe sanitation.

Water–land–poverty nexus (NWLP) analysis at the household and district level

In our analysis we used a logit regression (LR) model, combining explanatory variables extracted from secondary data on households and administrative units, to understand NWLP at the household and district level. The LR model examines the probability (P) of a person being in poverty given the socio-economic status, assets of households, and cropping patterns, crop productivity, land tenure and size of holding, water sources and irrigation patterns, and level of infrastructure development of the administrative district. We extracted the data for this analysis from the National Sample Survey Organisation (NSSO) 61st round survey in 2004/5, which contains 32,230 household records in 280 districts in 16 states or union territories of India. We outline some of the important observations below.

Agriculture-related economic activities are still the dominant form of livelihood of the rural population. Reduction of rural poverty can be accelerated in a two-track approach: the pro-poor agriculture growth interventions can still play a major role in reducing poverty in the agriculture-operator households, and improving skills and enhancing opportunities for employment in the non-farm sector can reduce poverty among the agricultural labourers.

In the IGP in India, growth in agriculture is significantly related to the growth in yield of food grains, and crop diversification. The existing grain yields range from 0.9 t/ha in the central region in Bihar to 4.5 t/ha in the western Punjab and provide a strong indication of the magnitude of the scope for growth in agricultural productivity. Districts with a high percentage of the area growing fruits and vegetables are also significantly associated with low HCR.

Rainfall, access to irrigation, irrigation using groundwater, and WP of irrigated agriculture are all significant in explaining the variation of HCR in the IGBs, with households having large areas of irrigated crops being significantly less poor. The incidence of poverty of households without land and without irrigation are similar, but decrease with better access to irrigation.

Better access to groundwater, in general indicating a reliable irrigation supply, is also associated with low poverty. The positive interaction of WP and the extent of the irrigated area also show that by increasing access to irrigation, increases in productivity can accelerate poverty reduction in areas with low levels of water endowment.

Considering the size of the landholding shows that a large number of marginal and small landholdings are associated with high levels of rural poverty across districts. Among the poor households, 33% have no access to land and a further 54% have only marginal to small landholdings. Moreover, many of these small landholdings are highly fragmented and therefore seem to be a major reason for low inputs and productivity. These areas are the hotspots of rural poverty in the eastern IGBs.

Access to infrastructure such as road, markets and electricity also has a major influence in reducing poverty. Districts with higher road density are associated with low HCR. Lack of markets is also a main determinant in the regions with high rural poverty.

Water–land–poverty nexus: lessons from case studies in India, Pakistan and Nepal

We also conducted case studies in India, Pakistan and Nepal to examine the NWLP at the household level, using questionnaire surveys. The survey was based on three separate questionnaires responding to the village and household profile and to fragmentation of the cultivated landholding. In India, we conducted the survey in three least progressive districts of Bihar, Vaishali, Darbhanga and Munger. In Pakistan, the case study site was

Rechna Doab and in Nepal, we conducted the study in four villages of Morang and Sunsary districts. We summarize the inferences from the case studies below.

Property rights and the status of water resources

In India, more than two thirds of the respondents purchased agricultural water through informal trade mechanisms, with 68% buying water from local groundwater markets, while 20% have their own pump and only 6% make use of public canal water. In Pakistan, however, each farmer had an individual allocation of a share in the canal network. In Nepal, use of public canal water was almost universal with only minimal ownership of tubewells.

Water volume allocation in property rights schemes

In India, about two thirds of the buyers and all the water owners make exclusive use of a specific water source without combining it with any other additional source. In the case of canal irrigation, however, only 20% depended solely on the canal supply. In Pakistan, the majority of tubewell owners cover only a small amount of the landholdings through this source. In Nepal, all the farmers depend only on a single source of water, either canal water or groundwater.

Farmers' perspective towards agricultural water use

Indian farmers claim that lack of credit is the major reason for low productivity, closely followed by insufficient supplies of water, and unpredictable weather. In Pakistan and Nepal, insufficient water supply is by far the most important constraint for higher productivity. Farmers in all three countries identified improved water availability as a major factor contributing to increased productivity. Farmers are very willing to set up a groundwater market to meet their water needs, but are not able to make financial contributions for such a venture. In India, 70% of respondents agree to setting up a co-operative tubewell system. Most respondents also gave high priority to the introduction of new technologies and specialized training on methods to improve production.

Conclusions and recommendations

The waterscape of the Indus and Ganges plains is like a palimpsest. It is a landmass that was thickly populated and intensively cultivated over 2000 years ago, and carries the imprints of countless waves of hydraulic tradition. Mastering and manipulating snow-fed perennial streams and rivers for watering fertile soils has remained the secret of the IGBs' endemically high population-carrying capacity. Unfortunately, past fortunes have turned for the worse. Today India, Pakistan, Nepal and Bangladesh, the four countries that share the contiguous Indus and Ganges basins have more poor people than any other region in the world and are amongst the worst laggards in the global human development index and human poverty index. Its vast, fertile plains remain overwhelmingly rural and agricultural. Though agriculture contributes only about 20–25% to the national GDPs, agriculture as a means of livelihood provides 78% of total employment in Nepal, 63% in Bangladesh, 66% in India and around half in Pakistan. With population pressures bursting at the seams and now exceeding one billion people, the availability of cultivable land per capita, through which people eke out a living, has shrunk to around 0.1 ha. With the exception of Nepal, the internally renewable fresh water resources are much below the threshold limit and stand at 325, 688 and 1149 m^3/capita/year respectively for Pakistan, Bangladesh and the Indian parts of the Indus-Ganges plains. It is certainly no exaggeration to say that the basins of the Indus and the Ganges basins are "under extreme pressure".

We summarize below some of the important recommendations that arise from the work we have presented, which will help to improve sustainability of the resources, enhance agricultural productivity, and alleviate poverty.

(1) *Regional perspective:* From both the hydrological and socio-ecological perspectives, the opportunities, constraints, strengths and weaknesses of the Indus Basin and western Ganges Basin ("northwestern region") are markedly different from eastern Ganges Basin (eastern Indian states, Nepal Terai and Bangladesh, "eastern region"). Water and land resources are intensively utilized in northwestern region and are poorly utilized in the eastern region. Development frameworks and policies need to be regionally differentiated to ensure long-term sustainability in the northwestern region and substantially improve the productivity and economic dynamism in the eastern region.

(2) *Use and abuse of water resources:* Most of the available water resources are utilized within the Indus Basin leaving little scope for expansion; the basin is closed. The strategic options are for improving crop productivity by moving towards high-value, diversified and precision agriculture, and to release some water from the irrigation sector to meet the growing demands for domestic, industrial and environmental flows. Net discharge from the Ganges Basin (37%) accounts for more water than any other use. Additionally, the basin also has a massive and presently unutilized static groundwater resource of about 7800 km^3. The Ganges Basin countries may immediately initiate sound groundwater development and use policies that will help increase productivity, reduce climatic vulnerability and also alleviate poverty.

(3) *Improving agricultural productivity*: The average WP for the rice–wheat rotation system in the Indus-Ganges plains is low at US\$0.131/m^3 whereas the "bright spot" areas (about 5% of basin area) have high water productivity of US\$0.190/m^3. If the basin average value could be increased to the "bright spot" level, the basin could theoretically save 31% of agricultural water consumption with the same production or alternatively increase production by 31% with the same water input. This shows a great opportunity for improving productivity through a set of technical and policy interventions.

(4) *Potential interventions for improving water productivity:* As access to water resources is relatively poor in large parts of the region, the critical crop water requirements are inadequately satisfied and crops experience different degrees of water stress. Farmers need to know more about these critical growth stages so that they can improve their management of specific crops in different regions through proper irrigation scheduling. Adoption of resource conservation technologies such as reduced/zero tillage, laser land levelling, raised bed planting and direct seeding of rice need to be promoted better to realize their benefits on a wider scale. Policy measures like regulatory delay in transplanting of paddy and breaking the water–energy nexus on the pattern of the *Jyotirgram* scheme will improve groundwater governance. Multiple-use water systems in the hilly areas of the basin and the eastern region will improve productivity and alleviate poverty of the small and marginal farmers.

(5) *Policy and institutional changes:* Energy infrastructure and policies in the eastern region need to be comprehensively and immediately improved so as to empower

the millions of small and poor farmers operating at a very low level of agricultural productivity. The groundwater boards and authorities in all the four countries need to be strengthened to move further from their present role of resource estimation to planning, development and management.

(6) *Managing the water–land–poverty nexus:* Reduction of pervasive rural poverty in the region can be accelerated in a two-track approach: pro-poor interventions to increase agricultural yields will play a major role in reducing poverty in the agriculture-operator households, and improving skills and enhancing opportunities for employment in the non-farm sector can reduce poverty among agricultural labourers. Providing access to irrigation can help achieve substantial improvements in productivity, diversification and thus reduction in poverty. Higher access to groundwater is associated with low poverty. Setting up of co-operative tubewells managed by small groups of farmers may be a viable option for helping large numbers of marginal farmers with smallholdings. Areas with small and fragmented holdings are hot spots of rural poverty in the eastern region and need immediate policy changes to stimulate consolidation. Access to infrastructure such as roads, markets, electricity and quality inputs can also have a major influence in reducing poverty.

Acknowledgements

The authors acknowledge with gratitude the financial and scientific help provided by the Consultative Group on International Agricultural Research (GCGIAR) Challenge Program on Water and Food (CPWF) and the International Water Management Institute (IWMI). Dr Simon Cook, Basin Focal Project Co-ordinator and Dr Vladimir Smakhtin, Theme Leader, provided great help and guidance. We thank Dr Myles Fisher for editing and bringing the paper to its present form.

Note

1. Small and marginal farmers will not have to obtain a permit if the well is proposed to be sunk for exclusively personal purposes (Section 6.1).

References

Abeywardana, S. and Hussain, I., 2002. Water, health and poverty linkages: a case study from Sri Lanka. *Asian Development Bank regional consultation workshop on water and poverty*. Colombo: IWMI.

Aggarwal, P.K., *et al.*, 2004. Adapting food systems of the Indo-Gangetic plains to global environment change: key information needs to improve policy information. *Environmental Science and Policy*, 7 (6), 487–498.

Ahmad, M.D., *et al.*, 2007. Water saving technologies: myths and realities revealed in Pakistan's rice-wheat systems. Research report 108. Colombo: IWMI.

Ahmad, M.D., Turral, H., and Nazeer, A., 2009. Diagnosing irrigation performance and water productivity through satellite remote sensing and secondary data in a large irrigation system of Pakistan. *Agricultural Water Management*, 96 (4), 551–564.

Amarasinghe, U.A., *et al.*, 2005. *Spatial variation in water supply and demand across river basins in India*. Research report 83. Colombo: IWMI.

Amarasinghe, U.A., *et al.*, 2007. *India's water future to 2025–2050: business as usual scenario and deviations*. Research report 123. Colombo: IWMI.

Amarasinghe, U., Shah, T., and Singh, O.P., 2008. *Changing consumption patterns: Implications on food and water demand in India*. Research report 119. Colombo: IWMI.

Anwar, T., Qureshi, S.K., and Ali, H. 2004. Landlessness and rural poverty in Pakistan. *The Pakistan Development Review*, 43 (4).

Babel, M.S. and Wahid, S.M., 2008. *Freshwater under threat: vulnerability assessment of freshwater resources to environmental change* [online]. Nairobi: United Nations Environment Programme. Available from: www.unep.org/pdf/southasia_report.pdf [Accessed 30 July 2010].

Biswas, B., *et al.*, 2006. Integrated assessment of cropping systems in the eastern Indo-Gangetic plain. *Field Crops Research*, 99 (1), 35–47.

BBS, 2006. *Sectoral need-based projections in Bangladesh 2006*. Dhaka: Bangladesh Bureau of Statistics.

Cai, X.L. and Sharma, B.R., 2010. Integrating remote sensing, census and weather data for an assessment of rice yield, water consumption and water productivity in the Indo-Gangetic river basin. *Agricultural Water Management*, 97 (2), 309–316.

Chakraborti, D., *et al.*, 2004. Groundwater arsenic contamination and its health effects in the Ganga-Meghna-Brahmaputra plain. *Journal of Environmental Monitoring*, 6 (6), 75N–83N.

Chhetry, D., 2001. Understanding rural poverty in Nepal [online]. *In:* C. Edmonds and S. Medina, eds. *Defining an agenda for poverty reduction: proceedings of the first Asia and Pacific forum on poverty (volume II)*. Manila: Asian Development Bank, 293–314. Available from: http://www.adb.org/Documents/Books/Defining_Agenda_Poverty_Reduction/Vol_1/chapter_26.pdf] [Accessed 28 July 2010].

Eastham, J., *et al.*, 2010. *Water-use accounts in CPWF basins: simple water-use accounting of the Ganges basin*. BFP05. Colombo: CPWF Working Papers, Basin Focal Project Series.

Erenstein, O., Malik, R.K., and Singh, S., 2007. *Adoption and impacts of* zero tillage *in the irrigated rice- wheat systems of* Haryana, *India* [online]. New Delhi: Centro International de Mejoramiento de Maíz y Trigo (CIMMYT) and the Rice-Wheat Consortium for the Indo-Gangetic Plains. Available from: http://www.rwc.cgiar.org/DownloadServer.asp?p=187&f=2002&n=Erenstein%20et%20al%202007%20Impact%20ZT%20Haryana.pdf [Accessed 28 July 2010].

Gosain, A.K. and Rao, S., 2007. Impact assessment of climate change on water resources of two river systems of India. *Jalvigyan Sameeksha*, 22 (1), 1–20.

GoI, 2005. *Report on third census of minor irrigation schemes (2000–01)*. New Delhi: Minor Irrigation Division, Ministry of Water Resources, Government of India.

Habib, Z., 2004. Pakistan: Indus Basin water issues [online]. *South Asian Journal*, 6 (6). Available from: www.southasianmedia.net/Magazine/Journal/6_pakistan_indus.htm [Accessed 30 July 2010].

Hobbs, P.R. and Gupta, R.K., 2003. Resource conserving technologies for wheat in rice-wheat systems. In J.K. Ladha, *et al.*, eds. *Improving the productivity and sustainability of rice-wheat systems: Issues and impact*. ASA Special Publication 65. Madison, WI: ASA, 149–171.

Hosterman, H.R., *et al.*, 2009. *Water, climate change, and adaptation: focus on the Ganges River basin*. Nicholas Institute Working Paper. Durham, N.C.: Nicholas Institute for Environment Policy Solutions, Duke University.

IPCC, 2001. Climate change 2001. *Third assessment report of the Intergovernmental Panel on Climate Change*. Cambridge: Cambridge University Press.

Jain, S.K., Agarwal, P.K., and Singh, V.P., 2007. *Hydrology and water resources of India*. Dordrecht: Springer.

Jain, S.K., *et al.*, 2009. A comparative analysis of the hydrogeology of the Indus-Gangetic and Yellow River basins. In: A. Mukherji, *et al.*, eds. *Groundwater governance in the Indo-Gangetic and Yellow River basins: Realities and challenges*. London: CRC Press, 43–66.

Jat, M.L., *et al.*, 2006. *Land leveling: a precursor technology for resource conservation*. Rice-Wheat Consortium Technical Bulletin Series 7. New Delhi: Rice-Wheat Consortium for the Indo-Gangetic Plains.

Jehangir, W.A., *et al.*, 2007. *Sustaining crop water productivity in rice–wheat systems of South Asia: a case study from Punjab Pakistan*. IWMI Working Paper 115. Colombo: IWMI.

Kam, Suan-Pheng, *et al.*, L.S. 2005. Spatial patterns of rural poverty and their relationship with welfare-influencing factors in Bangladesh. *Food Policy*, 30 (5–6), 551–567.

Kulkarni, A.V., *et al.*, 2007. Glacial retreat in Himalaya using Indian remote sensing satellite data. *Current Science*, 92 (1), 69–74.

Mikhail, M. and Yoder, R., 2009. *Multiple-use water service implementation in Nepal and India: Experiences and lessons for scale-up*. Colombo: International Development Enterprises, CPWF, and IWMI.

Moench, M., 1996. *Groundwater policy: Issues and alternatives in India*. Colombo: IWMI.

Molden, D., *et al.*, 2003. A water productivity framework for understanding and action. *In*: J.W. Kijne, B. Randolph, and D.J. Molden, eds. *Water productivity in agriculture: limits and opportunities for improvement*. Wallingford: CABI Publishing, 1–18.

Mukherji, A., 2006. Political ecology of groundwater: the contrasting case of water abundant West Bengal and water scarce Gujarat, India. *Hydrogeology Journal*, 14 (3), 392–406.

Mukherji, A., 2007. The energy-irrigation nexus and its implications for groundwater markets in eastern Indo-Gangetic basin: evidence from West Bengal, India. *Energy Policy*, 35 (12), 6413–6430.

Mukherji, A., *et al.* eds, 2009a. *Groundwater governance in the Indo-Gangetic and Yellow River Basins: realities and challenges*. London: CRC Press.

Mukherji, A., *et al.*, 2009b. Metering of agricultural power supply in West Bengal, India: Who gains and who loses? *Energy Policy*, 37 (12), 5530–5539.

Pant, D. and Gautam, K.R., 2008. *Literature synthesis: water, land and poverty nexus in IGB. Case of Nepal*. Draft report prepared for the IGB-Basin Focal Project.

Ravallion, M. and Sen, B. 1994. *How land-based targeting affects rural poverty*. Policy Research Working Paper 1270. Washington, D.C.: World Bank.

Repetto, R., 1994. The "second India" revisited: population, poverty and environmental stress over two decades. Washington, D.C.: World Resources Institute.

Rodell, M., Velicogna, I., and Famiglietti, J.S., 2009. Satellite-based estimates of groundwater depletion in India. *Nature*, 460 (7258), 999–1002.

Saaty, T.L., 1980. *The analytic hierarchy process*. New York: McGraw-Hill.

Scott, C. and Sharma, B.R., 2009. Energy supply and the expansion of groundwater irrigation in the Indus-Gangetic Basin. *International Journal of River Basin Management*, 7 (119), 119–124.

Shah, T., 2009. *Taming the anarchy: groundwater governance in South Asia*. Washington, D.C.: RFF Press.

Shah, T., *et al.*, 2009. Is irrigation water really free?: a reality check in the Indo-Gangetic Basin. *World Development*, 37 (2), 422–434.

Sharma, B.R., 2009. Impact of climate change on water resources and potential adaptations for agriculture in the Indus-Gangetic basin. *Abstracts 60th IEC meeting & 5th Asian regional conference on improving in efficiency of irrigation projects through technology upgradation and better operation & maintenance*, 6–11 December 2009, New Delhi. New Delhi: International Commission on Irrigation and Drainage, 301.

Sharma, B.R. and Ambili, G., 2010. Impact of state regulation on groundwater exploitation in the "hotspot" region of Punjab, India. *Towards sustainable groundwater in agriculture: An international conference on linking science and policy*. University of California, Davis, 15–17 June 2010. Available from: http://www.ag-groundwater.org/Materials/Ag-GW_2010_Abstract.Pdf

Sharma, B.R. and McCornick, P.G., 2006. Domestic policy framework on adaptation to climate change in water resources: Case study for India [online]. *In: Working together to respond to climate change: Annex I expert group seminar of OECD global forum on sustainable development*, 27–28 March 2006, Paris: OECD. Available from: http://www.iwmi.cgiar.org/tmp/H039635.pdf [Accessed 28 May 2010].

Sharma, B.R., *et al.*, 2010. *Water poverty in the northeastern hill region (India): potential alleviation through multiple- use water systems- cross-cutting learning from Nepal hills*. International Water Management Institute–National Agriculture Innovation Project (IWMI-NAIP), Report 1. New Delhi: IWMI.

Sikka, A., Scott, C.A., and Sharma, B.R., 2003. Integrated basin water management in the Indo-Gangetic basin. Challenge Program on Water and Food workshop, Dhaka, Bangladesh, 21–22 December 2003. Unpublished. Available from International Water Management Institute, New Delhi, India.

Tangri, A.K., 2000. *Integration of remote sensing data with conventional methodologies in snow melt runoff modeling in Bhagirathi river basin, U.P. Himalayas*. Technical report. Lucknow, India: Remote Sensing Application Centre.

UNDP, 2008. *Human Development Report 2007/2008. Fighting climate change. Human solidarity in a divided world*. New York: United Nations Development Programme.

WBSEB, 2006. *Yearbook*. Kolkata, India: West Bengal State Electricity Board (WBSEB).

WIDD, 2004. *Groundwater resources of West Bengal: an estimation by GEC-1997 methodology*. Kolkata, India: Water Investigation and Development Department (WIDD), Government of West Bengal.

The Karkheh River basin: the food basket of Iran under pressure

Mobin-ud-Din Ahmad and Mark Giordano

Development of the Karkheh's water resources has contributed in important ways to Iran's food security and underpinned the livelihoods of both basin farmers and urban consumers. However, the linkages between poverty and agricultural water use in the basin are now weak at best. Furthermore, there is now little if any additional water to develop. As a result, future water policy will need to increasingly focus on management and allocation of existing resources rather than development of new sources of supply, and poverty reduction strategies may need to aim outside agricultural water solutions.

Introduction

Iran is a land-abundant and water-short country. It has 1% of the world's population and 1.1% of its land, but less than 0.4% of the world's freshwater. The country already uses 74% of its annual total renewable freshwater, a figure placing it far into any definition of a water-scarce state. The vast majority of current water use, 93%, is utilized for agricultural production. The food needs of a rapidly growing population and strategic policy goals to move the country towards food self-sufficiency, from its recent position as one of the world's largest agricultural importers, will only put further pressure on water resources in the coming decades. Adding to these pressures will be even faster growth in demand for industrial and domestic water for an urbanizing population and likely increased recognition of the values of environmental flows.

Given this situation, the agricultural water challenges for Iran are in the first instance related not primarily to the development of new water resources, but rather to discovering ways to utilize more effectively existing resources for current needs, managing the competition for water between sectors and determining the role of agricultural water use in future poverty-reduction strategies. As in all countries, these challenges must also be considered in the light of environmental services naturally provided by water. Perhaps especially important for Iran, the goals of agricultural water management must also be viewed in the larger economic and geo-political objectives and context of the country.

The water challenges for Iran's Karkheh River basin are in some senses exemplars of the water challenges facing the entire country and similar regions around the world.

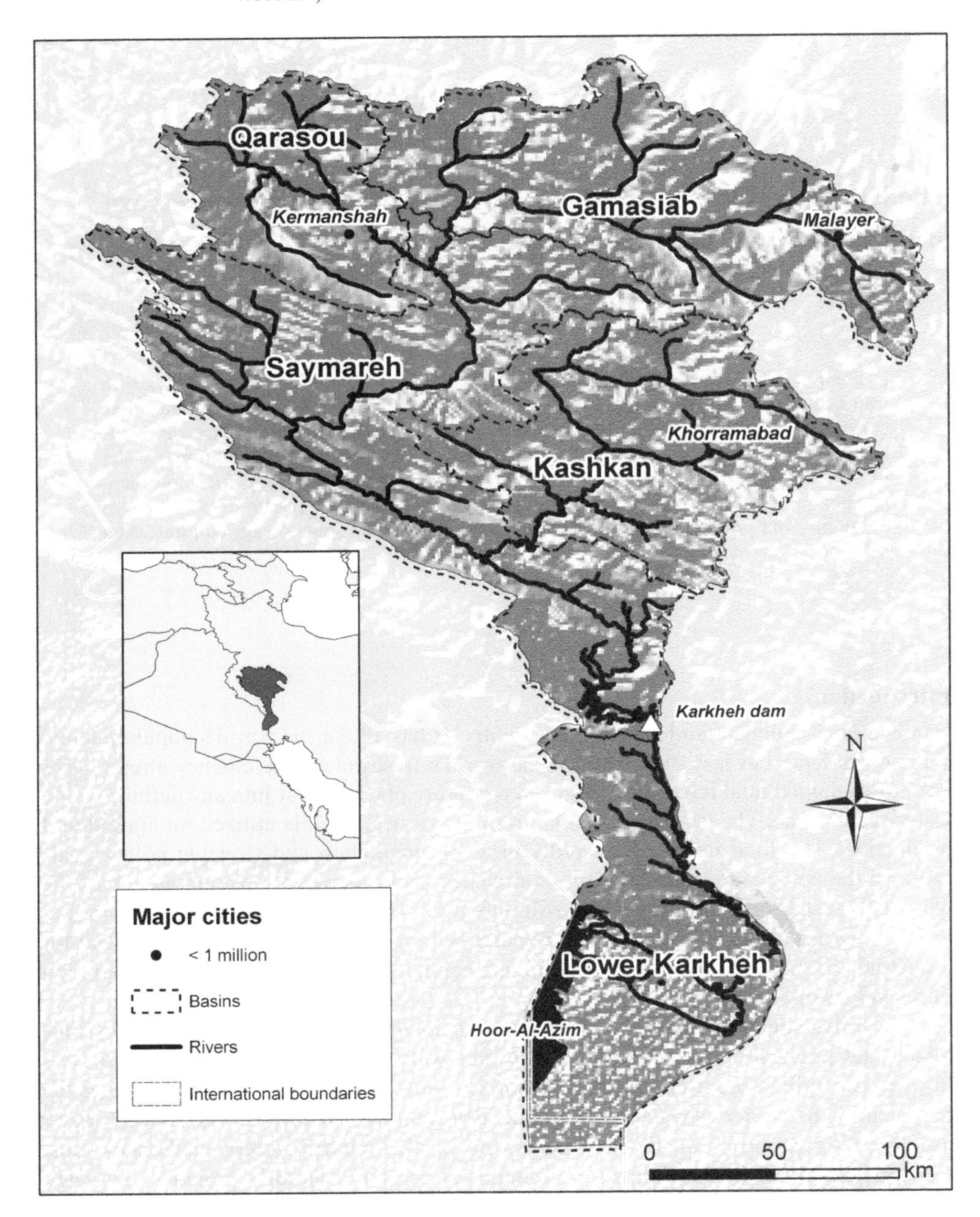

Figure 1. Location of Karkheh River basin in Iran.

The Karkheh Basin has a long history of water use and engineering, dating back to Mesopotamian times. The basin is now known as the "food basket of Iran" and is one of the main areas for the production of strategically important wheat in western Iran (Figure 1). Wheat production is facilitated by an irrigation system making up 9% of the country's total network. However, non-irrigated areas are also important sources for the production of both grains and, in particular, livestock products. In some cases, this production has resulted in degradation of lands and contributed to erosion problems impacting the operation of dams for hydropower and irrigation. As a key agricultural region, there is pressure to keep up agricultural production. The area has growing industry, is the home of important oil fields in the south and is experiencing a rapid demographic shift from agriculture towards cities. The use of the river also has important implications for environmental functions, perhaps especially in the Hoor-al-Azim wetlands on the border with Iraq.

This paper synthesizes the results of a two-and-a-half-year study commissioned by the Challenge Program on Water and Food (CPWF) on agricultural water use, water productivity and poverty in the Karkheh Basin. The paper first looks at the physical side of the basin, providing estimates of water use and water productivity across scales. It then examines the human side, providing the first ever basin-scale estimates of poverty in the Karkheh as well as a description of the institutional environment in which poverty, and water use, exist. From this content, we derive policy recommendations related to water use, water productivity and poverty alleviation.

Description of the Karkheh Basin

The Karkheh River was part of the flourishing ancient civilizations of Mesopotamia, and irrigation has been practised for millennia. The Karkheh Basin covers a total area of about 5.08 million ha with a 2002 population of 3.5 million, 35% of whom reside in rural areas. The population is expected to grow to 4.5 million by 2025 with around 25% still outside of cities (Marjanizadeh *et al.* 2009). Administratively, the Karkheh Basin is distributed among seven provinces and 32 districts. Hydrologically, the basin is divided into five sub-basins of the main rivers (Gamasiab, Qarasou, Kashkan, Saymareh and Karkheh) and 16 sub-catchments according to flow-gauging stations (Figure 1).

Nearly two thirds of the basin lies in the mountains (altitude 1000m–2500m), so that both surface and groundwater resources are replenished from winter snowfalls in the high Zagros ranges. Agriculture and human settlements are mainly in the valleys of the upper basin and in the hyper-arid plain, where the river eventually terminates in the Hoor-Al-Azim, a transboundary wetland shared with Iraq. River water becomes progressively more saline as it flows through the newly constructed Karkheh Dam to the downstream Hoor-al-Azim wetland with electrical conductivity reaching well above 3 dS/m during the dry months.

The climate of the basin is mainly semi-arid, with large variations in the average annual precipitation between the southern and northern regions. Annual rainfall in the southern part of the basin (the Lower Karkheh area south of the Karkheh Dam) is about 150 mm while in the northern part (Upper Karkheh) it rises to 750 mm (1961–90 average). About half the precipitation falls in the winter months of January–March and the summer months of June–September receive less than 2% of the total. In the Upper Karkheh, even in the hot months the maximum temperature fluctuates from 23°C to 36°C. In the Lower Karkheh temperatures rise above 40°C during the summer months, resulting in high evaporative demand. In winter, mean daily temperature falls as low as 6°C in the Lower Karkheh and −2°C in the Upper Karkheh. As a result of low temperatures, much winter precipitation falls as snow in the Upper Karkheh.

Farming activities are principally livestock rearing on rangelands and rainfed agriculture in the Upper Karkheh, complemented by irrigated cropping. Livestock is tightly integrated into all farming systems, with cattle predominating in the lowland and sheep and goats in the uplands. The dominant crops in Upper Karkheh are wheat, maize, barley and chickpea. In the past 15–20 years, there has been increasing private and state-sponsored groundwater development across the region together with major development of surface irrigation downstream of the newly commissioned Karkheh Dam.

Water use

Assessing the availability of water resources has been one of the key foci of the Iranian government for all Iran's river basins. A comprehensive study was conducted by JAMAB (1999) for the Karkheh River basin with the main motivation of planning the development of available renewable water resources to expand irrigated lands, provide water to increasing populations and industry, control floods and produce hydroelectricity. Using the information from JAMAB (1999), Masih *et al.* (2009) estimated the water accounts of the Karkheh Basin for the water year 1993/4, represented schematically in Figure 2. The year 1993/4 represents the average water availability in the basin and is considered as a reference year for future planning and allocation of water resources.

Because its goal was partially planning water diversion for irrigation, the work of JAMAB was focused on water use related to irrigated areas. To add greater detail in terms of both the nature of water use and the distribution of that use, the present study also undertook an additional analysis to estimate actual evapotranspiration using surface energy balance systems (SEBS) (Su and Jacobs 2001, Su 2002, Su *et al.* 2003) for 2002/3, the period with most complete data (Muthuwatta *et al.* 2010). We assessed the spatio-temporal distribution of precipitation and actual evapotranspiration at 1 km grid-cell resolution with results aggregated to 16 sub-catchments as shown in Figures 3 and 4. Analysis behind this information reveals that, during the study period, the Karkheh Basin

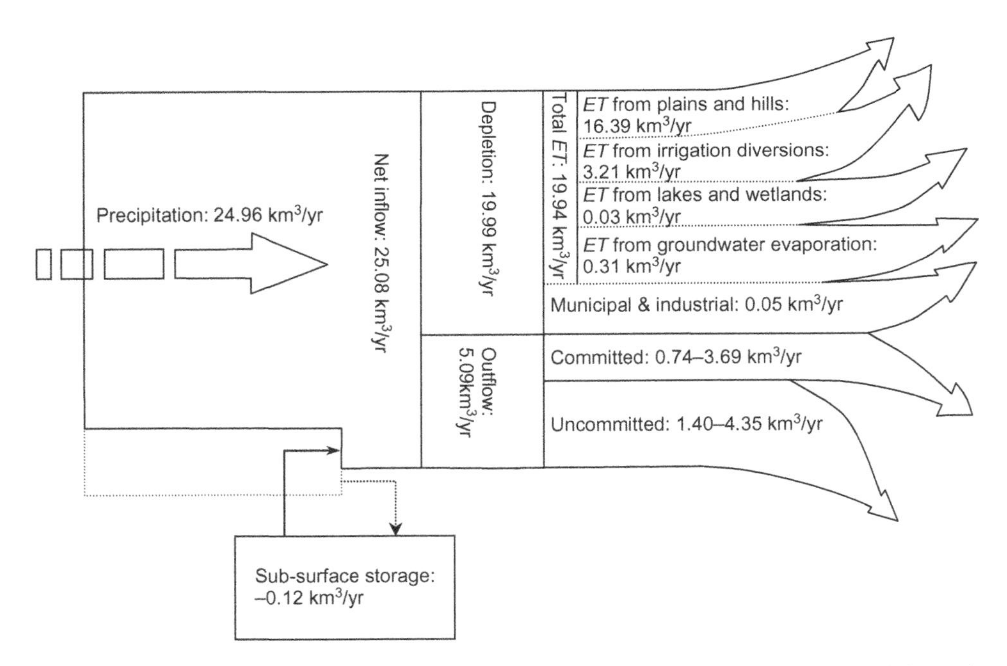

Figure 2. Schematic representation of the basin-level water accounting of the Karkheh Basin (1993–94).

Source: Adopted from Masih *et al.* (2008, 2009).

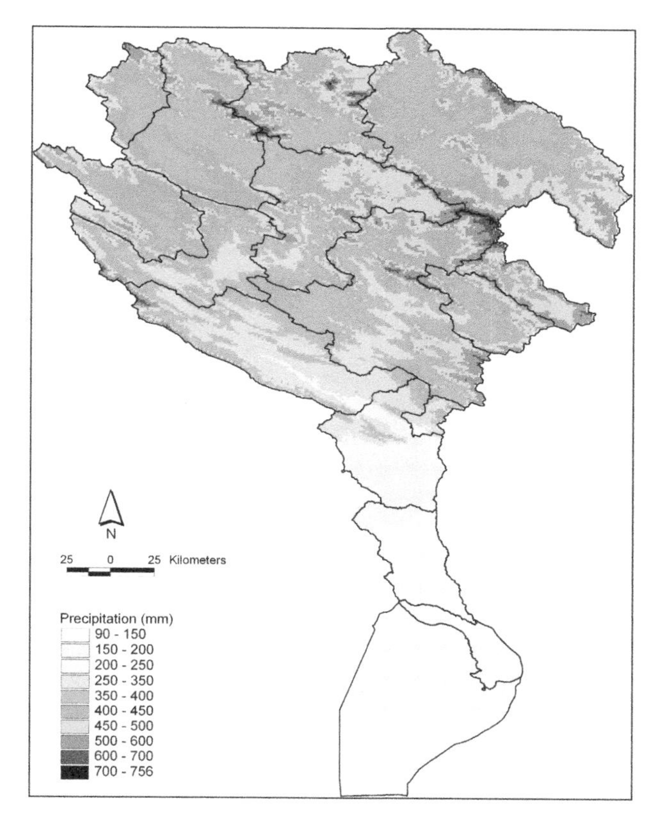

Figure 3. Spatial variation of annual precipitation (P) in the Karkheh River basin from November 2002–October 2003.

Source: Muthuwatta *et al.* (forthcoming).

received 18.5 km^3/year of inflows as precipitation out of which 16.7 km^3/year were outflows as evapotranspiration. Actual annual evapotranspiration (ET_a) varies from 41 mm to 1681 mm, with the highest values found in the Karkheh Dam and the lowest in the bare land/desert areas below the Karkheh Dam.

The analysis further reveals that, overall, irrigated areas (578,100 ha or 11% of total basin area) consume 14% of total basin precipitation while rainfed areas (1,163,000 ha or 23% of the total basin area) consume 20%. Rangeland and forests consume 18% and 11% of total rainfall respectively. However, there are important differences in water-use patterns between the upper and lower basin as shown in Figure 5. Rainfed crops and rangeland dominate the upper Karkheh while evapotranspiration from irrigated crops, wet soil, water bodies and bare land dominate the lower basin. Further highlighting the physical differences within the basin, two thirds (75%) of annual precipitation in the upper Karkheh is consumed as evapotranspiration whereas in the lower Karkheh potential evapotranspiration (ET_p) is three times higher than precipitation. High levels of water use in the lower Karkheh are made possible by irrigation supplied from dam releases and direct pumping from rivers. We again used information on water use in computing water productivity estimates in the next section.

While both sets of water accounts are useful for understanding the current state of water use in the Karkheh, they are insufficient for understanding the relationship between

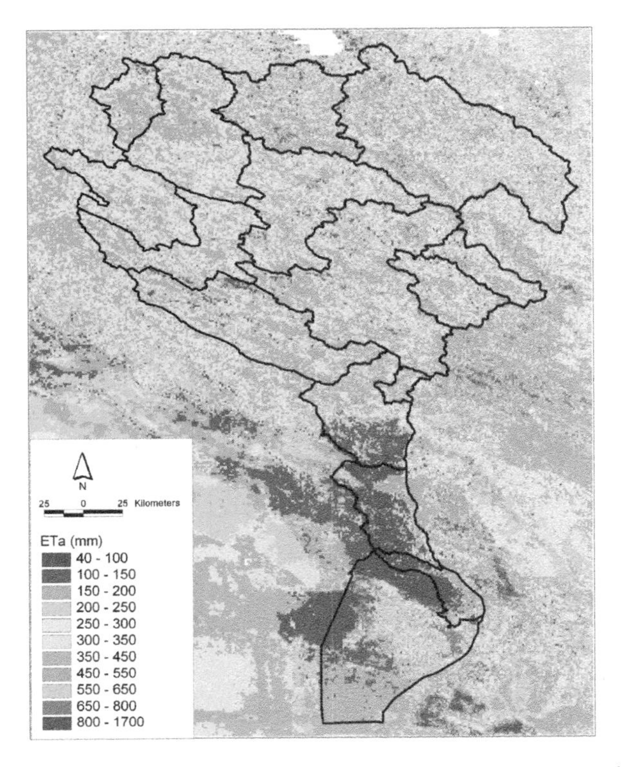

Figure 4. Annual actual evapotranspiration (ET_a) in the Karkheh Basin, November 2002–October 2003.
Source: Muthuwatta *et al*. (2010).

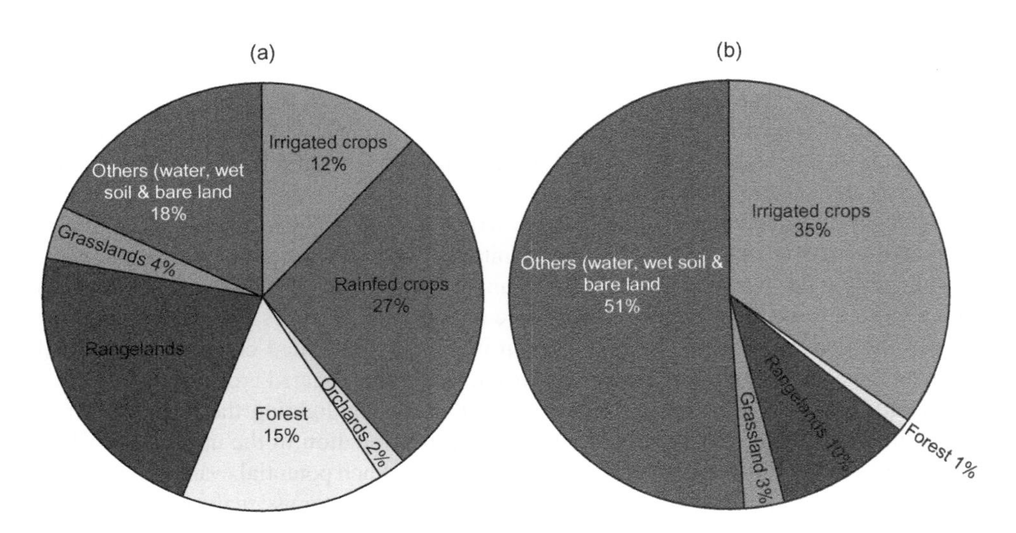

Figure 5. Percentage distribution of ET_a for different land classes for (a) the upper and (b) the Lower Karkheh Basin, November 2002 – October 2003.

these average figures, sustainability of use, and variability in supplies or for understanding the tradeoffs between changes in allocation and basin-scale water productivity change or poverty reduction.[1]

In terms of sustainability, an additional analysis conducted by JAMAB (2006a, 2006b) as part of the present study has shown that water resources are coming under increasing pressure mainly due to growing demand and recurrent drought. At the basin scale, a negative water balance of 0.144 km^3/year was reported for 2000/1, a drought year, indicating increasing stress on the groundwater storage in the basin. Groundwater withdrawals in Gamasiab and Qarasou sub-basins have already exceeded the safe limits (JAMAB 2006b) and further groundwater development has been prohibited in the plains of Malayer, Asadabad, Toyserkan and the Nahavand plains of Gamasiab sub-basin. Ongoing strategies to develop water resources in the Karkheh Basin are thus clearly coming into conflict with sustainable water use in the basin.

Existing work, including the 1999 JAMAB numbers used for planning, has tended to ignore the issue of flow variability. Long-term average flows at the key locations across the Karkheh river system are shown in Figure 6 for the period from 1961–2001 (Masih *et al.* 2008). Large seasonal variability is evident across the whole basin as is the similarity in high and low flow patterns. The maximum flow (at Paye Pole) of 12.59 km^3/year occurred in 1968/9 and the minimum of 1.92 km^3/year corresponds to the drought year of 2000/1. These large temporal variations indicate a high level of supply insecurity for current and further withdrawals for human use.

To quantify this insecurity, we analysed flow duration relative to mean annual availability of surface water as shown in Table 1. The analysis of the derived flow duration curves shown in Figure 7 clearly suggests that planning on the basis of mean annual flows can only provide a supply security of 35–50%. Furthermore, due to construction of the Karkheh

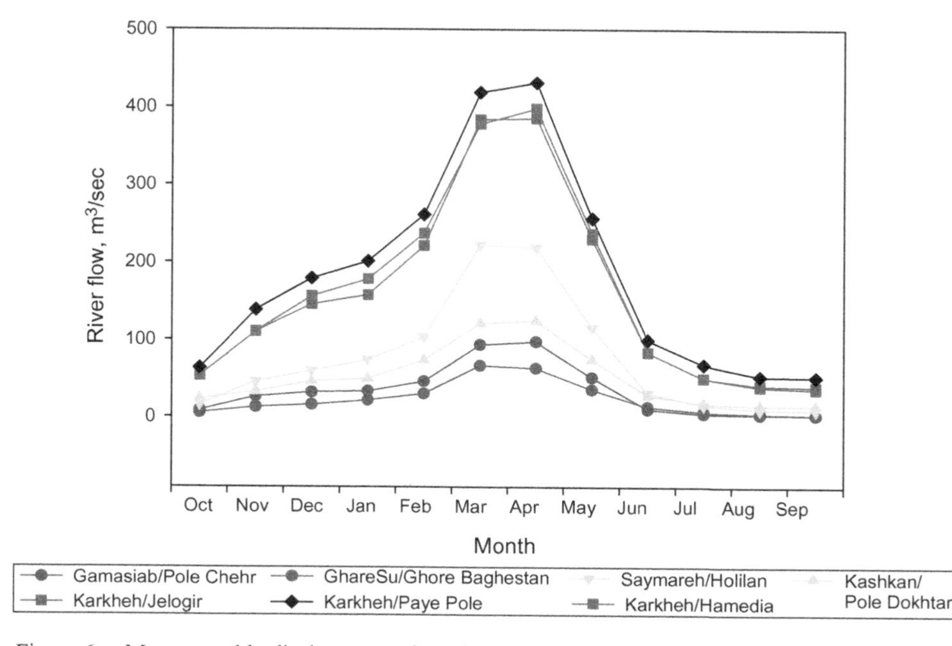

Figure 6. Mean monthly discharge at selected stations of the Karkheh River system, 1961–2001. Source: Masih *et al.* (2008).

Table 1. Various probabilities of annual river flows (in km^3/yr) at selected stations of the Karkheh River system. Based on the data for the period 1961–2001.

River/Station	Q5	Q10	Q25	Q50	Q75	Q90	Q95
Gamasiab/Pole Chehr	2.42	1.68	1.30	1.02	0.77	0.55	0.29
GhareSu/Ghore Baghestan	1.84	1.18	0.96	0.72	0.42	0.35	0.27
Saymareh/Holilan	6.04	4.25	2.98	2.34	1.50	1.17	0.87
Kashkan/Pole Dokhtar	3.08	2.46	2.06	1.65	1.11	0.85	0.78
Karkheh/Jelogir	8.96	8.23	6.19	4.84	3.56	2.60	2.23
Karkheh/Paye Pole	10.76	9.28	7.76	5.65	4.08	3.02	2.40
Karkheh/Hamediah	9.28	8.64	7.56	4.87	3.45	2.25	1.65

Source: Masih *et al* (2008).

Dam and downstream irrigation schemes, we anticipate that during below average/low flow years, conflicts will increasingly arise between the desire to retain water in the Karkheh Dam for hydropower generation and the need to supply downstream agricultural users with irrigation water. The issue of irrigation supply will likely be further exacerbated by problems of soil salinity. However resolved, both of these uses will also be accompanied by the diminished flows to riverine ecosystems and floodplains as well as to Hoor-al-Azim wetland further downstream (Masih *et al.* 2009).

Water productivity

Analysis of water productivity is essential to evaluate the performance of current water use at river-basin and other scales and to identify opportunities to improve the net gain from water. The gains can come by either increasing the productivity for a given quantity of water consumed, or by reducing the quantity consumed without decreasing production. A review of existing literature and understanding revealed that the information on water productivity in Iran in general, and the Karkheh Basin in particular, is limited (Ashrafi 2006).

At present, there are some estimates of field-scale physical crop water productivity. However, the basis of the estimates (i.e. productivity in terms of gross inflow, irrigation applied, or water consumed by evapotranspiration) are unclear. Water productivity estimates beyond the field scale also appear not to exist. This is the major bottleneck for water-policy makers to identify possible and viable options to enhance water productivity in a sustainable manner and to understand how changes in water use in one location may have impacts on productivity in other locations and across scales. The major goal of this component of the project was thus to fill these information gaps by providing explicit estimates of water productivity using physical measures at basin, sub-basin and farm scales and economic measures at sub-basin scale (see Ahmad *et al.* 2009 for more details). The analysis was conducted in two ways. First, estimates of physical water productivity at a range of scales were made based on field-scale measurements and surveys of farmers' responses, water availability and water consumption. Second, economic water productivity at the sub-basin scale was calculated mainly using secondary data from meteorological, hydrological and agricultural statistics.

For the entire Karkheh Basin, physical water productivity in terms of yield over gross water inflow for major crops and by irrigated and non-irrigated area are shown in Table 2. While comparisons across countries are problematic because of differences in basic physical conditions, we can say that the physical water productivity numbers for the Karkheh are

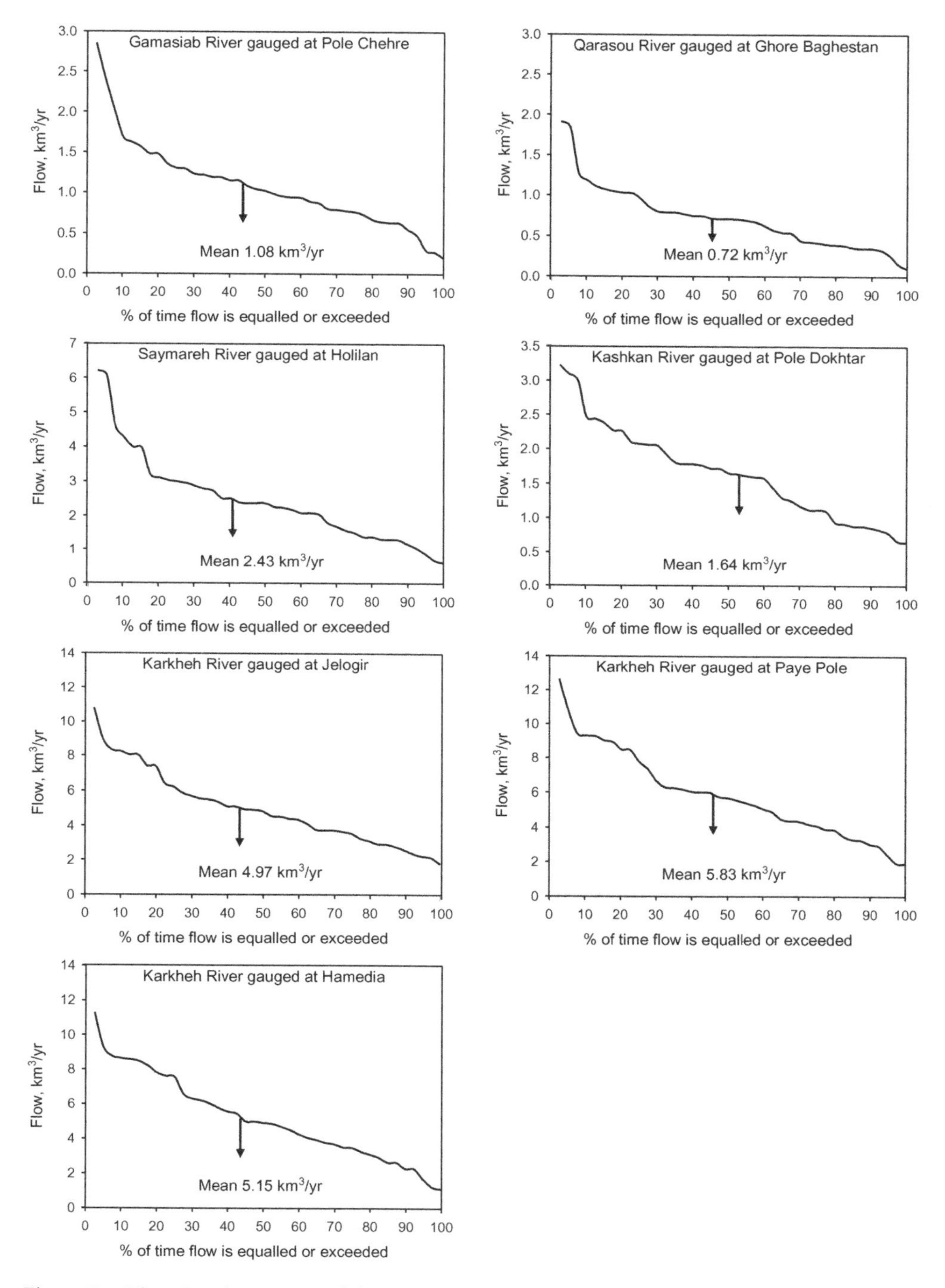

Figure 7. Flow duration curves of the mean annual flow at the selected locations of the Karkheh Basin (1961–2001) (note the difference in scales of the y-axis).
Source: Masih *et al.* (2008).

Table 2. Yield, water use and water productivity estimates for major crops of the Karkheh Basin and sub-basins, 2006.

Crops	Gamasiab	Qarasou	Kashkan	Saymareh	Lower Karkheh	Karkheh Basin
Wheat irrigated						
Yield	4860 ± 1300	4030 ± 970	3420 ± 1020	2680 ± 1070	2490 ± 1200	3320 ± 1510
Gross inflow	7550 ± 1530	6970 ± 1380	7250 ± 1960	6630 ± 1250	4500 ± 1140	6050 ± 1920
WP	0.65 ± 0.16	0.59 ± 0.16	0.48 ± 0.12	0.41 ± 0.15	0.55 ± 0.23	0.55 ± 0.20
Wheat rainfed						
Yield	1820 ± 570	1730 ± 550	1410 ± 580	1290 ± 480	1220 ± 620	1460 ± 580
Gross inflow	3100 ± 240	3190 ± 300	3580 ± 140	3650 ± 450	1840 ± 260	3320 ± 610
WP	0.59 ± 0.19	0.55 ± 0.18	0.39 ± 0.16	0.35 ± 0.12	0.69 ± 0.41	0.46 ± 0.22
Barley irrigated						
Yield	4050 ± 896	4250 ± 1060	2250 ± 350	1500 ± 710	1460 ± 830	2640 ± 1530
Gross inflow	6760 ± 1490	7270 ± 1530	5820 ± 290	5420 ± 60	4030 ± 1040	5320 ± 1800
WP	0.60 ± 0.08	0.61 ± 0.27	0.39 ± 0.08	0.28 ± 0.13	0.37 ± 0.18	0.470 ± 0.19
Barley rainfed						
Yield	1870 ± 590	1590 ± 650	1450 ± 500	1130 ± 507	900 ± 200	1410 ± 610
Gross inflow	3240 ± 240	3320 ± 390	3600 ± 130	3540 ± 340	1750 ± 280	3380 ± 480
WP	0.58 ± 0.18	0.49 ± 0.21	0.40 ± 0.14	0.32 ± 0.14	0.54 ± 0.17	0.43 ± 0.19
Maize irrigated						
Yield	8350 ± 1380	8270 ± 910	6000 ±	5750 ± 1320	6710 ± 1450	7440 ± 1560
Gross inflow	9820 ± 2420	11260 ± 5920	6570 ±	6720 ± 1800	10290 ± 4550	9990 ± 4400
WP	0.90 ± 0.28	0.88 ± 0.37	0.91 ±	0.88 ± 0.18	0.75 ± 0.30	0.84 ± 0.30
Chickpea rainfed						
Yield	590 ± 210	750 ± 260	430 ± 110	610 ± 150	–	620 ± 210
Gross inflow	1090 ± 650	1040 ± 250	1420 ± 746	1250 ± 540	–	1200 ± 530
WP	0.54 ± 0.44	0.76 ± 0.34	0.44 ± 0.32	0.57 ± 0.26	–	0.70 ± 0.84

Note: Units for yield, gross inflow and water productivity are kg/ha, m^3/ha and kg/m^3 respectively.
Source: Ahmad *et al.* (2009).

generally low by global standards. We discuss this further below in the context of economic water productivity. It at least suggests scope for improvement.

Analysis at the sub-basin scale can provide additional insights into where any efforts to increase physical productivity might best be targeted. As shown in Table 2, the basin averages mask variation in physical water productivity across sub-basins. The two upper sub-basins, Gamasiab and Qarasou, have the highest physical water productivity for all crops, while the Kashkan and Saymareh sub-basins have the lowest figures. These water productivity patterns are largely similar to yield patterns, highlighting water productivity–yield relationships.

Because of vast differences in physical conditions within a single basin like the Karkheh, one must be cautious in assuming that differences in water productivity across the basin indicate gaps that might be closed. To get a better feel for the potential to close the gaps, we analysed water productivity of individual farmers. The results for irrigated and dryland wheat are shown in Figure 8. They reveal substantial variation even between farmers within the same sub-basins. While the physical conditions on individual farms within a sub-basin of course differ, these figures give an indication that closable gaps do exist. For example, the difference between the top 10% of cases and average water productivity in each sub-basin is about 0.40 kg/m^3. If that gap were to be closed, wheat production could increase by approximately 1500 kg/ha with almost no increase in water use. While this is probably an overstatement of the realistic possibilities in the short to medium term, it does give an idea of the possible targets.

We did production-function analysis to understand key reasons for these gaps and what policy responses might therefore be possible. The results showed that access to irrigation, pumped water (from either groundwater or streams), seed rate and nitrogen use are the major factors governing wheat yield. This highlights the importance of water and other inputs in improving farm-scale yield and water productivity.

For irrigated areas, the analysis also reveals that farmers apply two to eight irrigations to wheat crops. Further segregation of data showed that in most cases, the highest yields can be attained by three to four irrigations. While there may be valid reasons for over-irrigation, this suggests that extension work for farmer education may allow increased production at lower levels of water input. Because all applied irrigation water is not consumed as evapotranspiration, the impact of the savings would not be proportional to the decreased application, but savings are possible.

For rainfed areas, the analysis shows that there is considerable scope for improving physical land and water productivity by exploring means of additional water application through supplemental irrigation wherever possible. Rainfed yields tend to be only half that of irrigated yields and much of the difference can be associated with water use. However, the gap of water productivity (as opposed to the gap in yield) between rainfed and irrigated areas is minimal, because of both lower yield and lower water use in rainfed systems. This stresses the need for comparing both land and water productivity when diagnosing overall system performance. The data also show that the land productivity gains with respect to marginal increases in water use is higher for rainfed systems than irrigated ones. For these reasons, and because of higher precipitation and lower irrigation requirements, in spatial terms more scope exists in the upper than lower Karkheh for improvements in productivity.

To move beyond a single crop (wheat in this case) to the best overall use for agricultural water requires a shift to the concepts of economic water productivity. Estimating economic water productivity is difficult as it requires data at similar scales on water use, production and economic value. As in most countries, data related to water use in Iran are collected within hydrological boundaries while production and economic data follow administrative

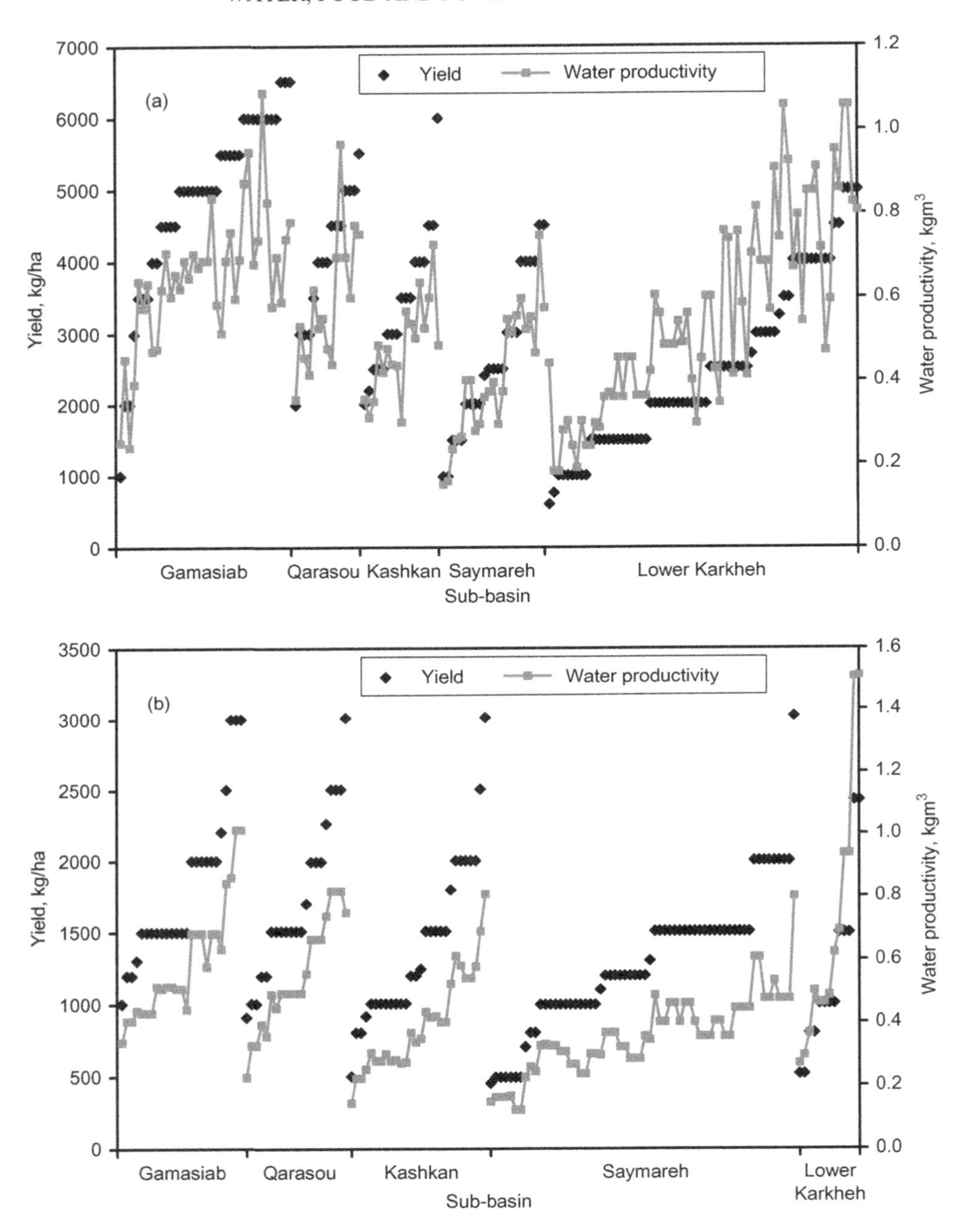

Figure 8. Inter- and intra-subbasin variation in water productivity of (a) irrigated wheat land and (b) rainfed wheat land. Water productivity is calculated in terms yield per unit of (a) gross inflow and (b) rainfall. The large differences between farmers within a sub-basin give an initial indication of the potential for improvement.

Source: Ahmad *et al*. (2009).

boundaries. We overcame this mismatch for the Karkheh by transforming district-level secondary data on agricultural, livestock and poultry production/economic value to the sub-catchment scale by spatial analysis (Ahmad *et al.* 2009). Finally, we aggregated gross value of production (GVP) and water consumption in terms of ET_a from different land uses to calculate water productivity at the sub-catchment scale. Results are presented separately for systems with and without livestock to highlight the importance of mixed crop-livestock systems.

The overall WP of rainfed crops was 0.051 $/m^3, ranging from 0.027 to 0.071 $/m^3 (Figure 9), whereas average irrigated crop WP is 0.22 $/m^3 ranging from 0.12 to 0.524 $/m^3. The coefficient of variation (CV) for irrigated crop WP was 0.45, which is almost double the CV of rainfed WP. The higher CV for irrigated crops compared with rainfed crops are largely attributed to large variation in cropping patterns between the different sub-catchments. In rainfed systems wheat and barley are the main crops whereas in irrigated systems farmers grow a mixture of crops such as wheat, maize, barley, sugar beet and vegetables. Rainfed WP declines from the upper to the lower Karkheh reflecting increased incidence of drought as one moves southwards. Patterns of precipitation across the upper Karkheh are similar, but rainfed WP shows quite large variability. This could be related to soils and other agronomic factors and requires further investigation. In contrast, the higher irrigated WP values are concentrated in middle and lower reaches of Karkheh except South Karkheh sub-catchment, which could be related to higher soil and water salinity. Although the highest irrigated WP is in Pole Zal sub-catchment, this could not be considered as a target value for improvement, given the high level uncertainties in the estimates of ET_a, GVP, and the smaller area of irrigation. Considering this, Jelogir, Pole Dokhtar, Ghore Baghestan and Doab in upper Karkheh and Abdul Khan and Hamedieh in lower Karkheh indicate the possible targets for medium-term interventions in neighbouring low-performing sub-catchments.

Vegetative WP (WP of crops, horticulture and forest, not considering the contribution of livestock in the system) was 0.097 $/m^3. However, there was wide variation across sub-catchments and values ranged from 0.004 to 0.36 $/m^3 at the Pole Zal and Hamedieh sub-catchment respectively (Figure 10).

The higher values occur where there is a higher proportion of irrigated lands in a sub-catchment whereas lower values reflect the dominance of other land uses (i.e. rangelands and forest and rainfed crops). It is important to note that the magnitude and distribution of agricultural economic WP changes substantially when livestock are included. The average value of overall vegetative and livestock WP becomes 0.129 $/m^3 (range 0.022–0.408 $/m^3). The impact of including livestock in WP calculation is greater in the upper Karkheh. The reasons include a high proportion of grass and rangelands as well as rainfed vegetation, which is an important grazing resource in addition to crop residues from irrigated and rainfed crops.

Poverty

Prior to this project there were few studies and little data available on the extent and distribution of poverty in Iran. Moreover, to our knowledge there were no poverty studies in Iran using anything other than administrative boundaries as the unit of analysis. As a result, it was necessary for the project team to both develop Karkheh-specific data sources and undertake an original analysis to describe poverty in the Karkheh Basin (Assadzadeh 2006, Assadzadeh unpublished data).

Iranian team members worked with the Statistical Center of Iran to obtain data from household income and expenditure surveys (HIES) for the years 1983, 1993 and 2004. The HIES is the most complete and consistent survey in existence within Iran related to poverty analysis. For the first time, the 2004 survey included partially geo-referenced data, which allowed us to reorganize the data away from traditional administrative boundaries and conduct analysis specifically on the Karkheh Basin. As some indication of the most

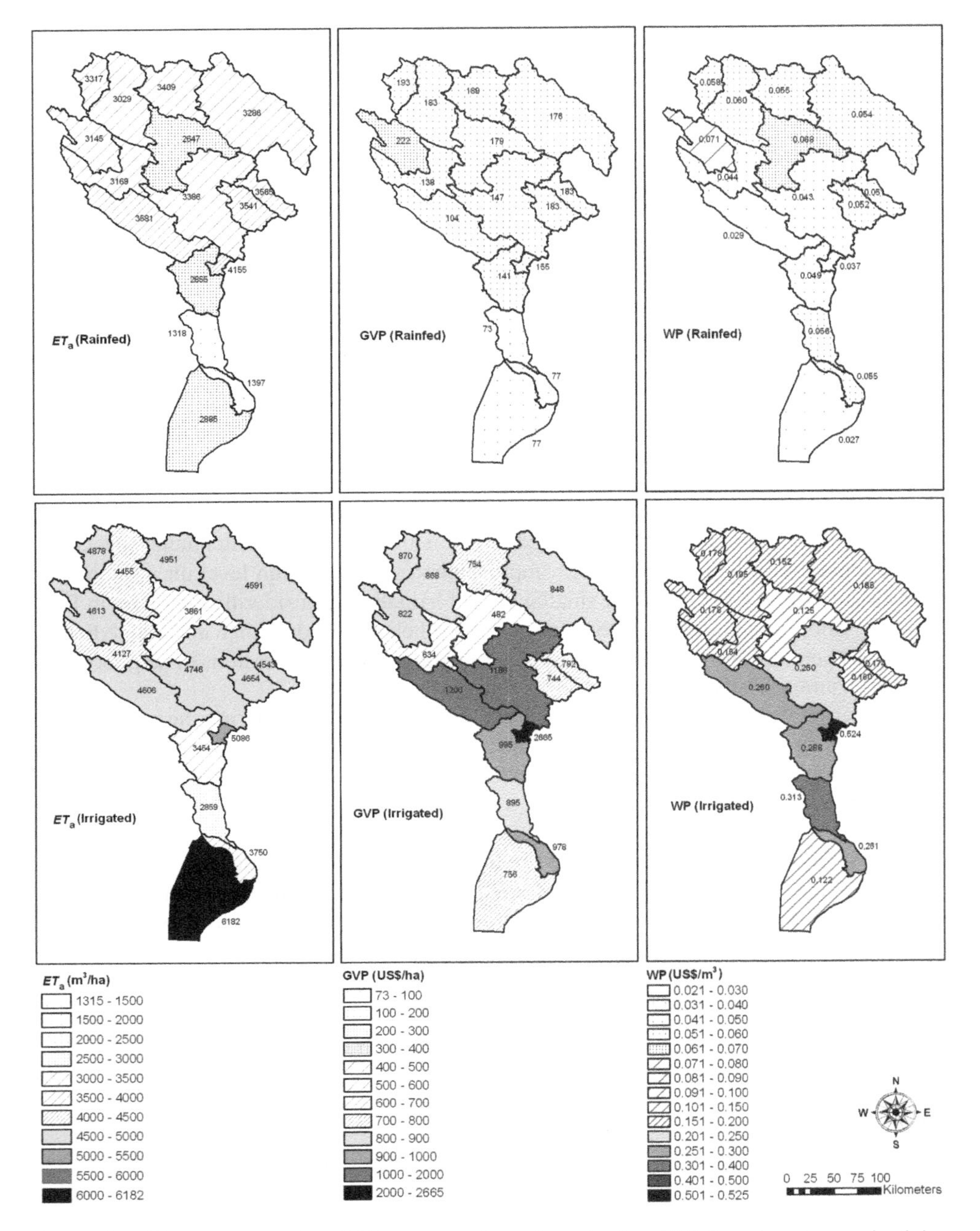

Figure 9. Water consumption (ET_a), gross value of production (GVP) and water productivity (WP) from rainfed and irrigated crops for 2002/3 in Karkheh River basin. The values in each sub-catchment represent the average values for ET_a, GVP and WP. (Note: 1 US$ = 8281 Iranian Rials in 2003).

Source: Ahmad *et al*. (2009).

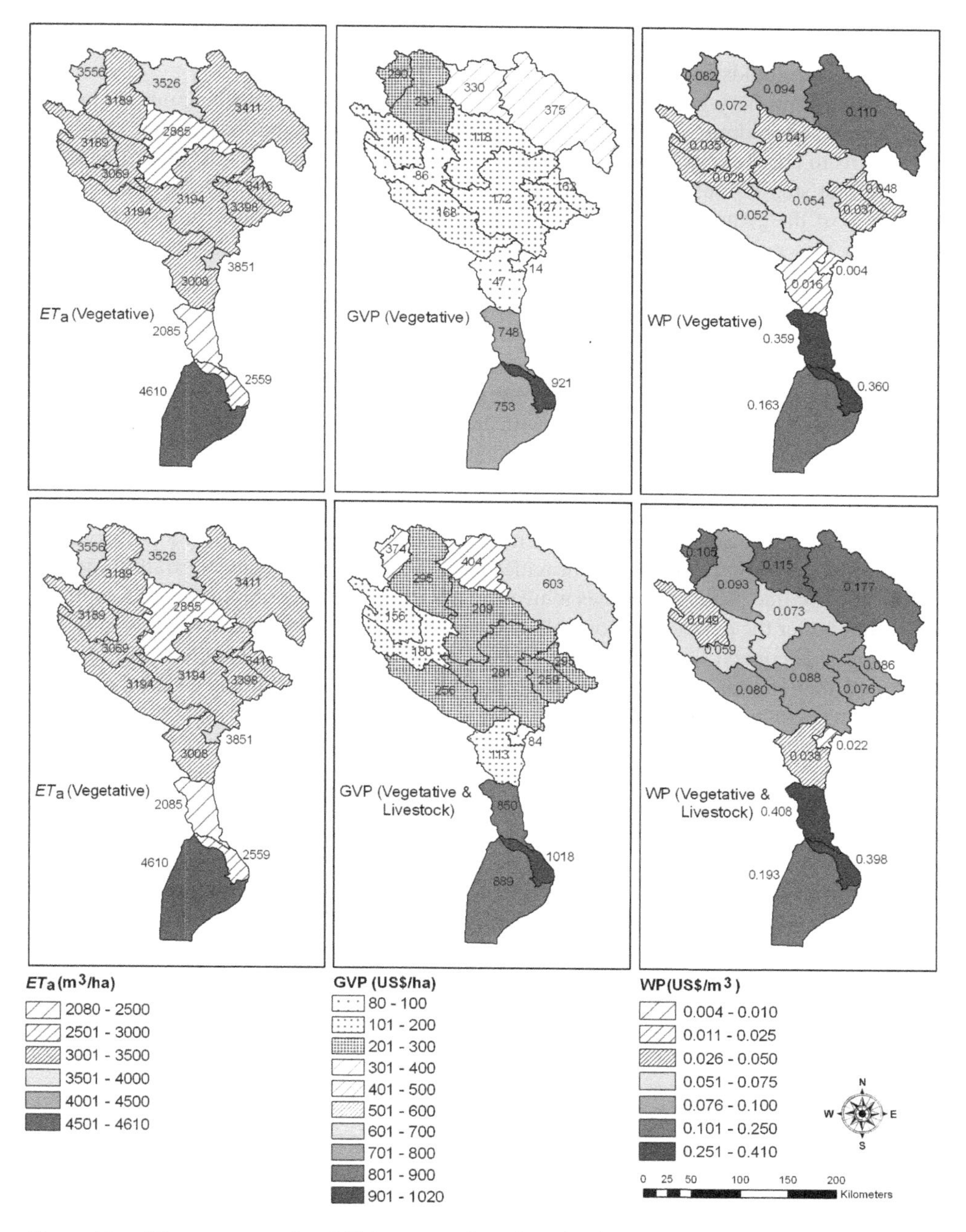

Figure 10. Water consumption (ET_a), gross value of production (GVP) and overall water productivity (WP) from vegetative and livestock for 2002/3 in Karkheh River basin. The values in each sub-catchment represent the average values for ET_a, GVP and WP. (Note: 1 US$ = 8281 Iranian Rials in 2003).

Source: Ahmad *et al.* (2009).

recent surveys' scope, it contains data for some 24,000 households across Iran and more than 2000 households for the Karkheh.

While our primary goal was analysis of poverty within in the Karkheh, our data organization and procedures facilitated simultaneous analysis of the entire country. This also allowed us to put poverty in the Karkheh in the context of poverty for the whole country.

The first step in measuring poverty is defining a poverty line. To do this for Iran, we made use of the estimates of the cost of diet provided by Rahimi and Kalantary (1992). Based on this information, separate poverty lines were constructed for both rural and urban areas. Based on the HIES, we then calculated poverty estimates using standard metrics including: the headcount, income gap, poverty gap ratio and the Foster, Greer, and Thorbecke measures. For the purposes of this synthesis, we present only the two most widely used measures, headcount and income gap.

The headcount ratio (HCR) is one of the most widely used poverty measures. HCR is simply the proportion of the population with income lower than the poverty line. Its primary advantage is that it provides an easy-to-understand metric of the extent of poverty within a given population. One major disadvantage of the HCR is that it does not take into account the severity of poverty among the poor. Thus someone with income 99% of the poverty line is counted the same as someone with income equal to only 10% of the poverty line. The income gap ratio (IG) measure provides the average proportionate shortfall of income below poverty line. Thus it provides a measure not of the number of poor people, as in the HCR, but rather the depth of the overall poverty problem in terms of income shortfall.

To provide context for the Karkheh results, we first highlight the results at the national level and their trends as shown in Table 3. The most striking finding is the large and rapid changes in poverty levels, which rose sharply from 1983 to 1993 before falling even more sharply by 2004. Poverty is believed to have been relatively low in early part of this period in part as a result of policy responses to the Iran–Iraq war. During that period, the government maintained tight controls over the economy and attempted to provide minimum subsistence to every household through a wide-ranging rationing system. By 1993, economic adjustment and liberalization policies adopted by former president Hashemi Rafsanjani resulted in inflation rates of 40%, eroding income and throwing large numbers of people into poverty. By 2004, a stabilization policy had become effective, helped also by high oil prices due to conflict in the Persian Gulf, and the country's economy was growing at 5–6% per year with inflation falling to 15%.

Table 3. Incidence of rural and urban poverty in terms of headcount ratio (HCR) and income gap ratio (IG) in Iran, 1983–2004.

	Rural		Urban	
Year	HCR	IG	HCR	IG
1983	0.391	0.384	0.187	0.342
1993	0.525	0.411	0.296	0.328
2004	0.334	0.362	0.236	0.331
		% change in poverty		
1983–93	34.3	7.2	58.3	−4.4
1993–04	−36.5	−11.9	−20.3	0.9
1983–04	−14.7	−5.6	26.2	−3.4

Source: Assadzadeh 2006.

Table 4. Incidence of rural and urban poverty in terms of headcount ratio (HCR) and income gap ratio (IG) in the Karkheh River basin, Iran, 1983–2004.

	Rural		Urban	
Year	HCR	IG	HCR	IG
1983	0.488	0.373	0.227	0.337
1993	0.593	0.376	0.404	0.347
2004	0.283	0.318	0.267	0.301
	Percentage change in poverty			
1983–93	21.6	0.8	78.0	3.5
1993–04	−52.3	−15.4	−33.9	−13.5
1983–04	−42.0	−14.7	17.6	−10.5

Source: Assadzadeh 2006.

The magnitude of these changes on poverty levels was higher in rural areas than urban, perhaps in part because the level of poverty was significantly higher in rural areas to begin with. While rural poverty is still higher in rural areas, the gap between the two is now substantially reduced.

As shown in Table 4, the overall trends in the Karkheh Basin are similar to those for Iran as a whole. However, the decline in rural poverty has been even more rapid than the decline in urban poverty. As a result, levels of rural poverty in the Karkheh are now lower than the levels of urban poverty. Furthermore, while the Karkheh Basin was poorer than Iran as a whole in 1983, the reductions in poverty after 1993 have been even more rapid with the result that the Karkheh now has lower than average poverty rates in both the rural and the urban sectors.

There is, however, substantial variation in poverty rates across the basin as shown in Figure 11 for rural areas. Poverty rates in the south of the basin are more than double those of some of the more northerly districts and above those of the national as a whole. This contrasts with frequent assumptions that the incidence of rural poverty is highest in upland areas.

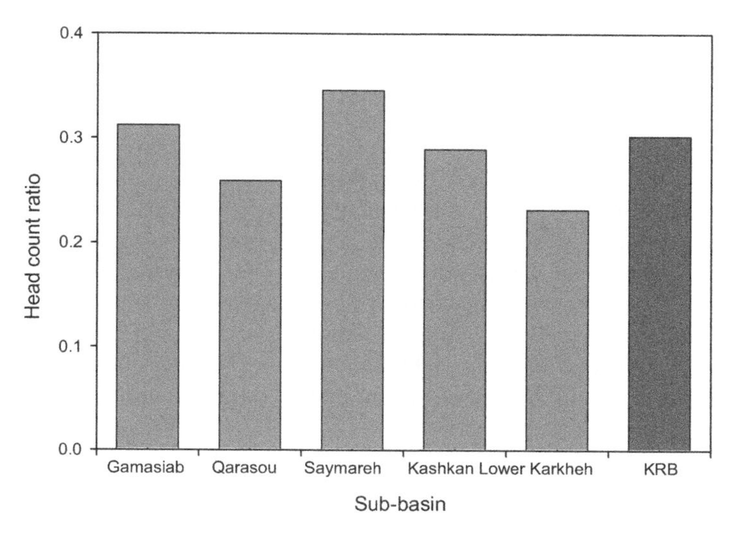

Figure 11. Sub-basin breakdown of the extent of poverty in the rural Karkheh River basin, 2004.

Institutions and the strategic policy environment

As in most countries, Iran develops and manages its water resources through a set of institutions. At the national level, there are three primary ministries in charge of water resources. These are the Ministry of Energy and the Ministry of Jihad-e-Agriculture (A. Assadzadeh unpublished data).

The Ministry of Energy (MOE) is responsible for energy supply and therefore water via hydropower production. While hydropower plays a major role in the Karkheh management, the role of the MOE goes beyond just hydropower production and includes all major hydraulic works including dams as well as primary and secondary irrigation canals and drainage. Within the MOE there is a Water Affairs Department responsible for overseeing and coordinating planning, development, management and conservation of water resources. There are also 14 regional water authorities (RWA), including one for the Karkheh, which report to the MOE and are responsible for feasibility studies of water project and project management.

The Ministry of Jihad-e-Agriculture (MOA) is responsible for rainfed and irrigated crop development and thus indirectly influences crop water use. It is also in charge of subsurface drains, tertiary and quaternary canals as well as on-farm development and irrigation techniques. In addition, the MOA is involved with the setting of agricultural prices and production policies, which influence farmers' cropping decisions, water use and water productivity as well as watershed management and rural development. Another set of organizations, both governmental and non-governmental, works to alleviate and prevent rural poverty.

While there is certainly institutional overlap and imperfection, the key point, especially in comparison with many of the other CPWF basins, is that a set of functioning institutions does exist both to manage water and to address poverty. The primary institutional issues for the Karkheh thus have more to do with the setting and funding of policy priorities rather than the creation of new institutions.

This is perhaps exemplified by examining the physical and policy environment within the Karkheh over the last 100 years. This development can be divided into four phases (Marjanizadeh *et al.* 2009). During the first phase (1900–50) and before, wheat and barely were the dominant crops based on rainfed cultivation, although there is some ancient evidence of irrigation. Starting in the second phase (1950–80), the region shifted slowly towards irrigated agriculture and more than 10% of the area came under irrigation. Starting in the third phase (1980–2000), groundwater began to grow in relative importance as surface supplies became scarcer. Due to the self-sufficiency programme launched by the government in the fourth phase (2000 onwards), wheat cultivation was given priority. As a result, the cultivation of other crops with less water demand and higher water productivity (in terms of units of water or value) were not considered as options.

Although these policies have helped close the self-sufficiency gap, natural resources, including water in the basin, came under extreme pressure. Thus trends are not sustainable and more concerted efforts will be needed for the management of land and water resources if the Karkheh River basin is to continue contributing its share in meeting the country's overall food requirements. This highlights the interplay between physical outcomes and the policy environment and the role of each in finding future solutions to the basin's water problems.

Understanding Iran's current water policy environment and choices related to water and poverty also requires an understanding of the country's broader national and strategic challenges and foreign policy challenges, in particular as related to its agricultural and foreign policy focus on self-sufficiency. The reasons for a self-sufficiency policy come from Iran's tumultuous past, two ongoing wars on the country's eastern and western flanks, and the fear that food supplies from the outside, in particular from the west, could be cut off. Food self-sufficiency was in fact already a goal of the Iranian Revolution. The downside of the food-self-sufficiency policy is that it encourages the production of grains in water-scarce areas that, from an economic and probably water productivity standpoint, would be better imported. Exacerbating the problem is a grain subsidy system aimed primarily at the welfare of urban residents. This subsidy system does not specifically target the poor but does encourage waste and pushes against Iran's own self-sufficiency goals.

Water, poverty and productivity: current policy issues

As everywhere, the linkages in the Karkheh Basin between water, agriculture and, in particular, poverty are complex. The control of water resources in the Karkheh has played a major role in the both the expansion of agricultural area and an increase in average yields over the last few decades. This has clearly contributed to Iran's strategic goal of food self-sufficiency.

At the same time, the increase in food production has not generally been associated with changes in rural or urban poverty levels in the Karkheh. Poverty increased and then decreased sharply over the 20-year period ending in 2004 following a pattern similar to the country as a whole. For the Karkheh, this is despite a relatively steady increase in area harvested from 786,000 ha in 1983 to 1,272,000 in 1993 to 1,707,000 in 2004 (Figure 12). Clearly the direct driving forces behind poverty change in Iran in general, and in the Karkheh in particular, have more to do with larger economic, political and international relations issues than with agricultural production and, by association, water use. This same finding holds true for rural Iran and the rural Karkheh Basin where poverty trends

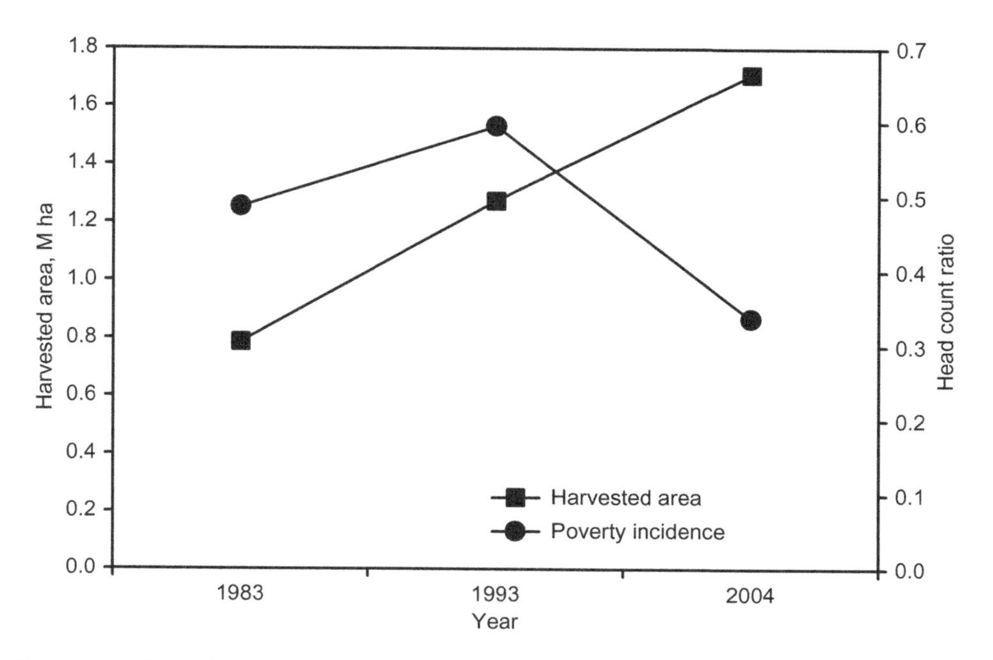

Figure 12. Trends in poverty and agricultural production, Karkheh Basin, Iran.

have been similar and one might hypothesize a closer connection between water use and poverty levels.

This is not to say that water, food and poverty are not linked in the macro environment, however, in other, more indirect ways. In a market economy, increased food production translates into lower prices and has poverty impacts on the urban poor, impacts often estimated in other regions as large as or larger than the impacts in rural areas. In Iran, the prices of many consumer staples are government controlled and so any price impact is more directly related to the cost of government subsidies. At least in theory, the decrease in implicit prices reduces the costs to the exchequer, which can free government resources for other poverty-reduction programmes. While the Government of Iran and private organizations have established many poverty-reduction programmes, the extent to which they have been facilitated by increased food production is unknown and is likely unknowable.

In the Karkheh Basin, rural poverty levels are already relatively low, and agricultural workers are also not amongst the poorest of rural residents. Furthermore, the absolute number of rural residents is falling and the proportion of rural residents in the overall population is declining rapidly. These demographic changes, the scarcity of water in Iran generally, the country's substantial other assets, together with evidence of the drivers of past poverty, suggest that factors other than water determine poverty. Measures that are unrelated to agricultural water are therefore likely to be the most effective direct solution to remaining problems of rural poverty in the Karkheh Basin and elsewhere in the country.

As an example of the possibilities, we can consider Iran's policy of providing subsidies to domestic energy, primarily petrol, users. These subsidies only indirectly impact water use and agriculture. However, their magnitude highlights how a change from a policy that does nothing to improve productivity could be used to improve the lives of the poor. Petrol in Iran sells at state-controlled prices at only a small fraction of world market prices. The result is high levels of inefficiency in use as well as smuggling. Because Iran's petroleum refining capacity is limited and despite the fact it is one of the world's largest oil exporters, it now has to import nearly half its petrol, which it sells internally at subsidized prices. These subsidies amount to more than 19% of gross domestic product (GDP) (Victor 2009, World Bank 2010). Based on our analysis of the income gap measure of overall poverty in Iran, a targeted transfer of slightly more than half this amount would move everyone in Iran now below the poverty line above the poverty line. While this might sound simple, the recent riots over the imposition of fuel rationing show the political difficulty in making any change to domestic energy policy. The same is would likely be the case for a change in food pricing policies. Nonetheless, the example highlights how changes in existing policy structures could be used to target poverty, likely much more effectively and with lower trade-offs than changes in agricultural water management or allocation.

In a related way, our analysis of poverty trends highlights how much impact the overall economic environment can have on both rural and urban poverty levels. Policies to foster overall growth, including industry and services, in a stable macro-economic environment again seem to be a more promising way to reduce poverty than changes in agricultural water management. However, such policies may actually increase competition for water between the agricultural and urban sectors and require careful overall planning to ensure that vulnerable rural groups are not disproportionately disadvantaged.

This discussion should not be taken to mean that there is no work to be done in agricultural water management. In fact, our findings reveal that there are substantial gains to be made through improvements in WP. In the short to medium term, this means focusing on improvements in physical WP, primarily the physical quantity of crop output per unit of water input so as to improve the use scarce water resources for national priorities of food security. Reduction in irrigation requirements would also help in controlling declining levels of the groundwater table. In the longer term, and if the international environment changes, the focus may be shifted towards increases in economic WP by moving water away from lower productivity grain production and towards higher value activities in and outside the agricultural sector including hydropower generation and urban uses.

A scenario analysis illustrated the tradeoffs between food production goals, environmental protection and equitable distribution of production gains over the basin (Marjanizadeh 2008, Marjanizadeh *et al.* 2010). The analysis showed that continuation of the government's wheat-self-sufficiency policy, implemented because of the threat of international sanctions, will place heavy strain on water systems and likely lead to further degradation of the downstream Hoor-al-Azim wetland. On the other hand, strict adherence to environmental flow standards will lead to lower wheat outputs and missed production targets. The scenario analysis also showed that a compromise between food production, environmental goals and equity is possible, and a further decline of water resources because of increased food demand is not inevitable.

It should be noted that the intention of the modelling effort was to show the use of scenario analysis as a tool in thinking about the future development of the Karkheh Basin. The scenario quantification was based on lumped estimates of hydrological and agricultural variables. While the model provides a good first estimate, physically based hydrological modelling could add more detailed and precise estimates of the hydrological behaviour of the basin (and the interaction of the surface and groundwater in particular). Further, a more detailed modelling of the income distributional effects of the scenarios would enhance the analysis. While these activities were beyond the scope of the project, they would form a logical next step for further work in the basin.

Recommendations

The direct driving forces behind poverty change in Iran in general, and in the Karkheh in particular, have more to do with larger economic, political and international relations issues than with agricultural production and, by association, water use. Measures outside the agricultural water sector are likely to be the most effective direct solution to remaining rural poverty problems in the Karkheh Basin and elsewhere in the country.

In the short to medium term, agricultural water policy should instead be focused on improvements in physical water productivity, primarily the quantity of wheat output per unit of water input, so as to improve the use of scarce water resources for given national food security priorities.

In the longer term, and if the international environment changes, the focus of agricultural water policy may be shifted towards increases in economic water productivity by moving water away from lower-productivity grain production and towards higher-value agricultural and other activities including hydropower generation and urban uses.

Acknowledgements

This paper is a synthesis of the results of a CPWF-funded Basin Focal project in Karkheh River basin in Iran. References to the published material behind this paper have been made wherever possible and readers may refer to those works and their authors for more details. The work was carried out through co-operation between the International Water Management Institute (IWMI), Soil Conservation and Watershed Management Research Institute (SCWMRI) of Iran, the Agricultural Engineering Research Institute (AERI) of the Agricultural Research and Education Organization

(AREO), Ministry of Jihad-e-Agriculture, Iran and Tabriz University, Iran. We would like to thank Seyed Abolfazl Mirghasemi for providing spatial data and help in field visits, Dr Sharam Ashrafi and Dr Nader Heydari for their help as CPWF basin coordinators, Dr Zia-ud-din Shoaei, additional staff from IWMI-Iran and the office's host Dr Arzhang Javadi, AREO, and the Ministry of Energy, Iran. The authors would like to acknowledge the entire Karkheh Basin project team including Iranian collaborators and consultants, CPWF researchers and basin coordinators, and a set of dedicated PhD students for their valuable contributions which were directly and indirectly used in this paper. We also thank Dr Hugh Turral, Dr Myles Fisher and an anonymous reviewer of this paper for valuable feedback which helped improving the quality of this manuscript.

Note

1. Reallocation planning requires data on in stream flows. However, lack of data prohibits further downscaling of the analysis presented already to the level of tertiary and quaternary catchments. For instance, in the case of surfacewater measurements there were 50 stream flow gauging stations installed after 1950 out of which only 25 are continually measured and some of these have missing data. Similarly, while there are some 22 tributary streams discharging into the main rivers of Karkheh Basin, discharge data at the outlet of their catchments is not available for half of them. Information on rainfall–runoff relationships is also lacking in the basin in general. Hydrological models can be used to partially remedy this problem. Models can be calibrated for catchments where discharge and other data (such as climate) are available. One key challenge, however, is the extrapolation of hydrological information from gauged to ungauged catchments (Sivapalan *et al.* 2003). This is a key issue in Karkheh Basin. A methodology has been developed for Karkheh Basin and applied (Masih *et al.* forthcoming). It will serve as a tool for conducting reach-by-reach analysis of water balances that can then be used to analyse water productivity or poverty trade-offs under different allocation scenarios. Similarly, because of the complex geology of the Karkheh Basin, the interactions between surface and groundwater are not well understood and require detailed investigation.

References

Ahmad, M.D., *et al.*, 2009. Mapping basin-level water productivity using remote sensing and secondary data in Karkheh River Basin, Iran. *Water International*, 34 (1), 119–133.

Ashrafi, S., 2006. Productivity and sustainability of land and water resources. *Land, water and environmental management in the Karkheh River Basin: Challenges and opportunities workshop.* SPII Campus, Karaj, Iran, 24 May 2006.

Assadzadeh, A., 2006. *Poverty analysis in the Karkheh River Basin.* CPWF Basin Focal Project Pre-forum meeting, November 2006, Vientiane, Laos.

JAMAB, 1999. *Comprehensive assessment of national water resources: Karkheh River Basin.* Tehran: JAMAB Consulting Engineers Co. in association with Ministry of Energy, Iran (in Persian).

JAMAB, 2006a. *Water balance report of Karkheh River Basin area: Preliminary analysis.* Tehran: JAMAB Consulting Engineers Co.

JAMAB, 2006b. *A report on the development of ground waters and utilization thereof in Karkheh drainage basin: an analysis of specifications.* Tehran: JAMAB Consulting Engineers Co.

Marjanizadeh, S., 2008. *Developing a "best case scenario" for Karkheh River Basin management (2025 horizon): a case study from Karkheh River basin, Iran.* Ph.D. thesis. University of Natural Resources and Applied Life Sciences of Vienna (BOKU).

Marjanizadeh, S., *et al.*, 2009. *From Mesopotamia to the third millennium: the historical trajectory of water development and use in the Karkheh River Basin, Iran.* Working Paper 135. Colombo: IWMI.

Marjanizadeh, S., de Fraiture, C., and Loiskandl, W., 2010. Food and water scenarios for the Karkheh River Basin, Iran. *Water International*, 35 (4), 409–424.

Masih, I., *et al.*, 2008. Understanding hydrologic variability for better surface water allocations in Karkheh Basin, Iran. *XIIIth World Water Congress on Global changes and water resources: confronting the expanding and diversifying pressures*, 1–4 September 2008, Montpellier, France. [Available from: http://www.worldwatercongress2008.org/resource/authors/abs823_article.pdf]

Masih, I., *et al.*, 2009. Analysing streamflow variability and water allocation for sustainable management of water resources in the semi-arid Karkheh River Basin, Iran. *Physics and Chemistry of the Earth*, 34 (4–5), 329–340.

Masih, I., *et al.*, 2010. Regionalization of a conceptual rainfall-runoff model based on similarity of the flow duration curve: a case study from the semi-arid Karkheh Basin, Iran. *Journal of Hydrology*, 391 (1–2), 188–201.

Muthuwatta L.P., *et al.*, 2010. Assessment of water availability and consumption in the Karkheh River basin, Iran using remote sensing and geo-statistics. *Water Resources Management*, 24 (3), 459–484.

Rahimi, A. and Kalantary, A., 1992. *Economic appraisal of subsidy*. Tehran, Iran: Institute of Commerce Research and Studies (in Persian).

Sivapalan, M., *et al.*, 2003. IAHS decade on predictions in ungauged basins (PUB), 2003–2012: Shaping an exciting future for the hydrological sciences. *Hydrological Sciences Journal*, 48 (6), 857–880.

Su, Z., 2002. The surface energy balance system (SEBS) for estimation of turbulent heat flux. *Hydrology and Earth System Sciences*, 6 (1), 85–99.

Su, Z. and Jacobs, C., 2001. *ENVISAT: actual evaporation. BCRS report 2001*: USP-2 Report 2001. Delft: National Remote Sensing Board (BCRS).

Su, Z., *et al.*, 2003. Assessing relative soil moisture with remote sensing data: theory, experimental validation, and application to drought monitoring over the North China Plain. *Physics and Chemistry of the Earth*, 28 (1), 89–101.

Victor, D., 2009. *The politics of fossil-fuel subsidies*. Geneva: International Institute for Sustainable Development.

World Bank, 2010. World Bank open data initiative [online]. Available from: data.worldbank.org/data-catalog [Accessed 10 May 2010].

Vulnerable populations, unreliable water and low water productivity: a role for institutions in the Limpopo Basin

Amy Sullivan and Lindiwe Majele Sibanda

There are nearly seven million rural people in the Limpopo Basin, many of whom are poor. They rely mainly on unreliable rainfall to support their mixed crop/livestock smallholder production systems. The poor have few avenues out of poverty and are especially vulnerable to disease and continued inequitable distribution of land and water resources. Infrastructure development and other investments target these communities within and institutional support is uneven and under capacitated. Addressing the needs of these populations and achieving sustainable development and livelihood security will take strengthened institutions working closely with the rural poor to meet their needs in the face of ongoing economic, political and climatic change.

Introduction

The Limpopo River basin in southern Africa, one of 63 internationally shared river basins on the continent, encompasses rich natural resources and biodiversity. The basin has 11 main sub-basins, covers 414,800 square kilometres (Wolf *et al.* 1999) and has a population of just over 14 million. Four major metropolitan areas fall within or near the basin – Bulawayo, Gaborone, Polokwane and Johannesburg-Pretoria – which together comprise about half the total basin population.

Portions of four countries make up the basin, with Botswana to the west, Zimbabwe to the north, Mozambique downstream to the east, and South Africa to the south. The latter covers nearly half the basin area, but none of the basin countries has more than 16% of its national surface area within the basin (Figure 1).

Geography

The main river course runs for nearly 1750 kilometres from the source to the mouth. Altitudes range from over 1600 m in the Drakensberg Mountains of South Africa, to sea level where the basin meets the Indian Ocean near Xai-Xai in Mozambique.

The basin has warm summers and mild winters, but temperatures vary widely, depending on altitude and proximity to the coast. Daily mean temperatures in the south and west of the basin range from 5°–10°C in the winter with frosts, and from the mid to the upper 30s

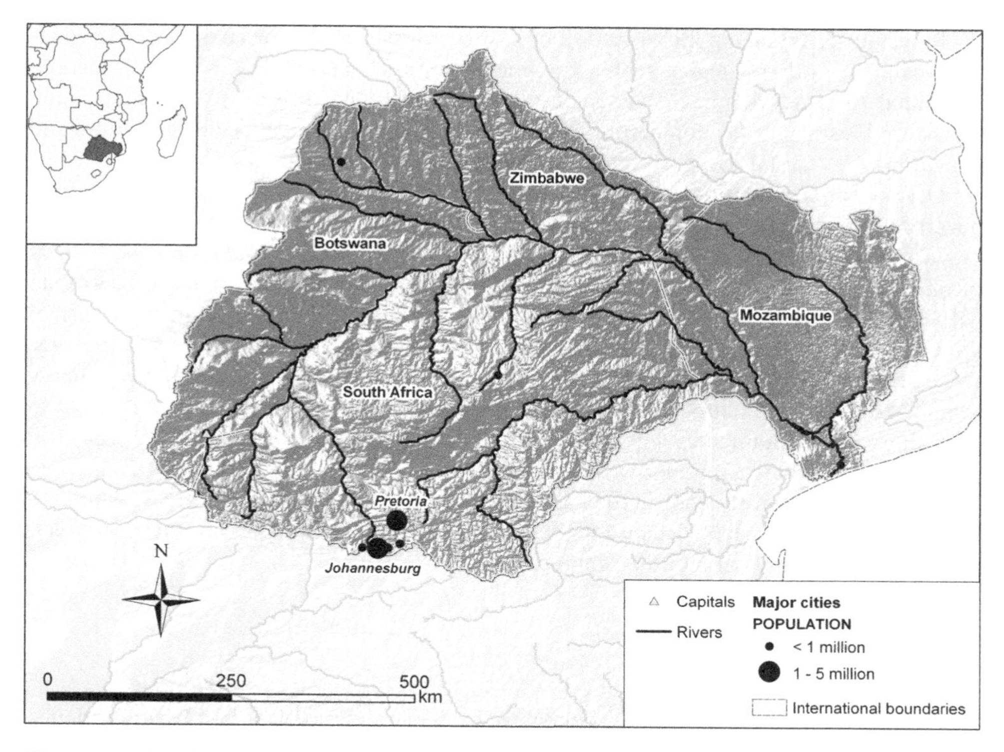

Figure 1. The Limpopo River basin.

in the summer. The northern and eastern parts of the basin are generally warmer because of their low altitude and the influence of the Indian Ocean.

A large portion of the basin falls within the Savannah Biome, known locally as the "bushveld". Its vegetation is characterized by a grassy ground layer and a distinct upper layer of woody plants. A major factor delimiting the biome is the limited rainfall, which, coupled with fires and grazing, allows the grass layer to remain dominant.

The colonial heritage

Countries of the Limpopo Basin have long and well-documented colonial histories, largely shaped by the Dutch, English and Portuguese. For nearly four centuries, policies, institutions, economies and settlements were all directed toward exploiting the natural and the human resource base of the indigenous populations for the benefit of the colonizers. Great wealth was generated in rural areas from mining and agriculture, but little was returned to develop either the human resource or rural infrastructure.

Poverty and policy

Poverty in the Limpopo Basin is concentrated in rural areas where populations rely largely on the natural resource base for their livelihoods. Poverty is higher amongst households headed by women than in those headed by men, mainly due to gender bias in access to productive resources. Low smallholder productivity and poor access to credit exacerbate the high levels of poverty. Unemployment, poor access to water and low levels of education prevail, leading to low income, compromised health, and few economic opportunities.

In response, national policies in all four basin countries have focused on poverty reduction, using various strategies such as public sector reforms, environmental protection, integrated rural development, human resource development, disaster management, land and natural resources development (including water resources, small-scale agriculture and irrigation), HIV and AIDS prevention, and empowerment of women.

Despite implementation of policy reforms and interventions in all four countries, both poverty and inequality remain high in the basin. Where there has been rapid economic growth, its benefits have been spread unevenly across the basin population. The highest incomes are enjoyed by those in formal sector employment, but they make up less than half of the labour force. The remainder are less well off, being engaged in the informal sector and subsistence agriculture, or unemployed. Low incomes, and hence high poverty rates, are concentrated in the rural areas, where economic growth has stagnated for many years. A number of factors contribute to this situation, including limited alternatives to agriculture for economic development in rural areas, which ultimately limit investment in education, health care and other basic services.

Historically, the rural poor in the basin were relegated to water-scarce areas. Although it is often difficult to establish causality between water scarcity and poverty, the association is quite clear in many parts of the Limpopo.

Vulnerable populations

Population distribution

While Botswana, Mozambique and Zimbabwe each has about one million basin residents, the relative proportion of the national population living in, and relying upon, the Limpopo Basin varies widely across the four countries, from 59% in Botswana to less than 10% in Mozambique and Zimbabwe (Table 1). Almost 60% of the basin's rural population lives in South Africa, followed by Mozambique (15%), Zimbabwe (13%) and Botswana (8%).

Resources

A general characteristic of the basin is that it lacks outstanding features, in that the river itself does not serve as the focus for any large city or major industry. The river mouth is not a harbour, nor is the river navigable in any important sense. Thus, while at certain points the river serves as a boundary between South Africa, Botswana and Zimbabwe, it does not, *per se*, generate a great deal of local activity.

Table 1. Distribution of population in the Limpopo Basin by country.

Country	Population of country in basin	Fraction of country's population in basin	Country share of total basin area	Fraction of country area in basin	National rural population as % of total in 1998	National rural population density (WDI 2008)*
	million	*%*	*(%)*	*(%)*	*%*	*/km²*
Botswana	1.0	59	19	14	31	230.8
Mozambique	1.3	7	21	11	62	307.9
South Africa	10.7	24	45	15	50	125.1
Zimbabwe	1.0	9	15	16	66	253.7

Source: Compiled from United Nations Development Program (2003).
* Data for 2003 for comparison.

Settlement patterns across the basin have been heavily influenced by mining and the availability of water. The inequitable distribution of land, water and economic resources left over from four centuries of European settlement manifests itself in pockets of extreme wealth in urban centres, and high productivity of commercial farms, juxtaposed with much larger pockets of poverty, particularly in rural areas, which are characterized by dispersed settlements that have low annual rainfall on degraded land.

Countries within the Limpopo Basin depend on its natural resources for agriculture, tourism and industry (mainly mining), which account for up to 50% of their GDP (WDI 2008). Unfortunately, countries with abundant natural resources, such as oil and minerals, are much less likely than others not so endowed to include the poor in their economic growth strategy and policies (Sachs and Warner 1995). For example, the resource-rich countries of central Africa are among the poorest in the world. The Limpopo Basin is no exception: although mining is a major economic activity, rural poverty is widespread. It is particularly deep on both sides of the Zimbabwe–Mozambique border in the northeast of the basin, and in the areas of South Africa formerly designated as homelands.

In these areas, nearly 80% of the rural people live on less than US$1 per day and unemployment is very high, with education, health care, infrastructure and access to clean water correspondingly low. Mozambique has the highest national population below the poverty line of a dollar a day among the four basin countries, with nearly 38%. Zimbabwe has 36% and Botswana has 33.3% of its entire population living on less than a dollar a day. A 2005 study showed that 40% of black South Africans lived on less than $2 per day while poverty had increased from 36% in 1995 to 47% in 2000 in the Limpopo Province within the basin (Hoogeveen and Özler 2005).

Economic history

Divestment away from smallholder agriculture and investment in large-scale commercial enterprises has changed the face of farming in parts of southern Africa in the twentieth century. Remnants of that change, combined with many other factors, are still visible in statistics about the size and composition of each country's agricultural labour force. Agriculture still employs a majority of the population in both Zimbabwe and Mozambique, and nearly half (45%) of the population in Botswana (Table 2). That only 10% of the South African labour force was engaged in agriculture in 1998, and nearly 75% of them were men, which is a departure from the regional norm, reflects the highly commercialized and mechanized nature of the agricultural sector in that country.

Table 2. Size and composition of the agricultural labour force in the Limpopo Basin.

			Agricultural labour force			
Country	1998 population	Total labour force	Number	Fraction of total	Male	Female
	million	*million*	*million*	%	%	%
Botswana	1.6	0.67	0.30	45	43	57
Mozambique	16.5	9.59	7.72	81	43	57
South Africa	42.1	18.03	1.73	10	74	26
Zimbabwe	11.4	5.63	3.52	63	47	53

Source: Challenge Program for Water and Food (2003).

Men from rural communities in southern Africa have long been seen as sources of labour, and were often directly or indirectly coerced to work in mining, industry and other economic sectors. Their participation in the labour market introduced remittances to the rural areas, thereby contributing to reduced household reliance on agricultural activities.

The combination of rural communities left with limited available labour and expulsion from communal lands changed the role of agriculture in household survival strategies across large parts of the South African and Zimbabwean portions of the Limpopo Basin. Subsistence farming remains a vital livelihood activity in large areas of the basin but the value of small-scale agricultural production has decreased. Poor land-use management has led to degradation of the natural resource base and women, who often found themselves as the sole provider present in the household, were left to farm the land with little or no technical or economic support.

Not all the poor depend on agriculture for their livelihoods, but the majority in Mozambique and Zimbabwe do. The actual number fluctuates across basin states, depending on a number of regional factors. Other livelihood options include social grants, remittances, trading, home-based businesses, and sporadic local employment, which suggests that agriculture cannot be a ladder out of poverty for all. Nevertheless, despite the weight of historical, economic and hydrological factors against them, there are still smallholder farmers in rural communities. Their sources of water are boreholes, shallow hand-dug wells, small reservoirs, rivers, streams and irrigation canals. However, they lack adequate road access to economic centres, water and sanitation systems, and municipalities with the capacity to deliver services.

Like the majority of the rural people, basin residents have low household incomes and purchasing power, while high levels of illiteracy and HIV/AIDS exacerbate the situation (South Africa, Department of Agriculture, 2006). Given uneven development and distribution of wealth, the current conditions reflect institutional unwillingness and inability to target strategic interventions to address poverty, HIV/AIDS, food insecurity, environmental degradation and competing interests. This results from the history of political and economic policies (apartheid, forced migration, labour extraction and civil war) that have shaped the development trajectory of the basin. Their profound effect on settlement patterns and resultant poverty is difficult to overestimate. We discuss this aspect in more detail below.

Income distribution

Mozambique is by far the poorest of the basin countries, and the one whose residents are most likely to be malnourished (Table 3).

Table 3. Gross domestic product (GDP), income, distribution and nutrition in the Limpopo Basin.

Country	GDP per capita	Population <$1/day	Gini index	Income share of lowest 10%	Income share of highest 10%	Undernourished people
	2005$	*%*	*%*	*%*	*%*	*%*
Botswana	7,423	33.3	63.0	0.7	56.5	23
Mozambique	376	37.9	39.6	2.5	31.7	54
South Africa	6,293	11.5	59.3	1.1	45.9	–
Zimbabwe	1,617	36.0	50.1	2.0	40.4	39

Source: Adapted from Hanjra and Gichuki (2008). South African statistics on nutrition not available.

In South Africa, inequitable distribution of income is well known, with small concentrations of extreme wealth contrasted with the poor majority, which is confirmed by a Gini index of 59.3 (Table 3). In Botswana, income distribution is more skewed than elsewhere in the basin, which reflects the dependence of the Botswanan economy on diamonds and the poverty in the arid rural areas — although the country is often cited as an economic success, its economic growth has bypassed the rural poor. Zimbabwe is much poorer and has a somewhat less skewed income distribution.

Estimates in rural Limpopo by Perret *et al.* (2005), show 70% of households living on less than US$227/month, and 40% living with less than US$125/month. Rural populations in Botswana survive on 37% of the average household income of their urban counterparts.

Rural poverty

Rural poverty can be assessed through a rural livelihoods approach, which examines other factors in addition to income. This approach suggests poverty is often correlated with certain vulnerable groups and that women and children are disproportionately over-represented among the ultra-poor (those living on less than $0.50/day). The ultra-poor represent the most alarming cases of deep monetary poverty, associated with isolation from the local social fabric and health problems such as HIV/AIDS. Living conditions in rural areas are much worse than in urban areas, and the rural poor are rarely heard at national policy level.

Persistent rural poverty in the Limpopo Basin has outlasted a variety of political and economic initiatives and needs a closer look. While data such as those from a human development index (HDI) or a water poverty index (WPI) are useful for comparison, they do not necessarily expose the combination of underlying causes that keep successive generations of rural families in poverty.

The HDI is a composite index based upon life expectancy, literacy and schooling, and standard of living. United Nations data from 2007 (United Nations Development Programme [UNDP] 2009) rank Botswana as the tenth most developed African country, and Mozambique as having the lowest HDI in Africa. South Africa falls below Botswana and data from Zimbabwe have not been available for several years.

Water poverty index (WPI) is an aggregate index, which describes relative scarcity of fresh water at a country level. The WPI from 2002 (World Resources Institute 2006, UNEP/GRID-Arendal n.d.) shows Mozambique as having a very low WPI (fresh water is abundant), Zimbabwe and South Africa as low (there is little scarcity of freshwater) and Botswana, which is the driest country in the basin, as having a medium WPI, that is, freshwater is moderately scarce.

Despite seasonal and periodic water scarcity in the Limpopo Basin, water availability is not a source of conflict between the four riparian states. Each country faces essentially the same major issue within its borders: how to prioritize development and allocate highly irregular natural resources, namely water and land, in the face of historical inequalities and great economic potential.

Water-consumptive land uses

The land uses that consume rainfall (Figure 2) are roughly proportional to the area of the basin that they cover: grassland covers nearly 57% of the basin land area and uses 54% of the rainfall, and rainfed agriculture covers 40% of the land and uses 40% of the water. Mining uses less than 5% of basin water but almost certainly impacts the quality of the

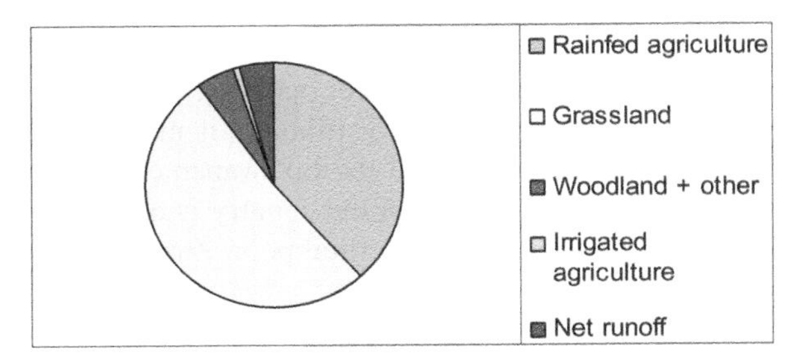

Figure 2. Land use/land cover in the Limpopo Basin.
Source: Adapted from Mainuddin *et al.* (2010).

water resources. Irrigated agriculture covers only 0.6% of the land area and diverts less than 1% of surface water, but is critical to those farmers who rely upon it.

According to water productivity analysis done under the Limpopo Basin Focal Project in 2010, most of the basin's water resources support shrub or grassland. Although agriculture occupies a relatively small portion of the basin area, large numbers of small-scale farmers are found throughout the basin, often on degraded soils with unreliable rains.

Irrigation is practised in some of the commercial and communal lands. While it usually has a much higher yield than rainfed cropping, there are vast differences between large-scale commercial irrigated operations and small-scale infrastructure. At least 300 small-scale irrigation schemes have been developed in the basin over the last 40 years but they now function poorly or not at all, after neglect, abandonment or management regimes that have changed with the profound political changes that have occurred.

The main land use in the Limpopo Basin, in terms of both the population and the land area involved, is mixed crop and livestock farming. The rural poor in Zimbabwe and Mozambique rely on the rainfall for both types of farming, a precarious livelihood. Their counterparts in Botswana and South Africa have the added option of social welfare grants, and are less likely to engage in agricultural activities, especially in drier areas. Agricultural production is either large-scale commercial production, or the small-scale mixed crop/livestock systems on communal land that dominate the smallholder sector.

The term "small-scale" often refers to the scale of economic investment as well as to the physical size of the operation, in the sector where the farmers are called "smallholders". At the lower end of the smallholder scale is subsistence agriculture, which contributes little to national agricultural outputs and primarily provides food for the household, but may provide food for sale when they have enough surplus or when they need money.

Subsistence agriculture

Subsistence farmers generally have limited access to input and output markets, but many in Zimbabwe have access to improved seeds through a well-developed distribution network (Bourdillon *et al.* 2007). Elsewhere, they rely on their own saved seed or on local sources of typically low-yielding varieties. Output marketing is often limited to local markets, or through parastatal institutions for cash crops such as sunflowers or cotton.

As elsewhere, subsistence agriculture in the Limpopo Basin is a low-input-low-output system that seeks to minimize risks caused by climate variability, and to make the most of limited resources. It is characterized by minimal use of inputs such as fertilizer and certified seed, and low levels of management. Its performance is restricted by insecure land tenure, low-level technologies, risky water supply, and limited access to other production resources, such as labour and cash. Moreover, the soils are often degraded and depleted of nutrients. Subsistence farmers in the basin typically crop small areas because they rely on hand-hoeing and family labour. This often leads to land preparation being delayed, increasing their risk of crop failure if rains arrive on time.

Livestock farming

Livestock production in the basin as a whole is economically and socially important. Subsistence farmers' livestock are typically managed in low-input systems of extensive grazing on poor quality feed in a variable climate. The stock access surface water or water provided via windmill, hand pump or other mechanism. They graze communal pastures during the day and are guarded in *kraals* at night. Milk production, which is used primarily to meet household needs, is low. On the other hand, large-scale commercial farmers with greater access to resources and secure tenure maximize productivity in intensive systems.

Figures from Botswana suggest that returns from livestock, mainly meat exports, contribute 2.6% of the 3% GDP contribution from agriculture (Republic of Botswana, Central Statistics Office 2004). Livestock numbers remained more or less constant over the decade to 2003, but fell by 30–40% in the drought year of 2004.

In South Africa, animal products contributed 45% to total agricultural production in the 1980s (Food and Agriculture Organization [FAO] 2004). Livestock outputs (meat, milk, eggs, skins, and so on) in Mozambique accounted for 25% of agricultural domestic product in the 1990s (FAO 2004). According to the famine early warning system network (FEWSNET), livestock numbers and production in Mozambique have been increasing since 1993 (FAO 2004). In 2003, Limpopo Province, which makes up the bulk of the Limpopo Basin in South Africa, had 1.23 million cattle (nearly 9% of the total national herd), of which 458,000 were in commercial areas and 775,000 in communal areas (Limpopo Province Freight Transport Databank). Nationally, about 60% of cattle are fattened for slaughter.

Livestock are crucial to smallholders' livelihood security, acting as a buffer against economic shocks. They also have cultural significance, because ownership of livestock is an indicator of wealth. Cattle are used to pay a bride price, to acquire and store wealth, to spread the risk in mixed farming systems, as draught power, and for meat and milk.

Nearly 70% of cattle, sheep and goats in southern Africa are kept by smallholders (FAO 2004), most of whom also keep chickens. In general, men own and manage cattle and goats, which are more important to the household. Women tend to own and manage the small livestock such as chickens and ducks. Smallholders mostly have local breeds of cattle that are generally well adapted to the basin conditions of high temperatures, low-quality diet, and ticks and other parasites, but their reproduction rates are poor, with a 24-month calving interval as opposed to 12–15 months in intensive systems (De Leeuw and Thorpe 1996).

Herd size and overall animal numbers are affected by the frequent droughts in the basin, which reduce fodder quantity and quality, and water availability. Cattle are typically kept in small herds of less than 10 head in Zimbabwe and four to 10 in Mozambique (International Fund for Agricultural Development 1996). In Botswana, herd sizes are relatively larger than in the other basin countries, with a 1996 estimate of just over 44 (FAO 2004). However, demographic pressure has decreased the number of households in Botswana that own cattle, while ownership of goats, sheep and chickens has increased (Low and Rebelo 1996).

Movement of stock as a drought-avoidance strategy is hampered by land tenure structures and also because severe droughts generally affect large areas. Yet farmers in many parts of the basin are reluctant to sell cattle to reduce stocking rates, preferring to maximize herd size as a safety net for use in times of drought (FAO 2004).

Crop farming

Rainfed cropping is the predominant crop production system in the basin, in spite of the high aridity, because of favourable temperatures and ample annual solar radiation. Potential production of crops is constrained mainly by rainfall and soil fertility. The supply risk of agricultural water is therefore a crucial issue for crop farmers in the Limpopo Basin. As the simulated rainfall pattern shown in Figure 3 suggests, the probability of crop failure due to drought is a constant threat.

The main rainfed crop is maize, with lesser amounts of wheat, sunflower, citrus, potatoes, and dry beans. Crop productivity in the basin reflects historical patterns, with yield differences caused by differences in production technologies and rainfall. South Africa is the only country in the basin in which maize yields have increased during the last 40 years. The average yield is about 3600 kg/ha but varies widely, from less than 1000 kg/ha in subsistence systems to over 8000 kg/ha in intensive commercial farms (Hanjra and Gichuki 2008). Average maize yields in the communal sector is only 250 kg/ha in Botswana and about 800 kg/ha in Zimbabwe (Hanjra and Gichuki 2008).

Given the widespread adoption of improved technologies and high yields in the commercial sector across all the basin, it is clear that it is not the availability of technology that limits smallholder production. The issue seems to be that smallholders do not have the access to markets and infrastructure that would allow them to use the technology that is available. A corollary is that technology alone is not the answer. Policies like water rights, land tenure, and subsides also play a vital role. We discuss some of these aspects below.

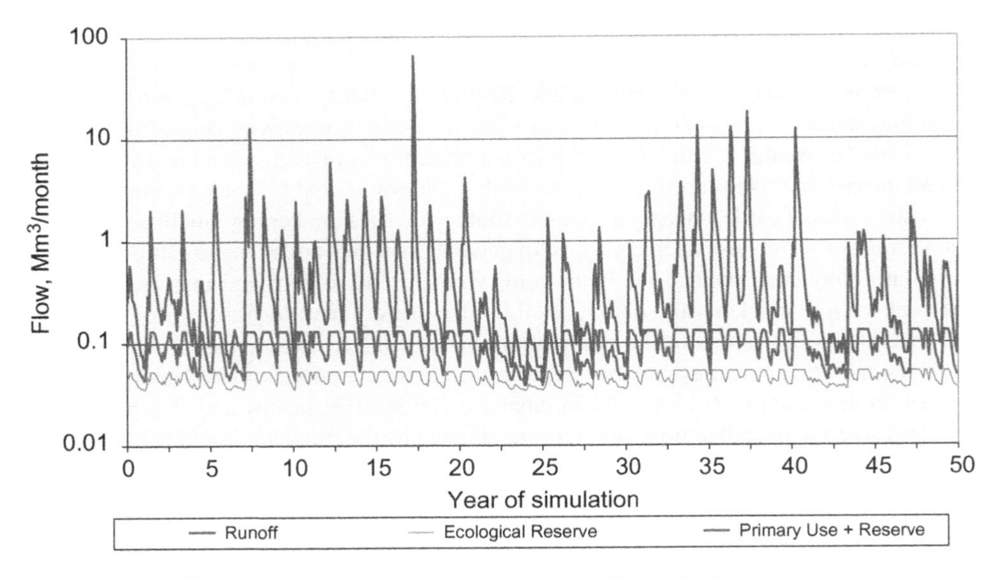

Figure 3. Runoff hydrograph – 50-year simulation – Crocodile sub-basin.
Source: Limpopo Basin Focal Project (2010).

Water in the Limpopo Basin

In a water-use account of the basin, Mainuddin *et al.* (2010) used an area of 412,900 square kilometres. Rainfall input is 230 km^3/yr with a mean annual discharge of 3.6 km^3 at the mouth, 1.6% of the total precipitation input. Rainfall varies from 200 millimetres to 1500 millimetres, with "much of the northern and western parts of the basin receiving less than 500 millimetres. Rainfall is highly seasonal, with 95 percent falling between October and April. The rainy season is short with the annual number of rain days seldom exceeding 50. The consequence is that the Limpopo and its tributaries have a very pronounced seasonal variation in flow, with negligible flows in the dry season" (Mainuddin *et al.* 2010). Rainfall varies considerably from year to year. Mean annual rainfall in the basin is less than 550 mm. Floods and droughts occur frequently, and intra-season variability makes farmers vulnerable to crop failure and livestock loss.

We divide water use into runoff, ecological reserve and primary water requirements. A hydrograph of typical river flow in Figure 3 illustrates how these water uses are prioritized in the Limpopo Basin. Within the water sector, agriculture is rarely recognised as a separate water use and is not prioritized. This is because, although rainfed agriculture (including livestock) supports most of the rural population, it carries a low water cost to the state so its role in rural development is not viewed as a water-related issue.

The ecological reserve or base flow is the water that must remain in the river system to preserve ecological functions. This need is met first from the most readily available water. The second water use to be met is known as "primary water requirements", which includes domestic and industrial uses. Agriculture accesses the water that makes up the difference between the primary use and the runoff (between the upper two lines in Figure 3). Yet the peaks in the upper line represent the rainfall that exits the basin quickly as flood flow, so that vast quantities of water are not captured and are unavailable to agriculture.

There are 250,000 ha of irrigation in the basin (FAO AQUASTAT 2004), which accounts for less than 1% of the total drainage area. There is increasing uncertainty about its environmental and social costs, however. There is no objective analysis of irrigation that identifies the environmental costs, the economic benefits and the complete list of beneficiaries.

Irrigated agriculture relies mainly on the "high-risk" water that comes in as an unpredictable and temporary excess flow. Supply risks can be mitigated to a limited extent by providing storage that converts some of the surplus flow into a more usable format. Converting surplus water (floodwater) does not, however, guarantee water for irrigators because shortages in the primary and ecological reserve supplies must be satisfied first.

It must be noted that water for domestic and industrial use is considered a priority in the basin, and so must be supplied from reliable sources. Irrigation water is given a lower priority and often comes from less reliable sources, leading to frequent water shortages.

Water demand

Table 4 shows current water distribution between sectors in each country. Water demand within the basin continues to grow. Rural development targets for water and sanitation delivery in all basin countries, as well as land and water reform, suggest that more equitable distribution of existing resources may be closer to a physical and economic reality than at any point in the past. Of course, competition for these scarce resources comes from powerful stakeholders such as commercial agriculture, mining and environmental interests, which will require decision makers to confront very difficult choices. According

Table 4. Water use by sector in the Limpopo riparian states.

Country	Agriculture (%)	Industry (%)	Domestic (%)
Botswana	48	20	32
Mozambique	89	2	9
South Africa	62	21	17
Zimbabwe	79	7	14

Source: SADC (1999).

to the Food and Agriculture Organization (FAO AQUASTAT 2004), at least two of the Limpopo's sub-basins in South Africa are over-subscribed; moreover, there are some inter-basin transfers.

Water availability

Per capita water availability in the basin, a theoretical amount calculated as the total amount of water available divided by the total population, varies widely, ranging from less than 900 m^3/person in Botswana to over 5800 m^3/person in Mozambique. In South Africa and Zimbabwe per capita water availability is about 1200 m^3. These figures are estimates and are useful only for comparison, but it is worth noting that they are likely to decline if climate change predictions hold true.

Ample per capita water availability suggests that water scarcity in the basin may be more economic than physical (Seckler *et al.* 1998). But even when there is sufficient water at a basin scale, it is often not available where poor people can make use of it, nor when they need it most. Nevertheless, water is relatively scarce in the basin, whose hydrology is affected by the extreme climate variability (Scott 2008). Rain falls during a short, intense rainy season, so that the majority of runoff from the basin occurs in short-lived flood peaks. As well as devastating floods, notably in 2000, there are long, severe droughts every 10–20 years throughout the region (FAO CLIMWAT).

Rainfall, evapotranspiration and runoff

Rainfall in the basin is unreliable, both within and between years, which exposes those relying on it to high risk of crop failure. Furthermore, the basin shows a large difference between potential and actual evapotranspiration (Figure 4), which indicates the level of water stress to which crops are exposed.

Upstream areas of the basin (in Zimbabwe, Botswana and part of South Africa) have low rainfall, which severely limits crop yields. However, the downstream and southern areas of the basin suffer relatively less water shortage and, in the South African parts of the southern basin, though not in the Mozambiquan parts, the cropping system benefits from higher rainfall and better-established farming systems.

With up to 80% of the rain falling between November and late February, rarely in more than 50 rainy days, the Limpopo and its tributaries have a very pronounced seasonal variation in flow, and low season flows are negligible. Several factors contribute to this fluctuating water availability, including rainfall, soil type, ground cover and slope. The hydrograph of typical river flow in Figure 3 illustrates how natural runoff (the upper line) can be highly erratic.

About 18% of the basin is classified as arid, with very hot summers, and just over 80% is semi-arid (Figure 5). Runoff varies across the basin, with relatively more water available in the southwestern portion of the basin.

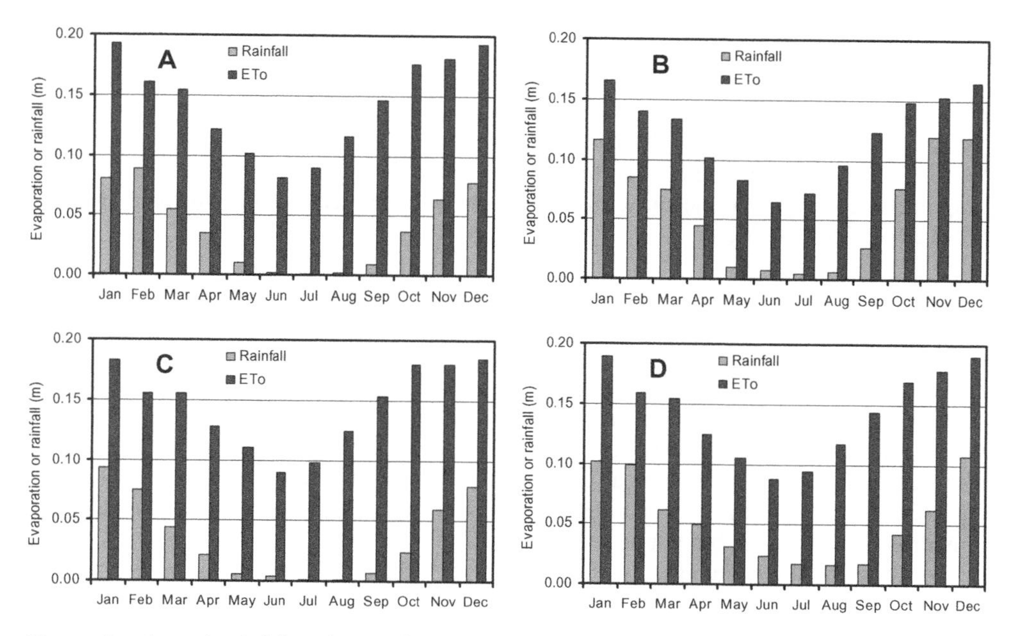

Figure 4. Annual rainfall and potential evapotranspiration at four points in the Limpopo Basin: (A) the Upper Olifants in South Africa; (B) the Lower Limpopo near Chokwe, Mozambique; (C) upstream at Oxenham Ranch in South Africa; and (D) the Tuli catchment in Zimbabwe.
Source: Mainuddin *et al.* (2010).

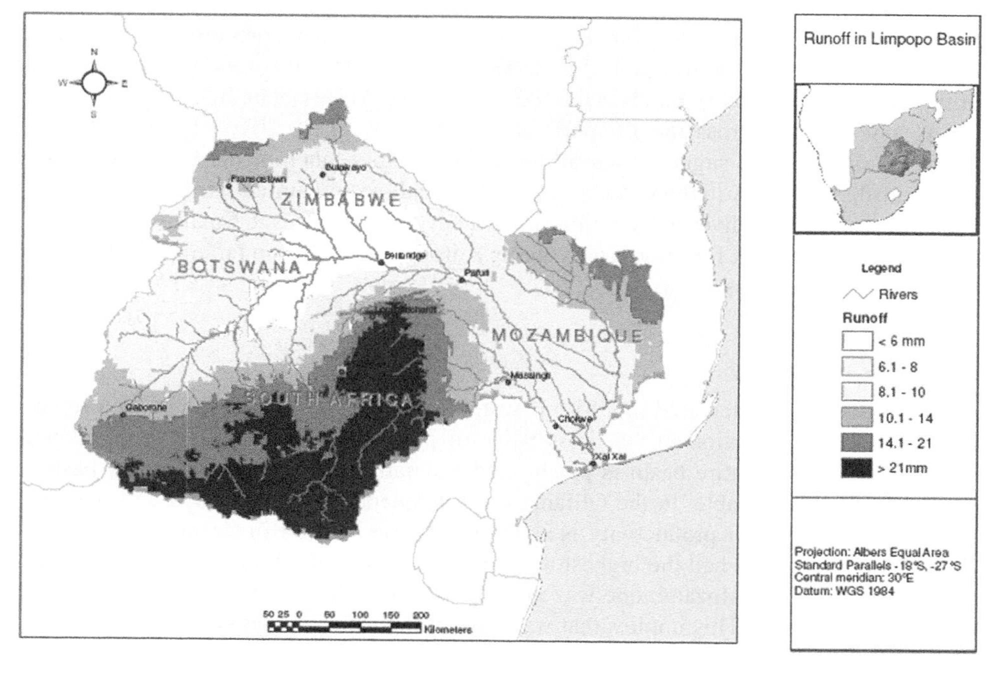

Figure 5. Limpopo Basin run-off map.
Source: Limpopo Basin Focal Project (2010).

Downstream of the confluence of the Shashi and Limpopo, there is a stretch of the river where, when the river is low, water that flows in does not flow out. Mean annual rainfall in this sector is 250–450 mm but evaporation is as high as 2100 mm/yr so that runoff from the surrounding area is negligible. The riverbed in this stretch is normally dry sand, varying from 20–200 metres in width. It is thought that when the inflow is low, there is so much evaporation of the sub-surface water from the sand bed that there is no downstream outflow (FAO 2004, Scott 2008).

Drought

As discussed above, the Limpopo Basin is susceptible to meteorological and hydrological droughts. A "meteorological drought" is defined as an interval of time, generally months or years, during which the *rainfall* at a given place falls considerably below the long-term average, normally also taking account of the balance between rainfall and evapotranspiration. A "hydrological drought" is a period of below-normal *stream flow* or depleted reservoir storage. Taken together, it is clear that meteorological drought may lead to hydrological drought, but not necessarily so.

For water resource planning, three aspects of drought are of interest:

- the duration of a particular drought;
- its severity; and
- the frequency of successive droughts.

The severity of a drought indicates the magnitude of the shortfall in the rainfall and is normally expressed as an aridity index (AI), which quantifies the precipitation deficits for multiple time scales by defining the ratio between mean annual precipitation and mean annual potential evapotranspiration (PET). An AI of 0.2 means that rainfall supplies 20% of the PET (Ezcurra 2006). Lands classified as arid have AIs less than 0.2.

AI assessments across the Limpopo Basin (Alemaw *et al.* 2010) are presented in Figure 6 below, which suggests that aridity is high across much of the Limpopo Basin. This contributes greatly to the vulnerability of the small-scale, rainfed farmers in the basin and helps explain why livestock continue to be a key to rural survival. Small-scale farmers can dispose of livestock for some return during a drought, unlike crop failure, which entails complete economic loss.

Low water productivity

Water productivity (calculated in US$/m^3 water applied) varies across the basin, primarily as a result of variation in the gross value of production of the major crops. Water productivity for the entire basin is patchy and generally low compared to other basins, and is also highly variable. In the Olifants sub-catchment, which is in the South African part of the basin, water productivity is as high as in the Nile Basin (from the 2005 data analysed, South Africa had the highest water productivity of the four riparian states), but water productivity in Mozambique was unexpectedly low despite the low water stress in that part of the basin. This implies that water is not the main limitation to productivity in Mozambique portion of the basin and can likely be addressed through means other than increased availability of water.

Analysis of gross value of production is challenging due to the limited availability of data at the scale of the administrative boundary in some basin countries. Nevertheless,

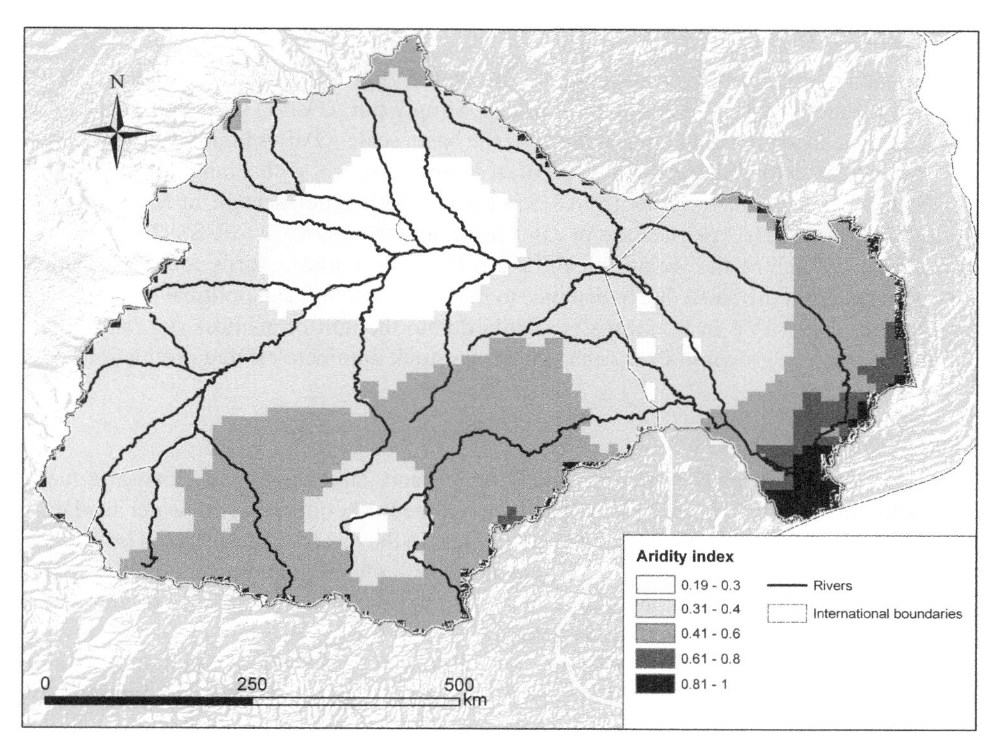

Figure 6. Aridity index map Limpopo Basin.
Source: Alemaw *et al.* (2010).

analysis suggested that the gross value of production of agriculture is the driving factor that determines differences in agricultural water productivity across the basin. The extremely variable water productivity in the basin is mainly due to the significant variation in the yields of different crop production systems. This variation in the gross value of production is caused not only by crop yield variation, but also by the wide fluctuations in market prices caused by the different market conditions in the riparian countries.

Higher agricultural water productivity (more than US$1.00/m^3, such as found in the Olifants sub-catchment) was obtained with crops such vegetables, fruit and tobacco that give a high net return. This suggests that better matching of crops to physical conditions could improve water productivity in this and other areas of the basin. Similar pockets of high-value crop production are interspersed with the generally very low gross value of production throughout the basin. These pockets mostly correspond with large-scale commercial farms that contribute to both national and regional food security, and also to rural employment. They perform relatively better than other areas as a result of higher inputs, better soils, better rainfall and better management practices. However, they use a disproportionate share of the land and water resources.

Poor farmers who cultivate on small-scale communal lands are most vulnerable to the risks of crop failure caused by the frequent droughts. Some of them farm in areas that are least endowed with fertile soils and good rainfall. Yields in these areas are invariably low, and market access is likely limited, resulting in lower gross value of production than in other areas.

Gross value of production

The productivity of crop-based agriculture is measured as gross value of production (GVP), which is a function of yield and market prices. Simply put, GVP is the value of the crop divided by the land area under production. At the basin scale, GVP is highly variable, ranging from US$19.5 to US$368/ha. It is greatest in South Africa, with a range from US$357 to US$368 per hectare. It varies from US$19.5 to US$187 per hectare in Mozambique, from US$35 to US$102 per hectare in Zimbabwe, and from US$50 to US$87 per hectare in Botswana. Much of the variation can be attributed to market returns for crops, which vary widely across the basin due to distance to market, economic and political stability, and crop quality. Low GVP in Botswana is mainly due to the unit of analysis (kg maize/ha). Not much maize is grown in Botswana, where livestock dominate rainfed production.

Climate change

Although water is already scarce in South Africa and Zimbabwe, and Botswana faces chronic water scarcity, it is very likely that there will be further decreases in water availability per person by 2025. Water availability and access in the basin will be affected by climate change and variability, and by the risks associated with planning and managing water and irrigation infrastructure. The latter have important implications for the agricultural water supply of both smallholders and commercial farmers in the basin.

In a pilot study by Jones and Thornton (2002), the weather simulation generator MarkSim (Jones *et al.* 2002) was used to generate 30-year runs of weather data for pixels of 18 kilometres. The generated and the current 30-year weather data and FAO data for all the agricultural soils within each pixel were input to the CERES maize model (Hoogenboom *et al.* 1994) for the variety Katamani Composite B, which is widely grown in the region. The modelled yields were used to generate the yield probabilities. In much of the region, probabilities of yields of 1.5 tons per hectare will be considerably less than current probabilities. Botswana and southern Zimbabwe will go from being risky at present to having almost no chance of producing maize yields from this variety in 2055. Mozambique will go from relatively risk-free production to decidedly risky.

Institutions in the Limpopo Basin

The Limpopo Basin is home to many statutory and traditional institutions that mediate access to natural resources. An emerging basin commission, four national governments, numerous provincial institutions, districts, wards and traditional authorities influence who has which sort of access to what land and water.

Despite a decade of stakeholder participation in integrated water resources management (IWRM), links and communication between decision makers at all levels and end users of the water resources remain weak. Furthermore, the lack of integration between the water, land and rural development sectors, each with its own goals, makes reducing rural poverty even more difficult. Insecure land tenure and water rights not only increase the vulnerability of the rural poor, but also discourage investment in development, and are a direct result of national statutory institutions.

Water governance and control mechanisms are mostly underdeveloped in all four riparian states and, in general, governance of water resources in the region lacks integration. Most countries have national ministries with the responsibility of planning, developing and managing water resources but these ministries are often separate from those that govern agricultural activities or other uses of natural resources, such as mining and tourism.

At the national level, because water resources are scarce, there is competition for water, with high-value interests such as mining and tourism having priority over agriculture. This further increases the vulnerability of farmers to drought and to the unpredictable climate by assigning them the water that remains after all other needs have been met. Moreover, the design of water-supply infrastructure and water allocation are both based on historical data that do not reflect recent or current changes in the rainfall patterns, a policy decision which further disadvantages agriculture. Little investment has been made to deliver available water to those in need, whether for domestic, subsistence or commercial uses.

The institutional environment across the basin is varied in its coverage, capacity and mandate to govern natural resource planning, use and conservation. In addition to sectoral issues (such as water, agriculture, mining and tourism), questions of scale, finance and public participation are particularly relevant within the basin. Current water-related institutions are mainly concerned with sectoral distribution of water for economic gains within nations, but the actual physical distribution is skewed away from the rural poor.

Relationships between countries

Wolf *et al.* (2003) named the Limpopo and other regional basins as being at risk of future conflict over water, but that assessment may not reflect the reality on the ground today. Current relations between the basin states are generally quite positive, with little outward disagreements. This may be due in part to the shared history of liberation struggles of Mozambique, South Africa and Zimbabwe. Nonetheless, Turton (2003) claims that basin hydrological data are contested, with each country promoting data that best represent its own interests.

There are a number of longstanding bilateral agreements between various basin countries. However, the establishment of the Limpopo Basin Permanent Technical Committee (LBPTC) in 1986 was the first recognized basin-wide initiative. The LBPTC focused on short-term water availability issues, division of basin waters, implementation of existing agreements, arrangements for common watersheds and other technical aspects. Recent progress has been made in transforming the LBPTC into the Limpopo Basin Commission (LIMCOM), with a secretariat established in Maputo, to develop a baseline study of the water resources in the basin. Three basin countries have ratified the agreement and Zimbabwe is expected to do so.

Since the relationship between livelihoods and water poverty is influenced by national or sectoral policies and strategies, it is important to highlight selected relevant policies and strategies implemented in each country to provide a basis for analysis.

National development policies

Persistent rural poverty within the basin is a result of history, bio-physical conditions as described above, and national priorities and policies as implemented by institutions. Each basin country has spent the last decades attempting to design and implement development policies to address poverty reduction while stimulating national economic growth, which remains a major challenge across the region.

Botswana

Botswana is not currently considered among the poorest African countries (World Bank 2006). Since the 1970s, Botswana's economy and society have changed from those of a poor country based on cattle rearing and subsistence agriculture to a more industrialized, urbanized, and developed one, due to the discovery of diamonds. Despite this transformation, poverty and inequality remain high, and the benefits of rapid economic growth have been spread unevenly across the country and between urban and rural populations.

The 1997 National Development Plan 8 (NDP 8) led to rapid economic growth based on exploitation of minerals and the use of revenues derived from mineral production for investment in economic and social infrastructure. The majority of beneficiaries were urban rather than rural dwellers. The ninth National Development Plan (2003–2009) targeted integrated development based on economic diversification, creation of employment and poverty reduction, public sector reforms, environmental protection, rural development and human resources development, including a campaign against HIV/AIDS. Reflecting the limited water available in the Botswana portion of the basin, the focus of the most recent development plan was not on rainfed agriculture.

Mozambique

Over the past decade in Mozambique, the government has created a number of development policies aimed at reducing poverty. The national programme for agricultural development in 1999 (PROAGRI) aimed at transforming the subsistence agriculture sector to market producers. PROAGRI was followed by the plan for the reduction of absolute poverty (PARPA) in 2001 and by PARPA-II in 2006. Their major objective was an integrated strategy to reduce the incidence of absolute poverty to 54% by 2003, and to 45% by 2009, through activities in agriculture, health, education and rural development. A complementary agricultural marketing strategy (ECA) was developed for 2006–2009, with the objective of promoting and improving agricultural marketing, inputs and agricultural services, to stimulate efficiency, equity and transparency by all stakeholders. While the government has made progress, it has yet to reach its targets.

The efforts of the Mozambique government to address water management issues led to the Water Act of 1995, which established basic principles of water management. It provided for improved use of available water resources for all purposes, through sustainable planning to meet the population's needs and economic development. The intended coverage level by the year 2000 was 40% of the national population. This target has not been reached but evidence of rollout of the Act can be seen across Mozambique.

South Africa

In South Africa, 1994 saw transition from racially separate development under apartheid to rural development programmes (RDPs) in a bid to reach the government's development goals. In spite of this partial success, the apartheid legacy remains: 10% of urban households and 39% of rural households still do not have access to basic sanitation services, and 26% of urban households and 23% of rural households do not have access to waste removal.

By 1997, South Africa had abandoned the RDP empowerment strategy for a growth, employment and redistribution (GEAR) strategy. The macroeconomic GEAR strategy included job creation, land reform, housing, services, water and sanitation, energy, telecommunications, transport, the environment and social welfare. GEAR also implied commitment to pro-market policies. The pro-market and welfare approaches have existed in parallel since then, especially in the rural development sector. Their co-existence has not led to significant reduction in rural poverty or improved service delivery to rural areas in recent years.

The government has delegated rural development to decentralized local government, but its concurrent welfare and cost recovery policies contradict the decentralization. Water and sanitation have received high priority and the government has made progress in providing free basic water, but that has meant less revenue to pay for provision of more services, which has hindered progress. South African policies are typically sound with solid rationale, but the human and institutional capacity to implement them, and deliver services to communities, remains an issue.

Zimbabwe

In Zimbabwe, the poverty alleviation action programme (PAAP) has been the main policy framework guiding programmes and strategies earmarked to alleviate poverty. It targets Zimbabwe's poorest rural areas through strategies that aim to improve livelihoods and access to resources, infrastructure and services, as well as increased knowledge. The present status of this and other programmes is difficult to assess, given the recent economic and political crises in Zimbabwe.

Water use and management in Zimbabwe is guided by policies at the national, district and council level. The Water Act of 1998 is the main policy instrument at the national level, guiding access, allocation, use, and management of water resources. Other policies that impact the livelihoods of people in catchment areas include the irrigation policy, drought mitigation policy, the livestock policy and the crop production policy. In addition to these national policies, traditional leadership may put informal institutions in place at the local level to suit the needs and requirements of communities, depending on conditions peculiar to each of them, for instance, the general prohibition of washing of clothes in rivers to preserve water quality.

Social and political turbulence

The last three centuries of natural resources management in the Limpopo Basin have seen colonial construction of nation-states, across existing social, cultural and physical boundaries, which changed the nature of the relationship between populations and the natural resources upon which they depended. Water and land, which used to be public goods under the domain of tribal authority, are now regulated by boundaries, borders and statutory laws dictating access and control. Long-standing agrarian communities have disintegrated as forced migration and the extraction of male labour for mining irreparably changed livelihoods and societies (Earle *et al.* 2006, Tewari 2009).

South Africa, Mozambique and Zimbabwe, with 93% of the Limpopo Basin's population and occupying 81% of its area, have seen dramatic political upheavals in past 40 years, which have profoundly affected the distribution of land, water and livelihoods. The period saw the end of minority governments in South Africa and Zimbabwe, and of the colonial administration in Mozambique. There was civil war in both Mozambique and Zimbabwe. Subsequent fast-track redistribution of farming land in Zimbabwe led to disastrous declines in productivity, 45% malnutrition and severe economic decline. The economy of Botswana, the fourth country of the basin, has been heavily influenced by the recent discovery of deposits of diamonds. The process of change in all four countries is therefore ongoing and the outcomes are uncertain.

Recent basin history cannot be told without acknowledging the integral relationship between access to land and water, and economic and political performance. Current land reform efforts in Zimbabwe and South Africa have had profound impacts on economies and populations at many scales, and have highlighted the need for institutional integrity in poverty reduction. Adding the burden of water and land reform to already over-burdened institutions serving large rural constituencies may raise frustrations and the pressure on governments to act decisively. The appropriate policies need to be implemented by well-supported institutions that have the mandate and necessary resources to engage directly with rural communities to improve their livelihoods. These two basin countries in particular continue to grapple with balancing the redress of past injustices while maintaining the economic productivity of the available natural resources.

With the spread of independence and democracy across the region in the last 30 years, increased attention was given to rural development through improving agricultural productivity. However, several decades of investment and interventions have not reduced poverty or improved the agricultural productivity of rural populations in the basin to the degree anticipated (Earle *et al.* 2006). As previously stated, agricultural productivity in the basin remains generally low, with small pockets of high productivity. Earle *et al.* (2006) suggest that a key reason for this is the failure of institutions at all levels to enhance and regulate people's access to water.

Lack of access to markets is another barrier that prevents poorer farmers from converting production into higher income. Market prices in the region also fluctuate greatly between countries due to regional social and political turbulence.

Dual economies

Independence and democratization brought great expectations, yet there have been few systemic changes to restructure the economies or the tenure systems that control the exploitation of natural resources. As a result, dual economic systems persist across the basin. On the one hand, there is the commercial sector, with clearly defined rights, which is well linked to markets, infrastructure and technology. On the other hand, there is the communal sector, with insecure rights to tenure, which is relegated to low-input activities with insecure livelihoods. Yet in all the basin countries, both are now under one system of governance, with one set of institutions and policies serving both groups. This clearly does not work. Institutions are likely to continue to be subject to political and financial pressure from powerful established commercial interests, which inevitably seek to protect their favoured status.

Changes in water resources management

African agriculture and management of water resources is being planned and implemented in ways not previously possible. Awareness among decision makers is high, monitoring and capacity building frameworks are in place, and progress is being made. Planning and development processes are being guided by state-of-the-art principles and driven by regional processes that value protection, conservation and efficient use of available natural resources.

Two regional initiatives can be considered quite successful. One, the Comprehensive African Agriculture Development Programme (CAADP), was initiated by the New Economic Partnership for Africa's Development (NEPAD). The other, regional implementation of IWRM, was initiated by the Southern African Development Community (SADC).[1]

IWRM has become influential at the regional and national levels in the riparian states along the Limpopo River, but it has yet to manifest itself as integrated water resources planning and delivery at subsidiary levels. While there is currently enough water to satisfy requirements, analysis shows that its distribution amongst domestic, environmental, agricultural, mining and tourism uses, varies widely and is not necessarily related to perceived value or benefits. Links between policy, planning and implementation have not lived up to expectations, with the will and capacity to implement lagging behind. Lack of access to water at specific places and times constrains rural development. This is especially true where the costs of water development are high in terms of infrastructure, institutions and markets, such as in the former homelands of South Africa.

An important shift in water resources management principles has been the move away from riparian water rights, especially in South Africa and Zimbabwe. The riparian principle, in force from the colonial period, guaranteed access rights to river water for non-commercial water use by communities, or individuals, who lived adjacent to the river. This principle has been largely replaced by permitted allocation. This exposes productive water users to potentially ponderous regulatory requirements and raises questions about governments' ability to regulate the activities of smallholders. Emphasis has been put on developing plans to manage water resources, and encouraging stakeholder participation in this planning process.

Changes in water resources management across the region aim for harmonization of approaches, and the entrenchment of a regulated system based on authorizing and licensing water uses from a water resource that is regarded as public property. With these regional moves toward IWRM, the traditional view of water as a free basic right poses a challenge for planning and budgeting. All agree that water is, or should be, a basic human right, but on the other hand it is an economic good. Policy makers have therefore adopted the stance that its development and delivery must be paid for by users, because, while the reserve is a guaranteed quantity of water for people and the environment, it still incurs costs. Moreover, water alone is not the problem. Strategic decisions must be made about how much water should be put where, and at what cost to which users.

Capacity building for stakeholder participation

The institutions in the basin have a great need for capacity building, especially at the local level. Although stakeholder participation is a principle of IWRM, current implementation is often unsatisfactory, with domination by elite actors and perpetuation of past inequities in local water-user associations and sub-catchment councils. Without adequate preparation and training, smallholders will continue to lose out in local negotiations for development and management of water resources. Yet improved stakeholder consultation alone will not address the historic inequities or physical water scarcity.

Decision makers must balance natural resources for continued economic growth through industry with secure water and land rights for rural populations. Water resources policies and governance are integral to national development, food security, political stability and stable ecosystems. Yet the links between scarce water and high levels of poverty we document here suggest the need for an integrated approach to water resources planning and development, which focuses on reducing the water-related vulnerability of poor populations and increases their access to secure livelihoods.

Conclusions

Numerous agricultural water management interventions have been introduced in the Limpopo Basin over the last 30 years with limited success. Poverty still persists. Many strategic intervention packages remain relevant to rural recipients in the basin, but targeting is a major challenge. This paper is a snapshot of the current situation and the recent past. These conclusions are therefore tentative and should be modified as nations within the region adjust their economic and social policies.

Vulnerable populations

The rural poor in the Limpopo Basin rely mainly on rainfed agriculture for their livelihoods, but often live in water-scarce areas and have few resources to invest in developing hydraulic property. Insecure land tenure and water rights increase their vulnerability and discourage investment in development. Yet, for those without the safety net of other options such as grants, agriculture remains their only possibility for livelihood. Improving agricultural productivity is therefore their only possible ladder out of poverty.

Other options include social grants, remittances, trading, home-based businesses and sporadic local employment. Without intervention, many are likely to abandon agriculture as a means of livelihood, which could negatively affect the productivity of the area. However, intervention to reduce rural poverty is extremely difficult because of the lack of integration between the water, land and rural development sectors, each of which has its own target or its own goals.

Given agriculture's role in rural development and poverty reduction, allocation of water for agriculture should be re-examined by each basin country in an effort to reduce supply risks for agricultural producers. This implicates the policy, budgeting and planning arms of government to prioritize risk reduction as an important step toward poverty reduction. It may be, however, that the approach of Botswana and South Africa of providing safety-net grants for the rural poor is the most viable option where the limited overall supply of water can be more productively used elsewhere. As long as this decision is taken transparently, with adequate consideration of the needs of all stakeholders, and the rural poor in particular are provided with adequate livelihoods, there could be little dispute. Further discussion of this aspect is outside the scope of this paper.

Unreliable water

Water is scarce and, to cope with drought and an unpredictable climate, farmers only receive the water left over after all other needs have been met. The design of the water supply system uses outdated data that do not incorporate recent or future changes in the natural resource base and climate. Reducing agricultural vulnerability is especially important given the high levels of rural poverty and the limited capacity of poor communities to deal with droughts and water shortages. Irrigation consumes less than 1% of rainwater in the basin but may have undetected environmental and social costs, such as increasing inequity. Objective analysis is required to identify variations in the costs, benefits, and beneficiaries of irrigation.

There is evidence to suggest that the existing irrigated area in the South African portion of the basin (198,000 hectares) already exceeds its potential by nearly 67,000 hectares (FAO AQUASTAT 2004). On the other hand, Mozambique and Zimbabwe have barely tapped their theoretical potential for irrigation. Policy makers and planners in those two countries can include further development of irrigated agriculture in their potential activities, whereas their counterparts in South Africa are under domestic political pressure to reallocate water away from licensed users to previously dispossessed recipients. Further irrigation development in South Africa is likely to be scrutinized.

Low water productivity

At the meso-scale, low water productivity probably results from poor infrastructure and linkages to markets, given that a wide range of productivity enhancing technologies are available in the region.

While all four countries have policies aimed at supporting small-scale farmers to increase their productivity and profits, the effectiveness of implementation varies considerably. The major issue hindering implementation of IWRM is secure financing to meet long-term goals and to build institutional and technical capacity (SADC 2007). Although IWRM focuses on water, other development sectors in all four basin countries have similar issues.

Widespread rural development and poverty alleviation would strengthen all the relevant sectors (water, agricultural, rural development, environment, education and health) so that the required multi-sectoral approach could be successful. Specific issues include improved market access through better infrastructure; increased rural education and ability to absorb education and capacity building and training efforts; commercial orientation of small-scale enterprises; and subsidized inputs when necessary.

Interventions

Smallholders in the basin, who depend on agriculture for their livelihoods, face a host of ecological and economic challenges. Intervention packages, ranging from technologies to institutions, could be tailored to address their priority needs, but it would take a concerted, combined and co-ordinated effort on the part of regional and national bodies to make their development a priority. All basin countries now have development targets to provide rural water and sanitation, as well as land and water reform, which suggests that a more equitable distribution of existing resources may be closer to reality than at any time previously.

Improved stakeholder consultation will help but will not address historic inequities or physical water scarcity. Decision makers must balance natural resources for continued economic growth through industry with secure water and land rights for rural populations. Water resources policies and governance are integral to national development, food security, political stability and stable ecosystems. Yet the links between scarce water and high poverty we suggest here should focus attention and investment on getting water to people.

On-farm interventions alone are insufficient, an enabling environment of policies, infrastructure, investment, and markets must exist for sustainability. Rural water pricing, permits, and cost-recovery policies must be carefully considered and synchronized with rural development policies so that they do not compromise the economic viability of smallholders.

Competition from powerful interests, such as commercial agriculture, mining and the environmental lobby, however, will require institutional integrity in poverty reduction to enable decision makers to make difficult decisions. While water scarcity in the basin is more economic than physical (Seckler *et al.* 1998), it is often not available where poor people can make use of it, so there must be trade-offs. Regional policy analysis and support bodies have roles to play as this process evolves across the region and basin countries.

Benefit sharing

There are examples of countries that transformed themselves by the equitable distribution of the riches derived by development of natural resources. For example, Sweden transformed itself from a poor to a rich nation in the last century by exploiting its forests and iron deposits and investing the derived wealth in initiatives like free education and health care. In doing so, it converted its natural resource wealth into human capacity, while it transformed resource extraction into industry and processing, which provided work for its skilled, well-educated labour force (de Vylder 1996). The wealthier basin countries, Botswana and South Africa, could emulate this model by investing returns from national resources in the further development of their rural populations. Ensuring free education, health care, basic services and infrastructure would increase the capacity of rural populations to help drive economic development.

In contrast, countries without natural resources only have their human resources to develop and they have to do this by focusing on the education and creation of a highly productive, skilled labour force. In such cases, typified by Switzerland, Japan and Hong Kong, the countries' economies then attract investment and grow. A key issue is the political will to distribute the benefits, such as health care and pensions. Cultural homogeneity and historical opportunity also appear to be important.

Basin-level planning and management can be taken sequentially or holistically. A sequential approach has each entity planning for its portion of the water, and how best to meet its needs and maximize its potential. Holistic planning allows for an approach commonly referred to as benefit sharing (Sadoff *et al.* 2002, Turton 2008). Benefit sharing treats the resource as a whole and encourages actors to explore how best to generate overall benefits, and share costs and benefits in a mutually agreed-upon manner. It calls for planning and development agents from riparian countries to think beyond the water flowing across their borders, and think of a cascade of benefits whereby the actual volume of water represents one of many possible benefits. Examples of benefit sharing from water have recently emerged from the Andes basins and the approach is gaining traction at a transboundary scale (Turton 2008).

The benefit-sharing approach to river basin management suggested by Turton has been practised, in small ways, within the Limpopo Basin. This approach to maximizing benefits to be shared equitably could ease current pressures on the resource and act as a buffer against the climate, economic and political changes so common in the region. At the basin scale, the LIMCOM could play a vital role in overall basin assessment, monitoring and planning for the purpose of benefit sharing.

On a cautionary note, it must be said that LIMCOM is currently dominated by water sector professionals, which does not bode well for integrated planning and development of basin water resources. Furthermore LIMCOM has yet to reach its potential as a multinational co-ordinating authority.

Recommended roles for institutions in the Limpopo Basin

Future development of the Limpopo River basin will depend heavily on the effectiveness of the institutions charged with meeting competing needs.

In production technologies

This study has shown high productivity potential in the Limpopo Basin when natural resources, infrastructure, market linkages and technologies are present. Rural populations relying on agriculture in the basin should adapt their production systems to prevailing conditions (wet or dry) and take advantage of existing and emerging technologies. For example:

(1) The Consultative Group on International Agricultural Research (CGIAR) and SADC are promoting conservation farming research and implementation within the basin, mainly for dry areas;
(2) Rainwater harvesting techniques, from household to field level, should be combined with value-chain and finance facilities to improve productivity. These strategies may include small reservoirs and other water storage techniques; and
(3) In areas with ample water, focus needs to be on market linkages, value chains, and increased competitiveness.

There is potential for:

(1) Increasing water-use efficiency at the farm level and reducing runoff from the system. This must be balanced against the effect on the amount of water available for storage. These issues are site-specific and require further research to determine the best mix of approaches and technologies for increasing productivity while improving water-use efficiency of the whole system;
(2) Improving agricultural productivity in the basin by enhancing and regulating people's access to water;
(3) Improving water productivity by better matching of crops to physical conditions, especially in Mozambique where water is less scarce but productivity is lower than in the rest of the basin; and
(4) Assessing the current livestock production strategies in relation to natural resource availability to increase productivity in this sector. Livestock plays a crucial role in rural livelihoods in the basin, yet is often overlooked as a strategy to reduce poverty.

In irrigation productivity

Irrigated agriculture supports a small proportion of the rural poor in the basin but there is scope in most areas to increase the productivity of irrigated crops:

(1) Institutional collapse is cited as a frequent cause of failure of interventions in the basin. Therefore, increased attention needs to be paid to capacity building for local level water management;
(2) Water supply risks for irrigation schemes need to be calculated and analysed before scarce financial resources are invested in schemes that are likely to run out of water;
(3) Several hundred small-scale irrigation schemes in the basin, mainly in South Africa and Zimbabwe, are in need of rehabilitation. South Africa has started such an effort, and the lessons learned should be derived and used to expand this activity to other areas;
(4) There is limited potential for new irrigated areas in Botswana. This should be explored, most likely in conjunction with high-value crops or animal fodder to support existing production activities; and
(5) There is considerable room for expanding irrigation in the wetter parts of the Mozambican basin, but natural hazards (floods) and institutional issues need to be carefully considered before making investments.

In multiple-use systems

Considerable attention has been given to multiple-use systems in the basin over the last five years. The term describes the real-life water use regime of millions of rural poor and suggests that, as conditions evolve, households use multiple water sources to meet multiple water needs. If agriculture is to provide a ladder out of poverty, water is a necessary precondition:

(1) Water policy and infrastructure development for the rural poor need to favour the previously dispossessed, without diminishing the economic capacity of existing users;
(2) Since water delivered to the poor will be used to meet their needs, as prioritized by them, the burden of infrastructure costs and water pricing policies must not be borne solely by them because not all of their water use will generate income; and
(3) Given the reality of water distribution across the basin, investment in multiple-use systems should target areas where water scarcity is not going to prevent or inhibit future growth.

In the policy environment

The policy environment for natural resources management in the four basin countries has progressed over the last 15 years, but further attention is needed in the following areas:

(1) Ratification and capacity building of LIMCOM to engage in basin-scale water resources planning, as well as monitoring water use and quality;
(2) Development of a framework to secure tenure rights to land and water for rural populations in the basin, one that will encourage investment and sustainability;
(3) Pricing and prioritizing the use of water by smallholders to ensure their economic viability;
(4) Adopting a multi-sectoral approach, rather than an IWRM-alone approach. The IWRM concepts of water as an economic good and user-pays must be balanced with the concepts of the water, land and rural development sectors so as not to over-burden end users and make their economic ventures unviable;
(5) Paying specific attention to the commercial and communal sectors of the basin so that each can reach the goals set for it. For the commercial sector, the goals are regional food security, job creation and avoiding mass migration of the poor to cities. For the communal sector, the goals are rural development, poverty reduction and livelihood security, which will also help avoid mass urban migration;
(6) Coordinating efforts on the part of regional and national bodies to make their development of intervention packages a priority;
(7) Getting the necessary support to the correct recipients at the appropriate time, which takes careful planning. Institutions responsible for rural development should therefore identify those who can use agriculture as a ladder out of poverty, assess their situation, and address their specific needs. This could start with:
    (a) Mapping the communities and households in Botswana and South Africa that rely on agriculture as a primary livelihood activity so that they can be targeted for best bet intervention packages to help them climb out of poverty. The rural residents in those two countries, who have a wider range of livelihood options available to them, including social grants, may not need the agricultural interventions so badly.

(b) Mapping conditions in Mozambique and Zimbabwe so that best bet technologies, in terms of water availability, soils and livelihood systems, can be targeted at the correct audience with the necessary support systems. Smallholder agriculture is crucial to the livelihoods of the majority of rural residents in those two countries because households rely heavily on agriculture for food, income and livelihood security;

(8) Prioritizing water resources development at national level by updating approaches to determining ecological reserves, storage capacity and supply risk so that they are based on current averages. This is particularly important in the face of climate change scenarios for the region; and

(9) Directing training and technology at the actual decision makers at the household level. Policy makers and practitioners wastefully continue to underrate women's roles as decision makers and implementers in rural livelihood systems.

As we suggested above, each of the four basin countries faces essentially the same major issue: how to prioritize development and allocation of highly irregular water and contested land in the face of historical inequalities and great economic potential. The key will be for governments to design efficient and effective ways in which to reach target audiences with the most suitable technologies to improve their livelihoods. This calls for a deeper understanding of rural populations and the motivation behind their choice of livelihood activities as well as stronger and more effective institutions to help them meet their needs.

Acknowledgements

The authors wish to acknowledge all those who contributed to the success of the Limpopo Basin Focal Project. The Challenge Programme for Water and Food sponsored and coordinated the projects with support from the EU and IFAD. The LBFP was co-led by the Agricultural Research Council of South Africa (ARC) and the Food, Agriculture and Natural Resources Policy Analysis Network (FANRPAN). Regional partners in research and policy platform activities included the Global Water Partnership–Southern Africa (GWP-SA) and the International Water Management Institute (IWMI). The following institutions led work package activities under the project: the University of Malawi, Poly Technique; ARC; the University of Botswana; IWMI; and FANRPAN. Additional research was undertaken by University of Eduardo Mondlane in Maputo Mozambique, the University of Pretoria and the University of Zimbabwe. This synthesis report is the result of over two years synthesis and analysis of mostly existing data. Therefore acknowledgements go to those researchers whose previous outputs are the basis of this work.

Note

1. The SADC Revised Protocol on Shared Water Courses (2000) provides guiding principles for water sector governance within the SADC region, including the four Limpopo Basin states, all of whom have signed the protocol.

References

Alemaw, B.F., Scott, K., and Sullivan, A., 2010. *Water availability and access in the Limpopo Basin*. Challenge Programme on Water and Food (CPWF) Working Paper: Basin Focal Project series, BFP-L01. Colombo: CGIAR CPWF.

Bourdillon, M.F.C., Hebink, P., and Hoddinott, J., 2007. Assessing the impact of high-yield varieties of maize in resettlement areas of Zimbabwe. *In*: M. Adato and R. Meinzen-Dick, eds. *Agricultural research, livelihoods, and poverty*. Baltimore: Johns Hopkins University Press and Washington: International Food Policy Research Institute, 198–237.

CPWF, 2003. Limpopo Basin profile: strategic research for enhancing agricultural water productivity. Advanced Conference Edition. Colombo: CPWF.

De Leeuw, P.N. and Thorpe, W., 1996. Low-input cattle production systems in tropical Africa: an analysis of actual and potential cow-calf productivity. *All Africa Conference on Animal Agriculture*, 1–4 April, 1996 Pretoria. *In: Conference handbook and volume of abstracts*. South African Society of Animal Science, 3.2.4.

de Vylder, S., 1996. *The rise and fall of the "Swedish model"* [online]. UNDP Occasional Paper 26. Available from http://hdr.undp.org/docs/publications/ocational_papers/oc26a.htm] [Accessed 15 July 2010].

Earle, A., *et al.*, 2006. *Indigenous and institutional profile: Limpopo River Basin*. Working Paper 112. Colombo: International Water Management Institute (IWMI).

Ezcurra, E., 2006. Natural history and evolution of the world's deserts [online]. *In:* E. Ezcurra, ed. *Global deserts outlook*. Nairobi: United Nations Environment Program, 1–26. Available from: http://www.unep.org/geo/news%5Fcentre/pdfs/GDOutlook.zip. [Accessed 5 November 2009].

FAO, 2004. *Drought impact mitigation and prevention in the Limpopo River basin: a situation analysis*. Land and Water Discussion Paper 4. Rome: FAO. Available from: http://www.fao.org [Accessed 8 December 2009].

FAO AQUASTAT, 2004. Rome: FAO. Available from: http://www.fao.org/nr/water/aquastat/main/index.stm [Accessed 24 September 2009].

FAO CLIMWAT. Rome: FAO. Available from: http://idn.ceos.org/KeywordSearch/Metadata.do?Portal=GCMDandKeywordPath=%5BKeyword%3D'BDP'%5DandNumericId=10766andMetadataView=FullandMetadataType=0andlbnode=mdlb1 [Accessed 19 March 2009].

Hanjra, M. and Gichuki, F., 2008. Investments in agricultural water management for poverty reduction in Africa: case studies of Limpopo, Nile and Volta River basins. *Natural Resources Forum*, 32 (3), 185–202.

Hoogenboom, G., *et al.*, 1994. *Crops models: DSSAT version 3.0. International Benchmark Sites Network for Agrotechnology Transfer*. Honolulu: University of Hawaii.

Hoogeveen, J. and Özler, B., 2005. Not separate, not equal: poverty and inequality in post-apartheid South Africa. Working Paper Number 739. Ann Arbor, MI: William Davidson Institute, University of Michigan.

International Fund for Agricultural Development (IFAD), 1996. *Review of research findings applicable to the dry communal areas of Zimbabwe*. Report 0657-ZI rev.1 of south eastern dry areas project and smallholder dry areas resource management project. Rome: Africa II Division, Programme Management Department, IFAD.

Jones, P.G., *et al.*, 2002. *MarkSim: a computer tool that generates simulated weather data for crop modelling and risk assessment*. Cali, Colombia: Centro Internacional de Agricultura Tropical.

Jones, P.G. and Thornton, P.K., 2002. Spatial modeling of risk in natural resource management [online]. *Conservation Ecology*, 5 (2), 27. Available from: http://www.consecol.org/vol5/iss2/art27 [Accessed 14 May 2009].

Limpopo Basin Focal Project, forthcoming. Synthesis Report. CPWF Project 62. Colombo: CPWF.

Limpopo Province Freight Transport Databank. Available from: http://www.ldrt.gov.za/wp-content/Limpopo_Databank/index.html [Accessed 28 November 2009].

Low, A.B. and Rebelo, T.G., 1996. *Vegetation of South Africa, Lesotho and Swaziland*. Pretoria: Department of Environmental Affairs and Tourism.

Mainuddin, M., *et al.*, 2010. *Water-use accounts in CPWF Basins: simple water-use accounting of the Limpopo Basin*. CPWF Working Paper: Basin Focal Project series BFP06. Colombo: CGIAR CPWF.

Perret, S., Anseeuw, W., and Mathebula, N., 2005. Poverty and livelihoods in rural South Africa. Investigating diversity and dynamics of livelihoods. Case studies in Limpopo. Unpublished Project report: WKKF / University of Pretoria / CIRA Tera, no. 17/06.

Republic of Botswana, Central Statistics Office, 2004. Untitled [Electronic version]. Gabarone: Botswana. http://www.cso.gov.bw/images/stories/Agric/livestock_table5.doc [Accessed 2 December 2009].

Sachs, J.D. and Warner, A.M., 1995. Economic reform and the process of global integration. *Brookings Papers on Economic Activity*, 1 (1), 1–118.

SADC, 1999. *Integrated water resources development and management in the Southern African development community*. Round Table Conference. Gaborone, Botswana: Southern

African Development Community Secretariat. Available from: http://www.sadcwscu.org.ls/programme/rtc/rtc-II.htm [Accessed 22 January 2009].

SADC, 2007. *Regional water strategy*. Gaborone, Botswana: Infrastructure and Services Directorate, Southern African Development Community Secretariat.

Sadoff, C., Whittington, D., and Grey, D., 2002. *Africa's international rivers: an international perspective*. Washington, D.C.: World Bank.

Scott, K., 2008. The impact of extreme climate variability on water resource development planning. *International Forum on Water and Food*, November 2008 Addis Ababa, unpublished proceedings.

Seckler, D., *et al.*, 1998. *World water demand and supply, 1990 to 2025: scenarios and issues*. IWMI Research Report 19. Sri Lanka: IWMI.

South Africa, Department of Agriculture, 2006. Linking information to action. Food insecurity and vulnerability information management system: FIVIMS. *Newsletter: Fighting hunger*. Available from: http://www.nda.agric.za [Accessed 27 October 2009].

Tewari, D.D., 2009. A detailed analysis of evolution of water rights in South Africa: an account of three-and-a-half centuries from 1652 AD to Present. *Water SA*, 35 (5), 693–710.

Turton, A.R., 2008. A South African perspective on a possible benefit-sharing approach for transboundary waters in the SADC region. *Water Alternatives*, 1 (2), 180–200.

UNDP, 2003. *Human development report 2003. Millennium development goals. A compact among nations to end human poverty*. New York: UNDP.

UNDP, 2009. *Human development report 2009* [online]. New York: UNDP. Available from: http://hdr.undp.org/en/statistics/data [Accessed 24 October 2009].

UNEP/GRID-Arendal, n.d. Water poverty index, by country in 2002, *UNEP/GRID-Arendal Maps and Graphics Library*. Available from: http://maps.grida.no/go/graphic/water-poverty-index-by-country-in-2002 [Accessed 10 July 2010].

WDI 2008. *World development indicators 2008*. CD-ROM. Washington D.C.: World Bank.

World Resources Institute, 2006. World Resources Institute. 2006. EarthTrends - Freshwater indices: water poverty index. Available from: http://earthtrends.wri.org/searchable_db/index.php?theme=2&variable_ID=1299&action=select_countries [Accessed 10 July 2010].

Wolf, A., *et al.*, 1999. International river basins of the world. *International Journal of Water Resources Development*, 15 (4), 387–427.

Wolf, A., Yoffe, S.B., and Giordano, M., 2003. International waters: identifying basins at risk. *Water Policy*, 5 (1), 29–60.

World Bank, 2006. *World Bank report 2006: equity and development*. Washington D.C.: World Bank.

The Mekong: a diverse basin facing the tensions of development

Mac Kirby, Chayanis Krittasudthacheewa, Mohammed Mainuddin,
Eric Kemp-Benedict, Chris Swartz and Elnora de la Rosa

Population is growing in the relatively unregulated Mekong River basin, and demands for hydropower and food are increasing. The basin has prospered but the poorest have not shared the benefits. Agricultural production is keeping up with rising food demand, but capture fisheries are unlikely to increase production, threatening the supply of animal protein in people's diets. National governments decide water issues unilaterally, with weak transnational institutions and limited public participation. Growing pressures, exacerbated by climate change, will likely increase tensions over access to water, reinforcing perceptions of institutional failure and stimulating demands for improved governance.

Introduction: the physical and human setting

The Mekong River basin is one of the most dynamic, productive and diverse river basins in the world. Decades of civil strife have largely "saved" the basin from the disruption of natural flow patterns that has beset most major transboundary river systems in the world, where water impoundments and diversions have been more intensive. As a result, the Mekong Basin, in relative terms, continues to enjoy exceptionally rich aquatic biodiversity and a heavy reliance on the river's environmental services – fishing, farming and grazing – for peoples' livelihoods.

The Mekong Basin varies from high montaine plateau at its source; through tropical forested mountainous upper middle sections; to densely settled, agricultural lower middle regions; to wide, flat irrigated floodplains of the Delta (Figure 1).

The rainfall is strongly seasonal. The dry season, from November to May, is particularly intense in northeast Thailand, which suffers from seasonal water shortage. In the dry season, the river flow is modest, but increases nearly tenfold in the wet season from June to October. At Chiang Saen, where the Mekong leaves the upper basin and enters the lower basin, the dry season flows are about 2.5 km^3/month and the peak wet season monthly flows vary from about 10.0 to about 20.0 km^3/month. At Phnom Penh, in southern Cambodia, the flows are about 6.0 to about 10.0 km^3/month in the dry season, and about 60.0 to 90.0 km^3/month at the peak of the wet season.

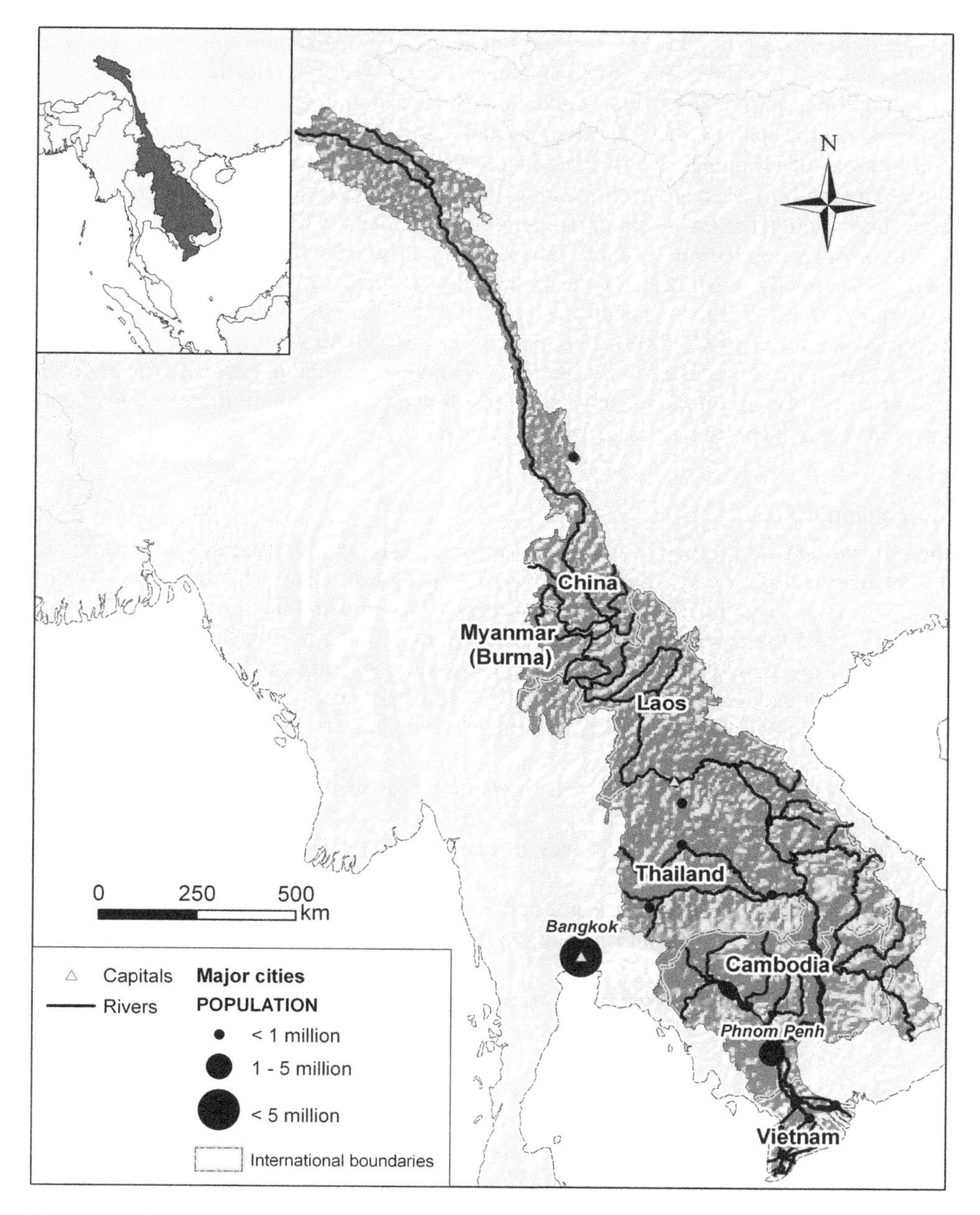

Figure 1. The Mekong River basin.

In the dry season, the Tonle Sap River drains the Tonle Sap ("Great Lake") to the Mekong. In the wet season, the Mekong rises above the level of the Tonle Sap, and pushes water up the Tonle Sap River, reversing its flow for about five months of the year. The Tonle Sap swells greatly during this period, and there is a great production of fish. Annual fish production is correlated with the magnitude of the annual flood (Mainuddin *et al.* 2009). The flood pulse and fish production is less dramatic elsewhere, but is nevertheless important throughout much of the lower Mekong. The basin supports the third largest inland fishery in the world (after the Amazon and Bangladesh). It is the most important source of animal protein for millions of people, including many of the poor (Hortle 2007).

Agriculture, particularly rice production, is the dominant land use in northeast Thailand, central and southern Cambodia, and the delta region of Vietnam. There are smaller areas of cropping in Lao PDR in the flatter areas near the Mekong and in the central highlands of Vietnam in the east of the basin. There is some irrigation in many of these areas, though the main area of irrigation is in the delta (MRC 2003).

The basin's population is about 60 million, most of whom are rural poor with livelihoods directly dependent on the availability of water for the production of food. Agriculture, along with fishing and forestry, employs 85% of the people in the basin, many at subsistence level (MRC 2003). The pressure on the natural resource base, particularly water resources, has increased in recent decades and has resulted in new patterns of development within the six riparian countries. Whilst living standards have increased markedly across the basin, there remain large areas of poverty.

Poverty and water

There is widespread poverty in the lower Mekong region that varies in its characteristics and intensity between and within countries. The people in Cambodia and Lao PDR are among the poorest in the world, while in northeast Thailand and the provinces of Vietnam that are part of the basin, many people suffer from severe poverty. Because livelihoods in the basin are strongly dependent on water resources, water and poverty are closely linked through access to cultivable land, as well as to fish (Kristensen 2001). Chaudry and Juntopas (2005) describe the key factors governing poverty in the Mekong as:

- Increasingly insecure tenure and rights of access of the poor to natural resources, such as land, forests and rivers;
- The predominance of subsistence-based agricultural practices, particularly in rice production;
- A regional inland fisheries sector under increasing pressure from many factors;
- The eradication of upland farming systems upon which minority peoples depend; and
- Processes for hydropower and infrastructure development that adversely impact those with the least stake, or voice, in national development projects.

Seeking to understand how wealth and poverty are perceived by communities, de la Rosa and Chadwick (2008) reviewed wealth-ranking studies. Wealth ranking is a participatory approach in which rankings are based on the views of the community residents, who generate their own criteria. They found a consistent set of seven categories that identified household well-being: food security, landholding, shelter, livestock, productive assets, disposable income, and income and debt. They collected indicators under each of these categories and combined them with water-specific data, covering water quality, water quantity and access. Using these indicators, water poverty was mapped using two different methods: a direct aggregation method based on whether districts are above or below the median value for the different indicators and an alternative method, developed for this project, that makes use of Bayesian inference (Kemp-Benedict *et al.* 2008). The two methods gave similar results, and enabled the identification of hotspots (Figure 2).

Subsequently, household surveys were carried out in a selection of the hotspot areas characterized by a combination of low wealth indicators and high levels of water stress.

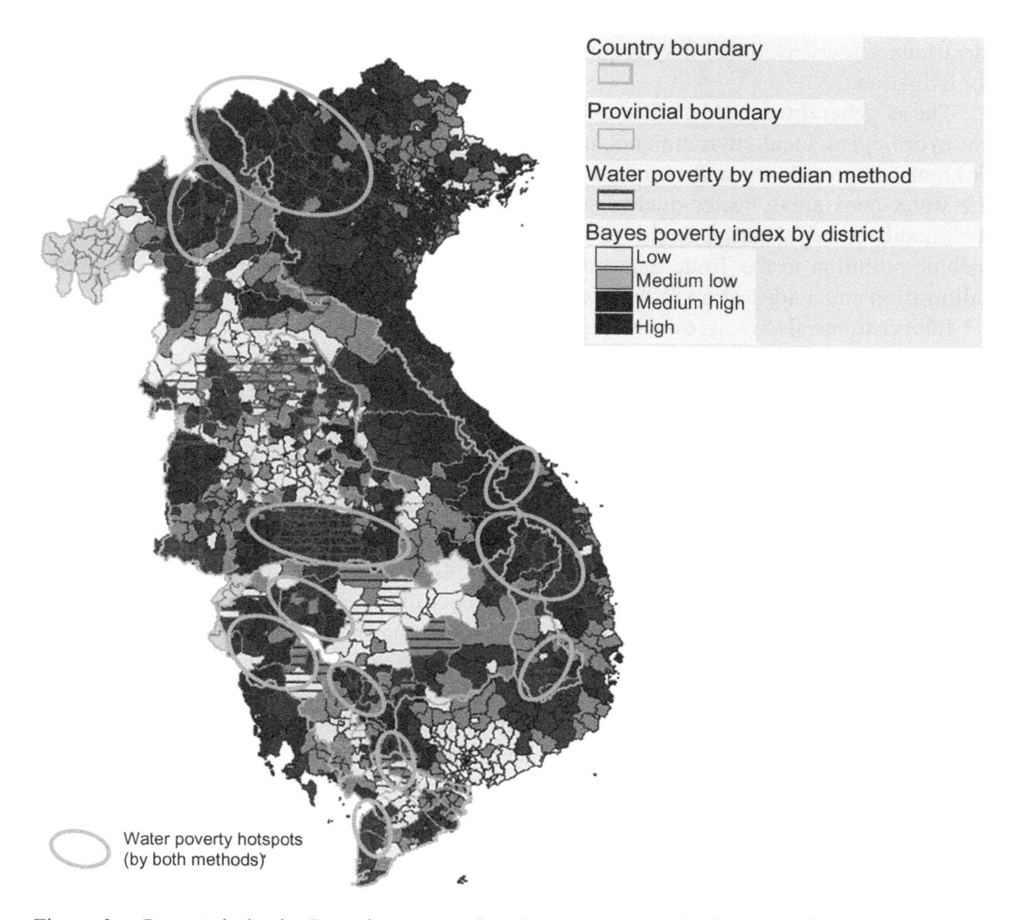

Figure 2. Poverty index by Bayesian approach and water poverty district by median value.

Based on the household surveys (Stockholm Environment Institute 2008a, 2008b, 2008c), it was found that in the hotspot areas, even as poverty is decreasing overall in the Mekong Basin, the poorest households remained poor. Livelihood activities in these areas are closely linked with water, with the majority of villagers engaging in crop farming, fishing or shrimp production. The combination of poverty and water-dependent livelihoods suggests that improvements in water productivity could be an important route to poverty alleviation.

Consistent with this expectation, fish ponds, excavated in the rice paddies and supplied by water and fish naturally during the rainy season, provide a valuable and reliable source of nutrition and cash for farmers. Rainwater harvesting provides drinking water to virtually all households, although storage is often insufficient to last through the dry season. Cash crops irrigated with water from tubewells, canals and other water bodies provide many farmers with a reliable income, but high fuel costs undermine profits.

Unsurprisingly, it was found that wealthier households are better able to cope with water problems than the poor, mostly because they have more diverse livelihood strategies (multiple crops, livestock and remittances) rather than by preferential access to water infrastructure. Indeed, large water infrastructure projects were scored less effective than smaller-scale interventions by farmers. For this reason, extension of the electricity grid was seen as more useful than the creation of large irrigation schemes, although making electricity widely available may require hydropower dams, which can also provide water for irrigation.

These generalities aside, different areas have different problems caused by different hydrologies, local environments, and communities' livelihoods strategies. Floods and droughts are common, and people have adapted to them. For this reason, in most of the water-poor areas, water quality seems to be the most important issue, as it affects livelihoods, food security, health and income of the poor. Problems of water quality include pollution in the Tonle Sap, groundwater quality in northeast Thailand, and water salinization and acidification in the Mekong Delta of Vietnam.

Interventions also focus on water quality, and include "hard" measures such as the provision of electricity, water supplies and sanitation, and "soft" measures such as monitoring of water-quality standards, as well as training in the appropriate use of fertilizers, pesticides and herbicides. Local solutions, such as the collection of rainwater by households, and giving high priority to smaller-scale, locally controlled irrigation projects appear to be more successful than large, externally imposed projects.

There is a major question in developing interventions to reduce poverty. Modelling results suggest (but do not prove) that while risk-spreading strategies can increase household and community resilience, they can also nullify the benefit of some interventions. The result is that an intervention, even when successful, may have only a small impact on overall quality of life. Moreover, for any given intervention, there can be a range of outcomes, such as different outcomes across the population of a village, different outcomes in different villages, or different outcomes in different years in the same village. Whichever happens, the end result is that outcomes are uncertain, so that instead of seeking or expecting a consistent outcome, the goal should be to make positive outcomes more likely and negative outcomes less likely. Over time, such interventions will allow households and communities to build up assets rather than drawing on them to buffer shocks.

There have been continuing efforts to reduce poverty in the basin through various interventions such as aquaculture management (Van Brakel *et al.* undated, Haylor 2001), forestry management (ADB 2003), and rice-shrimp farming (Brennan *et al.* 2002). There has not yet been a comprehensive assessment of basin-wide poverty and its links with the water and land resources, however, far less how they can be used to alleviate poverty, and what the consequences on biodiversity and the natural environment will be.

Hydrology and water resources

The flow for the sub-catchments of the Mekong (Kirby *et al.* 2006, Kirby *et al.* 2010) were modelled. Shown here are the flow modelling for an upstream and a downstream location (Chiang Saen and Phnom Penh), and the Tonle Sap catchment.

The flow at Chiang Saen shows a pronounced seasonal pattern, with some baseflow (Figure 3). The middle reaches of the Mekong preserve the flow pattern observed at Chiang Saen, with the volumes increasing as tributaries add to the flow. When the Mekong is at its peak flow, its level is above that of the Tonle Sap River, which drains the Tonle Sap. Hence water flows up the Tonle Sap River and is stored in the lake. This reverse flow reverts to normal flow when the Mekong River recedes, and the Tonle Sap River then drains the water stored in the lake plus additional water from runoff within the Tonle Sap catchment (Figure 4). The storage of water within the lake is of great importance to local fisheries and livelihoods.

At Phnom Penh, the Tonle Sap River joins the main stem of the Mekong. Flow at this point combines the influences of the floods in the reach from Kratie to Phnom Penh and the reversing flow of the Tonle Sap (Figure 5). The draining of the Tonle Sap back to the

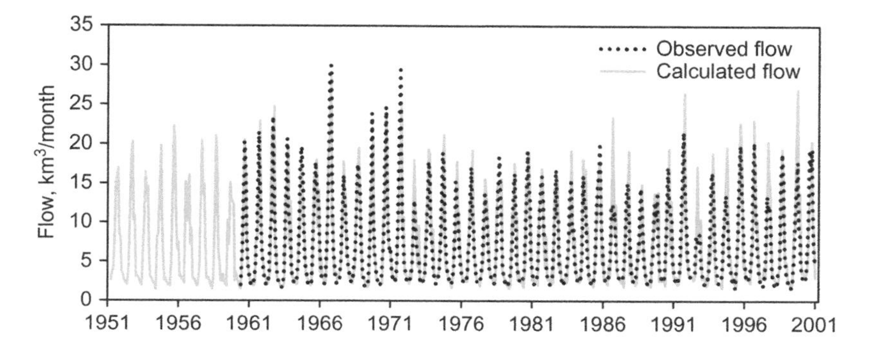

Figure 3. Flow from the upper Mekong at Chiang Saen for 1951 to 2000.
Source: Kirby *et al*. (2010).

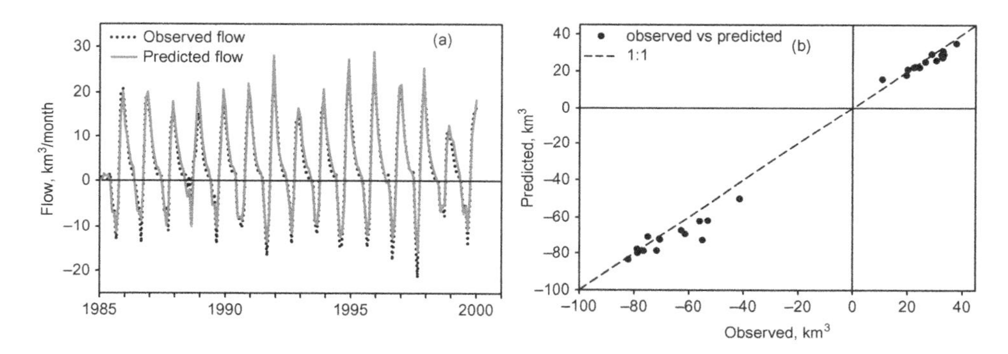

Figure 4. Comparison of observed and modelled flows in the Tonle Sap River 1985 to 1999. (a) hydrograph of the Tonle Sap River flows; (b) observed and estimated total annual outflows and inflows into the Tonle Sap Lake from the Tonle Sap River. The flows in the figure were modelled using observed Mekong flows at Kratie and independently modelled catchment runoff, and compared to observed flows in the Tonle Sap River.
Source: Kirby *et al*. (2010).

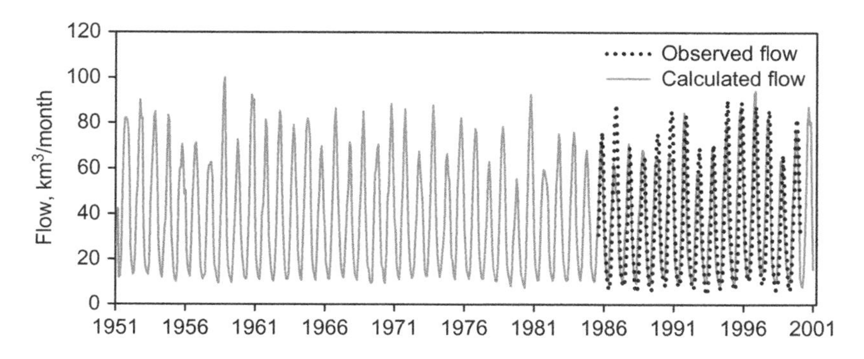

Figure 5. Flow in the Mekong at Phnom Penh for 1951 to 2000.
Source: Kirby *et al*. (2010).

Mekong in the dry season results in greater dry season flows than there would otherwise be. Flows from Phnom Penh to the mouths of the Mekong in the delta in Vietnam are, in aggregate, similar to those at Phnom Penh, but are divided amongst several main channels.

The distribution of the different water uses across the basin is shown in Figure 6, and is based on land use – land cover from the advanced very high resolution radiometer (AVHRR) satellite data (Kirby *et al.* 2010). The figure shows the different behaviour of the runoff-generating upper and eastern part of the basin, and the agriculture-dominated

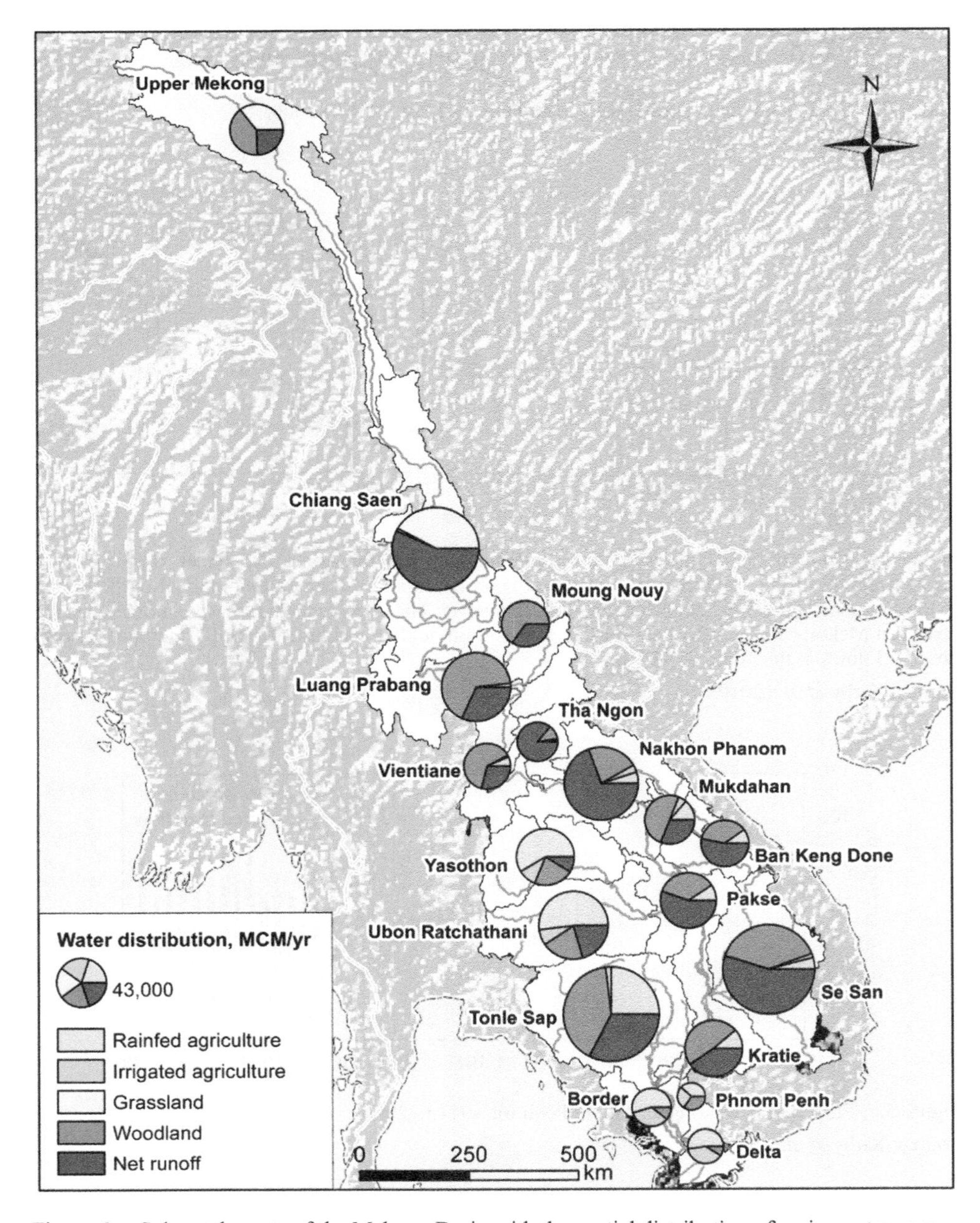

Figure 6. Sub-catchments of the Mekong Basin with the spatial distribution of major water uses.

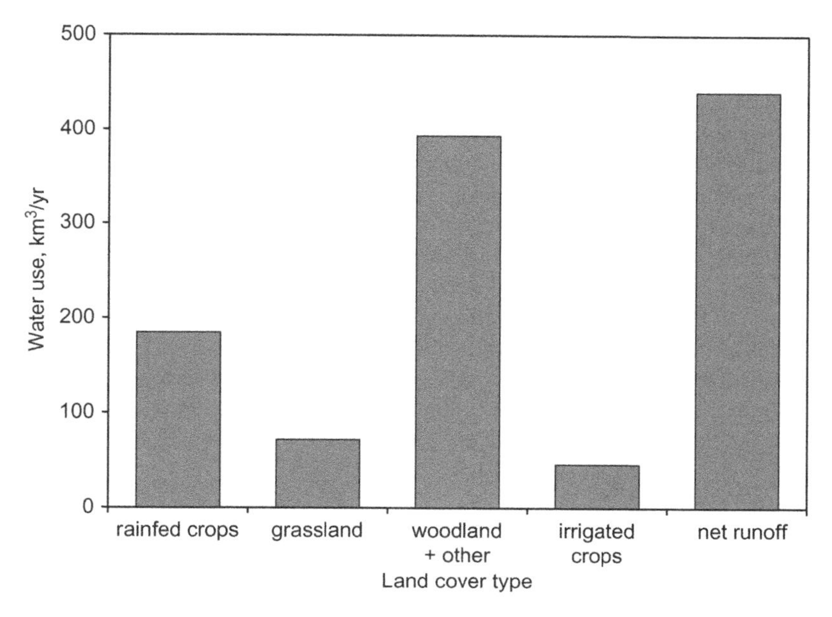

Figure 7. Summary of major water uses in the Mekong Basin.

middle-western parts of the basin in Thailand. Irrigation is a major water use in most parts of the basin. The figure depicts the water uses in each catchment, and the distribution of water uses across the basin. The net runoff from the whole basin is shown in Figure 7. The mean annual input by precipitation to the Mekong Basin totals about 1200 km^3. Net runoff is made up of the runoff remaining after all the water uses in the basin have been satisfied, and includes all other storage changes and losses. Net runoff from the basin is about 441 km^3 or about 37% of the total precipitation input. Forest and woodland is the most extensive land use, covering 43% of the basin. Its water use is correspondingly high, with a mean annual water use of about 390 km^3, or 33% of the total precipitation, or about 52% of the water consumed by the various land uses (i.e. the latter figure excludes net runoff). Urban water demand and use is very small relative to the other uses.

Irrigated agriculture covers about 6% of the basin. The estimated mean annual water use by irrigated agriculture is about 46 km^3 or 4% of the rainfall and 6% of the total water use (excluding net runoff). The majority of the irrigated water use is from crops irrigated from surface water resources. Grassland covers 22% of the basin, almost all in the upper basin, and consumes about 72 km^3 (10%) of the water used.

Thus, the main issue in the Mekong Basin is not water availability, except seasonally in certain areas such as northeast Thailand, but the impact of changed flows on ecology, fish production, access to water and food security. Changes in the natural flow regime may alter the environment of fisheries in the Tonle Sap and elsewhere. Altered low flows may impact salinity intrusion in the delta, thus altering the balance of rice and shrimp production, which in turn may affect food security and incomes. It is now time to turn to three potential threats to the natural flow regime.

Climate change

Several studies on climate change suggest that in several regions the dry season may lengthen and intensify, and that the rainy season may shorten and intensify. Thus both seasonal water shortages and floods may be exacerbated, as may saltwater intrusion into the delta (Hoanh *et al.* 2003, Snidvongs *et al.* 2003, Chinvanno 2004). Kiem *et al.* (2008) consider that extreme floods are more likely, whereas there is a reduced likelihood of droughts and low flow periods, assuming extraction remains at current levels. Eastham *et al.* (2008) also demonstrate that greater runoff, river flow and hence flooding are projected with climate change, but that the magnitude of the changes is uncertain, with little change also being within the range of projections. Ishidaira *et al.* (2008) also found that climate change is expected to result in greater runoff and river discharge but, from about 2050 onwards, climate-change-induced changes to vegetation (particularly greater vegetation in the Himalayas) would lead to a partial reversal of the increased flows.

The floods in the Mekong destroy life and property on the one hand, while on the other they are vital to many ecosystems and to fish production and hence food resources. The anticipated changes to climate and hence flow are expected to affect agriculture and food production greatly, and exacerbate the problems of supplying the increase in food required to feed growing populations (Hoanh *et al.* 2003, Snidvongs and Teng 2006, Eastham *et al.* 2008).

Dam development

Development of a large dam in the upstream areas would reduce the peak flows in the wet season and increase flow in the dry season. Reduction of flow peaks will reduce the risk of flooding and increased flow in the dry season will enable irrigation development. In terms of quantity of water the change is not great (Podger *et al.* 2004, Kirby *et al.* 2006), but the effect on the ecology and environment of the river could be very important. Change of the natural flow regime in the river, particularly low flows and flood onset timing, would affect fish production by changing migration triggers and fishing opportunities (Baran 2006, Baran *et al.* 2007a). Barlow *et al.* (2008) show that blocking of migration routes by dams on the mainstream of the Mekong may cause major damage to fisheries. Millions of poor people in Cambodia depend on the fisheries of the Tonle Sap for their livelihoods and would be affected by these changes. Increased dry-season flow could also reduce the intrusion of salinity in the Delta, affecting the management of shrimp farming, rice farming and other cropping.

Irrigation development

The population of the Mekong is expected to increase from the current 65 million to perhaps more than 82 million in 2030 (based on the medium variant projection, United Nations [UN] Population Division 2006). The increase in population with apparent rice consumption (including losses from field to market) at 150 kg per person per year (cf. Minot and Goletti 2000) means about 6 million extra tonnes of rice would be required. Kirby and Mainuddin (2009) examined the potential impact of this increase on water demand for irrigation. As a worst case, they assumed that the whole of the increased production will come solely from irrigation at the current level of productivity, and none is due to improvements in management or variety. The irrigation requirement of a rice crop varies from 1.5 to 3 m/yr or even more for soils which drain rapidly. In the latter case much of the water would return to the system for re-use. Assuming 2 m/yr is the net water requirement, and at the current average yield of 3 t/ha, some 20,000 km^2 is required to grow 6 million tonnes, then the net diversion requirement would amount to 40 km^3, which is about 8% of the current discharge to the sea.

As noted by Kirby and Mainuddin (2009), this rough estimate is, on the one hand, an overestimate of the likely water requirement because irrigation will not be the sole factor in increased production, while on the other it is an underestimate since production increases in rice would be supplemented by increased water demand for other crops and livestock. On balance, it would appear that the water demand of required increases in agricultural production is modest relative to the total volume of water in the Mekong. Locally, especially in the drier northeast Thailand, the impact could be greater. Podger *et al.* (2004) also estimated that the impact of a high agricultural development scenario on flows in the Mekong would be modest. However, the impact on the ecology and the environment is yet to be fully understood and could be important.

Agricultural productivity

Mainuddin and Kirby (2009a) examined agricultural productivity in the Mekong Basin. They noted that the average yields of rice in the world and Asia were 4.04 and 4.09 t/ha respectively, in 2004 (IRRI 2008). Rice yields were, in general, lower in the lower Mekong Basin, with only the delta in Vietnam meeting the average (based on figures from 1993–2004). Yields in Lao PDR and the central highlands of Vietnam were about 3 t/ha, whereas in Cambodia and northeast Thailand they were only 1.5–2 t/ha. The lower productivity of rice in these areas probably has many causes including lower rainfall, longer annual dry period, poorer soil, cultivation of local varieties with low yields, low fertilizer applications, and inadequate management. Lack of water in the dry season is a major production constraint for rainfed lowland rice and is particularly severe in northeast Thailand. It also affects large areas of rice cultivation in Lao PDR and Cambodia (Fukai 2001). Failure to adopt high-yielding varieties is also a factor in Lao PDR, Thailand and Cambodia (Kono *et al.* 2001, Linquist and Sengxua 2001, Hasegawa *et al.* 2008).

Mainuddin and Kirby (2009b) examined the agricultural water productivity (WP) (production per unit of water, rather than per unit of land as above). Increases in water productivity are required if the world is to meet growing food needs without increasing use of the world's water resources. Zwart and Bastiaanssen (2004) reported average WP for rice as 0.60–1.6 kg/m^3. Toung and Bouman (2003) give a very similar range of 0.4–1.6 kg/m^3 for lowland rice. Mainuddin and Kirby (2009b) found that WP of rice for Cambodia (maximum of 0.33 kg/m^3) and Thailand (maximum of 0.30 kg/m^3) were even lower than the minimum reported by these authors. WP of rice for Vietnam (maximum of 0.77 kg/m^3) and Lao PDR (maximum of 0.58 kg/m^3) fall within the range reported by Zwart and Bastiaanssen (2004) and Toung and Bouman (2003).

In all regions, however, rice yields and WP increased from 1993 to 2004, with the increase being greater in Lao PDR and Vietnam (Mainuddin and Kirby 2009a, 2009b). This increase was mostly because of the cultivation of high-yielding varieties in some areas and high fertilizer use (Hasegawa *et al.* 2008). Yields could be increased further even where water is limited by replacing low-yielding local varieties with high-yielding varieties, increasing application of fertilizer, improving management practices, and increasing application of herbicides and pesticides (Kono *et al.* 2001, Pandey 2001, Rickman *et al.* 2001, Schiller *et al.* 2001, Fukai and Kam 2004). Linquist and Sengxua (2001) showed that applying fertilizer alone increased yields by 134% and 107% in the Champassak and Saravane provinces of Lao PDR.

The yields and WP of other crops such as maize, cassava and sugarcane follow patterns similar to that of rice, except for Thailand. The yield of sugarcane in Thailand at 59 t/ha was second only to the Vietnam delta at 61 t/ha (compared to 23 for Cambodia and 30 for Lao PDR). Although agriculture in the lower Mekong Basin is dominated by rice cultivation, in recent years cultivation of other crops has increased in Lao PDR and the central highlands of Vietnam (Mainuddin and Kirby 2009a). This is most likely the result of bringing land under cultivation that otherwise lay fallow or was under forest, but some of it could be the result of double cropping. The return from non-rice crops per hectare was much higher than for rice. At the basin level, the 18% of the area under crops other than rice generates 37% of the total (rice plus other) income. The results suggest that cultivation of more non-rice crops can increase the income of the farmers and this will have an important impact on reducing poverty. In the long term, therefore, a shift toward integrated, diversified agriculture with a more balanced cropping pattern throughout the year is most likely to be more profitable than the present strong dependency on wet-season rice. Plans for related infrastructure and investments in developing agricultural markets, institutional development, and support services should therefore be a prime consideration in a planning process for the whole region (Kristensen 2001).

The analysis of Mainuddin and Kirby (2009a, 2009b) shows that there is important spatial variation of productivity within the riparian countries, particularly among non-rice crops, and that there is considerable scope for increases in the basin, particularly in Cambodia, Thailand and Lao PDR. Agricultural productivity is rising in the basin, and it seems that the demands for increased food due to increase in population can be met. There is even potential for the basin to maintain current levels of rice exports in the future if current trends continue. This could be achieved by modest increases in the areas irrigated in the dry season, coupled with increases in productivity. It also seems that the increases can be achieved without placing great demands on water resources. The principal issues will be food distribution, not agricultural production.

Fish – a key resource

Fisheries are a major factor in the well-being and livelihoods of the 65 million people who live in the lower Mekong Basin (MRC 2003). Some two thirds of the basin's population are involved in Mekong fisheries, at least part time or seasonally. Not only do they derive some of their livelihood from fisheries, they also depend on fish and other aquatic animals for food security (MRC 2003). Fish and other aquatic animals are the most important sources of animal protein for many people, in particular the rural population in the lower Mekong Basin (van Zalinge *et al.* 2003).

Mainuddin *et al.* (2009) reviewed the fisheries studies in the Mekong, and concluded that there are major uncertainties in estimates of fisheries production and cash value in the lower Mekong Basin. Catch surveys underestimate the production, whereas consumption surveys generally result in higher estimates of production. Nonetheless, a range of values can be derived. The highest estimates are 42 kg/capita/year in Lao PDR and 65 kg/capita/year in Cambodia. Many of the consumption-based estimates are only for the year 2000, however.

Fisheries production is dominated by capture fisheries in Cambodia (where it is concentrated around the Tonle Sap and the Mekong), Lao PDR and Thailand. In Vietnam, aquaculture dominates production, and is concentrated around the main rivers in the delta and along the coastal strip.

The uncertainties over estimates of production make other conclusions tentative, but it appears that production from capture fisheries increased relatively little from 1995 to 2005 in all four lower Mekong countries. A large reported increase in Cambodia in recent years appears to be a change in estimation methods rather than a true increase in production. In aquaculture, there is a clear, large increase in production in the Mekong Delta region of Vietnam since about 2000. Much of the increased production is exported (Ministry of Fisheries 2005, Starr 2007, Cuyvers and Van Binh 2008).

The value of the fisheries is, like the production, somewhat uncertain. Baran *et al.* (2007b) cite various estimates that place the value to the year 2000 as much as US$2 billion. These estimates clearly do not account for the recent great rise in aquaculture in the delta. Baran *et al.* give aquaculture production at 260,000 tonnes, whereas it is currently at least 1 million tonnes. The overall value is therefore US$3 billion or more. Other estimates, including those using other consumption figures, place the overall value somewhat lower. The value is probably not changing greatly with time, although the range of estimates and poor data mean that this conclusion is tentative. Aquaculture in Vietnam is rapidly increasing in value, matching the increase in production, and in 2003 was worth around US$1 billion (Ministry of Fisheries 2005).

The demand for fish will inevitably rise in the future, partly as a result of increasing population in the region and partly as a result of increasing incomes. Over and above this, there may also be a continuing rise in the export of fish products.

The increasing demand appears unlikely to be met through an increase in production of capture fisheries, but the current rapid growth of aquaculture, if it can be maintained, appears capable of meeting the demand. There are no quantitative estimates of the limits to the growth of this industry, however, nor whether it will pose risks for the capture fisheries by taking small fish fry as feed for aquaculture fish (Tuan and Mai 2005). Therefore, whether the current growth of aquaculture can be maintained is unclear. Rice-fish farming may also contribute to increased production, but again the impact appears not to have been quantified.

The lower Mekong fisheries face threats to production from changed water availability, quality, and barriers to fish migration, impact of climate change and overfishing. If the increased demand is to be met, these threats must be managed in such a way that developments do not reduce the production of fish, especially capture fish.

Thus, while there are both great concerns over the future of fisheries and some positive signs, one must conclude that there is insufficient information to make confident predictions. There is an urgent need to undertake studies in the Mekong of quantities and trends of fish production, both capture and aquaculture, how these are likely to respond to the various potential impacts that might change flow regimes, and whether the sector can continue to supply the current level of animal protein to the Mekong people. Even if there are increases in production (presumably from aquaculture), will it help local food security, or will it go to export? The studies called for here are as much, if not more, about economics and institutions as they are about fish biology and ecology.

Institutional setting

Here the literature on the institutional setting relevant to the Mekong is reviewed, and a few key points that arise are addressed.

Key point one: tensions are likely, and some are already evident

Studies of co-operation and conflict in global basins point to the Mekong as a basin with potential for conflict (Wolf *et al.* 2003, Yoffe *et al.* 2003). Stahl (2006), on the other hand, analyses transboundary conflicts and agreements in major river basins around the world, and concludes that the Mekong has a pattern of moderately positive co-operation, yet without much concrete action. This was an analysis of past tensions, and thus is not inconsistent with the above predictions. Basins where rapid change (either biophysical, such as dam development, or institutional) outpaces the institutional capacity to absorb the change are at risk of conflict (Wolf *et al.* 2003). In the Mekong, the pace of change and unilateral developments suggest that there may be political tensions within the next five to 10 years. Institutional factors are probably more important than biophysical factors. Amongst biophysical factors, water scarcity is the greatest indicator of potential conflict (Hensel *et al.* 2006), although Gleditsch *et al.* (2006) conclude that water scarcity and drought are almost unrelated to conflict.

In the Mekong, these global studies are borne out. Tensions are apparent, particularly over dam development and its impact on the poor people who are directly affected (Lawrence 2009) and on the tensions between dam development and fisheries (Foran and Manoram 2009, Friend *et al.* 2009). At the same time, these tensions are unlikely to be principal drivers of international conflict.

Key point two: public participation is limited and information is not shared

Many studies in the Mekong conclude that water governance is narrowly concentrated, primarily in the national governments and their agencies, and that it is necessary to broaden the sharing of information, decisions and benefits. Lebel and Sinh (2007) document the narrow politics of water governance and the lack of democratic participation for key issues in the Mekong, including irrigation development (Molle 2005, 2007), floods (Lebel and Sinh 2007), and hydropower expansion (Dore *et al.* 2007, Graecen and Palettu 2007). Even the development and use of hydrological models (Sarkkula *et al.* 2007) and scientific knowledge more generally (Contreras 2007) are explored as issues in which limited participation may lead to unequal exercise of power and outcomes that do not benefit the poor. None of the authors regards this as inevitable; science and models can also provide the information for alternatives and rational decisions. Dore *et al.* (2007) provide the most explicit call for greater engagement, through multi-stakeholder platforms.

Associated with the calls for greater participation are calls for greater transparency and accountability. Badenoch (2002) argues for enhanced governance practices, with increased transparency and accountability, and greater public involvement through multi-stakeholder dialogues. Lebel *et al.* (2004) propose "Nobody Knows Best" as a heuristic for forest management in southeast Asia. In Nobody Knows Best, all sources of knowledge and perspectives are welcomed, giving, for example, poor forest dwellers with their different values and uses a voice in forest management. The principles would also apply to water management (cf. Foran and Lebel [2006], who characterize current water governance as a more limited State Knows Best perspective). The Mekong Program on Water Environment and Resilience (M-POWER) dialogue "Informed and Fair" (Foran and Lebel 2006) recommended that information on developments should be freely available, that there should be more dialogue-based and cross-sectoral planning, that benefits should be shared and the disadvantaged should be compensated.

Middleton and Tola (2008), citing Agrawal and Gibson (1999), point out, however, that community participation is somewhat idealized in the development literature. Communities such as the Tonle Sap village studied by Middleton and Tola (2008) are made up of many actors with many agendas, internal differences and local politics. Furthermore, local communities cannot engage in basin-wide management, which must be assumed by larger institutions. Middleton and Tola (2008) argue for both local institutions and basin-wide management organizations.

Increased participation requires the commitment of local officials. Heyd and Neef (2006) describe how the espousal of greater openness and stakeholder inclusion in Thai water planning since the 1990s has led to little difference on the ground, partly because government officials are not disposed to devolve power.

Key point three: current institutions are weak and too focused on development

Many papers study the roles of institutions, especially the Mekong River Agreement and the Mekong River Commission, which are the main transboundary institutions in the basin. The general conclusion is that they are too weak to ensure sharing of information, decisions, and benefits. Hirsch *et al.* (2006) describe a major study into the tensions between national interest and transboundary water governance, focusing on the Mekong River Commission. They describe the current legal and institutional framework as too weak either for transboundary or for national governance. The Commission is too weak and uncertain in its directions and which interests its serves; in particular, it does not embrace a large diversity of stakeholders, and, being captive of a smaller range of issues and influenced by the National Mekong Committees, it fails to tackle many major issues head-on.

Sneddon and Fox (2006) argue that the Mekong River Agreement, being amongst states, limits the debates to transboundary environmental issues, and renders less visible issues within states, such as the Pak-Mun Dam, which affected fishing livelihoods in the Mun River in Thailand. The agreement also limits participation, excluding non-state entities. In Sneddon and Fox's view, anti-development forces such as the anti-Pak-Mun movement and the nature of the Mekong itself offer powerful counter-narratives to the perspective implicit in the Mekong River Agreement of the Mekong as a resource amenable to development.

Rena (2005) and the International Rivers Network (IRN) (2002) likewise argue that the Asian Development Bank (ADB) has a narrow focus on development, essentially hydropower dams and other large infrastructure development in the basin. They worry that many environmental and community concerns are overlooked. Jusi (2006) also argues that the lack of effective governance and consultation by the ADB leads to inadequate decision making and project implementation. In addition, the general benefit to the Lao economy of large dams often does not trickle down to the poor immediately affected by the development. Goh (2004) also argues that the ADB's Greater Mekong subregion programme and the Association of Southeast Asian Nations (ASEAN)'s Mekong Basin development programme are focused on development, and enthusiastically supported by China as it competes for influence in the region. Dore *et al.* (2007) point out that in the quest for development of hydropower in Yunnan, insufficient attention is given to environmental and social factors, and there is limited consideration of alternatives.

Lebel *et al.* (2006) conclude that in many flood-prone areas in developing nations institutions remain weak, and institutional reform to cope with floods has largely failed. They argue for a systematic approach to diagnosing institutional capacities and identify critical gaps before floods occur.

Key point four: national interests dominate

China, regionalization and the role of states dominate many discussions about the Mekong. China is in a position of great power, both because of its place on the river and because of its economic might, and its planned dam developments are unchecked and largely uncriticized by the lower basin countries (Osborne 2004). Both China and Thailand seek to develop hydropower to meet their growing energy demands (Dore *et al.* 2007, Graecen and Palettu 2007), even though these may be based on poor predictions of demand (Dore *et al.* 2007, Graecen and Palettu 2007). As a result, they are developing mainstream dams and aiding the development of dams in Lao PDR (McDonald *et al.* 2009). The Mekong River Commission is in a weak position to deal with national interests, at least partly because it is sponsored by the states. Furthermore, China is not a full party to the Mekong River Agreement. China shows increasing engagement but only at the level of technical co-operation and a co-operative regionalism focused on economic growth, the benefits of which are not necessarily shared equitably (Sokhem and Sunada 2008). The increasing regionalism of the Mekong area is primarily state-led and focused on economic development, and will not automatically lead to common prosperity.

There is considerable downstream opposition to upstream development (particularly hydropower), coming mainly from sectors of society that do not share the benefits and from non-governmental organizations (NGOs). Ojendal *et al.* (2002) show that while in principle poor people may desire dam development and the benefits it brings, in practice the benefits go elsewhere. There is little opposition from the national governments, which are beneficiaries of Chinese aid or access to hydropower and other forms of trade, and also may be engaged in similar forms of development often with Chinese help (Mehtonen 2008).

Hirsch *et al.* (2006) argue that many of the choices in the Mekong will ultimately be national political choices, rather than choices based on law, sustainable development principles or agreements amongst the countries of the Mekong.

Key point five: poor people will benefit from both regional and local governance

A common thread in much of the literature cited above is that benefiting the poor will require an alternative future, particularly public participation, strengthening of key institutions, and lessening the power of the states in favour of regional and local power, with less conflict and more participation. For many, a more powerful Mekong River Agreement and Mekong River Commission are necessary to deal with regional and transboundary issues. At the same time, local governance is seen as crucial to creating outcomes that benefit the poor.

Where to next?

It seems to us that enough is known and that it is time to act. Some actions are, of course, being taken, but not enough and not quickly enough. The Mekong River Commission is changing, but is still dominated by national interests. The rising pressure on the natural resource base, and the resulting tradeoffs over resources, are leading to greater tensions. The tensions will reinforce the perceptions of institutional failures and the demands for improved governance.

Climate change, with the greater floods and greater saline intrusion in the delta it will likely bring, will also further strain the capacity of institutions to cope, again reinforcing perceptions of institutional failures and demands for improved governance.

As noted above, Hirsch *et al.* (2006) argue that many of the choices in the Mekong will ultimately be national political choices. Mainuddin *et al.* (2009) noted that capture fisheries will not meet the likely future level of demand for fish in coming decades. The future of fish productivity in the Mekong will be determined primarily by political choices, barring a dramatic ecological change, such as might be caused by severe climate change with large changes to flows. The World Bank (2006) likewise notes that the future of capture fisheries in Bangladesh will depend on effective governance, including community participation. In both the Mekong and Bangladesh, the choices include: basin-wide choices, in particular of dams upstream and their effect on dry season flows and the timing and extent of floods; local choices on determining the conservation of wetlands; and access to the resource.

Thus, the authors of this study see that the short- to medium-term future will likely bring greater tensions and political conflict, greater benefits to some with losses by others. This will strengthen calls for greater public participation, with presumably more noticeable protests over key issues. Whether this will lead to increased participation, a greater sharing of benefits and improved livelihoods for the poor is an open question.

Conclusions

Much is known about the Mekong and the main issues: the rising pressure on the natural resource base leading to tradeoffs over resources between upstream and downstream interests, urban and rural areas, upland and lowland communities, sectors (notably between fisheries and hydropower), subsistence-based livelihoods and activities oriented towards industrialization, and civil society interests and formal resource agencies. These tensions are likely to increase with growing population, increasing development and resource use, especially hydropower and the growing demand for food. These tensions will reinforce the perceptions of institutional failures and the demands for improved governance.

Although there are biophysical constraints to water use and food production, especially the probable limit of capture-fish production being somewhere around current production, the key factors in future development and poverty alleviation appear to have more to do with institutions. Political choices will govern the future development of the Mekong.

Poverty is falling in the Mekong, though the benefits are not shared equally, and many rural people remain very poor. Many view development, of infrastructure, production, energy, and especially hydropower, as the way to reduce poverty further. Others vehemently criticize this view for failing to share information, decisions, and especially the benefits, and point to the poor that are disadvantaged and not compensated by development, especially to those disadvantaged by hydropower development.

Virtually all studies agree that greater public participation in decision making is required, although many add that other factors are necessary for full sharing of benefits. These include strengthened laws and the espousal of public participation by local officials.

Within this broad picture, policy and planning faces some key information gaps. These include:

- Means for improving low agricultural productivity, especially in northeast Thailand and Cambodia;
- How to preserve capture fisheries and sustainably develop an alternative supply of approximately the same production just to maintain the current level of fish in the diet;
- The impact of changed river flows and flooding due to climate change on the environment and fisheries;
- Institutions to share ownership of and benefits resulting from water resources management and the production that results from it; and

- Water quality, ranging from pollution of the surface water resources to the problems with potentially greater use of arsenic-contaminated groundwater in parts of Cambodia and the Delta.

The research landscape to address these gaps, however, appears to us highly fragmented. There are many research projects, involving many institutions, but there is no overall sense of direction. Perhaps this does not matter, perhaps the research marketplace, like markets everywhere are supposed to, produces the optimum result. Perhaps, on the other hand, there are too many overlaps, such as the development of many hydrology models, many of which appear not to be used. A key consideration for research in the region is whether greater gains could be made with better co-ordination.

Acknowledgements

The Mekong Basin Focal Project was funded by the Challenge Program on Water and Food and by the CSIRO Water for a Healthy Country programme. The authors thank several anonymous reviewers for their helpful comments on an earlier version of the paper.

References

Agrawal, A. and Gibson, C.C., 1999. Enchantment and disenchantment: the role of community in natural resource conservation. *World Development*, 27 (4), 629–649, cited by Middleton and Tola (2008).

ADB, 2003. *Technical assistance for poverty reduction in upland communities in the Mekong Region through improved community and industrial forestry*. Manila: Asian Development Bank.

Badenoch, N., 2002. *Transboundary environmental governance: principles and practice in Mainland Southeast Asia*. Washington, D.C.: World Resources Institute.

Baran, E., 2006. *Fish migration triggers in the Lower Mekong Basin and other tropical freshwater tropical systems*. Mekong River Commission (MRC) technical paper 14. Vientiane, Lao PDR: MRC.

Baran, E., Starr, P., and Kura, Y., 2007a. *Influence of built structures on Tonle Sap fisheries*. Phnom Penh: WorldFish Center.

Baran, E., Jantunen, T., and Chong, C.K., 2007b. *Values of inland fisheries in the Mekong River Basin*. Phnom Penh: WorldFish Center.

Barlow, C., *et al.*, 2008. How much of the Mekong fish catch is at risk from mainstream dam development? [online] *Catch and Culture*, 14 (3), 16–21. Available from: www.mrcmekong.org/download/programmes/fisheries/Catch_Culture_vol14.3.pdf [Accessed July 2010].

Brennan, D., *et al.*, 2002. *An evaluation of rice-shrimp farming systems in the Mekong Delta* [online]. Report prepared for the World Bank, Network of Aquaculture Centres in Asia-Pacific, World Wildlife Fund and Food and Agriculture Organization of the United Nations Consortium Program on Shrimp Farming and the Environment. Published by the Consortium. Available from: library.enaca.org/Shrimp/Case/Vietnam/FinalVietnam.pdf [Accessed 2 August 2010].

Chaudry, P. and Juntopas, M., 2005. *Water, poverty and livelihoods in the lower Mekong Basin*. Basin Development Plan working paper version 3. Vientiane, Lao PDR: MRC.

Chinvanno, S., 2004. *Information for sustainable development in light of climate change in Mekong River Basin* [online]. Bangkok: Southeast Asia START Regional Center. Available from: http://203.159.5.16/digital_gms/Proceedings/A77_SUPPAKORN_CHINAVANNO.pdf [Accessed December 2006].

Contreras, A., 2007. Synthesis: discourse, power and knowledge. *In*: L. Lebel, *et al.*, eds. *Democratizing water governance in the Mekong region*. Chiang Mai: Mekong Press, 227–236.

Cuyvers, L. and Van Binh, T., 2008. *Aquaculture export development in Vietnam and the changing environment: the case of Pangasius in the Mekong Delta* [online]. CAS discussion paper 59. Antwerp: Centre for Asian Studies, University of Antwerp. Available from: webh01.ua.ac.be/cas/PDF/CAS59.pdf [Accessed July 2010].

Dore, J., Xiaogang, Y., and Yuk-shing Li, K., 2007. China's energy reforms and hydropower expansion in Yunnan. *In*: L. Lebel, eds. *Democratizing water governance in the Mekong region*. Chiang Mai: Mekong Press, 55–92.

Eastham, J., *et al.*, 2008. Mekong River Basin water resources assessment: impacts of climate change [online]. Canberra: CSIRO. Available from: http://www.clw.csiro.au/publications/waterforahealthycountry/2008/wfhc-MekongWaterResourcesAssessment.pdf [Accessed May 2009].

Foran, T. and Lebel, L., 2006. *Informed and fair: water and trade futures in the border regions of mainland southeast Asia*. Unit for Social and Environmental Research (USER) working paper WP-2007-02. Chiang Mai: USER, Chiang Mai University.

Foran, T. and Manoram, M., 2009. Pak Mun dam: perpetually contested? *In*: F. Molle, T. Foran, and M. Kakonen, eds. *Contested waterscapes in the Mekong: hydropower, livelihoods, and governance*. London: Earthscan, 55–80.

Friend, R., Arthur, R., and Keskinen, M., 2009. Songs of the doomed, the continuing neglect of capture fisheries in hydropower development in the Mekong. *In*: F. Molle, *et al.*, eds. *Contested waterscapes in the Mekong: hydropower, livelihoods, and governance*. London: Earthscan, 307–332.

Fukai, S., 2001. Increasing productivity of lowland rice in the Mekong region. *In*: S. Fukai and J. Basnayake, eds. *Increased lowland rice production in the Mekong region*. Australian Centre for International Agricultural Research (ACIAR) proceedings 101. Canberra: ACIAR, 321–327.

Fukai, S. and Kam, S.P., 2004. Improved crop production under water constraints. *In*: V. Seng, *et al.*, eds. *Water in agriculture: proceedings of a CARDI international conference research on water in agricultural production in Asia for the 21st century*. 25–28 November 2003, Phnom Penh, Cambodia. ACIAR Procedings 116. Canberra: ACIAR, 182–195.

Gleditsch, N.P., *et al.*, 2006. Conflicts over shared rivers: resource scarcity or fuzzy boundaries? *Political Geography*, 25 (4), 361–382.

Goh, E., 2004. *China in the Mekong River Basin: the regional security implications of resource development on the Lancang Jiang*. Institute of Defence and Strategic Studies (IDSS) working paper 69. Singapore: IDSS.

Graecen, C. and Palettu, A., 2007. Electricity sector planning and hydropower in the Mekong Region. *In*: L. Lebel, *et al.*, eds. *Democratizing water governance in the Mekong region*. Chiang Mai: Mekong Press, 93–125.

Hasegawa, T., *et al.*, 2008. A model driven by crop water use and nitrogen supply for simulating changes in the regional yield of rainfed lowland rice in Northeast Thailand. *Paddy Water Environment*, 6 (1), 73–82.

Haylor, G., 2001. *Poverty reduction and aquatic resources*. Support to Regional Aquatic Resources Management (STREAM). Available from: http://aquacomm.fcla.edu/2467/1/Haylor-PovertyReduction_opt.pdf [Accessed August 2010].

Hensel, P.R., McLaughlin Mitchell, S., and Sowers, T.E., 2006. Conflict management of riparian disputes. *Political Geography*, 25 (4), 383–411.

Heyd, H. and Neef, A., 2006. Public participation in water management in northern Thai highlands. *Water Policy*, 8 (5), 395–413.

Hirsch, P., *et al.*, 2006. *National interests and transboundary water governance in the Mekong*. Sydney: Australian Mekong Research Centre, University of Sydney.

Hoanh, C.T., *et al.*, 2003. *Water, climate, food and environment in the Mekong basin in Southeast Asia. Final report* [online]. Amsterdam: IVM Institute for Environmental Studies. Available from: http://www.geo.vu.nl/~ivmadapt/downloads/Mekong_FinalReport.pdf [Accessed December 2006].

Hortle, K.G., 2007. *Consumption and the yield of fish and other aquatic animals from the lower Mekong Basin*. MRC technical paper 16. Vientiane, Lao PDR: MRC.

IRRI, 2008. *IRRI world rice statistics* [online]. Los Baños, Philippines: International Rice Research Institute. Available from: www.irri.org/science/ricestat/data/may2008/WRS2008-Table03.pdf [Accessed 2 August, 2010].

IRN, 2002. *China's upper Mekong dams endanger millions downstream*. Briefing paper 3. Berkeley, CA: IRN.

Ishidaira, H., *et al.*, 2008. Estimating the evolution of vegetation cover and its hydrological impact in the Mekong River basin in the 21st century. *Hydrological Processes*, 22 (9) , 1395–1405.

Jusi, S., 2006. The Asian Development Bank and the case study of the Theun-Hinboun hydropower project in Lao PDR. *Water Policy*, 8 (5), 371–394.

Kemp-Benedict, E., Chadwick, M.T., and Krittasudthacheewa, C., 2008. The Bayesian methods for livelihood, water and poverty analysis. *2nd CPWF International Forum on Water and Food*, 9–14 November 2008, Addis Ababa, Ethiopia, Volume II, 144–147. Available from: http://www.ifwf2.org [Accessed August 2010].

Kiem, A.S., *et al.*, 2008. Future hydroclimatology of the Mekong River Basin simulated using the high-resolution Japan Meteorological Agency (JMA) AGCM. *Hydrological Processes*, 22 (9), 1382–1394.

Kirby, M., *et al.*, 2006. Basin water use accounting method with application to the Mekong Basin. *In*: S. Sethaputra and K. Promma, eds. *Proceedings of the International symposium on managing water supply for growing demand*, 16–20 October 2006, Bangkok, Thailand. International Hydrological Programme (IHP) technical documents in hydrology 6. Jakarta: IHP, United Nations Educational, Cultural and Scientific Organization (UNESCO), 67–77

Kirby, M. and Mainuddin, M., 2009. Water and agricultural productivity in the lower Mekong Basin: trends and future prospects. *Water International*, 34 (1), 134–143.

Kirby, M., Mainuddin, M., and Eastham, J., 2010. *Water-use accounts in CPWF basins: Simple water-use accounting of the Mekong Basin*. Challenge Program on Water and Food (CPWF) Working Paper: Basin Focal Project series, BFP02. Colombo: Consultative Group on International Agricultural Research (CGIAR): CPWF.

Kono, Y., *et al.*, 2001. A GIS-based crop-modelling approach to evaluating the productivity of rainfed lowland paddy in North-east Thailand. *In*: S. Fukai and J. Basnayake, eds. *Increased lowland rice production in the Mekong region*. ACIAR Procedings 101. Canberra: ACIAR, 301–318.

Kristensen, J., 2001. Food security and development in the Lower Mekong River Basin: a challenge for the Mekong River Commission. *Paper presented at the Asia and Pacific forum on poverty: reforming policies and institutions for poverty reduction*. 5–9 February 2001. Manila: Asian Development Bank. Available from: http://www.wcainfonet.org/servlet/BinaryDownloaderServlet?filename=1062597220185_food.pdf&refID=103579 [Accessed August 2010].

Lawrence, S., 2009. The Nam Theun 2 controversy and lessons for Laos. *In*: F. Molle, *et al.*, eds. *Contested waterscapes in the Mekong: hydropower, livelihoods, and governance*. London: Earthscan, 81–114.

Lebel, L., *et al.*, 2004. Nobody knows best: alternative perspectives on forest management and governance in southeast Asia. *International Environmental Agreements: Politics, Law and Economics*, 4 (2), 111–127.

Lebel, L., Mikitina, E., and Manuta, J., 2006. Flood disaster risk management in Asia: an institutional and political perspective. *Science and Culture*, 72 (1–2), 2–9.

Lebel, L. and Sinh, B.T., 2007. Politics of floods and disasters. *In*: L. Lebel, *et al.*, eds. *Democratizing water governance in the Mekong region*. Chiang Mai: Mekong Press, 37–54.

Linquist, B. and Sengxua, P., 2001. Nitrogen management for the rainfed lowland rice systems of Laos. *In*: S. Fukai and J. Basnayake, eds. *Increased lowland rice production in the Mekong region*. ACIAR Proceedings 101. Canberra: ACIAR, 179–190.

McDonald, K., Bosshard, P., and Brewer, N., 2009. Exporting dams: China's hydropower industry goes global. *Journal of Environmental Management*, 90 (supl.3), S294–S302.

Mainuddin, M., Kirby, M., and Chen, Y., 2009. *Fisheries productivity and its contribution to overall agricultural production in the lower Mekong River Basin*. Colombo: CPWF.

Mainuddin, M. and Kirby, M., 2009a. Agricultural productivity in the lower Mekong Basin: trends and future prospects for food security [online]. *Food Security*, 1 (1), 71–82. Available from: http://www.springerlink.com/openurl.asp?genre=article&id=doi:10.1007/s12571-008-0004-9 [Accessed 6 January 2009].

Mainuddin, M. and Kirby, M., 2009b. Spatial and temporal trends of water productivity in the lower Mekong Basin. *Agricultural Water Management*, 96 (11), 1567–1578.

MRC, 2003. *State of the basin report: 2003*. Phnom Penh: Mekong River Commission.

Mehtonen, K., 2008. Do the downstream countries oppose the upstream dams? *In*: M. Kummu, *et al.*, eds. *Modern myths of the Mekong: critical review of water and development concepts, principles and policies*. Helsinki: Helsinki University of Technology, 161–172.

Middleton, C. and Tola, P., 2008. Community organizations for managing water resources around Tonle Sap Lake: a myth or reality? *In*: M. Kummu *et al.*, eds. *Modern myths of the Mekong:*

critical review of water and development concepts, principles and policies. Helsinki: Helsinki University of Technology, 149–160.

Ministry of Fisheries, 2005. *Vietnam fisheries and aquaculture sector study: final report* [online]. Available from: http://siteresources.worldbank.org/INTVIETNAM/Resources/vn_fisheries-report-final.pdf [Accessed July 2010].

Minot, N. and Goletti, F., 2000. *Rice market liberalization and poverty in Vietnam*. Research report 114. Washington D.C.: International Food Policy Research Institute.

Molle, F., 2005. *Irrigation and water policies in the Mekong region: current discourses and practices*. Research report 95. Colombo: International Water Management Institute.

Molle, F., 2007. Irrigation and water policies: Trends and challenges. *In*: L. Lebel, *et al*., eds. *Democratizing water governance in the Mekong region*. Chiang Mai: Mekong Press. 9–36.

Ojendal, J., Mathur, V., and Sithirith, M., 2002. *Environmental governance in the Mekong hydropower site selection processes in the Se San and Sre Pok Basins*. SEI/REPSI Report series 4. Stockholm: Stockholm Environment Institute.

Osborne, M., 2004. *River at risk: the Mekong and the water politics of Southeast Asia*. Paper 2. Sydney: Lowy Institute for International Policy.

Pandey, S., 2001. Economics of lowland rice production in Laos: opportunities and challenges. *In*: S. Fukai and J. Basnayake, eds. *Increased lowland rice production in the Mekong region*. ACIAR Proceedings 101. Canberra: ACIAR, 20–30.

Podger, G.M., *et al*., 2004. *Modelled observations on development scenarios in the lower Mekong Basin*. Mekong Water Resources Assistance Strategy. Vientiane, Lao PDR: World Bank.

Rena, S., 2005. Saving the Tonle Sap Lake the ADB way: building a large-scale modern harbour, the "gateway" to environmental and social ruin. *In*: *The ADB and policy (mis)governance in Asia: focus on the gobal South*. Bangkok: Chulalongkorn University Social Research Institute, 45–53.

Rickman, J.F., *et al*., 2001. Direct seeding of rice in Cambodia. *In*: S. Fukai and J. Basnayake, eds. *Increased lowland rice production in the Mekong region*. ACIAR Proceedings 101. Canberra: ACIAR, 60–65.

de la Rosa, E. and Chadwick, M.T., 2008. *Wealth ranking study*. Report. Colombo: Challenge Program on Water and Food.

Sarkkula, J., *et al*., 2007. Mathematical modelling in integrated management of water resources: magical tool, mathematical tool or something in between? *In*: L. Lebel, *et al*., eds. *Democratizing water governance in the Mekong region*. Chiang Mai: Mekong Press, 127–156.

Schiller, J.M., *et al*., 2001. Constraints to rice production system in Laos. *In*: S. Fukai and J. Basnayake, eds. *Increased lowland rice production in the Mekong region*. ACIAR Proceedings 101. Canberra: ACIAR, 3–19.

Sneddon, C. and Fox, C., 2006. Rethinking transboundary waters: a critical hydropolitics of the Mekong basin. *Political Geography*, 25 (2), 181–202.

Snidvongs, A., Choowaew, S., and Chinvanno, S., 2003. *Background paper. Impact of climate change on water and wetland resources in Mekong River Basin: directions for preparedness and action* [online]. Regional Centre report 12. Bangkok: Southeast Asia START. Available from: http://www.iucn.org/themes/climate/wl/documents/regional_waterstudies/southeast-asia-final.pdf [Accessed December 2006].

Snidvongs, A. and Teng, S.-K., 2006. *Global international waters assessment [GIWA], Mekong River* [online]. GIWA regional assessment 55. Bangkok: United Nations Environment Programme. Available from: http://www.giwa.net/publications/r55.phtml [Accessed December 2006].

Sokhem, P. and Sunada, K,. 2008. Modern upstream myth: is a sharing and caring Mekong possible? *In*: M. Kummu, M. Keskinen, and O. Varis, eds. *Modern myths of the Mekong: critical review of water and development concepts, principles and policies*. Helsinki: Helsinki University of Technology, 135–148.

Stahl, K., 2006. Influence of hydroclimatology and socioeconomic conditions on water-related international relations. *Water International*, 30 (3), 270–282.

Starr, P., 2007. Catfish processing takes off in delta as global demand soars [online]. *Catch and Culture,* 13 (2), 16–17. Available from: http://www.mrcmekong.org/download/programmes/fisheries/Catch_Culture_vol13.2.pdf [Accessed July 2010].

Stockholm Environment Institute, 2008a. *Household level investigation on water poverty and livelihoods: 1. Tonle Sap case study*. Colombo: CPWF.

Stockholm Environment Institute, 2008b. *Household level investigation on water poverty and livelihoods: 2. Northeast Thailand (Si Sa Ket) case study*. Colombo: CPWF.

Stockholm Environment Institute, 2008c. *Household level investigation on water poverty and livelihoods: 3. Mekong Delta case study*. Colombo: CPWF.

Toung, T.P. and Bouman, B.A.M., 2003. Rice production in water scarce environments. *In*: J.W. Kijne, *et al*., eds. *Water productivity in agriculture: limits and opportunities for improvement*. Wallingford: CABI Publishing, 53–67.

Tuan, V.A. and Mai, B.T.Q., 2005. Trash fish use as a food source for major cultured species in An Giang and Dong Thap provinces, Mekong Delta. *In*: *Vietnam Regional workshop on low value and "trash fish" in the Asia Pacific region*, 7–9 June 2005, Hanoi, Viet Nam. Bangkok: Asia Pacific Fishery Commission, 211–222.

UN Population Division, 2006. *World population prospects: the 2006 revision population database*. Available from: esa.un.org/unpp/index.asp?panel=1 [Accessed 12 December 2008].

Van Brakel, M.L., Muir, J.F., and Ross, L.G., (undated). *Modelling of aquaculture-related development, poverty and needs in the Mekong Basin*. Stirling: Institute of Aquaculture, University of Stirling.

Wolf, A.T., Yoffe, S.B., and Giordano M., 2003. International waters: identifying basins at risk. *Water Policy*, 5 (1), 29–60.

World Bank, 2006. *Bangladesh country environmental analysis* [online]. Bangladesh development series paper 12. Dhaka, Bangladesh: World Bank. Available from: www.worldbank.org.bd/bds [Accessed 3 August 2010].

Yoffe, S.B., Wolf, A.T., and Giordano, M., 2003. Conflict and cooperation over international freshwater resources: indicators of basins at risk. *Journal of the American Water Resources Association*, 39 (5), 1109–1126.

van Zalinge, N., *et al*., 2003. The Mekong River system. *In*: R.L. Welcomme and T. Petr, eds. *Proceedings of the second international symposium on the management of large rivers for fisheries: sustaining livelihoods and biodiversity in the new millennium*. 11–14 February 2003, Phnom Penh, Cambodia, Volume 1, 333–355. Available from: http://www.fao.org/docrep/007/ad525e/ad525e00.htm [Accessed August 2010].

Zwart, S.J. and Bastiaanssen, W.G.M., 2004. Review of measured crop water productivity values for irrigated wheat, rice, cotton, and maize. *Agricultural Water Management*, 69 (2), 115–133.

Water, agriculture and poverty in the Niger River basin

Andrew Ogilvie, Gil Mahé, John Ward, Georges Serpantié, Jacques Lemoalle, Pierre Morand, Bruno Barbier, Amadou Tamsir Diop, Armelle Caron, Regassa Namarra, David Kaczan, Anna Lukasiewicz, Jean-Emmanuel Paturel, Gaston Liénou and Jean Charles Clanet

Livelihoods in the Niger River basin rely mainly on rainfed agriculture, except in the dry extreme north. Low yields and water productivity result from low inputs, short growing seasons, dry spells, and excessive water. The overlap of traditional and modern rules impedes secure access to water and investments in agriculture by generating uncertain land tenure. Improved agriculture and water management require technical, sociological, and regulatory changes to address the wider causes of poverty. Illiteracy and poor water quality, both correlated with high infant mortality, are pressing problems. Rapidly increasing population, climatic changes and dam construction contribute to rural vulnerability.

Introduction

The Niger River basin is the largest basin in West Africa. It covers six agro-climatic zones in nine countries (Figure 1) and presents a cross-section of the complex development issues of West African societies. The main livelihood is traditional, low input, rainfed farming, but ranges from nomadic pastoralism in the north, irrigated and fertilized agriculture in parts of Mali, Niger and Nigeria and fishing throughout. The Inner Delta in Mali is one of the largest wetlands and multi-use systems in Africa at three million hectares with over one million herders, fishermen and farmers. Much of the population in the basin suffers from extreme, chronic poverty and remains vulnerable to droughts and malnutrition. The rich ethno-linguistic diversity, recent independence and ongoing political insecurity further complicate the development of the basin.

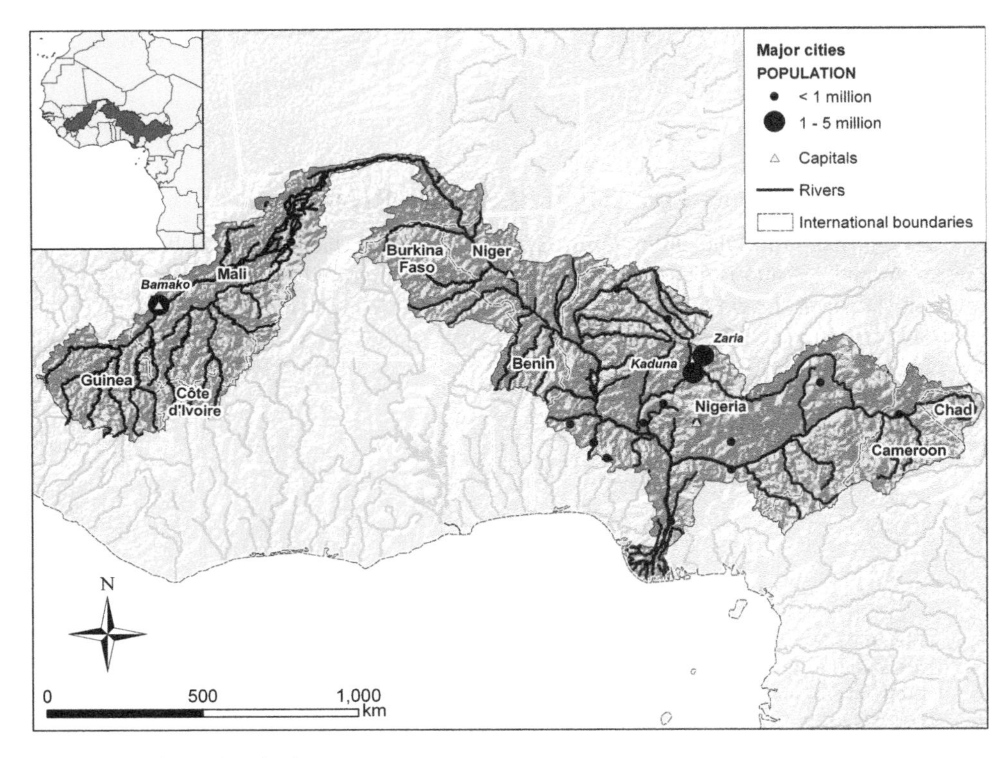

Figure 1. Niger River basin.

This paper reports the links between water, food and poverty in the Niger Basin. We diagnose the hydrologic and agronomic systems and identify how better agricultural water management may reduce vulnerability in the basin. We examine statistical relations between water and poverty, and analyse major future threats and opportunities, as well as the critical influence of institutions on water and agricultural development. Faced with increasing food and water insecurity as a result of climatic, demographic and land use changes, we discuss issues of water productivity in the basin.

Characteristics of the Niger River basin

Geography and hydrology

The Niger River is 4200 km long, the third longest in Africa after the Nile and Congo/Zaire. It rises in the mountains of Guinea and Sierra Leone (50 km^2 are in Sierra Leone) before flowing northeast towards the Sahara, and the vast flood plain of the Inner Delta in Mali. After the Inner Delta it flows southeast, is joined by the River Benue and finally reaches the Atlantic Ocean through the Niger Delta in Nigeria.

Spread across ten West African countries, the basin is 2,170,500 km^2 in area, the ninth largest in the world (Showers 1973). The northern section of the basin, extending across the Sahara desert into Algeria, is hydrologically inactive. The *active* basin, which we call hereafter the basin, covers 1,272,000 km^2 (Table 1) and nine countries: Benin, Burkina Faso, Cameroon, Chad, Côte d'Ivoire, Guinea, Mali, Nigeria and Niger. They are all members of the Niger Basin Authority (NBA).

Table 1. Countries of the Niger Basin.

Country	Area	Proportion of basin within country	Proportion of country within basin
	km²	%	%
Benin	44,967	3.5	38.7
Burkina Faso	86,919	6.8	31.5
Cameroon	86,381	6.8	18.4
Côte d'Ivoire	23,550	1.9	7.3
Guinea	98,095	7.7	39.9
Mali	263,168	20.7	20.9
Niger	87,846	6.9	7.4
Nigeria	562,372	44.2	61.5
Chad	19,516	1.5	1.5
TOTAL Active basin	1,272,814	100	

Source: Marquette (2008).

Demography

Population of the basin was estimated at 94 million in 2005 (CIESIN/CIAT 2005), of which 71% live in Nigeria. Due to a high fertility rate, populations of most countries in the basin increased by 50% between 1990 and 2005 (Tabutin and Schoumaker 2004) and the growth rate of the population is currently estimated at 3.2% (Bana and Conde 2008, Guengant 2009). Demographers estimate, according to the lowest scenario, that the population of the basin will double by 2050, but, if the fertility rates remain constant, the population could increase fourfold by 2050 (Figure 2). This could jeopardize current and future development.

Population density in the basin is four to five times greater than the national averages, as people concentrate along their lifeline, the Niger River. The population is 64% rural.

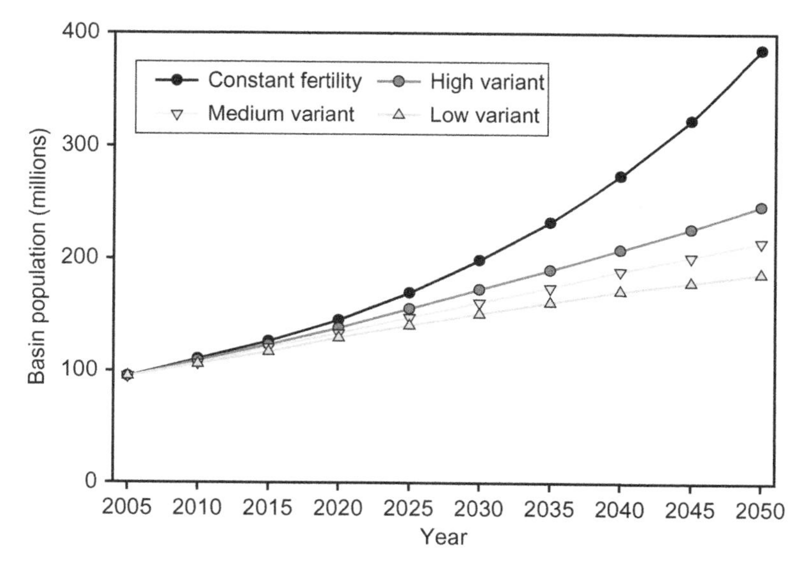

Figure 2. Evolution of Niger Basin population 2005–2050.
Source: based on United Nations Population Division (2006).

However, this is changing rapidly and by 2025 the majority may be urban. Urbanization is fuelled by a massive rural exodus, as well as a century-old migration from the inland to coastal areas and is sustained by recent political and climatic crises. The population is young (44% are under 15 years of age) and largely illiterate (with 35% overall literacy rates and only 18% for women).

Economy

When ranked by gross domestic product (GDP) (purchasing power parity, per capita), all nine countries of the Niger Basin fall in the bottom quarter of national incomes. Agriculture represents a large part of the Niger Basin GDP, with crops making up 25–35%, livestock 10–15%, and fishery 1–4%. The main livelihood/agricultural systems in the basin include dry- and wet-season cropping, pastoral systems, crop-livestock systems and fishing. The major crops are yams, cassava, rice, groundnuts, millet, sorghum, plantains, cocoa beans, maize, sugarcane and cotton.

The basin countries have important mineral resources, including gold, bauxite and uranium. Nigeria is the region's largest oil and gas producer, with 3% (36 billion bbl) of the world reserves, mostly in the Niger Delta (CEDEAO 2007). Installed hydroelectric capacity is 6185 GWh, less than 21% of the basin's potential. As in many parts of Africa, the Niger Basin suffers from a huge deficit in transport infrastructure, which undermines economic growth and regional integration.

Poverty

The United Nations Human Development Index, a composite ranking based on national income, life expectancy and adult literacy rate, ranks all of the Niger Basin countries in the lowest quintile of countries (Table 2) (UNDP 2007). Life expectancies (on average 50 years) are in the bottom 15% of all countries worldwide (Aboubakar 2003, Bana and Conde 2008).

Niger Basin childhood mortality rates (death prior to the age of five) of up to 250 per 1000 live births are two to three times higher than those in neighbouring countries in northern and southern Africa (Balk *et al.* 2003, Guengant 2009). After respiratory diseases, water-related diseases, namely malaria and diarrhoeal diseases are the largest causes of child mortality (UNICEF 2008, OMS 2006, ECOWAS-SWAC/OECD 2008). HIV infection rates are 1.1–7.1%, significant but less than in southern Africa.

The proportion of people living below the poverty line (US$1.25 per day) is high throughout the basin and is especially acute in Burkina Faso (70.3%), Guinea (70.1%) and Niger (65.9%) (World Bank 2009). There are an estimated 138 million poor in the basin countries, most of whom are rural. Table 2 provides a snapshot of the development status for these countries according to an array of commonly applied poverty metrics.

Sociology and institutions

The ethno-linguistic diversity in the basin is one of the richest in the world with over 400 vernacular languages and five official languages. Though half of these could disappear by 2050, the sheer number of them restricts the circulation and dissemination of information and innovations.

Traditional customs, influenced by animist culture, continue to define local activities and practices (Clanet 1994). Partly from the inability of central government administrations to implement their directives, village and land chiefs maintain considerable influence and

Table 2. Poverty and water situation indicators for countries of the Niger Basin.

Country	GDP[1]	Population below poverty line[2]	Life expectancy at birth	Under-five mortality rate	TARWR[3]	WPI[4]	Basic human needs index[5]	SVI[6]	HDI[7]	HDI world rank	Gini Co-efficient[8]
	$US 2007	*%* 2007	*years* 2007	*%* 2007	*m3/yr/capita* 2005	2002	*L/day* 2000	*index* 2004	*index* 2007–8		2007
Benin	1500	47.3	55.4	19.1	3820	39.3	15	0.584	0.437	163	36.5
Burkina Faso	1200	70.3	51.4	15.0	930	41.5	17	0.658	0.370	176	39.5
Cameroon	2300	32.8	49.8	14.9	17,520	53.6	33	0.640	0.532	144	44.6
Chad	1600	61.9	50.0	20.8	4860	38.5	11	0.618	0.388	170	–
Côte d'Ivoire	1800	23.3	47.4	19.5	4790	45.7	28	0.584	0.432	166	44.6
Guinea	1000	70.1	54.8	15.0	26,220	51.7	26	0.562	0.456	160	38.1
Mali	1200	51.4	53.1	21.8	7460	40.6	6	0.585	0.380	173	40.1
Niger	700	65.9	55.8	25.6	2710	35.2	20	0.725	0.374	174	50.5
Nigeria	2200	64.4	46.4	19.4	2250	43.9	24	0.621	0.470	158	43.7
OECD mean[9]	37,500	n/a	78.3	0.52	39,090	39.7	n/a	n/a	0.939		10.4
Non OECD	10,900	n/a	66.1	6.76	26,800	23.6	n/a	n/a	0.686		18.6

Note: [1] Gross domestic product at purchasing power parity, per capita.
[2] US $1.25/day.
[3] TARWR = Total actual renewable water resources.
[4] WPI = Water poverty index (100 = lowest poverty; 0 = highest poverty).
[5] 50 L is the commonly accepted minimum; Gleick (1996).
[6] SVI = Social vulnerability index (higher index is more vulnerable).
[7] HDI = Human development index (higher index is more developed).
[8] Gini coefficient: 100 = complete inequality, 0 = complete equality.
[9] Organisation for Economic Co-operation and Development (OECD) mean based on the 27 high-income countries as defined by the World Bank (2009).
Source: Ward *et al.* (2009).

power (Jacob 2005). Internal political tensions and peripheral rebellions also undermine central governments and their development efforts.

Water availability and access

Rainfall and agro-climatic zones

Rainfall depends on the Atlantic monsoon between May and November and gives a wet season and dry season. In the most southern part of the basin in Nigeria, the wet season is subject to a period of reduced rainfall: the second, short dry season. Climatic zones vary from hyper-arid to sub-equatorial and annual rainfall fluctuates from over 4000 mm in southern Nigeria/Cameroon to less than 400 mm (with no rain in some years) on the fringes of the Sahara desert in northern Mali and Niger (Figure 3).

In the Sahelian climate of the far north, even short-season crops cannot be grown reliably, but in the rest of the basin, rainfall is broadly sufficient for rainfed agriculture. Rainfall is spatially and temporally variable, however, causing water excess and droughts, which are more problematic for agriculture than low annual rainfall (Mahoo *et al.* 1999 cited by Rockstrom *et al.* 2002). In the north of the basin, short (two-to-four-month-long) wet seasons restrict the growing season, while in the south, both dry spells during the wet season and excess rainfall can cause crop failures.

During the 1970s and 1980s isohyets over the whole basin shifted south by about 150 km with devastating droughts across West Africa (Conway *et al.* 2009) and the Sahel in

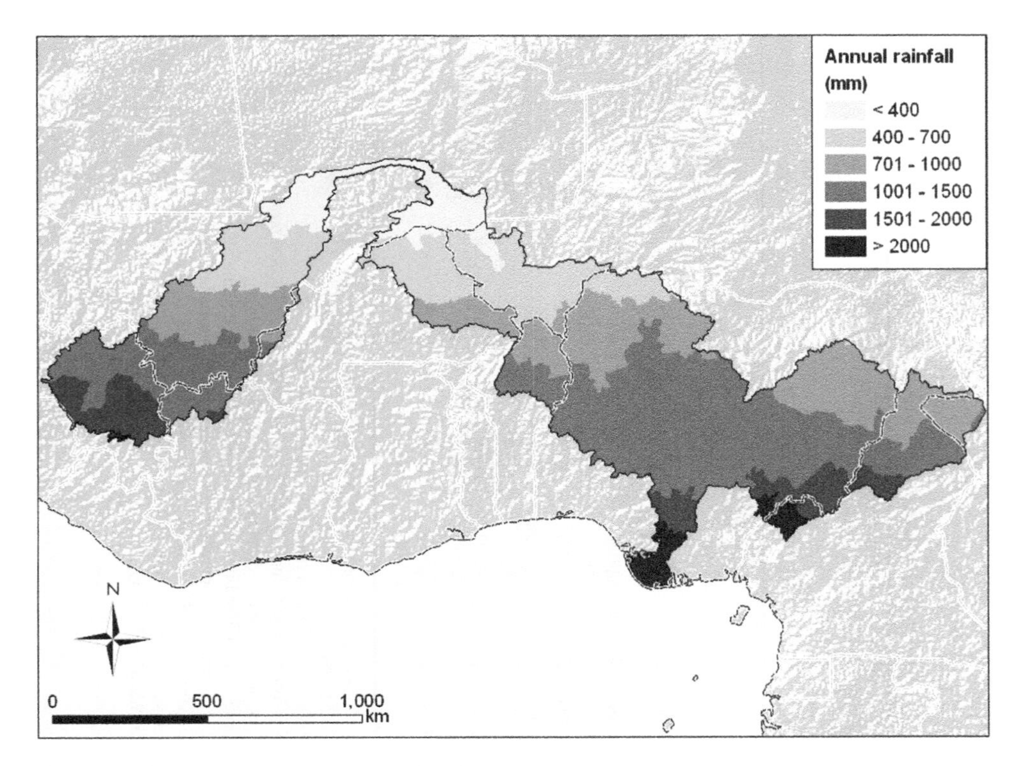

Figure 3. Niger River basin: annual rainfall.
Source: Mahé *et al.* (2009a).

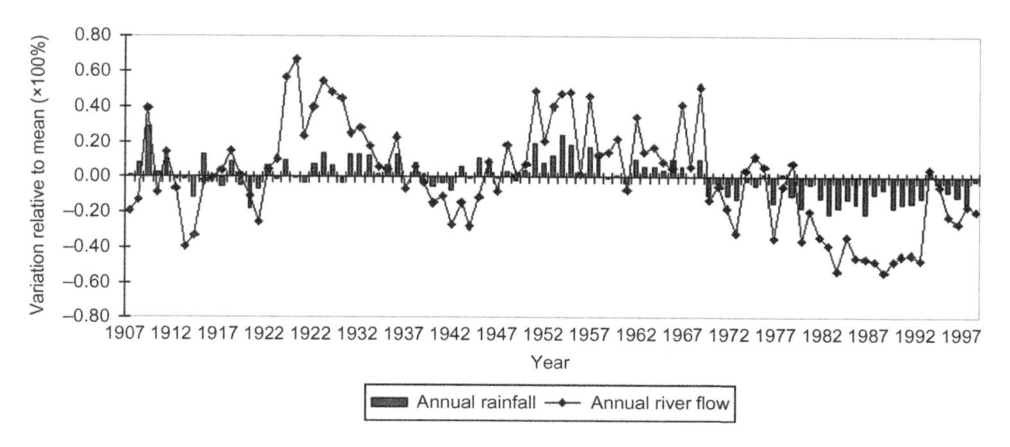

Figure 4. Rainfall-runoff in the Soudano-guinean part of the River Niger in Mali and Guinea. Source: Mahé *et al.* (2009a).

particular. Rainfall has increased since 1994, but remains erratic with more periods of severe drought (Mahé and Paturel 2009), which continue to reduce water availability in the basin. Rainfall–runoff graphs over the last century in the upper Niger Basin reflect the major droughts (Figure 4).

Evaporation

Potential evapotranspiration (ET_p) is high across the basin, especially in the north, due to advection from the Sahara. This dries up areas of inland drainage and causes high losses in water reservoirs. When rainfall (P) exceeds ET_p, the surplus infiltrates the soils, recharges groundwater and, excluding supplementary irrigation, defines the cropping season. In the north of the basin, this only occurs during the months of July and August. When rainfall is less than ET_p, November–April across the basin, plants exhaust soil water reserves and there is neither runoff nor infiltration.

Flows

The annual mean flow into the Inner Delta from Guinea and Mali is 46 km^3, but the mean outflow at Taoussa is only 33 km^3. The delta, whose area can reach 30,000 km^2 during the flood, changes the hydrology of the Niger by delaying the flood by two to three months, and reducing the flow 24–48% in dry or wet years respectively (Mahé *et al.* 2009b). The middle section of the Niger receives six tributaries from Benin and Burkina Faso and the mean annual flow entering the lower Niger in Nigeria is 36 km^3. With the contribution of its main tributary, the Benue River in Nigeria, and heavy rainfall, the mean annual discharge at the mouth exceeds 180 km^3.

The Niger River is highly seasonal with fairly high interannual variation. Runoff in all the sub-basins has lessened substantially since 1970, reflecting the decreased rainfall (Figure 4). Runoff proportionally fell most in the upper basin (upstream of the Inner Delta), caused directly by reduced contribution from rainfall and indirectly by less groundwater recharge and related baseflow. In the lower basin, including the Benue River basin, runoff was less affected because the rainfall decreased less. In contrast, in the Sahelian parts of the basin, runoff coefficients increased, partly due to reduced rainfall and soil compaction effects but mainly to increased agriculture and reduced cover by natural vegetation, leading to higher flood peaks, erosion, sediment transport and dam silting (Mahe *et al.* 2005). These variations in climate and river regimes have important implications for designing the projected dams in the basin as well as on the water available for agriculture.

Groundwater resources

Groundwater can provide a valuable water supply for dry season agriculture and quality drinking water. Studies have shown that under the Soudano-Guinean climate (Guinea, Mali, Côte d'Ivoire and Cameroon), baseflow from groundwater makes an important contribution.

There are large aquifers in sedimentary strata in Mali, Niger, Chad, Nigeria and Cameroon, and discontinuous aquifers in the Guineo-Sudanese and the Sudano-sahelian zones of Guinea, Mali, Côte d'Ivoire, Burkina Faso and Niger. Groundwater recharge is variable and depends on the geology, topology, climate and crop cover. Recharge rates vary from 20 mm/year in the Sokoto region of Nigeria (Adelana *et al.* 2006) to 136 mm/year in Katchari in northern Burkina Faso (Filippi *et al.* 1990). Around Niamey, Niger groundwater recharge increased fivefold to 25 mm/year when natural savannah was replaced by millet crops (Leduc *et al.* 2001). Extrapolation to other parts of the basin is uncertain and resources remain largely unknown.

Withdrawals are poorly quantified, but are estimated to be less than 5 km^3/year in Mali and Niger (WWAP 2009). There are proposals to develop groundwater and intensification may be worthwhile, but the impact must be carefully monitored to ensure that the amount pumped is sustainable.

Water balance

At the scale of the whole basin, evapotranspiration is the main component of the water balance. The mean rainfall of 690 mm over the total basin gives a gross input of 1500 km^3/yr. Flow at the confluence of the Benue and Niger rivers is 183 km^3, and discharge at the mouth is about 200 km^3 (blue water), which indicates high availability of water in the lower basin. Consumptive withdrawals for humans, livestock and industry are low and groundwater storage may be assumed constant. The difference, 1200–1300 km^3 or 80% of the basin water resources, is therefore evapotranspired. Mainuddin *et al.* (2010) indicate that grassland is the dominant water use compared with woodland and rainfed agriculture, but this is subject to classification difficulties. Water evapotranspired from irrigation is extremely low and highlights the importance of exploiting green water effectively.

Crops, livestock and fisheries

Rainfed and irrigated agriculture

There are over 2.5 million ha of arable land in the Niger Basin, of which only 20% are exploited. Although the Niger Basin possesses one of the largest wetlands, and has 27 large dams (ABN and BRLi 2007) and over 5000 small dams (Cecchi 2009), irrigation remains poorly developed and 85% of the cultivated area is rainfed. Agriculture has therefore adapted to rainfall and cropping zones roughly follow the isohyets. In the extreme north, rainfall is just sufficient for seasonal pasture. As one moves further south there is millet and sorghum, then banana, plantain, cassava, yam and finally rice in the south as well as in irrigated areas in Inner Delta in Mali, and in Niger and Nigeria.

Subsistence agriculture represents 78% of total agricultural production (Niasse 2006) and dominates all forms of rural activities. It remains an itinerant agriculture with extensive characteristics: low mechanization, lack of inputs (except Nigeria, which possesses fertilizer factories, thanks to its petroleum [Serpantié 2009a]), and with much of the labour provided by women and children. This agriculture is currently the only option available to farmers facing climatic uncertainties, inadequate support and weak commercialization possibilities (Serpantié and Lamachère 1989). On the positive side, agriculture has met the increased demand in food, as daily per capita production has been stable for 25 years at 2000 kcal. The projected demographic increase, however, is likely to make it difficult to reach a desired level of 2500 kcal/capita/day (CEDEAO 2007).

The Niger Basin Authority reports that 265,000 ha are under full control irrigation, of which 117,000 ha are in Mali, 46,000 in Niger and 84,000 ha in Nigeria. These countries have invested in large perimeters and full control irrigation over the decades, but traditional systems such as recession flooding, lowland and free flooding still dominate in terms of surface area. Farmers and donors now increasingly attempt to control water supply better, notably due to recent droughts. While the NBA favours large-scale developments to reach one and a half million hectares by 2025 according to its investment plan, many donors currently favour small-scale irrigation, preferably privately funded and owned. A number of small dams already exist in Burkina Faso, Mali, Côte d'Ivoire and are being actively developed as part of public, private and non-governmental organization (NGO) projects.

There is a vast land potential for irrigation. However, surface waters could not supply the amount of total irrigable land. Current agricultural withdrawals in the dry season already have an impact on the Inner Delta wetlands of Mali (De Noray 2003, Zwarts *et al.* 2005), on the Niger Delta in Nigeria (NDES 1999, Uyigue and Agho 2007) and on production of hydroelectricity of the Kainji Dam in Nigeria. Faced with food crises and climatic changes, donors are funding the expansion of irrigation, mostly in Nigeria and Mali. This could push agricultural withdrawals from 9 km^3 to nearly 30 km^3 by 2025, more in Mali than in Nigeria as the latter consumes less water per unit area. New dams are being built, notably to extend dry season irrigation in the Office du Niger in Mali. However, new dams to support low flows risk further reducing flows in the Inner Delta, lessening the extent of the flood, affecting livelihoods of a million herders, fishers and traditional rice growers as well as the wetland ecosystems (notably hippopotamus and manatees in the both the river and Inner Delta). As a result, irrigated agriculture will need to improve its water efficiency and the economic gain derived from the water used to justify the heavy investments it demands.

Livestock

Scattered over more than 1.5 Mkm^2 distributed over nearly 13° latitude, the 50 M herders of the basin maintain 138 M livestock units (Diop *et al.* 2009a). The north–south distribution of species is a function of their resistance to drought and their ability to exploit natural rangelands. Camels constitute the dominant form of breeding in the north of the basin and may be found to 13°N. At lower latitudes, zebu (*Bos indicus*) cattle follow to 8°N, the northern limit of trypanosomiasis. Below this latitude, only *Bos taurus* cattle and smaller ruminants tolerant of the endemic survive. There are small ruminants, mostly goat and sheep, across the basin.

There are two major livestock production systems in the basin: nomadic herders, who live on Sahelo-Saharan fringes and who have large herds of zebu cattle, and sedentary

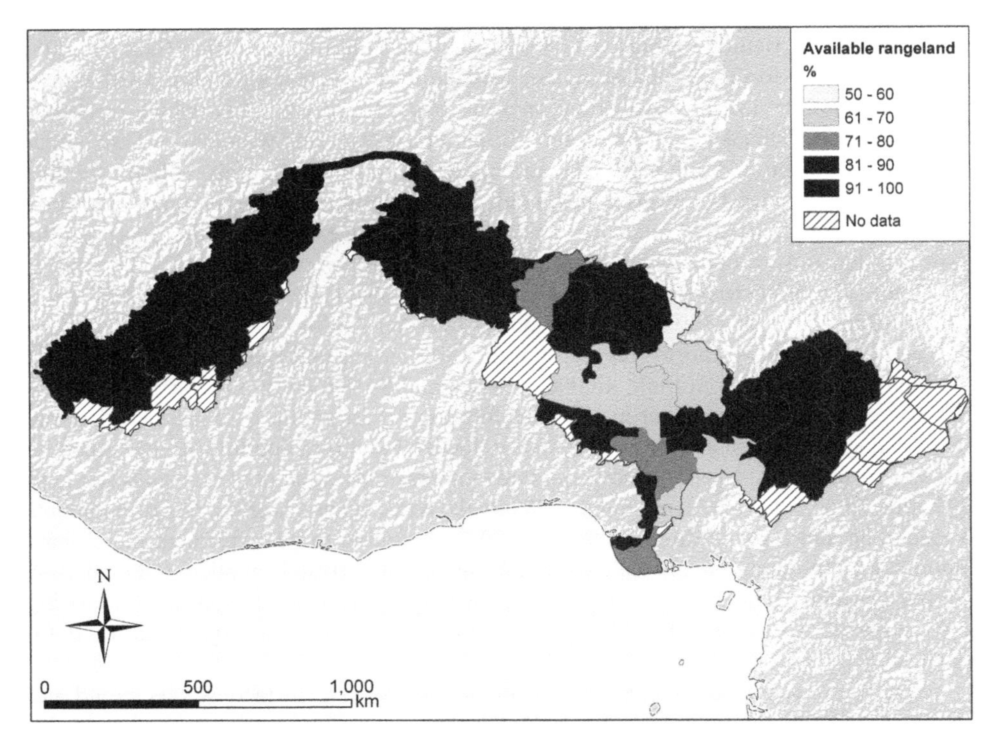

Figure 5. Amount of rangeland (total area-cultivated area) available in the Niger Basin.
Source: Serpantié (2009b).

agro-pastoralists, who typically own fewer than 10 small ruminants and some larger bovines. Nomadic herds make up 5–15% of livestock units and can travel over 800 km in their annual transhumance (Diop *et al.* 2009a). Livestock is an important livelihood strategy in the basin, and provides food, income and cash reserves. Livestock numbers have increased in all countries over the last 30 years and are expected to continue to rise, responding to the projected population increase and the increased demand for meat and dairy products in local diets (Diop *et al.* 2009b).

Livestock production remains predominantly extensive and can be severely impacted by the development of agriculture, which restricts the nomads' access to grazing land and water points. Despite possible synergies between farmers and herders, whose animals graze crop residues and return manure/fertilizer, conflicts are increasing. Contrary to what is often reported, natural rangeland is widely available throughout the basin (Figure 5). The upper basin, the Sahelian zones, the Inner Delta and the less populated eastern part of Burkina Faso have the most available rangeland and are therefore most capable of supporting increased livestock numbers. Even in central Nigeria, where rangeland is scarcer largely due to higher population density, 50–70% of the land is not cropped and therefore potentially available for grazing. Resolution of the conflict requires the correct implementation of institutional support for nomad pastoralists.

Fisheries

Fishing is mainly concentrated around the large floodplains of the Inner Delta or in the Sélingué, Kainji, Jebba and Lagdo reservoirs. The fishers in the basin may be divided between full-time fishers, mostly belonging to ethnic groups recognized as fishers, agro-fishers who spend a part of the year growing crops, and occasional or part-time fishers for whom fishing is a secondary activity. When they have access to land, the "full-time" fishers also practise rainfed or flood recession cultivation. Of 100,000 professional fishers in the basin, supporting roughly 900,000 people, 62,500 are in the Inner Delta and 13,000 in the large reservoirs (Morand *et al.* 2009). Total fish catch in the basin (not including the estuarine delta) is about 240,000 t/yr (Neiland and Béné 2008), with a value of almost US$100 million. Fish are an important source of animal protein in Africa, ranging from 40% in Nigeria to 49% in Cameroon (Food and Agriculture Organization [FAO] 2009). Total demand for marine and inland fish is estimated at 730,000 tonnes, assuming an average fish consumption of 7.7 kg/capita/year (the value for Africa).

According to recent surveys on the Inner Delta and the lakes of Kainji, Lagdo and Sélingué, lack of nets and canoes, the variability of the hydrology and the poor strength of their fishers' associations are secondary constraints to their livelihoods. They rank first food shortage, and second the lack of access to health care, to good quality household water, to school, or to credit. Lack of access to land occurs specifically in some regions where the fishers are considered to be new settlers. Poor enforcement of both traditional and modern (legal) fisheries regulations by communities or the state make both fishers and fish stocks vulnerable. Difficult access to markets in some regions of the basin and the competition from marine fish trade lead to low prices. The projected demographic increase as well as the construction of dams and water withdrawals will exert increasing stress on fishers, notably through increased competition for space, conflicts with the other activities (herders, farmers), and degraded environments. The effect of climate change on all rainfall regimes (IPCC 2007) may further increase the vulnerability of fishers.

Fish culture in ponds, around irrigated perimeters, and in cages in reservoirs can constitute a valuable solution to perturbed fisheries. By gathering the fishing communities in a reduced number of sites, it may also allow for better access to markets and increased recognition. Shifting from hydrologically highly variable natural systems to regulated systems can also remove uncertainty and reduce risk. The communities presently involved in fishing are poorly prepared to manage this new activity, however. Furthermore, migration may result in loss of rights (especially for access to land). Nigeria is one of the few African countries, with Zimbabwe, Ghana and Egypt, where some fish culture has developed, mostly as small- and medium-scale enterprises. Two thousand fish farms, covering 60,000 ha, produce 80,000 tons of fish per year, and the number is increasing rapidly (Brummett *et al.* 2008, FAO 2009).

National or pro-poor policies have not, up to now, taken into account the fisheries sector, partly because its importance has not properly been evaluated. New policies (NEPAD 2006), along with the creation of new infrastructure, such as ice plants, and fishing harbours around new water bodies, may give better access to markets, better prices and better living conditions for the families.

Agricultural water productivity

Rainfed water productivity

We calculate water productivity (WP) in rainfed agriculture using dry yield cereal production as the numerator and total rainfall or "evapotranspirable water" (ETW) (Serpantié 2009b) as the denominator. Total rainfall provides a measure of water supplied to rainfed agriculture on a given area, while ETW corresponds to the fraction of rainfall actually

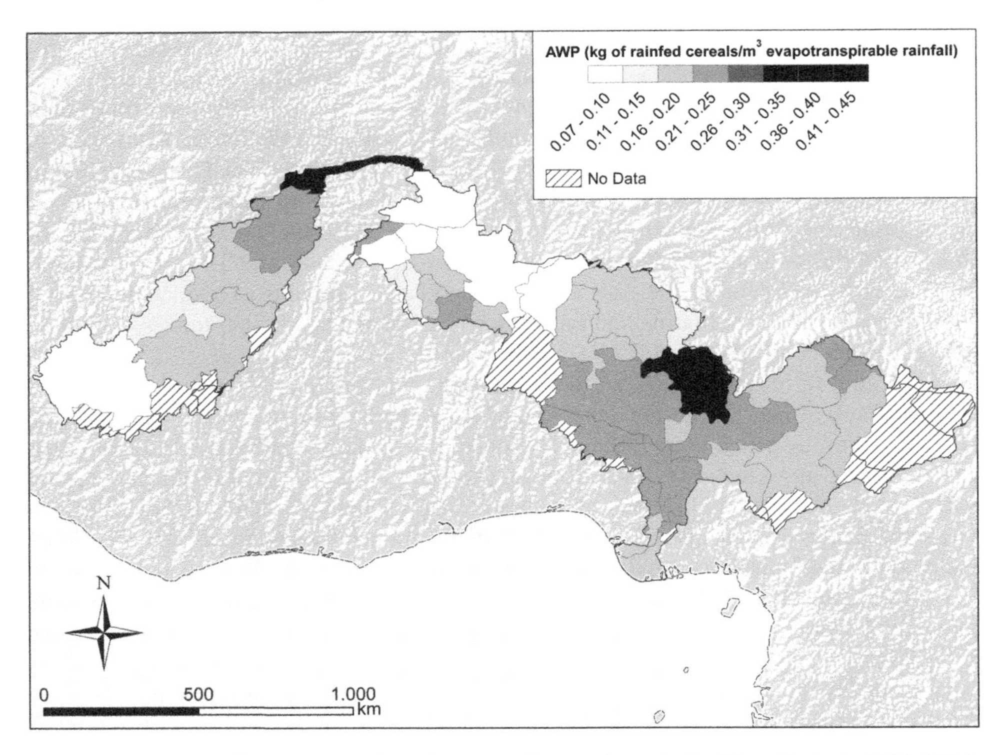

Figure 6. Water productivity (WP) based on ETW for provinces in the Niger Basin in 1999. WP is the ratio of annual average provincial yield for rainfed cereals in kg/ha to the average annual ETW in m^3/ha.
Source: Serpantié (2009b).

available to the plant and excludes rain that falls when the plant cannot exploit it or in excess of the demand (drainage). ETW therefore better reflects agricultural performance in strictly rainfed conditions and results (Figure 6) appear closer to the yield maps (Figure 7), except in very dry or very humid climates. ETW also allows a better comparison of water productivity in different climates, as only water directly useful to rainfed agriculture is taken into account. It does not, however, take into account the additional rainfall that could potentially be harnessed through rainwater harvesting.

Using both methods, rainfed water productivity is around 0.1 kg/m^3, around 10 times lower than in temperate agriculture. Kaduna state in Nigeria has the highest WP because of its fertilized and more intensive agriculture, while north of the 800 mm isohyet, the low WP of Niger is due to its low yields. In southwest Guinea, where the dominant crop is irrigated lowland rice, low WP is due to the low-yielding fonio (*Digitaria exilis*).

There are four main strategies to improve WP of rainfed agriculture: increase the area of rainfed production; increase the productivity of rainfed agriculture independently of water by the use of fertilizer and crops with longer growth cycles; increase water-use efficiency where water is scarce (essentially north of the 700 mm isohyet with millet and sorghum); and increase plant tolerance both to drought and to excess water. Solutions such as rainwater harvesting may extend the cropping season and reduce exposure to drought and to dry spells, which affect rainfed agriculture more often and more widely than low total rainfall. By reducing the risk of crop failure, supplementary irrigation could also encourage

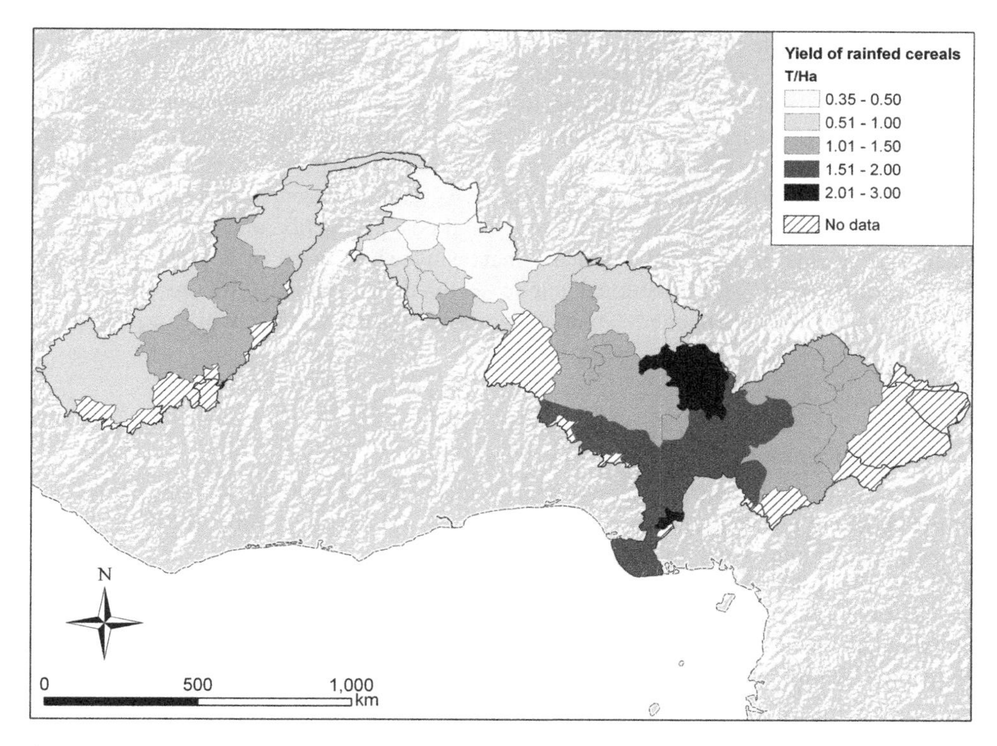

Figure 7. Average annual yield of rainfed cereals for provinces in the Niger Basin in 1999. Source: Serpantié (2009b).

farmers to increase fertilizer use (Rockstrom *et al.* 2002), which is currently marginal. Adopting supplementary irrigation, however, requires investment and capacity building, and may only be feasible where storing and distributing water can be done at costs coherent with the farmers' income and the production's value (i.e. high value or high productivity crop).

Water productivity in irrigation

We calculated WP for irrigated crops by using sample yields and dividing by estimates of water applied (Barbier *et al.* 2009). WP for irrigated rice is relatively low, 0.14–0.67 kg/m^3 similar to FAO's (2008) 0.05–0.6 kg/m^3. Values in the dry season can be as low as 0.06 kg/m^3 especially when counting distribution losses.

Excessive withdrawal, wasted water and low yields all contribute to low WP of rice. Improvements are possible but countries do not perceive the need to increase water productivity. The recent wetter period, and construction of dams to provide dry-season flows further reduces the perception. Current priorities instead are better satisfaction of crop water needs to obtain maximum yields, guarantee food security and create jobs.

Plans to extend the Office du Niger irrigated area to several hundred thousand hectares by 2025 may reduce water wastage as the higher density of plots will decrease distribution losses. Dams built to enable dry season production will increase water consumption due to higher evapotranspiration and may undermine WP increases. Market gardens in the dry season, which produce high-value crops with less water wastage, have much higher WP than rice. Small-scale private irrigation in which users pay for the fuel to pump water is more water efficient, but its expansion also implies more dry season withdrawals. There is a strong case to develop small-scale irrigation in a sustainable and equitable way.

Whatever type of irrigation, substantial increases in yield and water productivity are possible without excessive investments, as farmers gain experience, increase fertilizer input, and the industry becomes more organized. Agricultural groups must be supported to improve provision of inputs and seeds, storage and post production and reduce wastage of water. Rice yields in Mali and Niger are increasing while those in Nigeria are falling. Nigeria suffers from a dysfunctional public sector due to which full-control irrigation perimeters have been abandoned. Nigeria now prioritizes the expansion of rice production on to lowlands, which yield less but are profitable.

Livestock water productivity

It takes a large dataset to calculate livestock WP (Diop *et al.* 2009c). We have to account for the variety of products and services that livestock provide (meat, milk, fertilizer, animal traction, leather and skins), to estimate the water consumed by the animals and evapotranspired by the fodder and crop residues. We could not obtain homogeneous data at same spatial scales for the basin, so we calculated WP using kg of livestock for 1999 divided by a theoretical average water consumption for animal feed using rain use efficiency (RUE) (Serpantié 2009b).

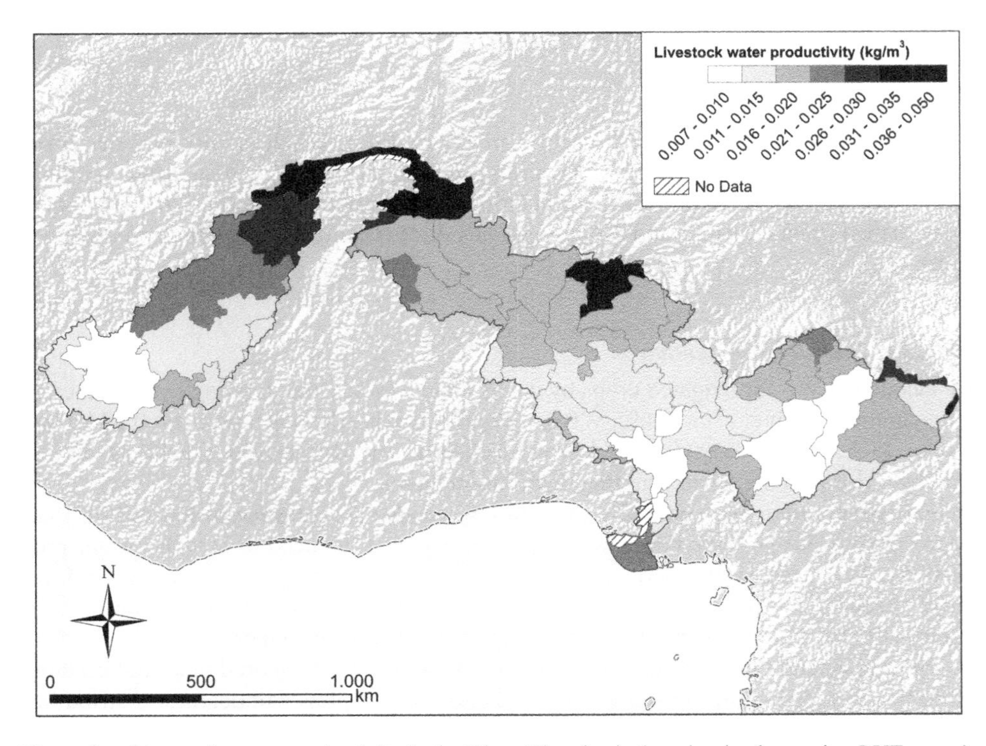

Figure 8. Livestock water productivity in the Niger River basin (maximal values using RUE range). Source: Serpantié (2009b).

LWP of livestock (0.002 to 0.05 kg per m^3 water) is low compared to WP of rainfed and irrigated crops, which is in accord with the place of herbivores in the trophic chain. The best water productivities are in the Sahel and the lowest in the zones above 1200 mm rainfall (Figure 8). WP could also be calculated using energy or dollar value to reflect the superior value of meat over many crops.

Options to improve productivity and water productivity include better animal health and vaccination coverage, selective breeding, commercialization of crop residues, fodder and animal byproducts and increased planting and selection of shrubs and trees for use as fodder. Institutionally, legislation is required to support nomadic herders, first by designating transhumance routes and water access (Clanet and Ogilvie 2009).

Water productivity in fisheries

Traditional fishing does not consume water, so that we use a marginal concept in which water management, such as water level in a dam, can be related to a change in the fish catch. Fisheries are vulnerable to river flows, which are affected by changing rainfall but also by dams and their management. In the Inner Delta, inflow controls the flood level, which in turn is related to the fish catch. A decrease of the inflow to the delta of 1 m^3/s, equivalent to 13 Mm^3 in the July–September flood discharge, decreases the fish catch in the next year by about 28 tonnes (Laë 1992, Morand *et al.* 2009). This gives a marginal WP for the Delta fishery of 0.002 kg/m^3, at the lower end of the range for livestock.

Interpreting water productivity

WP implies that water quantitatively limits production everywhere, which it clearly does not. Where rainfall is sufficient, or more (broadly in the basin >800mm for maize, >700mm for sorghum and millet (Serpantié 2009c)), water ceases to be a productive factor and may even be detrimental in excess. Improvements in yield and WP then depend on other factors such as soil fertility. This implies that the low WP of southern Nigerian crops should not be interpreted uncritically that agriculture is not performing or exploiting water well. Where water does limit production, cautious interpretation is also required. The high WP of crops in northern Mali reveals an efficient use of water but does not imply a good agricultural investment due to the very low and insecure yields. To interpret WP one needs to take account of both the numerator (production) and the denominator (water resource).

Water losses in agriculture may be beneficial to other parts of the system (Seckler *et al.* 2003). Excess water contributes to groundwater recharge, river flow, ecosystem services or even climate regulation (Monteny and Casenave 1989). In the Office du Niger, 44% of the water losses return to the Fala stream (Barbier *et al.* 2009). Stakeholders wishing to reduce water losses to improve WP must therefore consider the current uses and value of the drained water. New methodologies may assist in calculating WP effectively in multi-use systems at varying scales. Moreover, standard definitions for the numerator and denominator in WP (Molden *et al.* 2003), especially for rainfed crops and livestock production, would increase possibilities to compare WP values and formulate recommendations (Bessembinder *et al.* 2005).

Institutional analysis

The continued authority of traditional chiefs on one hand and the influence of former colonial powers on the other create a complex and fragmented institutional context in the Niger Basin. The heavy dependence on transboundary flows in countries such as Niger (91%) adds to this complexity (WWAP 2006).

At the regional level, the Economic Community Of West African States (ECOWAS), which gathers all basin countries except Chad and Cameroon, plays a major role in the regional process towards the implementation of integrated water resource management (IWRM). ECOWAS adopted a regional action plan for IWRM in 2000 (RAP-IWRM/WA) and created the permanent framework for co-ordination and monitoring IWRM in West Africa (PFCM WA). This structure supports the Niger Basin Authority and basin countries in implementing transboundary IWRM and elaborating national IWRM plans.

At the basin level, the NBA, created in 1980, is responsible for co-ordinating equitable development of water resources in the basin according to the principles of IWRM. The NBA adopted the Water Charter project 2008 for a legal and regulatory framework and has devised an Action Plan for Sustainable Development (PADD) that is linked to an Investments Program.

The institutional framework to manage natural resources in the basin remains at the national level, but is undergoing major reforms driven by international pressure. In contrast to the trend towards privatization in land tenure, reforms to water legislation have sought to assert state control over water resources and introduce mechanisms to administer and allocate water rights at the local level (Hodgson 2004). With time, decentralization will confer the main role in agriculture and water issues to new local institutions and increase the involvement of local people, who currently have little say (Bazie 2006).

Legal pluralism

Decentralization, participatory governance, and NGOs through water user associations introduce new structures and new rules at the local level. Together with economic, social and demographic changes, they encourage new practices, which, even if not official (Lund 2000), can contest the legitimacy of, and weaken, traditional authorities (Lavigne Delville 2005). This creates a legal pluralism (Caron 2009) in which rules based on different, or even contradictory, principles coexist. Arbitration becomes more complex, and can lead to problems of land and water governance (Figure 9) and conflicts (Lavigne Delville 2005, Cotula 2006). The result of this change dynamic varies between ethnic groups and communities, but in some cases legal pluralism leads to positive institutional innovations, such as recognition of women, youth, or minority groups who are often discriminated against under traditional law.

Impact on land tenure

Existing informal land tenure agreements are often denied by recent reforms to land tenure legislation, which create a dual and separate system (Hodgson 2004, McAuslan 2006). The reforms fail to recognize communal tenure as viable and introduce registration systems to secure land rights and their transactions. These procedures have not helped poor rural smallholders, women, and young people (McAuslan 2006, Toulmin 2008) who can not get title to their land, partly due to high costs and corruption. New participative systems of land titling, in line with the commitment to decentralization of the water legislation, are required to help to protect the customary land tenure rights of the poor. These require an innovative design rather than attempting to use imported systems.

As water rights are associated with land rights in most systems (Ramazzotti 1996, Caron 2009), insecure land tenure affects secure access to water, and restricts investment in agriculture (Figure 10). Legal pluralism, which creates insecurity in terms of definition, allocation and enforcement of land rights, is then a cause of stagnant agricultural productivity and rural poverty. The literature suggests that institutions, by affecting

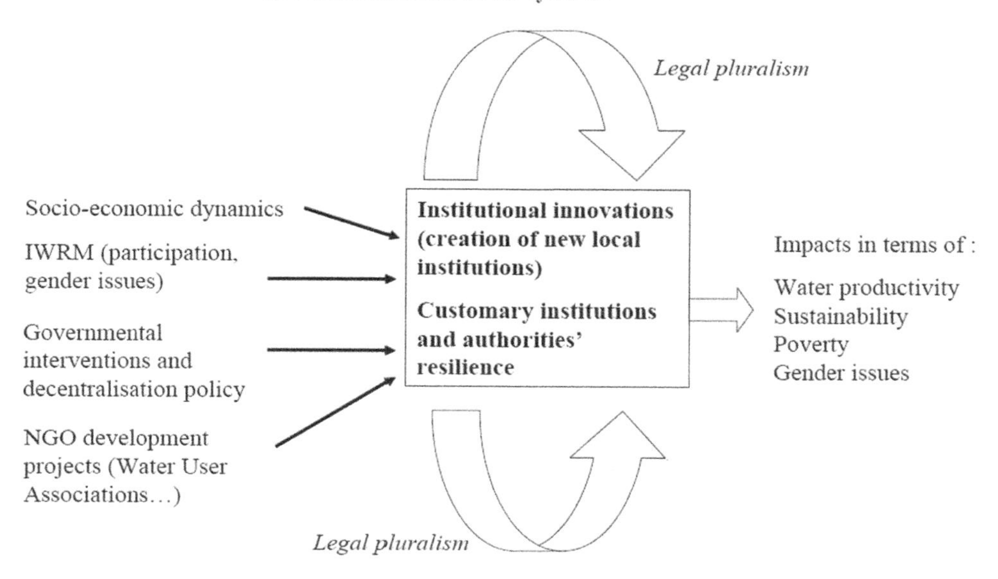

Figure 9. Influence of legal pluralism on water productivity.
Source: Caron (2009).

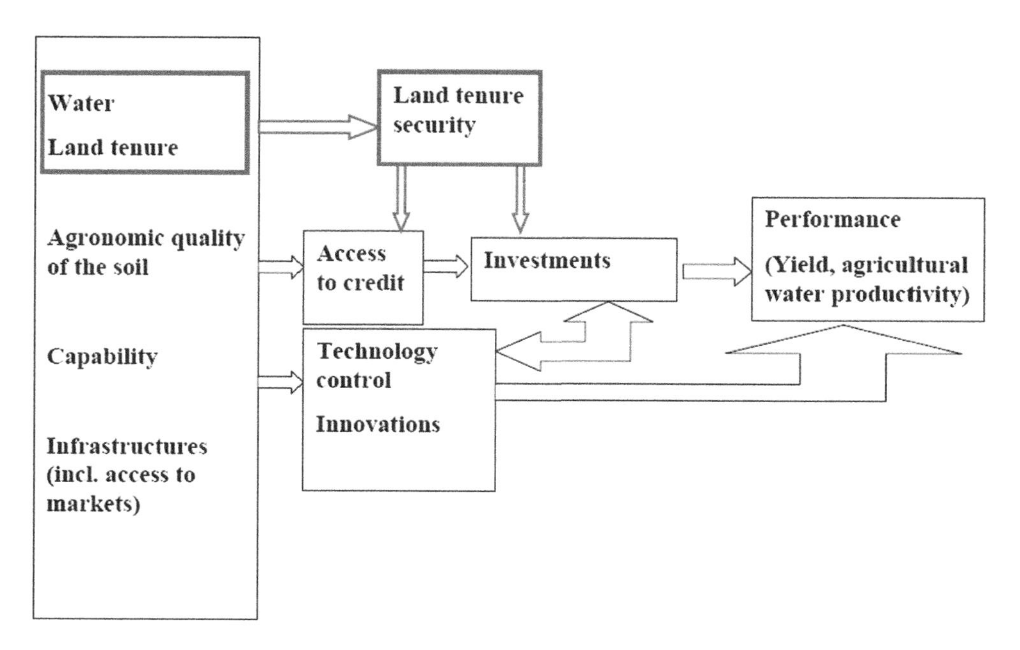

Figure 10. Factors affecting agricultural water productivity.
Source: Caron (2009).

stakeholders and their behaviour (Ostrom 1990, North 2005), play a major role in the dynamics of both development and poverty (Meisel and Ould Aoudia 2007). Here, by affecting land tenure security and the behaviour of farmers, institutions influence the performance of agriculture in the basin, which remains low despite considerable hydrologic and agronomic potential.

A case study of the Talo Dam and associated irrigated perimeters in Mali confirmed the importance of property rights to promote efficient, equitable and sustainable water use (Cotula 2006, Caron 2009). The decentralized land-allocation process was inadequate in securing land rights for women and young people, despite being an objective of recent law reforms such as the Malian *Loi d'Orientation Agricole* (2005). Cattle breeders were also marginalized, losing access to water and land without compensation (Caron 2009).

Institutional indicators

Figure 11 summarizes information on the security of land tenure (IPD 2006, UNDP 2007, OECD 2009). The institutional indicators and their analysis are qualitative and subjective, and moreover were not available for Guinea nor at the sub-national scale. Property rights for agricultural land are mainly traditional and informal in all eight of the states in the basin for which there are data. Except in Cameroon and Côte d'Ivoire, the scale of collectively owned land is high or very high. Security over traditional land property rights and transactions is low, except in Chad and Nigeria.

Further analysis using component multiple analysis of the institutional characteristics of the basin countries also highlights the lack of homogeneity between the countries, despite tendencies to group them as "informal-fragmented" (Meisel and Ould Aoudia 2007). Nigeria contrasts with other countries (Niger, Benin, Burkina Faso and Cameroon) possessing good internal security and subject to strong exogenous pressures. Chad is low due to the relative absence of both public freedoms and the autonomy of the civil society.

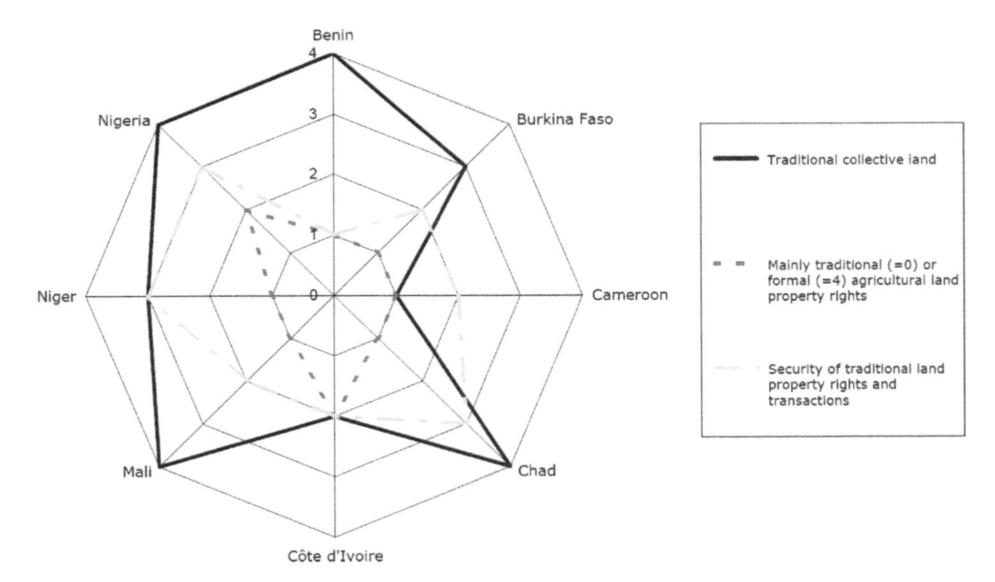

Figure 11. Traditional property and security of rights transactions in agricultural sector. From 0 (very low) to 4 (very high).
Source: Caron (2009).

Water poverty analysis

Water poverty

Water poverty occurs when people are either denied dependable water resources or lack the capacity to use them. Water scarcity is commonly thought to arise due to physical or economic constraints, though Molle and Mollinga (2003) distinguish three further causes of scarcity: managerial, institutional and political scarcity, reflecting the complex nature of water poverty.

The lack of a comprehensive metric that reliably captures the multi-factorial characteristics of water poverty has led to a raft of measurement techniques, each with advantages and disadvantages (Table 2). Such indicators have become increasingly widespread and favoured by decision makers, as they provide a more legible, though often simplified view of the reality on the ground. The added simplicity facilitates communication and comparison, but reduces objectivity and representativity. The widely used Falkenmark "water stress index" (Falkenmark *et al.* 1989) defines a threshold of 1700m^3 of renewable water resources per capita per year, under which a country is deemed to suffer from water scarcity. All countries except Burkina Faso exceed this threshold; however, the indicator fails to capture spatial and temporal variations, crucial in a country such as Mali, which ranges from a sub-humid to a hyper-arid climate. A more comprehensive measurement of water poverty is the water poverty index, which notably takes into account communities' abilities to access and use water but suffers from the use of arbitrary weights and which must ideally be generated at a local rather than a national or a regional scale (Sullivan and Meigh 2003). Variables that measure a relatively mono-dimensional and objective situation (Molle and Mollinga 2003) (e.g. childhood mortality rate) may instead offer the closest depiction of the real situation in these communities.

Relations between water and poverty

Indices may provide an overview of the poverty and water situations but they do not reflect the linkages between water and poverty. Composite indices intrinsically mask the importance of each factor, making interpretation of the potential causes behind water poverty and formulation of subsequent interventions difficult. To detect and analyse a hypothetical relationship between water and poverty, we estimated statistical relationships, using poverty maps and correlation coefficients. Significant correlations do not imply causality but point towards water resource factors that may influence poverty.

At the national scale, weak correlations between widely used water (Falkenmark's Water Stress Index, Water Poverty Index [WPI]) and poverty metrics (headcount ratio, Human Development Index [HDI], Social Vulnerability Index [SVI]) were identified for all African nations (excluding small island states). With little evidence for a strong association between a country's water situation and its development performance on the African continent, relations, if they exist, must be sought at a greater spatial resolution with more representative variables.

Poverty mapping

To account for a high proportion of subsistence livelihoods and a large non-market, hybrid economy, we used child mortality, child stunting and a composite wealth index as poverty metrics. The wealth index is country specific and cannot be compared internationally (Rutstein and Johnson 2004). Data were taken from the demographic and health surveys (Measure DHS 2008) and interpolated to estimate values in non-sampled regions.

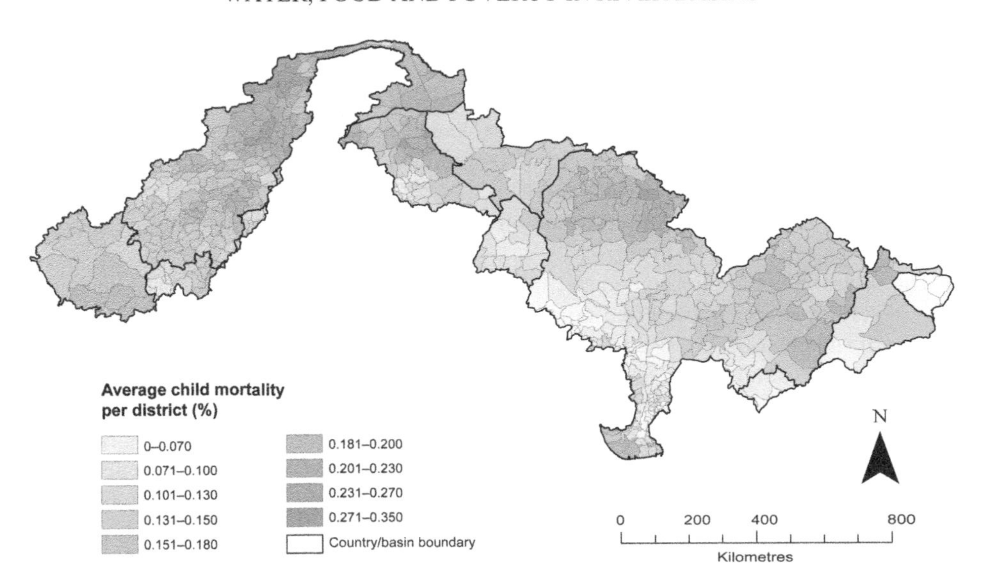

Figure 12. Estimated child mortality (proportion of children who die before age 5) across the Niger Basin (based on births recorded since 1980).
Source: Ward *et al.* (2009).

Figure 12 illustrates the spatial distribution of child mortality in 630 administrative districts across the basin.

The poverty indices were assessed for statistical correlation with an array of possible poverty determinants, both water and non-water related. This was undertaken in the first instance at a basin scale using geographically weighted regression (GWR) and at a sub-national scale using local indicators of spatial association (LISA) clusters and spatially explicit regression analysis. Ward *et al.* (2009) give further details on the methods and results. The analysis of spatially referenced child mortality, child morbidity and the wealth index identified three major poverty hotspots in the Niger Basin, situated in southern Mali and the Inner Delta, northeast Burkina Faso and northwest Nigeria (Figure 13). We expect communities situated in regions of intersecting hotspots for the three poverty metrics to face the greatest poverty and vulnerability challenges.

Links between poverty and water, poverty and agricultural productivity

Increased education and water quality, measured by the proportion of people drinking from unprotected water sources, were most clearly associated with decreases in poverty. These variables are significant and relatively stationary across the study area, and can therefore be addressed with whole-of-catchment-scale policies with less attention to regional differences. A statistical relationship between water quality and child health poverty measures seems consistent with the vital role given to water and sanitation in alleviating poverty (UNDP/SEI 2006). Insufficient access to clean water is known to impact on human health, through the development of water-borne diseases (e.g. diarrhoea, cholera) and water-washed diseases (e.g. scabies, trachoma) (Bradley 1974). Diarrhoea is the third cause of child mortality in West Africa after malaria and respiratory infections (ECOWAS-SWAC/OECD 2008) and new water-borne diseases such as Whipple disease are still emerging (Fenollar *et al.* 2009).

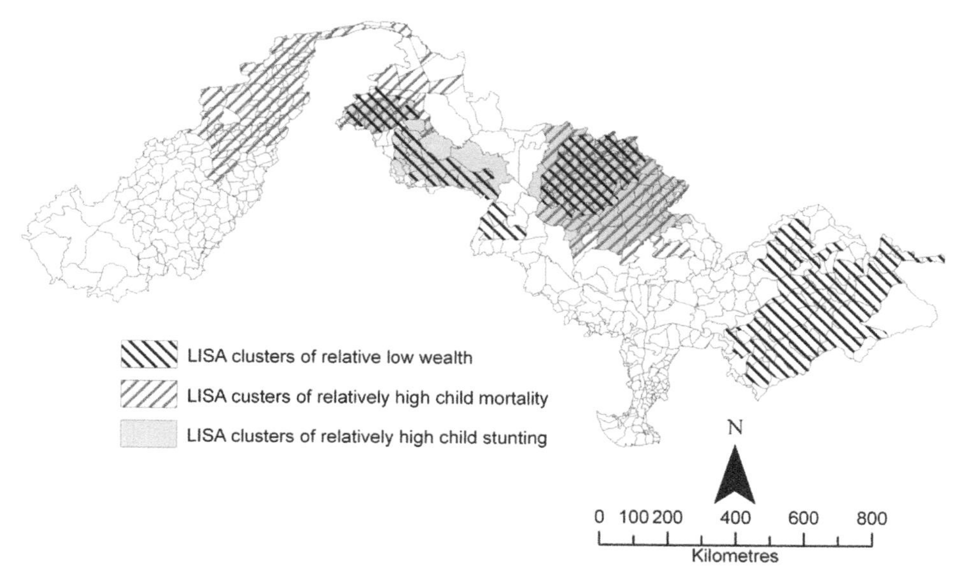

Figure 13. Overlapping map of significant statistical poverty hotspots.
Source: Ward *et al.* (2009).

The literature suggests that agricultural water management also has the potential to reduce rural poverty (Namara *et al.* 2010). We found weak correlations between agricultural water determinants and poverty variables. Total actual renewable water resources (TARWR) does not account for difficulties in accessing water and therefore only provides a theoretical indication of the water potentially available for agriculture. The metric also does not translate annual variations, crucial in countries that may experience drought and flooding in the same year (Rijsberman 2006). Though TARWR is a commonly used indicator, in the Niger Basin it does not accurately reflect the situation of a community with regard to water availability nor its poverty status.

Hussain and Hanjra (2004) argue that increased irrigation and proximity to dams provide a pathway out of poverty, indicating community opportunities and capacity to access and transform water into food. We found the relationship to hold in only some instances. The spatial regression analyses suggest either that irrigation's contribution to rural welfare is low in the Niger Basin, or that the spatial extent of irrigation is too limited at present to cause any detectable reduction in poverty at this scale of analysis. It may also be that the benefits of irrigation do not (yet) accrue to the people engaged in its practice, or that they do so at levels too small to register in our analyses. Variables demonstrated to be statistically non-stationary (i.e. their influence varies across the landscape) may be more appropriately addressed using a geographically targeted policy approach.

Agriculture-related indicators, including primary productivity, soil quality or livestock numbers, provided little explanatory power over poverty. A similar study in Malawi (Benson *et al.* 2005) found that a rise in maize yields actually resulted in increased poverty, presumably due to equity issues, with higher yields not benefiting local populations. Despite growth in agricultural productivity being expected to reduce poverty in the rural agriculture-dominated economies of West Africa (Thirtle *et al.* 2003), poverty prevails in areas of good soil quality, high productivity and sufficient available water.

Similar studies evaluating the significance of explanatory variables in poverty mapping have found limited correlations between poverty and agro-ecological or socioeconomic determinants (Hyman *et al.* 2005). This points to the complexity of transforming available water into adequate food production and a pathway out of poverty. Beyond reliable access to water, the ability to derive profit from water depends on several additional conditions such as access to land, labour, seeds, fertilizer, pesticides, tools and machinery, fuel, storage, transformation processes, roads, markets and political security. Hanjra *et al.* (2009) point to significant correlations between these variables and agricultural productivity, but variable interactions are critical in determining resultant poverty. Some of these factors may be stationary at the regional scale (such as roads and access to markets); others such as access to land may vary widely within the same village from one family or ethnic tribe to another, and thus require detailed analysis. Structural causes of poverty such as the positioning of individuals in the socioeconomic structure (Mulwafu and Msosa 2005) may have heavy influence.

Overall, it is difficult to isolate one contributing factor to poverty. Interactions between environmental, social and institutional factors are complex and an evaluation of poverty and its causes requires analysis at multiple spatial resolutions. Access to water for agriculture and productive purposes plays a crucial role in poverty alleviation but is not a sufficient condition and much depends on the capabilities and endowments (Chambers and Conway 1992) (e.g. level of training, diverse income sources, capital and support networks) of a given household or community. These, not simply the absence, presence or quality of water, determine whether they will fall or subsist in a state of poverty.

Threats and opportunities

Interventions

Successful interventions to reduce poverty have been introduced in certain areas over the years, but solutions to achieve sustained and widespread impacts are still lacking. Recommended interventions are highly contextual and require rigorous analysis for each watershed in the basin. They vary according to climatic and socio-economic conditions and according to the livelihood strategies and agricultural systems (Namara *et al.* 2009). Policy solutions must therefore rely on mixes of sequenced instruments tailored to address temporally and spatially diverse patterns of poverty.

Recommended interventions include developing infrastructure (wells, reservoirs), multiple-use systems (notably integrating livestock and fisheries with agriculture), adapting crop demands to water supply and vice versa (sowing dates, water harvesting, supplemental irrigation), drought-tolerant crops, fertilizer use and so on. Improvements in rainfed agriculture can have an important impact on poverty reduction and food security due to the large population dependent on it. Current farmer strategies to reduce risks due to rainfall deficits prevent intensification. Solutions to reduce soil-water stress and the risk of crop failure, such as rainwater harvesting and drought tolerant crops, are necessary if farmers are to invest in fertilizer and other inputs that are essential to boost yields.

Farmers also need to be linked to input and output markets and financial services, and have access to training and storage, but also to have secure access to land and water, possibly through communal land tenure arrangements. Mitigation strategies such as early-warning systems and storage options are also required to help reduce the impact of extreme events. Transparent, participatory governance is required to ensure water resources are developed in an equitable, participatory and sustainable way. Exogenously driven institutional dynamics must be viewed as a source of opportunities to develop new pertinent, culturally- and context-adapted regulation frameworks blending customary and Western law, which respond to the needs of rural poor, both men and women, who are the majority of landholders in all the member countries. The recognition of collective rights to water and land may help guarantee the access to water by the rural poor.

In certain areas of the basin, upstream and downstream conflicts coming from the development of the Niger River are inevitable and are expected to be intense, given the increasing demands for water from more users (population growth) and uses (industry, hydropower, ecosystem). IWRM is emerging in the basin but water resources are typically managed at a local or national scale. Nevertheless, as resources become scarcer, a transboundary, basin approach to water management appears both relevant and necessary. This is especially true in the upstream part of the river between Guinea and Niger, where rainfall decreases as the river flows downstream making northern populations extremely reliant on the river. In Nigeria, the substantial rainfall and contribution from the Benue River makes transboundary water management of the Niger River less of a priority, but the added presence of dams in Mali and Niger can be expected to impact negatively on water availability and hydroelectric production in western Nigeria.

Several tools such as Water Evaluation and Planning (WEAP), Modèle Intégré du Delta Intérieur du Niger (MIDIN) and rainfall/runoff modelling should be implemented by stakeholders as predicting tools, along with companion modelling tools. In addition to regulatory measures to determine and restrict maximal abstractions, measures such as payment for environmental services should be considered. These could come from hydroelectric production financing interventions upstream to reduce water consumption and improve water availability.

Future threats

Agriculture in the Niger River basin notably suffers from high spatial and temporal variations in rainfall, poor soil fertility, inadequate communication and transport infrastructure, as well as a critical lack of a cohesive, transparent social and institutional context conducive to agricultural investment. In addition to these constraints, the basin faces an array of development challenges in the years to come. Of these, three appear to be vital, due to the severe and widespread difficulties they may cause. These are population increase, climate variability and intense river development (dams, abstractions). Surveys in the basin confirm the growing concern of the rural poor over these issues (Mills *et al.*, 2009).

Projected population increase in the basin could well jeopardize current and future development efforts. The basin population estimated around 95 million in 2005 is expected to double by 2050 in the lowest scenario and could be multiplied by four if fertility remains constant. Current fertility rates exceed six to seven children per woman and as mortality has started decreasing, demographic increase rates now exceed 3% per year (Bana and Conde 2008). More worryingly, in countries like Mali (unlike Ghana) fertility is not decreasing, resulting in a progressive rise in the demographic increase rate. Future population trends will therefore depend on the speed of fertility decrease and the prevalence of pandemics such as HIV/AIDS (Guengant 2009). Clearly the additional demand on water and food resources to feed up to 300 million additional people added to the projected change in diets, climate change and water demand for industry and hydropower will lead to significant pressure on natural resources and ecosystems and increase vulnerability of rural poor communities.

Results from climate-change modelling present many uncertainties and contradictions. However, on average there is a trend towards: an increase in temperature; in variability and extreme events; a later start to the rainy season; more dry spells; and an overall increase in rain in the central part of West Africa and a decrease in the west (Mahé *et al.* 2009a). The effect on yields increases the uncertainties as we must account for increased evaporation, possible decreased rainfall and increased CO_2 fertilization, but there is little doubt that climate change will increase the strain on already-vulnerable agriculture.

Under the NBA investment plan, several large dams are due to be built, notably Fomi in Guinea, Taoussa in Mali and Kandadji in Niger. These offer important opportunities to triple the irrigated area of land, up to 400,000 hectares in Sahelian countries, but will have effects on both local populations and people downstream. Various scenarios have been studied in the literature, which all inherently impact on one element of the system. Trade-off analysis must be undertaken in consultation with local stakeholders to ascertain which element must be favoured – hydropower, irrigation, fisheries and ecosystems – and how to minimize negative impacts. Allocation of resources will inevitably be prioritized, putting vulnerable populations at risk. Expansion of the Office du Niger irrigation project for instance will result in a decreased flood in the Inner Delta, affecting traditional rice growers, herders and fisheries. Fish production could be reduced by 8500 t/yr.

Conclusions

This project aimed to acquire an interdisciplinary view of the relations between water, food and poverty in the Niger Basin. Water availability in the basin is not as low as common perception may suggest, but is subject to severe spatial and temporal variations. Rainfed agriculture is concentrated over less than two months in the north and more than nine months in the south of the basin, and is affected by repeated dry spells as well as severe rainfall deficits as in the 1970s and 1980s. Climatic scenarios predicting increased temperatures, higher variability, dry spells and extreme events as well as reduced rainfall in western parts of West Africa will increase the strain on an already vulnerable agriculture.

Livelihood strategies, though varied, rely predominantly on rainfed agriculture, which suffers from multiple problems, including low mechanization, lack of supplementary irrigation, low fertilizer input, inadequate support and the absence of commercialization. Institutional analysis reveals that the progressive introduction of new rules and structures (decentralization, IWRM, NGO projects) and the continued dominance of customary laws create a legal pluralism, leading to confusion and conflicts. Land tenure, which conditions secure access to water resources and investments in agriculture, is notably affected by the legal pluralism and the reforms that favour individualized tenure and land titles. Improvements in rainfed agriculture hold the greatest potential to tackle poverty thanks to the large population dependent on it, but notably require innovative, participative initiatives to protect the tenure rights of the poor.

Statistical analysis of correlations between poverty measures and water variables identified the importance of water quality in reducing poverty but highlighted the complex relation between agricultural water management and poverty. The pathway linking these is more complex, and resultant poverty depends on the interactions between environmental, social and institutional factors. Assessing the causes of poverty in the basin requires detailed and close-up analysis, and related interventions must be geographically targeted to account for these complicated interactions. In light of the current constraints and the impending challenges the basin will face as a result of demographic and climatic changes, concerted efforts are required by all basin stakeholders to improve agriculture and water management through technical, sociological and regulatory means.

Agricultural withdrawals already impact ecosystems and projected dam building to extend dry-season irrigation will exert further pressure on environments such as the Inner Delta, which provides for the livelihoods of over a million herders, fishermen and traditional rice growers. Trade-off analysis in consultation with local stakeholders must be undertaken especially in the upper basin to determine priorities and to seek to minimize negative impacts. Water productivity, though a partial measure whose calculation and interpretation must be refined, could in time assist stakeholders to highlight where additional value can be derived, especially as the Niger moves towards becoming a closed basin and water resources constitute a limiting factor.

Acknowledgements

This paper presents results from the Niger Basin Focal Project (BFP) project. We gratefully acknowledge the support provided by the Challenge Program Water and Food (CPWF) and thank the team of researchers from Africa, Australasia and Europe who collaborated on this project.

References

ABN and BRLi, 2007. *Evaluation des prélèvements et des besoins en eau pour le modèle de simulation du Bassin du Niger: rapport définitif. Niamey, Niger: secrétariat executif.* Niamey, Niger: Autorité du Bassin du Niger and BRL ingénierie.

Aboubakar, A., 2003. L'initiative du bassin du Niger (IBN): développement durable et gestion intégrée d'un grand fleuve. *Afrique Contemporaine*, 206 (2), 179–203.

Adelana, S.M.A., Olasehinde, P.I., and Vrbka, P., 2006. A quantative estimation of groundwater recharge in part of the Sokoto Basin, Nigeria. *Journal of Environmental Hydrology*, 14, paper 5, 1–16.

Balk, D., *et al.*, 2003. *Spatial analysis of childhood mortality in West Africa.* Calverton, MD and Palisades, NY: ORC Macro and Center for International Earth Science Information Network (CIESIN), Columbia University.

Bana, I. and Conde, N., 2008. *Perspectives de développement du Bassin du Niger* [online]. Niger Basin Authority. Available from: www.unwater.unu.edu/file/2.+ABN+Niger.pdf [Accessed 1 August 2009].

Barbier, B., *et al.*, 2009. *Productivité de l'eau d'irrigation dans le bassin du fleuve Niger.* Ouagadougou, Burkina Faso and Colombo: CIRAD/Institut International d'Ingénierie de l'Eau et l'Environment (2iE) and Challenge Programme on Water and Food.

Bazie, J.B., 2006. *Reconnaissance, identification et caractérisation des usagers de l'eau du bassin du fleuve Niger en vue de leur implication et participation effective au processus de la vision partagée, synthèse régionale.* Niamey, Niger: Autorité du Bassin du Niger.

Benson, T., Chamberlin, J., and Rhinehart, I., 2005. An investigation of the spatial determinants of the local prevalence of poverty in rural Malawi. *Food Policy*, 30 (5–6), 532–550.

Bessembinder, J.J.E., *et al.*, 2005. Which crop and which drop, and the scope for improvement of water productivity. *Agricultural Water Management*, 73 (2), 113–130.

Bradley, D.J., 1974. Water supplies: the consequences of change. *In*: *Human rights in health. (Ciba Foundation Symposium 23, Amsterdam).* Amsterdam: Elsevier-Excerpta Medica, 81–98.

Brummett, R.E., Lazard, J., and Moehl, J., 2008. African aquaculture: realizing the potential. *Food Policy*, 33 (5), 371–385.

Caron, A., 2009. *BFP Niger. Institutional analysis.* Montpellier and Colombo: IRD and CPWF.

Cecchi, P., 2009. *Les petits barrages en Afrique de l'Ouest et note sur les aménagements du Bani.* Montpellier and Colombo: IRD and CPWF.

CEDEAO (Communauté Economique des Etats de l'Afrique de l'Ouest), 2007. *Atlas web de l'intégration régionale de l'Afrique de l'Ouest/ The web atlas on regional integration of West Africa* [online]. Available from: www.atlas-ouestafrique.org [Accessed 1 October 2009].

CIESIN/CIAT, 2005. *Gridded population of the world version 3 (GPWv3): population density grids.* Palisades, NY: Center for International Earth Science Information Network (CIESIN), Columbia University, and Cali, Colombia: Centro Internacional de Agricultura Tropical (CIAT). Available from: sedac.ciesin.columbia.edu/gpw [Accessed 12 September 2008].

Chambers, R. and Conway, G., 1992. *Sustainable rural livelihoods: practical concepts for the 21st century.* IDS discussion paper 296. Brighton, UK: Institute of Development Studies.

Clanet, J.-Ch., 1994. *Le pastoralisme au Sahel central*. PhD thesis. Université de Paris IV – Sorbonne.

Clanet, J.-Ch. and Ogilvie, A., 2009. Farmer-herder conflicts and water governance in a semi-arid region of Africa. *Water International*, 34 (1), 30–46.

Conway, D., *et al.*, 2009. Rainfall and water resources variability in sub-saharan Africa during the twentieth century. *Journal of Hydrometeorology*, 10 (1), 41–59.

Cotula, L., 2006. *Land and water rights in the Sahel: tenure challenges of improving access to water for agriculture.* Issue Paper 139. London: International Institute for Environment and Development.

De Noray, M.-L., 2003. *Waza Logone: histoires d'eaux et d'hommes*. Gland, Switzerland and Cambridge, UK: World Conservation Union-International Union for Conservation of Nature.

Diop, A.T., *et al.*, 2009a. *Les productions animales dans le bassin du Niger: tendance évolutive – productivité de l'eau – optimisation*. Dakar, Sénégal and Colombo: ISRA/Ecole Nationale d'Economic Appliquée (ENEA)/CIRAD/Pôle Pastoral Zones Sèches (PPZS) and CPWF.

Diop, A.T., *et al.*, 2009b. *L'élevage et le pastoralisme dans le bassin du Niger*. Dakar and Colombo: LENRV/PPZS/CIRAD and CPWF.

Diop, A.T., *et al.*, 2009c. *Productivité de l'eau en Elevage et optimisation des ressources fourragères du bassin du fleuve Niger*. Dakar, Sénégal and Colombo: LENRV/PPZS/CIRAD and CPWF.

ECOWAS-SWAC/OECD, 2008. *Atlas on regional integration in West Africa, population series: communicable diseases* [online]. Available from: www.atlas-ouestafrique.org/spip.php?rubrique53 [Accessed 4 November 2009].

Falkenmark, M., 1989. The massive water scarcity now threatening Africa: why isn't it being addressed. *Ambio*, 18 (2), 112–118.

Fenollar, F., *et al.*, 2009. Tropheryma whipplei in fecal samples from children, Senegal. *Emerging Infectious Diseases*, 15 (6), 922–924.

Filippi, C., Milville, F., and Thiery, D., 1990. Evaluation de la recharge naturelle des aquifères en climat soudano-sahélien par modélisation hydrologique globale: application à dix sites au Burkina Faso. *Journal des sciences hydrologiques*, 35 (1), 29–47.

FAO, 2008. Water productivity [online]. *In*: *Unlocking the water potential of agriculture.* Rome: FAO. Available from www.fao.org/documents/show_cdr.asp?url_file=/DOCREP/006/Y4525E/y4525e06.htm [Accessed 31 July 2010].

FAO, 2009. *Fisheries and aquaculture information and statistics service* [online]. Rome: FAO. Available from: www.fao.org/figis/servlet/SQServlet?ds=Aquaculture&k1=COUNTRY&k1v=1 &k1s=159&outtype=html [Accessed 27 July 2009].

Guengant, J.-P., 2009. *Migrations internationales et développement* [online]. Recherches Interdisciplinaires et Participatives sur les Interactions entre les Ecosystèmes, le Climat et les Sociétés en Afrique de l'ouest (Colloque Ripiecsa). Bobo-Dioulasso, Burkina Faso: Institut de recherche pour le développement (IRD)–Agence inter-établissements de recherche pour le développement (AIRD). Available from: www.aird.fr/ripiecsa/index.htm [Accessed 1 October 2009].

Hanjra, M., Ferede, T., and Gutta, D.G., 2009. Reducing poverty in sub-Saharan Africa through investments in water and other priorities. *Agricultural Water Management*, 96 (7), 1062–1070.

Hodgson, S., 2004. *Land and water: the rights interface*. Legislative Study 84. Rome: FAO.

Hussain, I. and Hanjra, M.A., 2004. Irrigation and poverty alleviation: review of the empirical evidence. Colombo: IWMI. *Irrigation and Drainage Journal*, 53 (1), 1–15.

Hyman, G., Larrea, C., and Farrow, A., 2005. Methods, results and policy implications of poverty and food security mapping assessments. *Food Policy*, 30 (5–6), 453–460.

IPD, 2006. *Institutional profiles database* [online]. France: Ministry for the Economy, Finance and Employment (MINEFE) and African Development Fund (ADF). Available from: www.cepii.fr/ProfilsinstitutionnelsDatabase.htm [Accessed 1 October 2009].

IPCC, 2007. *IPCC Working Group 1. AR4 report* [online]. Geneva: Intergovernmental Panel on Climate Change. Available from: www.ipcc-wg1.ucar.edu/wg1/wg1-report.html [Accessed June 2009].

Jacob, J.-P., 2005. *Sécurité foncière, bien commun, citoyenneté. Quelques réflexions à partir du cas burkinabé* [online]. Etude RECIT no. 6. Ouagadougou, Burkina Faso: Institut universitaire d'études du développement (IUED). Available from: www.labocitoyennetes.org/documents/062_ETU6_SecFonBieComCit_BF_Jacob.pdf [Accessed 1 August 2010].

Laë, R., 1992. Influence de l'hydrologie sur l'évolution des pêcheries du delta central du Niger de 1966 à 1989. *Aquatic Living Resources*, 5 (2), 115–126.

Lavigne Delville, P. and Hochet, P., 2005. *Construire une gestion négociée et durable des ressources naturelles renouvelables en Afrique de l'ouest. Rapport final de la recherché* [online]. Paris: Groupe de recherche et d'échanges technologiques (GRET)/Changes in Land, Institutions and Markets (CLAIMS)/African Development Fund (ADF). Available from: www.foncier-developpement.org/analyses-et-debats/operations-foncieres/construire-une-gestion-negociee-et-durable-des-ressources-naturelles-renouvelables-en-afrique-de-l2019ouest/at_download/file [Accessed 1 August 2009].

Leduc, C., Favreau, G., and Schroeter, P., 2001. Long-term rise in a Sahelian water-table: the continental terminal in south-west Niger. *Journal of Hydrology*, 243 (1), 43–54.

Lund, C., 2000. *African land tenure: questioning basic assumptions* [online]. Dryland Issue Paper no. E100. London: International Institute for Environment and Development (IIED). Available from: www.iied.org/pubs/pdfs/9023IIED.pdf [Accessed 1 August 2010].

Mahé, G., *et al.*, 2005. The impact of land use change on soil water holding capacity and river flow modelling in the Nakambe River, Burkina Faso. *Journal of Hydrology*, 300 (1–4), 33–43.

Mahé, G., Lienou, G., and Paturel, M., 2009a. *BFP Niger: water availability and access*. Montpellier and Colombo: IRD and CPWF.

Mahé, G., *et al.*, 2009b. Water losses in the Inner Delta of the River Niger: water balance and flooded area. *Hydrological Processes*, 23 (22), 3157–3160.

Mahé, G. and Paturel, J.E., 2009. 1896–2006 Sahelian annual rainfall variability and runoff increase of Sahelian Rivers. *Comptes Rendus Geoscience*, 341 (7), 538–546.

Mahoo, H.F., Young, M.D.B. and Mzirai, O.B., 1999. Rainfall variability and its implications for the transferability of experimental results in semi-arid areas of Tanzania. *Tanzania Journal of Agricultural Science* 2 (2), 127–140.

Mainuddin, M., *et al.*, 2010. Water-use accounts in CPWF basins: simple water-use accounting of the Niger Basin. CPWF Working Paper: Basin Focal Project Working Paper No. 5. Series, BFP09. Colombo: CPWF.

McAuslan, P., 2006. Improving tenure security for the poor in Africa. Framework paper for the legal empowerment workshop-Sub-Saharan Africa. Rome: FAO. Available from ftp://ftp.fao.org/docrep/fao/010/k0781e/k0781e00.pdf [Accessed 1 August 2010].

Measure DHS, 2008. Country Datasets [online]: Benin (2001), Burkina Faso (2003), Cameroon (2004), Cote d'Ivoire (1998-9), Guinea (2005), Mali (2006), Niger (2006), Nigeria (2003). Calverton, MD: Macro International. Available from: www.measuredhs.com [Accessed 29 September 2008].

Meisel, N. and Ould Aoudia, J., 2007. *A new institutional database: institutional profiles 2006* [online]. Working Paper no. 46. Paris: Agence Française de Développement. Available from: www.afd.fr/jahia/Jahia/lang/en/home/publications/documentsdetravail/pid/2647 [Accessed 31 July 2010].

Mills, D., *et al.*, 2009. Vulnerability in African small-scale fishing communities. *Journal of International Development*, 26 (DOI: 10.1002/jid.1638)

Molden, D., *et al.*, 2003. A water-productivity framework for understanding and action. *In*: J.W. Kijne, R. Baker and D. Molden, eds. *Water productivity in agriculture: limits and opportunities for improvement*. Wallingford: CABI Publishing, 1–18.

Molle, F. and Mollinga, P., 2003. Water policy indicators: conceptual problems and policy issues. *Water Policy*, 5 (5–6), 529–544.

Monteny, B.A. and Casenave, A., 1989. The forest contribution to the hydrological budget in Tropical West Africa. *Annals of Geophysics*, 7 (4), 427–436.

Morand, P., *et al.*, 2009. *BFP Niger – Fisheries*. Montpellier and Colombo: IRD and CPWF.

Mulwafu, W. and Msosa, H.K., 2005. IWRM and poverty reduction in Malawi: a socio-economic analysis. *Physics and Chemistry of the Earth*, 30 (11–16), 961–967.

Namara, R.E., Barry, B., and Owusu, E., 2009. Niger basin focal project intervention analysis. An overview of the development challenges and constraints of the Niger Basin and possible intervention strategies. Montpellier and Colombo: IRD and CPWF.

Namara, R., *et al.*, 2010. Agricultural water management and poverty linkages. *Agricultural Water Management*, 97 (4), 520–527.

NDES, 1999. *Niger Delta environmental survey Phase I Report. Vol.1. Environmental and socio-economic characteristic (revised edition).* Lagos: Environmental Resources Managers Limited.

Neiland, A. and Béné, C., 2008. Review of river fisheries valuation in West and Central Africa. *In: Tropical river fisheries valuation: background papers to a global synthesis.* Studies and Reviews (1836). Penang, Malaysia: WorldFish Center, 47–106.

NEPAD, 2006. *Plan d'action pour le développement des pêcheries et de l'aquaculture en Afrique/ Action plan for the development of African fisheries and aquaculture.* Johannesburg: Nouveau partenariat pour le développement en Afrique.

Niasse, M., 2006. *Les bassins transfrontaliers fluviaux.* Atlas de l'intégration régionale d'Afrique de l'ouest, série espaces. Paris: Communauté Economique Des Etats de l'Afrique de l'Ouest -Club du Sahel et de l'Afrique de l'Ouest/Organisation de Coopération et de Développement économiques (CEDEAO-CSAO/OCDE).

North, D., 2005. *Understanding the process of economic change.* Princeton: Princeton University Press.

OECD, 2009. *Gender, institutions and development database 2009 (GID-DB).* Paris: Organisation for Economic Co-operation and Development. Available from stats.oecd.org/Index.aspx? DataSetCode=GID2 [Accessed 1 October 2009].

OMS, 2006. *Travailler ensemble pour la santé. Rapport sur la santé dans le monde 2006* [online]. Geneva: World Health Organization. Available from: www.who.int/whr/2006/fr/index.html [Accessed 1 August 2010].

Ostrom, E., 1990. *Governing the commons. The evolution of institutions for collective action.* Cambridge: Cambridge University Press.

Ramazzotti, M., 1996. *Readings in African customary water law.* Legislative Study, no. 58. Rome: FAO.

Rijsberman, F.R., 2006. Water scarcity: fact or fiction? *Agricultural Water Management*, 80 (1–3), 5–22.

Rockstrom, J., Barron, J. and Fox, P., 2002. Rainwater management for increased productivity among small-holder farmers in drought prone environments. *Physics and Chemistry of the Earth* 27 (11–22), 949–959.

Rutstein, S. and Johnson, K., 2004. *The DHS wealth index* [online]. DHS Comparative Reports no. 6. Calverton, MD: ORC Macro. Available from: www.childinfo.org/files/DHS_Wealth_ Index_(DHS_Comparative_Reports).pdf [Accessed 31 July 2010].

Seckler, D., Molden, D. and Sakthivadivel, R., 2003. The concept of efficiency in water-resources management and policy. *In:* J.W. Kijne, R. Barker and D. Molden., eds. *Water productivity in agriculture: limits and opportunities for improvement.* Wallingford: CABI Publishing, 37–51.

Serpantié, G. and Lamachère, J.-M., 1989. *Pour une connaissance des conditions de mise en oeuvre des aménagements du ruissellement.* Montpellier: Office de Recherche Scientifique et Technique d'Outre-Mer (ORSTOM).

Serpantié, G., 2009a. Dans quelles situations le concept de productivité de l'eau pluviale est-il pertinent: écosystèmes naturels: le cas des savanes? Montpellier and Colombo: IRD and CPWF.

Serpantié, G., 2009b. *Le concept d'efficience de l'eau pluviale est-il pertinent? Agricultures pluviales à base d'annuelles.* Montpellier and Colombo: IRD and CPWF.

Serpantié, G., 2009c. *L'utilité de l'eau pluviale sur le bassin du Niger.* Montpellier and Colombo: CPWF

Showers, V., 1973. *The world in figures.* Toronto: John Wiley.

Sullivan, C.A. and Meigh, J.R., 2003. Considering the Water Poverty Index in the context of poverty alleviation. *Water Policy*, 5 (5), 513–528.

Tabutin, D. and Schoumaker, B., 2004. The demography of Sub-Saharan Africa from the 1950s to the 2000s. A survey of changes and a statistical assessment. *Population-E*, 59 (3–4), 455–555.

Thirtle, C., Lin, L., and Piesse, J., 2003. The impact of research-led agricultural productivity growth on poverty reduction in Africa. Asia and Latin America. *World Development*, 31 (12), 1959–1975.

Toulmin, C., 2008. Securing land and property rights in sub-Saharan Africa: the role of local institutions. *Land Use Policy*, 26 (1), 10–19.

UNDP/SEI, 2006. *Linking poverty reduction and water management* [online]. Prepared on behalf of the Poverty-Environment Partnership. United Nations Development Programme and Stockholm Environment Institute. Available from: www.who.int/water_sanitation_health/ resources/povertyreduc2.pdf [Accessed 1 August 2010].

UNDP, 2007. *Human development report 2007/2008: fighting climate change.* New York: United Nations Development Programme.

UNICEF, 2008. *State of the world's children: the child survival* [online]. New York: United Nations Children's Fund. Available from: www.unicef.org/sowc08/docs/sowc08.pdf [Accessed 1 August 2010].

United Nations Population Division, 2006. *World population prospects: the 2005 revision*. New York, United Nations Department of Economics and Social Affairs. Available from esa.un.org/unpp/index.asp [Accessed 1 July 2008].

Uyigue, E. and Agho, M., 2007. *Coping with climate change and environmental degradation in Niger Delta of southern Nigeria* [online]. Benin, Edo, Nigeria: Community Research and Development Centre. Available from: www.priceofoil.org/wpcontent/uploads/2007/06/07.06.11 -Climate_Niger_Delta.pdf [Accessed 1 August 2010].

Ward, J., Kaczan, D., and Lukasiewicz, A., 2009. *A water poverty analysis of the Niger River basin, West Africa*. Report for Niger Basin Focal Project. Colombo: CPWF.

World Bank, 2009. *PovcalNet online poverty analysis tool* [online]. New York: World Bank. Available from: go.worldbank.org/NT2A1XUWP0 [Accessed 4 February 2009].

WWAP (World Water Assessment Programme), 2006. *The United Nations world water development report 2. Water: a shared responsibility*. Paris and New York: United Nations Educational, Scientific and Cultural Organization (UNESCO) and Berghahn Books.

WWAP, 2009. *The United Nations world water development report 3. Water in a changing world: a shared a responsibility*. Paris: UNESCO.

Zwarts, L., *et al.*, 2005. *The Niger, a lifeline: effective water management in the upper Niger Basin*. Lelystad, Netherlands, Sévaré, Mali, Amsterdam, Netherlands and Veenwouden, Netherlands: RIZA, Wetlands International, Institute for Environmental Studies, and Altenberg and Wymenga ecological consultants.

The Nile Basin: tapping the unmet agricultural potential of Nile waters

Seleshi Awulachew, Lisa-Maria Rebelo and David Molden

This paper provides an overview of poverty levels, hydrology, agricultural production systems and water productivity in the Nile Basin. There are opportunities to manage water better in the basin for use in agriculture to improve food security, livelihoods and economic growth by taking into account not only the water in the river, but also by improving management of the rain water. Crops, livestock, fisheries and aquaculture have long been important in the Nile but do not feature in the water discourse.

Introduction

The Nile River basin includes ten African countries (Table 1 and Figure 1), five of which are among the poorest in the world, with low levels of socio-economic development. In this paper, we document the Nile water resources, production systems and poverty levels. We then identify interventions in water management that can increase water productivity and alleviate poverty in the basin, and guide development and future research. The central hypotheses are that poverty is related to access to agricultural water; that poverty is related to water productivity irrespective of its origin; and that poor people are less able to manage water-related risks such as drought. The data support these hypotheses.

Over 70% of the basin's people depend on subsistence, rainfed agriculture (Ayalew and Awulachew 2008), which, because of the variable rainfall, causes food insecurity. The effects of frequent drought are worsened by war, HIV-AIDS, water-borne diseases and regressive policies. With the exception of Egypt, growth in gross domestic product (GDP) has been slow over the last 30 years. Land and water resources are not well utilized or managed, and are degrading rapidly. For example, soil erosion in the Blue Nile Basin is approximately 303 Mt/yr, with up to 45% delivered to rivers (Awulachew *et al.* 2008).

In contrast to countries upstream, Egypt has a productive, large-scale irrigation system, a diversified economy and moderate living standards. Elsewhere, poverty and lack of access to water are key concerns. Ethiopia must increase food production to feed its growing population, but long droughts have left more people food insecure and accentuated the country's inability to provide enough food. Because of its long civil war, Sudan can no longer provide food security to its growing population. Other basin countries face similar problems. The historic internal and external tension and instability in the region has

Table 1. GDP and population data of countries within the Nile Basin.

Country	Area in the basin	Total area	Total population	Population growth	GDP	GDP
	000 km²	*000 km²*	*millions*	*annual %*	*billion US$*	*per capita*
Burundi*	13	25	8.1	3.9	1.4	174
Democratic Republic of Congo (DRC)*	22	2313	64.8	3.2	11.1	171
Egypt	327	968	74.2	1.8	188.0	2450
Eritrea	25	122	5.1	3.6	1.7	328
Ethiopia*	365	1124	81.2	2.6	33.9	418
Kenya*	46	580	35.9	2.6	30.2	842
Rwanda*	20	24	9.8	2.5	5.0	512
Sudan	1979	2492	39.1	2.2	54.3	1388
Tanzania*	84	891	30.4	2.5	22.2	547
Uganda*	231	207	33.2	3.2	15.7	472

Source: Data from International Monetary Fund (IMF 2009).
Notes: Gross domestic product (GDP) total and per capita based on current prices. *Indicates countries with significant HIV-affected populations.

diverted the countries' meager resources to military expenditure. Infrastructure therefore remains inadequate, contributing to low productivity and high levels of poverty.

The benefits of the Nile need to be more equitably shared, which is difficult in the face of the unbalanced distribution of water and wealth. The Nile Basin Initiative (NBI), launched in 1999 as a mechanism to share the benefits of the Nile water, has made some progress, but is often tested. After 10 years of negotiation, the upstream countries Uganda, Rwanda, Tanzania, Ethiopia and Kenya signed a new Comprehensive Framework Agreement (CFA). The agreement was strongly opposed by Sudan and Egypt, which could create a rift between upstream and downstream countries. Alternatively, the agreement may be part of a long-term negotiation process that will ultimately benefit more people.

The poverty and water nexus

Poverty is a complex mix of political, economic, environmental, socio-cultural and historical factors. Here the focus is on the links between water, poverty and agriculture. In doing so, we distinguish between access to water (the ease of obtaining it) and water availability (the total amount of water available from all sources), because it is access to water and not its availability that most affects people's livelihoods. Limited access to water reduces agricultural productivity, aggravates poverty and limits the capability of poor farmers to cope with risks, such as drought.

We also distinguish between poverty and vulnerability – people's capacity to respond to the risks they confront in pursuit of their livelihoods – and the loss in wellbeing that the risk causes. People's response to risk, in turn, is determined by the options available to them in their livelihood assets, as well as by the strategic, policy and institutional environments that limit the negative impacts the risk might cause. Poverty is fixed in time and measured as economic wellbeing while vulnerability can happen at any time and is usually caused by particular events. Water risks and the incapacity to deal with them add to the vulnerability of the rural poor in the basin, while interventions to address the risks will build resilience, which will reduce poverty.

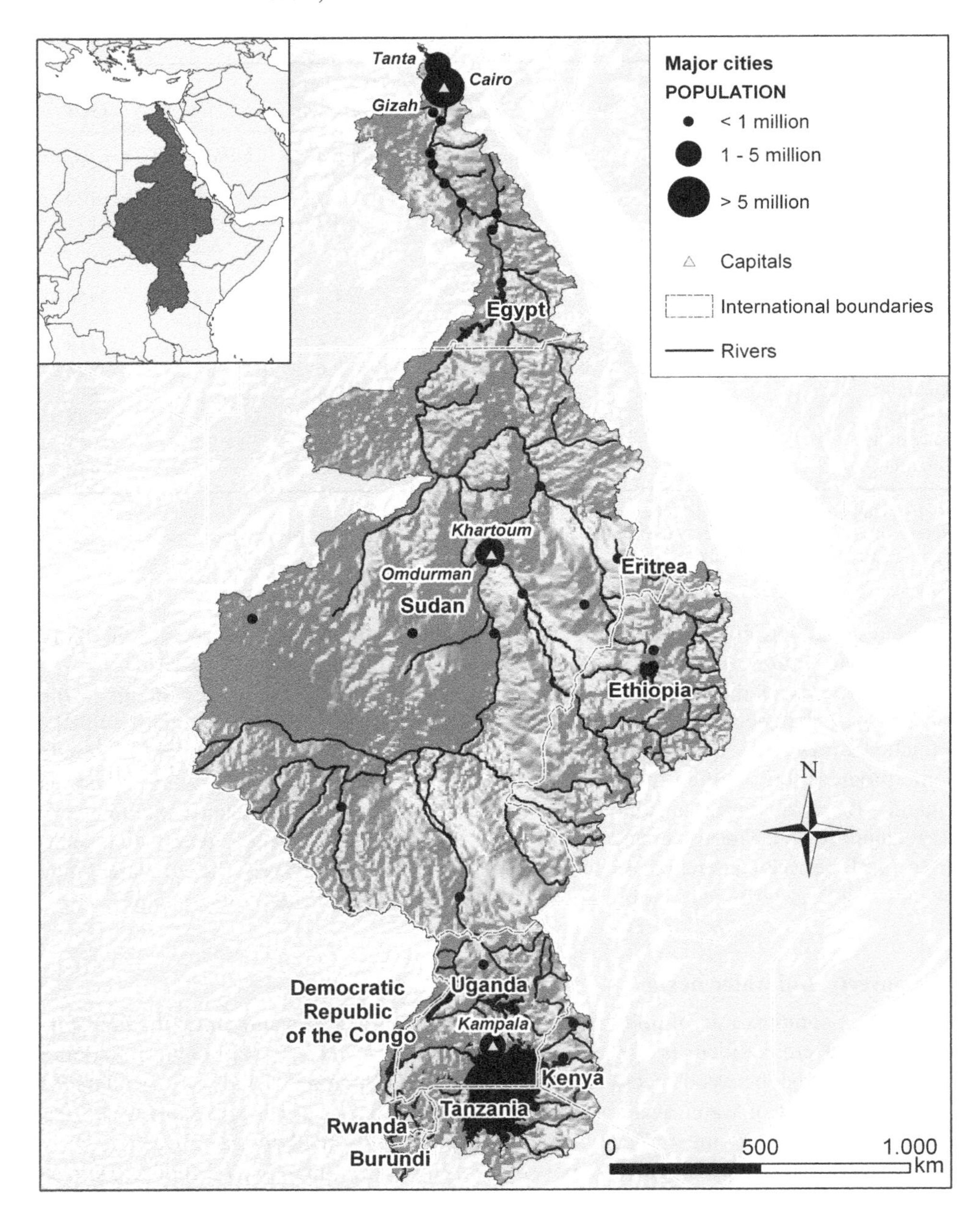

Figure 1. Map of the Nile Basin.

Poor households are those whose expenditure per adult equivalent falls below an absolute poverty line that reflects the monetary cost of meeting certain basic requirements of life, including both food and non-food items (income poverty) (ILRI 2002, PEAP 2004, Cook and Gichuki 2006, UBOS 2007). Food security, poverty levels and poverty inequality were used as indicators to map rural poverty in the basin. Levels of income poverty at sub-national level (Figure 2a) ranged from 17% in Egypt to over 50% in five of the 10

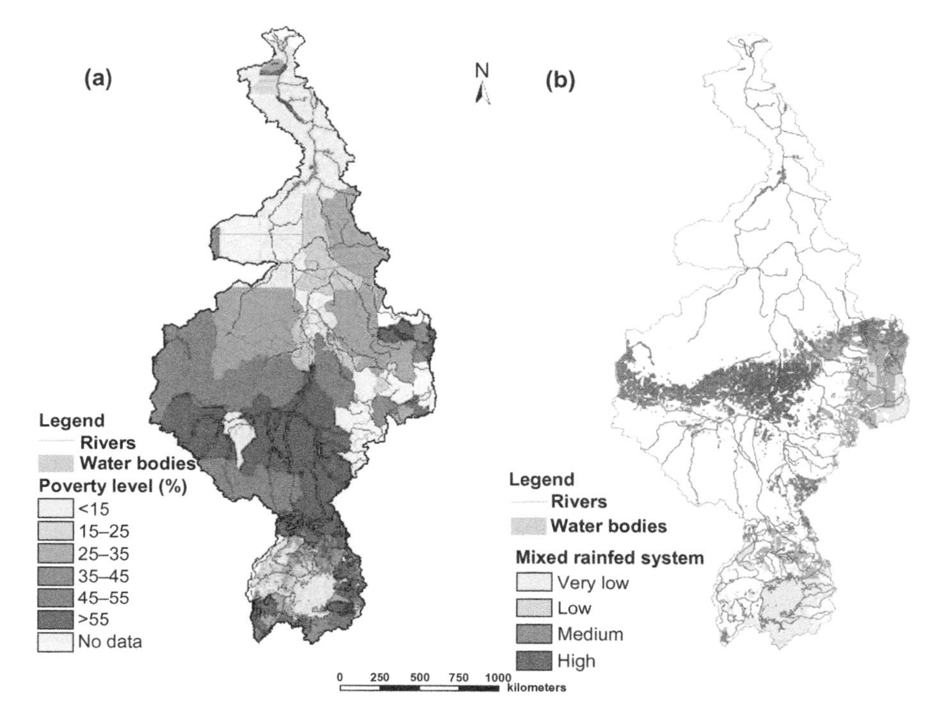

Figure 2. Poverty in the Nile Basin: (a) aggregate basin level; and (b) rainfed systems.

basin countries. Kinyangi *et al.* (2009) and Molden *et al.* (forthcoming) give details of the methodology and poverty profiles in various production systems.

We identified "hotspots" of poverty within each agricultural system (Figure 2b). Both pastoral and agropastoral (crops and livestock) rangeland systems have high levels of poverty (Figure 2). Sorghum and millet are the main food crops but are rarely marketed for income. Drought causes crop failure, and sudden food shortages. The central and southern parts of Sudan, the lakeshore region of northwestern Tanzania, and northern Uganda, where drought is common, have high incidence of poverty. Food insecurity is widespread in the rangeland systems of Sudan. Increased grain prices in times of drought worsen the situation and many households suffer food deficits. Options to reduce food-related poverty include increasing income from livestock by raising livestock water productivity through greater integration with crop farming.

We used three major factors: water availability and water accessibility; the biophysical resource base; and prevailing socio-economic conditions as proxy indicators of vulnerability (Thornton *et al.* 2006). We combined them in a probability function and mapped the data. There are several methods to measure water poverty (Sullivan 2002). The contribution of the location of water sources in agricultural systems, relative to the spatial demand of people and livestock were estimated. Figure 3 shows the framework developed to link vulnerability, water and poverty.

The first of the three components of water-related poverty is the water system, whose characteristics determine access to water and its availability for use in agriculture in rural areas, and for domestic/industrial uses in urban settlements. The second is the social condition of the system where water is allocated through institutional infrastructure. Third

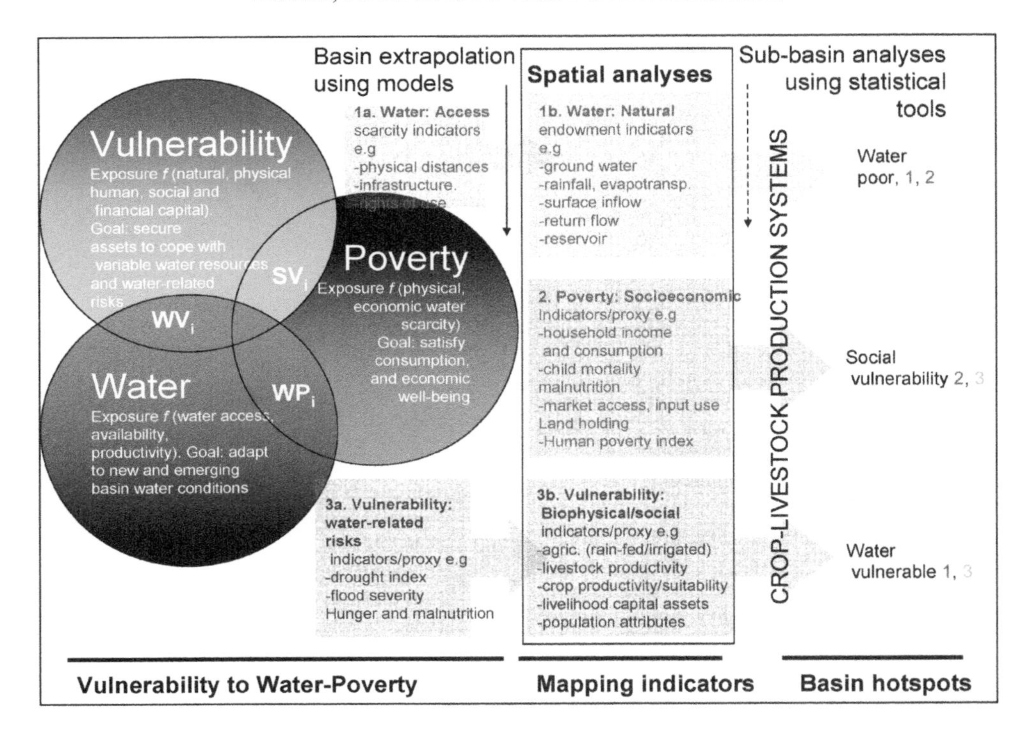

Figure 3. Conceptual framework linking vulnerability, water and poverty.

is the natural and human environment that modifies access to water. This framework links water, food and poverty by integrating water-related poverty and vulnerability, which can be explained through:

(1) Mapping poverty using household income and expenditure data. Poverty levels are higher in rural agricultural areas compared to urban areas. In the Nile Basin, with the exception of Egypt, rural poverty accounts for 90% of total poverty and nearly 80% of the poor depend on agriculture and farm labour as their main sources of livelihood;
(2) Characterizing the water-poor, people who are deprived due to physical and economic water scarcity. Most of the water-poor are in degraded and deforested rangelands, and in mixed rainfed systems that have poorly developed infrastructure, which limits access to water; and
(3) A dynamic livelihood system characterized by biophysical and social vulnerability and a weak base of capital assets, which indicates an incapacity to adapt to water stress. Areas with high vulnerability scores in rangeland and mixed rainfed systems are associated with low crop and livestock water productivity.

Kinyangi *et al.* (2009) and Molden *et al.* (forthcoming) provide full details and maps.

Within all agricultural systems there is low to medium risk of biophysical shocks due to water hazards. Mixed irrigated systems have low vulnerability while pastoral, agropastoral and mixed rainfed production systems are highly vulnerable. Poverty and access to water are linked through crop- and livestock-based livelihoods, while increasing the systems' water productivity can contribute to poverty reduction and food security. High levels of poverty in the basin are partly caused by dependence on rainfed agriculture and limited access to managed water.

Labour is the biggest income-generating asset for most people. Those with good access to water use it productively to grow food and for small-scale cottage industries. In contrast, poor access to water reduces a household's labour supply because of the time it takes to carry water.

Ill-conceived policy can be disastrous. Herders in the Nakasongola District of Uganda's cattle corridor invested in water harvesting (valley tanks). In the 2009 drought, local government authorities allowed this water to be used for domestic purposes and prohibited livestock watering, so that herders were forced to take their animals to Lake Kyoga for watering. Large concentrations of domestic animals at the lake soon depleted the feed and led to outbreaks of animal disease. In spite of having invested in water harvesting, herders lost 20–30% of their animals to disease, and the survivors lost weight and were in poor condition (Kinyangi *et al.* 2009).

Accounting for water availability and access

To identify interventions in water management that can improve poor people's access to water requires an understanding of the water available within the basin. The water resources and hydrology of the Nile are well documented (Hurst 1957, Shahin 1985, Sutcliffe and Parks 1999, Conway 2005, ENTRO 2006). The river rises in the high rainfall areas of the Equatorial Lakes Region (ELR) and the Ethiopian highlands. Rainfall is bimodal in the ELR and unimodal in the Ethiopian highlands, varying from over 2000 mm/yr at the sources to nil in the Sahara desert. Total rainfall in 2007 was estimated at 1745 km^3. Figure 4 depicts the rainfall distribution, potential evapotranspiration (PET), and rainfall (P) minus PET to describe runoff sources and identify deficit areas within the basin (Demissie *et al.* 2010).

The Nile consists of eight sub-basins: Lake Victoria; the Equatorial Lakes; the Sudd; Bahr el Ghazal; Sobat; Central Sudan; Blue Nile Basin; and the Atbara Basins (Conway and Hulme 1996). Downstream of Khartoum (the confluence of the Blue Nile and the White Nile) and the Atabara is a ninth 'main Nile' sub-basin (ENTRO 2006), while downstream of the Aswan High Dam (AHD), the tenth sub-basin consists of the Nile between the dam and the Mediterranean Sea, including the Delta. In the Appendix, Figure A1 shows the major tributaries, lakes, water balance and major dams constructed or planned by 2008.

The Blue Nile contributes 57% of the mean river flow at AHD, the White Nile 29% (48% of which is from the Sobat), and the Atbara River 14%. The Eastern Nile catchments (Sobat, Blue Nile and Atbara River) contribute 85% of the Nile flow at the AHD. Evaporation from the natural and manmade lakes and wetlands in the basin is important. The White Nile flow varies little because of the evening-out effect of lakes and wetlands, in contrast to the seasonal flow of the Blue Nile. The old Aswan Dam was built 200 years ago to decrease the sharp seasonality of the Nile flows. Subsequently, numerous dams and barrages were built in Egypt, Sudan, and more recently in Ethiopia (Appendix Table A1).

The AHD, with 162 km^3 of storage, is the largest reservoir in the basin. Average flow (1900–50) into Egypt at Aswan was 84.1 km^3/year. The 1959 Treaty of the Nile between Egypt and Sudan is based on this figure, with Egypt's allocation 55.5 km^3/yr and Sudan's 18.5 km^3/yr, and the remainder as flow to the sea. The Treaty makes no reference to upstream countries. Both Egypt and Sudan have based large-scale investments in irrigation

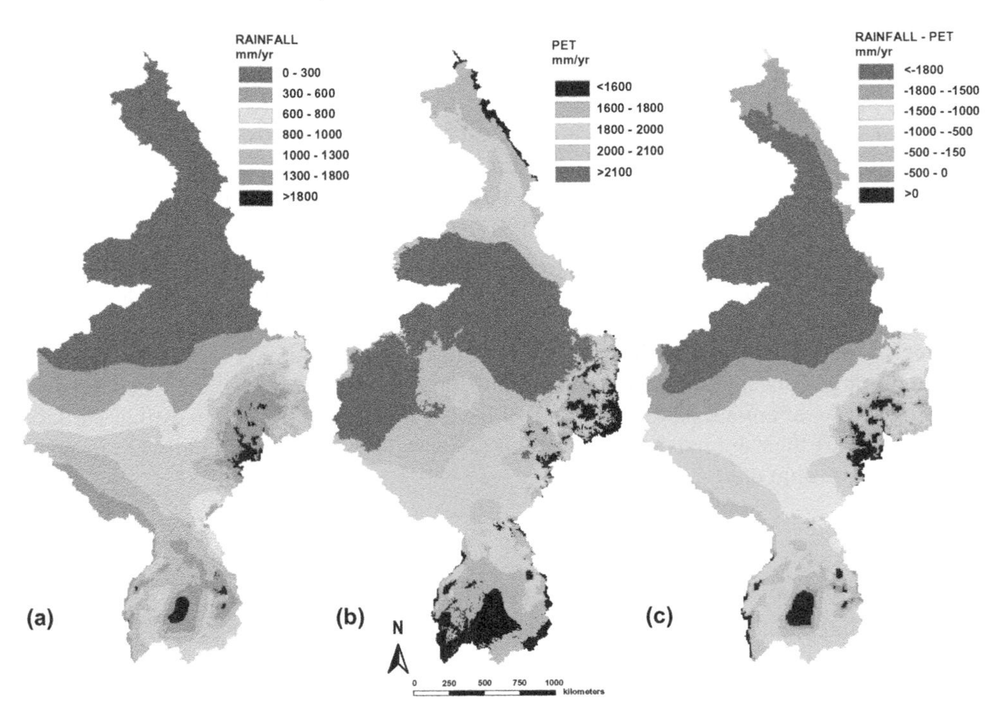

Figure 4. Water sources and sinks in the Nile Basin: (a) rainfall distribution; (b) potential evapotranspiration; and (c) potential runoff.
Source: IWMI 2009.

infrastructure on this figure. However, it is unclear how accurate this figure is, particularly as the Nile flow recorded at Aswan is highly variable. Figure 5 shows long-term mean annual flows at Aswan and releases at the AHD after construction. The average annual flow of the Nile between 1872 and 2002 was 88.2 km^3. It is, however, difficult to ascertain the precise amount of inflow into the AHD from public records, as the flow represents a composite number rather than an actual flow measurement (Sutcliffe 2009). The total average outflow from the AHD during its operational life through 2002 was 62.8km^3.[2] The flow is likely to change in the future with climate change and upstream development.

Calculating water availability from river flows, rain gauges and groundwater studies is straightforward, depending on the number and quality of measurements. In contrast, quantification of water access (the ease of obtaining water) in a large basin is not straightforward. People obtain water from many different sources including from rain, river, groundwater and storage structures, which vary from the small drinking water ponds in the plains of Kurdfan in Sudan, to the large AHD. Water availability in the basin, whether from rainfall, or from rivers, lakes, reservoirs, and groundwater varies across the basin, yet even where water seems plentiful, access may be difficult. Conversely, as in Egypt where rain is scant, people over millennia have developed the means to access the Nile's water.

The hydroclimatology of the Nile is temporally and spatially variable. The short rainy season (June to September) provides more than 80% of the Nile water. Rainfall and vegetation cover ranges from equatorial forest in the south, to semi-arid and arid climates in the central and lower parts of the basin. This variability is reflected in the agricultural

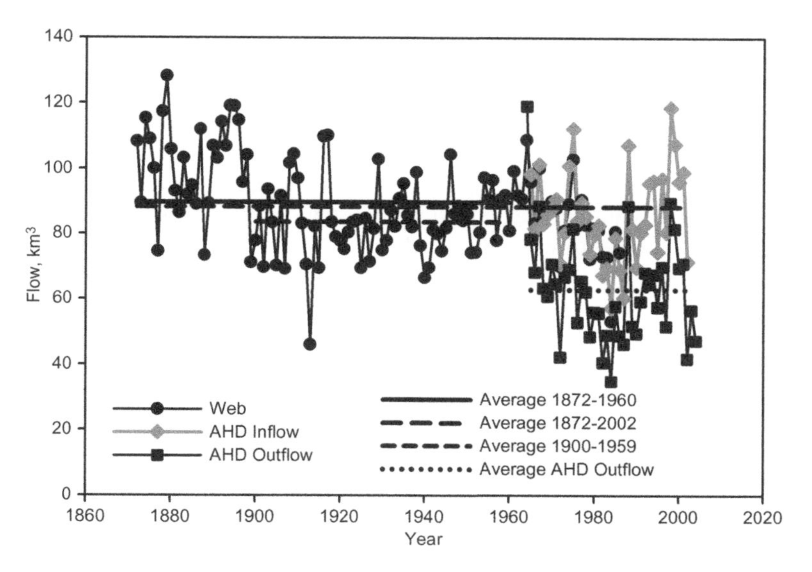

Figure 5. The long-term mean annual inflow and releases of the Nile at AHD.
Source: Various.

systems in the basin. The upper parts of the basin in Ethiopia, the ELR, and the middle part of Sudan have vulnerable rainfed systems. Improving access and reducing vulnerability in these areas requires management of water including the rainfall, water in the rivers, and groundwater. In contrast, irrigated agriculture dominates central Sudan and southern Egypt. While access to water in these regions is good, however, water productivity in the Gezira of Sudan is low due to institutional, policy, and market-related problems.

Water use by land cover

We used satellite remote sensing to provide data of combined land and water use, and indirectly to estimate access. For example, from satellite data one can identify irrigated areas where access to water is better than in rainfed systems. We analysed satellite data for 2007 to identify land-use patterns, precipitation, evaporation, transpiration, and dry matter production. We further used the water accounting framework (WA+, Molden 1997, Bastiaanssen and Perry 2009) in the analysis (Mohamed *et al.* 2009). We computed land and water accounting at a spatial resolution of 1 km and then aggregated them to identify 15 major land-cover classes in the basin (data not presented).

The Nile water balance for 2007

The annual water balance of the Nile was assessed as (Q_{out} = P-ET), using the outflow to the Mediterranean (Q_{out}) as the closing component. Values of Q_{out} in the literature differ: Molden (1997) estimated 14.1 km^3/yr for 1989/90, Oosterbaan (1999) gives nine to 11 km^3. The outflow can also be estimated as the difference between the AHD release and water consumption downstream. Data for outflows from AHD vary (55.5–73.6 km^3/yr) as do water uses in Egypt (31–68.3 km^3/yr) (Aquastat-FAO 1997, Hefny and Amer 2005), giving outflows of 9–30 km^3/yr. For 2007 and 2008, the outflow was 25.7 km^3 (including environmental flows) (Droogers *et al.* 2009), corrected to take account of rainfall interception by the different land cover classes (Gerrits *et al.* 2007). Nile outflow to the Mediterranean and other losses are uncertain and the potential to expand irrigation further thus remains unclear. A committed outflow to the Mediterranean is estimated at 9.8 km^3/yr for leaching salt in the delta and for aquatic ecosystems in the coastal region (Molden 1997), although this figure requires further analysis.

Water accounting results

The sole supply entering the basin is precipitation, which is either evaporated, transpired, or flows into rivers or groundwater (Table 2). Evapotranspiration accounts for >98% of the total precipitation, leaving 29 km^3/yr as outflow to the Mediterranean. Some of the outflow component may remain in the AHD reservoir as carryover storage and be spilled if the reservoir capacity is exceeded. Managed land use, that is, land that is used mainly for rainfed agriculture, uses 189 km^3/yr, and managed water used by irrigation, cities, and industry is 69 km^3, one third that of rainfed agriculture. River flow is only a small portion of the total rainfall amount. ET from pastoral lands is the greatest water consumer, followed by rainfed agriculture, then irrigation (Table 2).

Uncertainty in the ET and precipitation estimates gives uncertainty in the outflow term. This is, however, a critical figure for the basin as it gives an indication of how much more water could be developed and consumed through managed water use (irrigation, cities and industries). If we assume a committed environmental outflow at the mouth of 9.8 km^3/yr, the water available for use in irrigation, cities and industry is reduced to 76.6 km^3/yr, or approximately 4% of the total precipitation. The small amount of water available reflects a high degree of development with intensive diversion (57km^3 is diverted out of 77km^3 available). Moreover, the greatest potential source of water is rainfall before it reaches the river. While irrigation is often regarded as the major water user, forests and savanna consume (by ET) 21 times more water.

Table 2. Results of water accounting in the Nile Basin for the year 2007.

Inflows and outflows	km^3/yr	Water depletion	km^3/yr
Inflow (precipitation)	1745	Natural and pastoral	1458
Evapotranspiration	1716	Rainfed	189
Outflow	29	Irrigation, cities, industries	69
		Total	1716
		Depleted fraction	0.98
Blue water accounts			
Total blue water	98	Outflow plus ET by irrigation, cities, industries	
Environmental outflow	10	First order rough estimate	
Available for depletion	88		
Depleted	69		
Depleted fraction	0.78		

Climate change

While it is agreed that climate change will increase temperatures in the basin, different global circulation models (GCMs) give different predictions of rainfall (Strezepek *et al.* 2001, Conway 2005, IPCC 2007). The impacts on hydrology and in particular on runoff in the Nile are therefore uncertain. Elshamy *et al.* (2008) analysed rainfall and potential ET as predicted by 17 GCMs for the period 2081–98 to compute Blue Nile flow at the Ethiopia–Sudan border. Temperatures increased 2–5°C, and potential ET 2–14%. Precipitation changed -15% to +14%, with a mean of no change. Current inter-annual variability of Nile precipitation is determined by several factors, of which the El Niño-Southern Oscillation (ENSO) and the sea surface temperature over both the Indian and Atlantic Oceans are claimed to be the most dominant (Farmer 1988, Nicholson 1996). It is unclear what the influence of the predicted increased frequency and magnitude of ENSO events (IPCC 2007) will have on the Nile.

Agricultural and water productivity

After access to water, water productivity (the amount of water required to produce one unit of yield) is a key driver to improve livelihoods. We analysed the various production systems of crop, livestock, and fisheries in seven major regions of the basin, based on their dominance of farmers' livelihoods.

Mapping production systems

We used a classification of farming systems (Seré and Steinfeld 1996, Dixon *et al.* 2001) to describe and analyse opportunities and constraints for development (Otte and Chilonda 2002). Where relevant, we added the type of main crop to make an explicit link to food production (Notenbaert *et al.* 2009) (Figure 6). We grouped farming systems in the basin into irrigated agriculture, rainfed agriculture, livestock, fisheries, aquaculture and multiple-use systems. Mixed rainfed agricultural systems (crops plus livestock) and agropastoral systems are the largest agricultural land uses in the Nile Basin.

Crop productivity and farming systems

We calculated water productivity for the basin using secondary agricultural statistics and satellite data to estimate the standardized gross value production (SGVP) of each crop and the actual depleted water from cropped areas (Karimi and Molden personal communication, 2009, Molden *et al.* forthcoming). There is considerable variation in land and water productivity across the basin (Figure 7). Over almost two thirds of the basin, rainfall is insufficient to meet the water demand of crops so that water productivity is low. This results in the contrast between Egypt, which produces high yields of high-value crops such as wheat and maize, and Sudan, where both yields and value of rainfed crops is low. It also emphasizes the importance of irrigation in increasing water productivity.

Livestock productivity

The Nile Basin has large agro-pastoral and pastoral areas, which use approximately 45% of the total water used in the basin (Kirby *et al.* 2010). Most estimates of water use by livestock are either at a local scale or are based on regional or even global assumptions

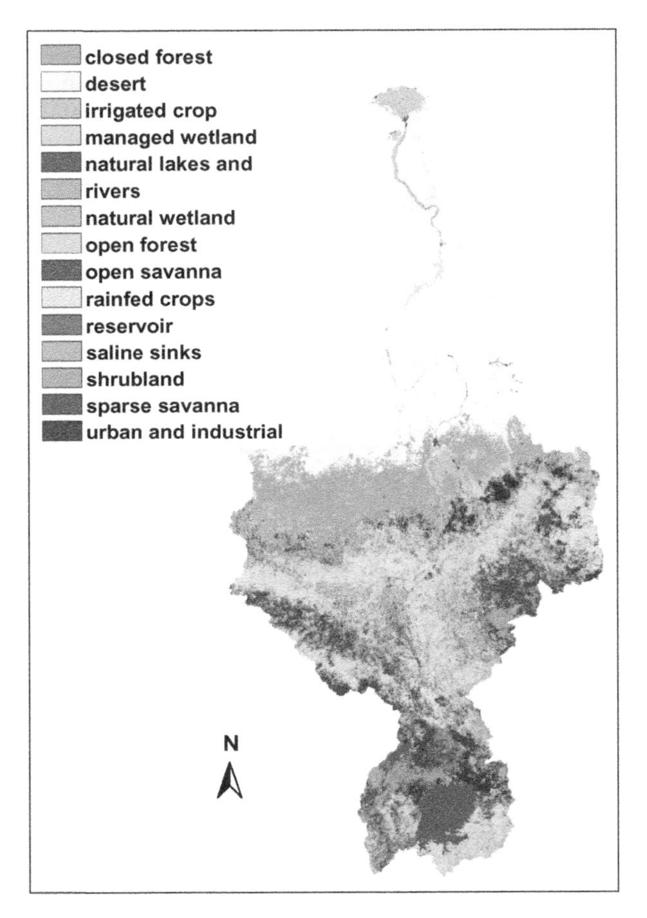

Figure 6. Global land-cover classes in the Nile Basin in 2008, defined according to the UN land cover classification system (LCCS).

Source: European Space Agency (2010).

for feed intake and water requirements. We used livestock densities and the availability of water and feed to calculate the ratio of water use by livestock compared with the total water used (Figure 8) and to derive livestock water productivity (LWP), the ratio of the outputs from livestock to the amount of water depleted in producing them (van Breugel *et al.* 2009). LWP largely depends on water productivity of the feed. Furthermore, because livestock are an integral part of many of the Nile's production systems, LWP needs to be accounted for as a component of agricultural water productivity (van Breugel *et al.* 2009).

At the basin level, water use by livestock is only a small proportion of total water depleted through evapotranspiration (Figure 9), although there are considerable differences across the basin. Lack of water limits the availability of feed in a large part of the basin's arid and hyper-arid areas. In other areas, however, livestock consume only part of the plant biomass and hence account for only that fraction of the water depleted through evapotranspiration. Because of low milk and meat production, water productivity of livestock in the basin is very low, but is highly variable. The feed resource is often degraded through injudicious grazing. A key intervention is rehabilitation by reseeding and grazing management, which is widely applicable at different geographical scales.

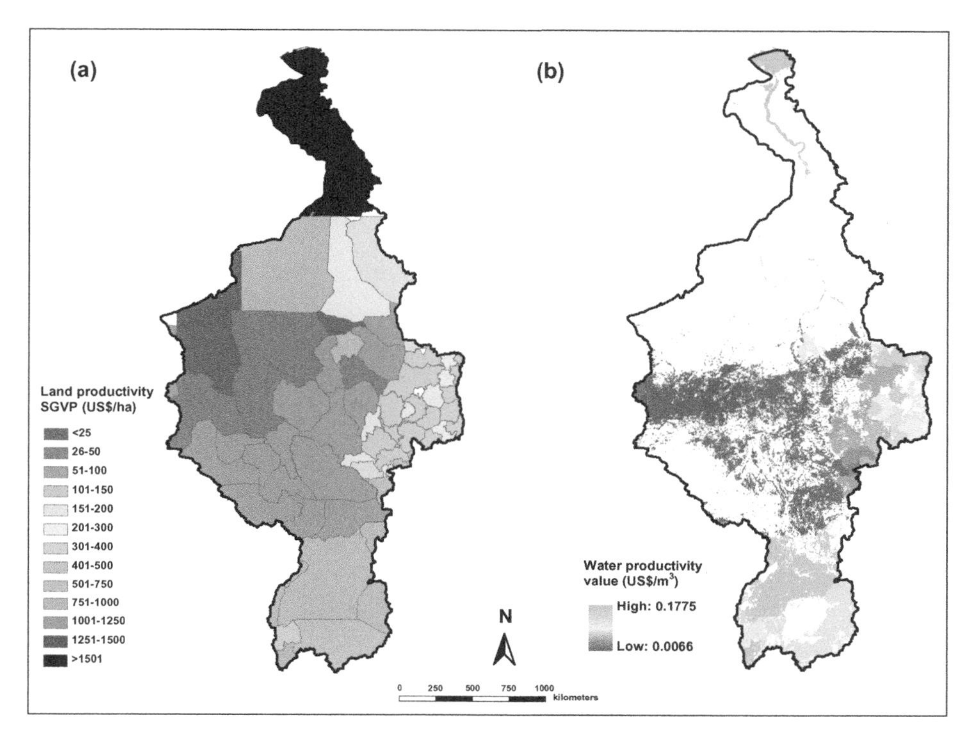

Figure 7. Productivity in the Nile Basin: (a) standardized gross value of land productivity (SGVP); and (b) water productivity based on SGVP/ET_a

Source: P. Karimi and D.J. Molden (personal communication 2009).

Fish production systems in the basin

Harvest fisheries are important for food security and employment in the Nile Basin. Fisheries are mainly in freshwater lakes, rivers and marshes, although Egypt and Sudan also have well-developed marine fisheries. Lake Victoria, shared by Tanzania (51%), Uganda (43%) and Kenya (6%), supports the biggest freshwater fishery in Africa, producing 1Mt/yr, although it is threatened by overfishing and pollution. Laws requiring larger net sizes are being implemented to protect the now-scarce Nile perch and reduce overfishing. Pollution causes algal blooms that deoxygenate the water, and invasion by water hyacinth has reduced fish numbers and productivity. Measures are needed to reduce pollution and to prevent further decline.

Aquaculture in the Nile Basin produced 700,400 tons of fish in 2007, valued at over US$1 billion (Table 3). Egypt is the largest producer with aquaculture increasing its share of the total fish produced from 16% to 56% between 1997 and 2007 (WorldFish 2009). Uganda is a distant second producer with Rwanda, Kenya and Sudan developing aquaculture using foreign aid.

Focus studies: opportunities for improvements

To illustrate the variability in water use by agriculture across the Nile Basin, we present seven focus studies that show how people use water and land resources, the threats and challenges they face, and the opportunities for improvement.

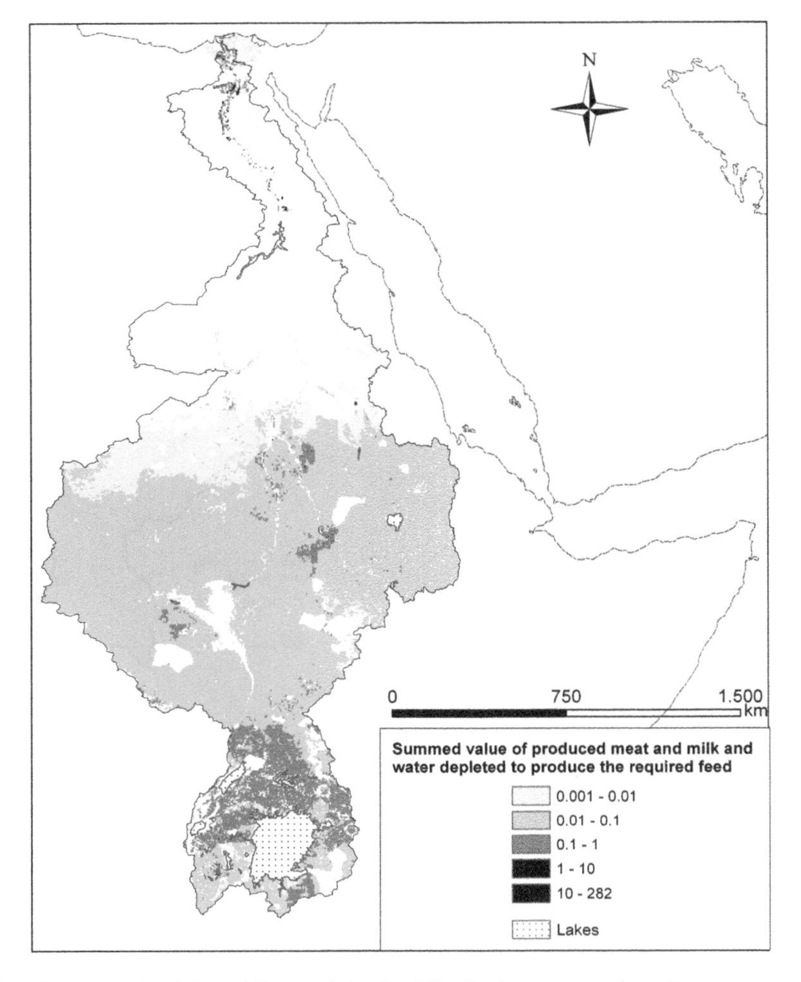

Figure 8. Water productivity of livestock in the Nile Basin, expressed as the monetary value of meat and milk produced for the water depleted to grow the animal feed.

Uganda's cattle corridor

Uganda's cattle corridor covers 84,000 km^2 where most people herd livestock. Poor households in the corridor confront multiple risks; they have insecure access to water and pasture as well as food deficits and disease (Peden *et al.* 2009). Moreover, they are usually far from water points, which they often share with livestock. Most household water is fetched by women. Many water sources dry up in the dry season, so that harvesting and storage of rainwater is a solution. Individuals who do so are perceived to be less poor as they are likely to have more livestock and to be food secure. In contrast, poor households have only small landholdings of often degraded land, are dependent on seasonal water and are also likely to be food insecure.

Rainwater harvesting with valley tanks requires more land than that typically owned by poor farmers. For it to be useful for the poor, community-based water and land management is needed, which requires appropriate institutional arrangements (Molden *et al.* forthcoming).

Table 3. Aquaculture production quantity and value in Nile Basin countries in 2007.

	Quantity		Value	
Country	*ton*	*% of total*	*USD 000*	*% of total*
Burundi	200	0.03	600	0.04
DRC	2970	0.42	7,435	0.56
Egypt	635,500	90.73	1,192,614	89.41
Kenya	4240	0.61	6311	0.47
Rwanda	4040	0.58	7327	0.55
Sudan	1950	0.28	3840	0.29
Tanzania	410	0.06	102	0.01
Uganda	51,110	7.30	115,662	8.67
Total	700,420		1,333,891	

Source: FAO (2007).

Lake Victoria

Lake Victoria, with an area of 68,800 km^2, is the largest lake in Africa. It is surrounded by a densely populated region supporting a variety of livelihoods. There is huge hydropower potential in the rivers feeding the lake. A key point is that only 15% of the inflow to the lake flows from it to the White Nile: the remainder is evaporated. The region's climate is unstable in a timeframe of centuries; it has recently changed from prolonged drought to less-than-normal rainfall, but drought is again likely within a decade (Swenson and Wahr 2009). The "Agreed Curve" specifies the permitted outflows depending on the level of the lake. To meet its increased demand for power, Uganda has released 55% more water than the Agreed Curve (Kull 2006), which, coupled with several years of low rainfall, reduced the lake to an 80-year low (Mubiru 2006). Low water levels have acute consequences for navigation and for fisheries because shallow waters in the coastal areas, which are important for many fish species, dry up. Co-operation between all concerned authorities is necessary to identify coherent solutions that ensure the sustainability of Lake Victoria.

Lake Victoria supports Africa's biggest lake fishery. Despite declining catches of several species, the fishery maintained a high variety until the 1960s. Many species became extinct by the early 1970s (WEMA Consultants 2009) and were replaced by the introduced Nile perch (*Lates niloticus*). There were initially good catches of Nile perch, but it is now over-fished and its population greatly reduced (Table 4). Paradoxically, numbers of several threatened endemic species have increased. The lake's fisheries are now threatened by pollution, illegal fishing and invasion by water hyacinth. Subsistence agriculture

Table 4. Lake Victoria species % and total landings from commercial fishing.

Fish species	2000	2005	2006
Nile Perch %	42	29	24
Dagaa %	40	48	54
Haplocromines %	–	13	13
Tilapia %	17	9	7
Other species %	1	<1	<1
Total Landings (000 tons)	620	804	1061

Source: LVFO (2006).

in the catchment on highly erodable land has increased the silt burden of inflowing rivers. Pollution comes from municipal and industrial discharge into rivers feeding the lake and on the shoreline and causes algal blooms, which deoxygenate the water and suffocate fish. With the riparian populations growing at rates among the highest in the world, the multiple activities in the lake basin have increasingly come into conflict.

The Sudd wetland

The Sudd is a wetland located along the White Nile River in southern Sudan occupied by the Nilote people. It is the largest inland wetland in Africa averaging 30,000 km^2 and expanding as large as 130,000 km^2 in the wet season. Agriculture, pastoralism, fishing and hunting are the major economic activities (RIS 2006). The Sudd also provides flood regulation, essential grazing lands for livestock and fisheries, and supports high levels of biodiversity.

There is debate about the water balance of the Sudd (Mohamed *et al.* 2006), mainly how much is lost to evapotranspiration and how much contributes to downstream flow (Figure A1). From 1905 to 1983 average annual inflows and outflows to and from the Sudd were 33.3 km^3 and 16 km^3 respectively (Sutcliffe and Parks 1999), so that about 50% of the inflow evaporates or transpires. The Sudd acts as a flood regulator in that increased inflow does not give a proportional increase in outflow (Howell *et al.* 1998), because floods increase the surface area so that there is more evaporation.

The 360-km-long Jonglei Canal was proposed to bypass the Sudd, reduce evaporation losses and provide more downstream flow. Only 240 km was dug when work was stopped in 1983 for technical, financial and political reasons (Hughes and Hughes 1992). The canal was expected to reduce the water level of the swamp by 10% during flood season and by 20% during the dry season, greatly reducing the area of the seasonal floodplain. If completed the canal would disrupt the wetland ecosystem, the seasonal movement of livestock, and the seasonal migration of great herds of African wildlife.

There are three Sudd vegetation communities used for grazing (Denny 1991): river-flooded grassland, which is the most productive for year-round grazing because the dead grass has high protein content; seasonally flooded grassland, which includes rain-flooded grasslands, seasonally inundated grassland, and rainfed wetlands on seasonally waterlogged clay soil, all three of which are heavily used by livestock; and floodplain scrub forest, at higher elevations on well-drained soils around the floodplains. The Nilotic pastoralists use a transhumance system to optimize the seasonal flooding and drying cycle. Livestock population is estimated at 1 million head (Birdlife International 2009), one of the highest cattle to human ratios in Africa (Okeny 2007).

Fisheries in the wetland are underdeveloped. In 1989, there were an estimated 7500 fishing canoes, with daily landings of 17–28 kg each and 151,000 people involved (Bailey 1989). Fishing in the Sudd is mostly subsistence, with commercial fishing becoming more important involving at least 1000 commercial fisherfolk in the early 1980s (UNDP 1983). Estimates indicate that southern Sudan could harvest 300,000 t/yr without depleting the fishery resource (Okeny 2007). Fishing falls short of its potential because of poor management, and inadequate policy and legal frameworks. The sector is further constrained by inadequate roads and electricity, and poor access to markets as well as the lack of cold storage facilities. Fish has to be sold or transported immediately, or be sun-dried or smoked.

Data on the Sudd livestock and fisheries are inadequate to guide management and policy interventions, but there are clear needs to promote efficient livestock management.

Lack of infrastructure and access to markets are problems for both livestock and fishing. If these were overcome, the Sudd could make a substantial contribution to Sudan's economy.

The Ethiopian highlands

The Ethiopian highlands consist of the Blue Nile itself (Abbay in Ethiopia) and two tributary sub-basins, Baro-Akobo and Tekeze-Atabara, which together contribute over 85% of the Nile flow at AHD (Awulachew *et al.* 2008, Figure A1). Poverty and environmental degradation are linked directly to population pressure and are the main issues in some parts of the highlands. Soil erosion is a major threat with total soil loss of 303 Mt/yr of which 102 Mt/yr comes from cultivated land (ENTRO 2006, Awulachew *et al.* 2008). About two thirds of soil eroded is from communal grazing and settlement areas. Approximately 45% reaches the stream system, and 33% remains as silt load at the Sudan border. In addition to this critical resource degradation (Woldeamlak 2003), agricultural productivity is low and poverty is widespread, made worse by recurrent drought.

Mean annual rainfall (1961–90) is 1220 mm with 70% falling June to September in a unimodal distribution peaking in July or August. ET_p is higher during the dry season (December to April) than in the wet season. Over 80% of the mixed farming systems in the highland areas are based on cereals followed by an enset root crops complex and coffee. Depending on rainfall, farmers can double crop and there is some shifting cultivation. There are mixed farming systems in the highlands of the upper Blue Nile and pastoral/agro-pastoralism in the lowlands (Figure 9). Mixed farming of cereal-based crops, enset root crops complexes and coffee crops compose one system. The largest proportion of cropping area (over 80%) is cereal-based cultivation, sub-divided

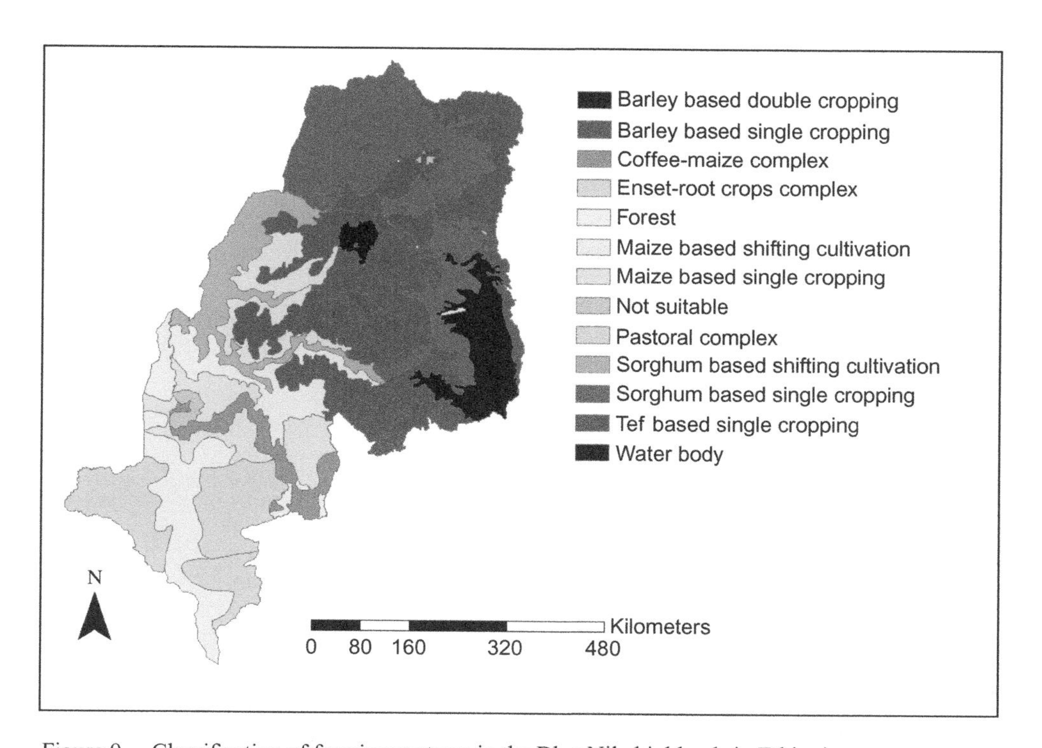

Figure 9. Classification of farming systems in the Blue Nile highlands in Ethiopia.

into single-cropping, double-cropping and shifting-cultivation systems. Wheat and barley dominate the highlands giving way to sorghum and maize at lower altitudes.

The major constraints for crop production are soil erosion, drought, poor agronomic management and lack of mechanization. Soil conservation, improved agronomy and mechanization have huge potentials to increase the productivity of agriculture (BCEOM 1998, NDECO 1998). Moreover, soil conservation would reduce sediment loads and lengthen the service life of reservoirs downstream. There is also potential to expand agriculture in the lowlands with irrigation. Despite limited potential of irrigation in Ethiopia, because the Blue Nile supplies 85% of the water at AHD, Sudan and Egypt are concerned that expansion of irrigation in Ethiopia will reduce the flows available for their use. This is a thorny issue of counting benefits and costs, who receives the benefits, and who pays for them.

The central belt of Sudan

Sudan's central belt extends from its borders with Chad, Libya and the Central African Republic in the west to Ethiopia and Eritrea in the east. The belt covers 75% of the country and 80% of the population (2007), with rainfed crops, irrigated crops and livestock keeping the main agricultural activities. Livestock and rainfed crops are equally important. Poverty, low productivity, under-performing irrigation systems and conflict are major challenges to development. The central belt contributes little runoff to the Nile so that downstream implications of changes in land use are unimportant. The belt contains 73% of Sudan's livestock, which graze rangelands (74%) and crop residues (21%) (Peden *et al.* 2009). Low and variable rainfall jeopardizes feed availability in the rangelands. Demand for animal drinking water exceeds availability in all areas except Khartoum and the Red Sea State with a deficit of 1 M m^3/day at the peak of the dry season (Table 5).

Table 5. Average daily rural drinking water availability, demand, and balance (000 m^3/day) in different states within Sudan's central belt, 2007.

State/Region	Available water	Average drinking demand	Peak drinking demand	Balance at average demand	Balance at peak demand
Red Sea	126.4	20.1	31.7	106.3	94.7
Khartoum	83.2	25.0	28.1	58.2	55.1
Gedarif	55.1	66.4	85.9	−11.3	−30.8
Kassala	44.0	61.4	86.7	−17.5	−42.7
Sennar	32.8	71.6	92.1	−38.8	−59.3
North Darfur	52.4	87.5	115.9	−35.0	−63.5
White Nile	48.2	118.8	156.8	−70.6	−108.6
Gezira	61.5	140.9	170.5	−79.4	−109.0
Blue Nile	19.1	151.9	203.4	−132.7	−184.3
South Darfur	51.1	187.2	235.6	−136.1	−184.5
West Darfur	29.5	172.3	229.3	−142.8	−199.8
Greater Kordofan	244.5	335.2	464.4	−90.8	−220.0
Total	847.9	1438.4	1900.5	−590.5	−1052.7

Source: Available water computed from data of the Ministry of Irrigation; Livestock in 2007 estimated from data of MoARF (2006).

Note: Requirements are calculated according to Payne (1990): average demand 25, 30, 4, 4 l/day for cattle, camels, sheep and goats; at peak summer months, respective values: 35, 65, 4.5, and 4.5 l/day. Human rural requirements are 20 l/day/person according to the Ministry of Irrigation.

Nomadic herders move animals to where feed and water is available, but during the dry season water points are unevenly distributed, so that animals are often unable to access the abundant feed resources that have already utilized water. Water productivity of livestock is very low because of the spatial imbalance between feed and drinking water.

Irrigation in Gezira

The Gezira Scheme is 8800 km^2 of irrigated agriculture that contributes to 3% of the Sudanese national GDP, produces about two thirds of Sudan's cotton exports, and considerable volumes of food crops and livestock for export. It also contributes to national food security and the livelihoods of the 2.7 million people in the area. The Gezira has a perverse impact on downstream flows in that the scheme's low irrigation efficiency returns more water to the Nile than it otherwise would. Both yields and cropping intensities are low. Operation and maintenance of the scheme was highly centralized and inefficient, although it has recently become more decentralized.

The Gezira scheme has undergone four different policy and institutional arrangements during the last four decades, marked by decreasing central management of operation and maintenance, introduction of charges for water, and liberalization in the choice of crops. Until liberalization in 1981, cotton was a mandatory crop, financed and marketed by the government. Subsequently, both the yields and the area sown to cotton fell (Gamal 2009), replaced by food crops like wheat, sorghum and groundnuts, which farmers finance and market themselves. Sorghum is popular as a feed crop with income from livestock in some cases exceeding that from crops.

The *Gezira Act* of 2005 required farmers to manage canal irrigation themselves, which has worsened their overall socio-economic condition (Gamal 2009). Water shortages could develop in Gezira if upstream schemes claim more water for irrigation, whether for increased area or for crops with higher water demand (Guvele and Featherstone 2001). The most important development issue for Gezira is the need to revitalize the scheme to reach its full potential.

The Nile in Egypt and the delta

Over millennia, Egypt has invested in managing the Nile waters to support a highly productive agricultural system. Egypt is the wealthiest country in the basin with the lowest levels of poverty (Table 1), and is therefore the most powerful country within the Nile Basin Initiative, where it wields great influence.

The Nile downstream of the AHD is completely regulated; water is released through a control system that serves irrigation, hydropower and cities. The capacity of Lake Nasser is 162 km^3, about twice the annual flow of the Nile, so that the discharge from the dam can be regulated throughout the year (see Figure 3). This security comes with a cost of evaporation losses of 10–14 km^3/yr, depending on the lake level, and up to 1 km^3/yr flowing to groundwater (Aly *et al.* 1993).

Egypt's irrigated agriculture performs well in terms of crop productivity, crop value and net returns. Water productivity of crops, livestock and aquaculture in the irrigated areas are high in spite of physical water scarcity (high use compared to water availability). Domestic needs have increased to approximately 8.8 km^3/yr thus decreasing the quantity of water available for agriculture, although water from irrigation and cities is reused downstream, leading to high efficiency of the system as a whole (Molden 1997). In the future there will be increasing pressure to serve cities, grow more food and meet growing environmental concerns.

Aquaculture in Egypt has boomed recently in the delta providing both dietary protein and economic return (WorldFish 2009). In 2006, aquaculture at Kafr el-Sheikh produced 295,000 tons of fish, of which an estimated 20% was sold within the area. Water use, measured by seepage and evaporation from ponds, was 1.12–3.61 m^3/kg fish, comparable in mass to crops, but much higher than crops in cash terms (Molden *et al.* forthcoming). While economically important, aquaculture's impacts on poverty are less clear. Over 330,000 people, 45% of the workforce, work in agriculture and fishing in Kafr el-Sheikh, and less than 10% in aquaculture (WorldFish 2009). Fish consumption per capita has increased dramatically without either imports or stressing the highly exploited capture fisheries, and at lower prices, which benefit consumers.

The environmental flow into the Mediterranean needs to be about 10 km^3/yr (Molden *et al.* forthcoming). Current outflow is estimated at 13 km^3/yr, but water balance estimates from remote sensing suggests the outflow is higher. If verified, this would represent water that could be productively used in Egypt or even further upstream.

Weaving water and food into a single development process

The previous sections have provided a general overview of the basin and introduced critical issues related to water and agriculture. In this section the focus is on themes that are crucial to development in the basin, and how interventions might transform them. Many reports consider only the narrow strip of the Nile River giving little attention to the entire landscape of the basin and the development opportunities for all the water and land resources. If one considers all the rain that falls on the basin as Nile water, poor people are not limited only to water opportunities within the riparian zone.

Water, derived from rainfall, albeit with varying runoff coefficients, is available in both upstream and downstream countries. As noted previously, water access is not the same thing as water availability. Most studies focus primarily on the water in the river without recognizing that it is access to water that makes a difference to people's livelihoods. It is emphasized here that rain is the main water resource in the basin and that the main depletion pathway is evapotranspiration (ET) from the landscape as a whole.

The current study also considers different management options for agricultural water, ranging from soil water conservation to large-scale irrigation, looking for opportunities within the whole landscape. The role of fish and livestock is considered as well as the importance of other ecosystem services to support livelihoods in different parts of the basin. Finally, it is recognized that policies and institutions are the driving force behind access and productivity, and that policies and actions outside of the river such as trade, or livestock management, influence the river itself.

Sensitivity of the basin water balance: where can gains be made?

The question of how much water is used in the Nile Basin and who uses it is important. The water accounting above detailed water supply, water use and production patterns within the basin. Based on land cover, rainfall, and a satellite-derived map of evaporation, 1745 km^3 of rainfall was estimated for 2007. Of this, ET of the natural land cover accounts for 1458 km^3/yr (85%) followed by ET from managed land use of 189 km^3/yr (11%), and managed water use of 69 km^3/yr (4%). It is unclear how much water in Egypt evaporates and seeps from Lake Nasser, or how much flows to the Mediterranean. The aggregate for 2007 varies from 13 to over 20 km^3.

Productivities of managed lands in the rainfed system of Uganda, Sudan or Ethiopia are low so that important gains are possible. Similarly, productivity of irrigated crops and water use efficiency (except in Egypt) are low and can be improved. Major gains could be made through shifting non-beneficial water uses such as non-productive evaporation to beneficial water uses such as crop transpiration. There appears to be scope for increased irrigation withdrawals, but the expansion depends on how the entire Nile is managed and this expansion is limited. The amount of additional irrigation is also not fully clear because data of actual water use are not available so that a full analysis is impossible. There is, however, certainly scope for improving the water productivity of rainwater and irrigation in some areas.

Potentials of increasing productivity and consequences

Agricultural productivity responds both to variations in crop yields and the production of high-value irrigated crops. It is lowest in low rainfall areas where there is no irrigation, with rainfed wheat in Sudan yielding less than 0.2 t/ha, but 1.6 t/ha in the Ethiopian highlands, while irrigated wheat in Egypt yields 6.5 t/ha. Rainfed maize in Ethiopia or Uganda yields 2 t/ha compared with 8 t/ha with irrigation in Egypt.

Rainfed agriculture, supported in some places by small-scale irrigation (SSI) and water harvesting systems, is the dominant form of agriculture in the upstream countries, whereas Sudan and Egypt are dominated by irrigated agriculture in large-scale irrigation (LSI) schemes. In the transition areas, the system is dominated by pastoralist/agropastoralist systems.

Rainfall management strategies include: (1) farm water management; (2) maximizing transpiration and minimizing soil evaporation; (3) collecting excess runoff from farm fields and using it during dry spells as supplementary irrigation; (4) drainage of water-logged farm areas without damaging vital environmental services, such as vertisol farm fields; (5) enhancing livestock productivity; and (6) using stream flows and groundwater through technological interventions to produce water for supplementary and full irrigation.

The choice of strategies and interventions is not straightforward and requires an understanding of the biophysical, technical, institutional, social and environmental factors that determine their sustainability. A comprehensive list of agricultural water management (AWM) interventions that are common in the basin for smallholders and that can enhance agricultural water access in rainfed, small-scale irrigation and livestock production systems was developed. The generic tabular matrix identifies AWM interventions for water control, pumping, conveyance, and application, which are customized to the sources of water (rainfall, surface water and ground water) including re-use and drainage (Molden *et al.* forthcoming). A multi-criteria analysis (MCA) was also developed as a spreadsheet for linking AWM technologies according to their suitability for different agricultural production. The spreadsheet identifies variables that are relevant for decision making, and that can be assigned weights according to expert opinion. The MCA brings together numerous variables and identifies factors to guide the identification of suitable interventions (Molden *et al.* forthcoming).

The impacts of interventions with respect to productivity, poverty and food security can be evaluated by various methods such as analysis of impact with and without interventions, as well as through modelling. It was not possible to obtain basin-wide data for impact studies, so the study was restricted to the Ethiopian highlands for which detailed data were available.

The average productivity of some farming systems, such as teff, is less than 1 t/ha. Moreover, regardless of the farming systems and the crop species, the productivity in the Blue Nile is lower than the national average, and much less than the potential of the crops.

Yields of some crops in the Ethiopian highlands can be vastly increased by using improved varieties and suitable agronomy (Erkossa *et al.* 2009). For example, planting on tie-ridges increases the yield of maize, sorghum, wheat and mung beans from 50 to over 100% compared with traditional planting on flat beds. Drainage of waterlogged vertisols can increase wheat yields by over 100%. Improved maize varieties increased yields from 2.6 to 5.8 t/ha. It is obvious that suitable management of water, combined with improved crop varieties and fertilizer can have big impacts on yields.

We also explored whether farmers' access to selected technologies, such as wells, ponds, river diversions and small dams, reduced poverty and, if so, which ones had the highest impacts. We calculated net present value of the selected technologies to assess which AWM technologies were worthwhile investments and suitable for out-scaling. We used a dataset from 1517 households from 29 *Kebeles* (peasant associations) in four regions of Ethiopia within the basin and some neighboring basins (Hagos *et al.* 2009).

AWM technologies increased per capita income by an average of US$82.00, and, moreover, reduced poverty by 22% compared to non-users. The impact on poverty depended on the type of AWM technology. Deep wells, river diversions, and micro dams gave 50%, 32% and 25% reduction respectively compared to the reference (rainfed) system. In contrast, *in situ* technologies such as soil and water conservation, which have important benefits against degradation, do little to reduce poverty.

Importance of livestock systems interventions

The importance of water use by livestock is not fully understood and is routinely underestimated. Six major livestock systems cover 60% of the basin, are home to 50% of the Nile's people, and deplete a large part of the water from rain through ET to produce the forages, pasture and crop residues consumed by livestock. We carried out a similar analysis to that described above to identify effective water management interventions for livestock systems. The improvement of livestock productivity, however, requires interventions beyond water management. Based on a livestock water productivity (LWP) framework developed for the Nile Basin (Peden *et al.* 2007), four basic strategies can help increase LWP. These are:

(1) Strategic sourcing of animal feeds and the selection of feed sources that require relatively little water to produce;
(2) Enhancing animal productivity and value through application of available animal science and marketing options;
(3) Conserving water resources through better land and water management associated with animal keeping; and
(4) Strategic spatial allocation of animal, land and water resources across landscapes to avoid overgrazing near watering points and under-utilization of feed far from watering points.

These four strategies often need to be applied simultaneously, and can increase LWP by more than 100% throughout most of the Nile Basin, although detailed approaches will vary from one place to another. For example, the interventions include the use of teff crop residues for oxen feed in the Ethiopian highlands; veterinary control of fasciolosis plus other diseases that limit livestock densities; moderate herd sizes in rangelands; and spatially optimal establishment of drinking water sites in central Sudan.

Importance of aquaculture in the basin

There is further scope for improving economic water productivity through enhanced aquaculture. Aquaculture in the Nile delta in the case study above and its expansion in other basin countries in the section on agricultural and water productivity has been discussed. Egypt has supported the development of aquaculture to promote farmers' livelihoods and to provide nutritional benefits to poor farming families. The programmes have been provided at minimal cost, and often free of charge. Egypt's advanced technical knowledge could be used to help support development of aquaculture in other basin countries through training and capacity development. Uganda's aquaculture export market has boomed over the past 10 years. The government promotes aquaculture to boost livelihoods and food security of farmers either through capture of floodwaters or use of groundwater in the northern and eastern areas of the country.

Protection of ecosystem services

A wide range of productive ecosystems exists in the Nile Basin, including the highland areas, rivers, lakes, wetlands and the delta. These (including the agricultural systems) provide a range of ecosystem services, that is, societal benefits from ecological processes (Gordon *et al.* 2009). The Nile Basin's ecosystems contribute to the production, retention and circulation of water from the highland areas with an annual rainfall of 2000 mm to Northern Sudan and Egypt with virtually zero annual rainfall. In addition to supplying water for irrigation, industry, hydropower, and human consumption, the basin's natural systems provide resources for food production, medicines, fuel and building materials. The ecosystem services support a range of livelihoods across the basin including rainfed and irrigated agriculture, livestock production and fisheries. They are also of aesthetic, cultural and heritage importance for a wide variety of Nile Basin communities. Maintenance of these ecosystems is critical for the continued provision of these services, and to ensure environmental sustainability across the basin.

Sustainable management plans for the entire Sudd wetlands will require the involvement of a stable government, multiple stakeholders, and policies that will address an all-embracing strategic plan. The strategic plan should span fisheries, oil industry demands, local population needs, and energy and infrastructure projects. Crop-based agriculture in the Sudd is primarily slash and burn, especially in areas that receive high numbers of returnees from the war. This practice, carried out in pristine areas that still have significant biodiversity, has negative impacts on the ecosystem.

Major ecosystem degradation occurs in upstream landscapes due to expansion of agriculture into marginal areas, deforestation and the resulting erosion. For example, out of the total 302.8 million tons of soil eroded in the Blue Nile catchments of the Ethiopian highlands, 45% leaves the landscape reaching the stream and about 90 million tons reach the border with Sudan annually (Awulachew *et al.* 2008).

The lake ecosystem around Lake Victoria has undergone substantial changes, which have accelerated over the last three decades. Recent pollution studies show that eutrophication has increased from human activities (Scheren *et al.* 2000). Policies for sustainable development in the region, including restoration and preservation of the lake's ecosystem, should therefore be directed towards improved land-use practices, and control over land clearing and forest burning. Diminishing water levels and pollution have acute consequences for several economic sectors, which depend on the basin's lakes. Variations in water level affect shallow waters and coastal areas, which are of particular importance for numerous fish species in certain stages of their life cycles. Cooperation between all concerned authorities is necessary to identify coherent solutions to ensure the sustainability of the lake's fisheries. The policies, legal, and institutional frameworks for the management of the Lake Victoria Basin are broadly guided by the treaty that established the East African Community in 2000. This came into force in July 2000 and designated the Lake Victoria Basin as a regional economic growth zone to be exploited jointly.

Three main strategies by which agricultural water management can deal with tradeoffs include: (1) improving water management practices on agricultural lands; (2) forming better linkages with the management of downstream aquatic ecosystems; and (3) paying more attention to how water can be managed to create multifunctional agro-ecosystems (Gordon *et al.* 2009). This can only be done if ecological landscape processes are better understood, and the values of ecosystem services other than food production are also recognized (Gordon *et al.* 2009).

Upgrading Gezira

The area of food crops grown in the Gezira Scheme is variable and responds to political and financial incentives, as well as to farmers' avoidance of risk. The amount of water consumed by these crops varies considerably, but their water productivity is low due to the low level of inputs, and the lack of appropriate institutional and support services, including water management. Nevertheless, there is a lot of potential to improve the water productivity in the Gezira. As discussed in the case study on the Gezira, it was found that every national-level policy had a deep impact on agricultural outcomes in the scheme and that farmers readily reacted to changes in incentives. The major actions needed include improving water-use efficiency, undertaking the institutional reforms that are necessary to support management of the Scheme, supply of inputs, and consolidation of the reforms to ensure that they continue in the long term.

Large-scale interventions

Large-scale interventions in water management are those such as large dams, canals and diversions that substantially influence the hydrology of the system and access to water. These entail important modifications of the water use for agriculture or for other purposes such as hydropower. We modelled the whole Nile Basin as one integrated system and analysed current, medium- and long-term scenarios considering irrigation, hydropower, the environment, and wetlands. While irrigation, environmental and wetland requirements are sensitive, hydropower demands, which are a non-consumptive use, were taken as unimportant in affecting overall water availability in the basin.

According to national plans, irrigation, including small-scale schemes, will increase from a current total of 5.5 Mha to 7.9 Mha in the medium term and nearly double to 10.6 Mha in the long term (Demissie *et al.* 2009). Figure 10 shows the existing development planned under three scenarios. While there is now little irrigation in the upstream countries, there are plans for more irrigation both upstream and downstream. Rwanda, for example, plans to expand its irrigated area from 2000 ha to 150,000 ha, Ethiopia from 15,900 ha to 1,216,100 ha, and Sudan and Egypt by 2,327,600 ha and 880,900 ha respectively. Note also that with advancing technologies and changes from conventional irrigation, the amount of expansion cannot be predicted accurately.

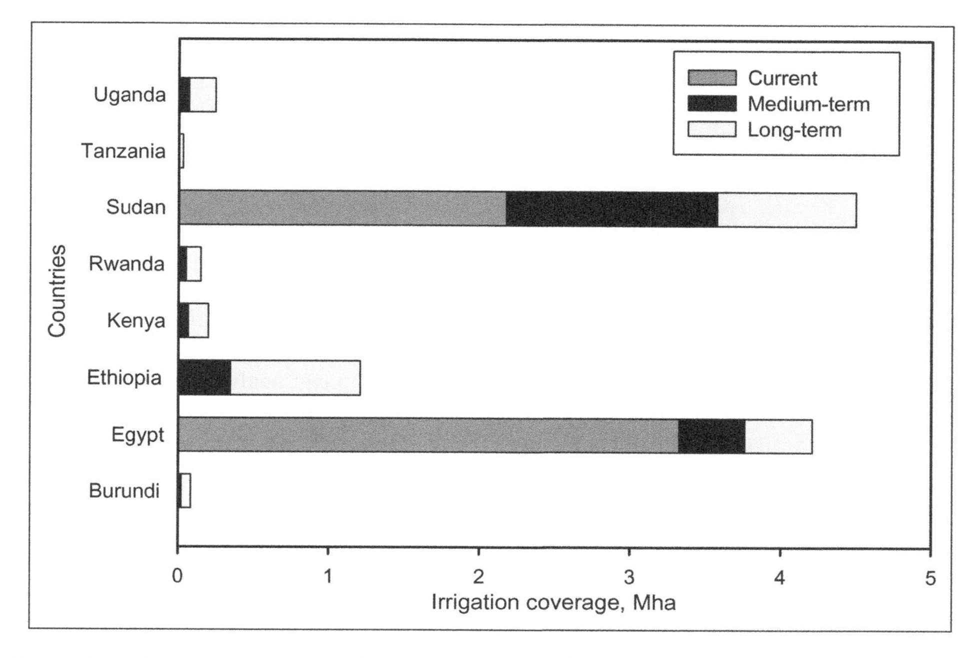

Figure 10. Aggregated incremental irrigation plans within the Nile Basin by country representing current, medium and long-term scenarios.

In contrast, it is important to note that such expansions are mainly related to surface water, and do not consider renewable ground water or management of rainwater.

We found that more large-scale irrigation is possible, but not at the levels planned. At the current level of water application and irrigation efficiency, and in the absence of reservoir management, the total water requirement for the long-term scenario above would be 127 km^3. For the medium-term scenario, the requirement is 94.5 km^3, still higher than the 84.1 km^3 (1900–50 average) and the 88.2 km^3 (long-term average) available (Demissie *et al*. 2009, Molden *et al*. forthcoming). To balance the demand with the amount of water available, possible mitigation measures include:

(1) Co-ordinated planning to expand irrigated land and manage the entire river to enhance overall gains in economic water productivity. Part of this planning requires transparent sharing of data. It was not possible to obtain data on existing flow patterns;
(2) Improved water productivity on currently irrigated lands. For example, overall production in the Gezira is far below achievable levels, with ET much less than it could be;
(3) Increased efforts to formulate real practices that save water including re-use that releases water for additional irrigation or urban uses;
(4) Increased storage capacity upstream and downstream storage translated to low evaporation areas;
(5) Enhanced carry-over storage;
(6) Managed cropping intensity and type of crops; and
(7) Adjusted expectations and ambitions to the river's capacity.

Choosing the right interventions

We identified multiple options, ranging from rainwater management to large-scale irrigation; from local gains in productivity and associated impacts on poverty to regional economic relevance, over diverse enterprises covering crop production, livestock and fisheries. No single intervention serves as a panacea, but there are numerous actions needed.

The Nile is a diverse area with different social, cultural and biophysical considerations. To help identify types of interventions, the Nile hydronomic zones were developed (Demissie *et al.* 2010, Molden *et al.* forthcoming). At a simple level, there are five major zones: irrigated areas; mixed rainfed systems in semi-arid areas; pastoral and mixed systems in the water-source zone; wetlands and other environmentally sensitive zones; and arid areas. A second level of zoning using principal component analysis provides a more disaggregated classification based on soils, topography and climatic considerations.

The zones consist of 19 distinct classes, which include different aspects of water availability, or indicate existing or potential water management, soil type, and environmental sensitivity, all of which allow better targeting of interventions. About 10% of the basin falls within the environmentally sensitive zone of wetlands and protected areas. In this zone, water development interventions should be treated with caution to ensure the maintenance of ecosystem services. The humid and wet-humid zones are the water source zones of the basin and account for less than 15% of the area. Taken from a water perspective, interventions in this zone need to consider not only upstream benefits but also downstream impacts. Since the identified zones have unique climatic and soil properties, water management interventions required to address issues in each zone should also be unique. In the remaining 85% of the land area, only local aspects of water management need be considered.

Ultimately, water governance will facilitate sustainable and productive development of the Nile's waters. The NBI was formed to improve cooperation between the basin's countries, and while it has made important progress, it has neglected several areas. It should give more attention to water and agriculture, and broaden its focus to include fisheries and livestock. There are numerous other institutions involved in water and agriculture across the basin, but there is a dire need for improved human and institutional capacity to implement programs to the benefit of the rural poor (Cascão 2009, Molden *et al.* forthcoming).

Conclusions

The focus of this study was to identify water management interventions that have a high potential to increase water productivity and alleviate poverty in different parts of the Nile Basin, and to inform future research and development efforts. Establishing a clear overall picture of the hydrology, agriculture and productivity, livelihoods, and poverty within the basin is not an easy task due to the basin's size and complexity, and is further complicated by limited data availability, access and quality.

Within the Nile Basin countries there are high levels of poverty and food insecurity. The majority of people depend on agriculture for their livelihoods, and agriculture plays an important role in the economies of all of the basin countries. In those countries where agriculture contributes the highest proportion to GDP and managed water is scarce, poverty is prevalent and GDP per capita is low. Rainfall is highly variable across the basin. Runoff zones constitute a mere 15% of the basin and the average runoff coefficient for the basin is less than 10%, one of the lowest in the world. When considering water resources in the basin, it is important to differentiate between water availability and water access. Water is accessible in northern Sudan and Egypt because of adequate infrastructure. Poverty and vulnerability across the basin are linked to access to water through crop- and livestock-based livelihoods, and increasing agricultural water productivity can potentially contribute to poverty reduction and food security.

The dominant production systems in the basin are rainfed, based on mixed crop-livestock, agropastoral, and pastoral. Irrigation is important in Sudan and Egypt. The semi-arid areas of rainfed systems have the lowest crop productivity, and the irrigated areas provide the highest. Rainfed cropping systems cover 7% of the basin and produce approximately 80% of the food (measured as dry biomass), while irrigated land covers 1.6% and produces 20%. Poverty mapping indicates that the highest levels of poverty within the basin are where people are highly dependent on rainfed agriculture and have limited access to managed water. Irrigated crop-livestock systems show low vulnerability while pastoral, agropastoral and mixed rainfed production systems exhibit high vulnerability to water-related hazards. In spite of its importance, water for agriculture is not well exploited in the rainfed systems.

The Ethiopian highlands contribute 85% of the Nile flow at AHD, and the ELR only 15%. The frequently quoted 84.1 km^3/yr entering the AHD is lower than the long-term mean flow of 88.2 km^3/yr. AHD loses 12 km^3–14 km^3 to evaporation with a total loss of 25.4 km^3 including seepage and spillway.

Rain is the ultimate source of green and blue water and a higher emphasis should be placed on ET from natural landscapes as the current main water use. The total rainfall of 1745 km^3 (2007) and its consumptive use shows that most water is consumed by natural land cover. There are important losses of water through non-beneficial uses, which with improved management practices can be transferred to beneficial uses. An example is cropped land on vertisols in the Ethiopian highlands, which become waterlogged, have low productivity, and contribute little to groundwater, with most water lost to evaporation. Draining these lands can save significant water, increase runoff coefficient and improve the productivity of the land.

We analysed water productivity for crops, livestock and aquaculture within the basin, examining ET, production systems and crop yields. We developed a comprehensive map of crop water productivity for the basin. Egypt has higher productivities while for crops elsewhere they are only about one third of the Egyptian levels.

The Sudd and other wetlands contain a huge untapped potential for future development of both commercial and subsistence fisheries. Although the Sudd is the biggest inland floodplain system in Africa, very little is known about either the wetland or the livelihood systems it supports due to the civil war and political instability in the region. Oil companies have begun to operate in the wetland, and there are plans for the development of various upstream water infrastructure projects, which will reduce inflows to the Sudd. Looking to the future, use of the wetland could either lead to prosperity and improved livelihoods, or be a point of conflict. Egypt has over 90% of the aquaculture in the basin, but there are opportunities elsewhere. Insufficient processing facilities and markets are at present key constraints to improved fish production in other parts of the basin. While wetland-related ecosystem functions are sensitive, it was observed that there is no measurable impact of hydropower development on water volume except modification of temporal variability.

Most governments prefer irrigation as a water intervention because of its capacity to raise productivity and reduce vulnerability. Yet not all the ambitious plans to expand irrigation in the long-term development scenario are possible with the available water resources, infrastructure and current water management.

Some less ambitious interventions such as supplemental irrigation using groundwater wells, micro dams, and river diversions for upgrading rainfed systems provide high returns in terms of poverty reduction, productivity gains and food security. Interventions need to be selected that reflect a range of options in terms of agricultural water management, that take into account the particular production systems and the prevailing biophysical, social and institutional conditions. Techniques such as hydronomic zone mapping can help to collate numerous biophysical factors into manageable zones in order to identify relevant interventions. Similarly, the multi-criteria analysis technique can help prioritize suitable technologies and practices and assist decision making.

Ultimately, water governance will facilitate sustainable and productive development of Nile waters. The NBI was formed to improve co-operation between the basin's countries. While the NBI has made important progress, it should give more attention to water and agriculture, and also focus on fisheries and livestock. While there are numerous other institutions involved in the management of water and agriculture across the basin, there is a still a dire need for improved human and institutional capacity to implement programmes for the benefit of the rural poor.

Acknowledgments

This paper presents findings from CPWF59, Nile Basin Focal Project, a project of the Consultative Group on International Agricultural Research (CGIAR) Challenge Program on Water and Food (CPWF) and implemented by the International Water Management Institute (IWMI) together with the Nile Basin Initiative (NBI), the NBI subsidiary Action Program of Eastern Nile Technical Regional Organization (ENTRO), the International Livestock Research Institute (ILRI) and the World Fish Center (WorldFish). The work was carried out and documented by contributors listed in the contributors list. In addition, the authors acknowledge the help and insights received from the NBI shared vision programme and its subsidiary action project managements. Many national systems such as Egypt's Ministry of Water Resources and Irrigation, Nile Water Sector Cairo, Nile Research Institute, National Water Research Center Cairo, South Sudan's Ministry of Water Resources, Makarere University Kampala, Ministry of Water Resources Department Entebbe Uganda, Ministry of Water Resources, Department of Hydrology of Ethiopia, FAO Uganda, and many secretaries, drivers, and farmers helped us plan and implement our field trip and programme meetings. We appreciate the leadership and direction provided by Simon Cook and the CPWF for sustainable support. Finally, the financial support from the CPWF and its donors is gratefully acknowledged.

Notes

1. With contributions from K. Conniff, Solomon S. Demissie, Teklu Erkossa, P. Karimi, J. Kinyangi, Y. Mohamed, A. Mukherji, A. Notenbaert, Don Peden and Paulo van Breugel.
2. "The composite record at Wadi Halfa/Kajnarty/Dongola is available from 1890, while the flows measured downstream at the Aswan reservoir commenced in 1869. These flows have been measured by various methods and standardized from time to time. They have been tabulated as 'Water arriving at Aswan' and also as 'Natural River at Aswan'; the latter is adjusted for water abstracted from the Blue Nile in the Gezira main canal, and from 1963 in the Managil canal; it allows for the regulation of the Sennar reservoir and the Aswan reservoir, but not for reservoir evaporation. From 1978 the Natural River flows has included estimated evaporation from the AHD and the Jebel Aulia reservoir, but not all the effects of upstream storage and abstractions" Sutcliffe (2009).

References

Aly, A.I.M., *et al.*, 1993. Study of environmental isotope distribution in the Aswan High Dam lake (Egypt) for estimation of evaporation of lake water and its recharge to adjacent groundwater. *Environmental Chemistry and Health*, 15 (1), 37–49.

Aquastat-FAO, 1997. *Aquastat-FAO.* Available from: http://www.fao.org/nr/water/aquastat/main/index.stm [Accessed October 2009].

Awulachew, S.B., *et al*., 2008. *A review of hydrology, sediment and water resource use in the Blue Nile Basin*. Working paper 131. Colombo: IWMI.

Ayalew, S.M. and Awulachew, S.B., 2008. *Intervention analysis literature review*. Unpublished data.

Bailey, R.G., 1989. An appraisal of the fisheries of the Sudd Wetlands, River Nile, southern Sudan. *Aquaculture and Fisheries Management*, 20, 79–89.

Bastiaanssen, W. and Perry, C., 2009. *Agricultural water use and water productivity in the large-scale irrigation (LSI) schemes of the Nile Basin* [online]. Wageningen, Netherlands: WaterWatch. Available from: http://www.waterwatch.nl/publications/reports/2009.html [Accessed December 2009].

BCEOM, 1998. *Abbay River basin integrated development master plan project. Phase 2, section iii. Agriculture*. Addis Ababa: Ministry of Water Resources, Government of Ethiopia.

Bird Life International, 2009. *Important bird area factsheet: Sudd (Bahr-el-Jebel system), Sudan*. Available from: http://www.birdlife.org [Accessed 27 May 2010].

Cascão, A.E., 2009. *Institutional analysis of Nile Basin Initiative: what worked, what did not work and what are the emerging options?* Colombo: CGIAR CPWF. Unpublished data.

Conway, D., 2005. From headwater tributaries to international river: observing and adapting to climate variability and change in the Nile basin. *Global Environmental Change*, 15 (2), 99–114.

Conway, D. and Hulme, M., 1996. The impacts of climate variability and future climate change in the Nile basin on water resources in Egypt. *Water Resources Development*, 12 (3), 277–296.

Cook, S. and Gichuki, F., 2006. *Analyzing water poverty: water, agriculture and poverty in basins* [online]. Basin Focal Project working paper 3. Available from: http://www.waterforfood.org/fileadmin/CPWF_Documents/Documents/Basin_Focal_Projects/BFP_restricted/Paper_3_14JY06.pdf [Accessed 22 July 2010].

Demissie, S.S., Awulachew, S.B., and Molden, D., 2009. *Intervention analysis in the Nile: large-scale interventions*. Colombo: CGIAR CPWF. Unpublished data.

Demissie, S.S., Awulachew, S.B., and Molden, D., 2010. *Biophysical classification for efficient water resources management: hydronomic zones of the Nile Basin*. Colombo: CGIAR CPWF. Unpublished data.

Denny, P., 1991. Africa. *In*: M. Finlayson and M. Moser, eds. *Wetlands*. Oxford: Checkmark Books, 115–148.

Dixon, J., *et al*., 2001. *Farming systems and poverty: improving farmers' livelihoods in a changing world*. Rome and Washington D.C.: Food and Agriculture Organization of the United Nations and World Bank.

Droogers, P., Immerzeel, W., and Perry, C., 2009. *Application of remote sensing in national water plans: Demonstration cases for Egypt, Saudi-Arabia and Tunisia*. Report 80. Wageningen, Netherlands: FutureWater.

ENTRO, 2006. *Cooperative regional assessment for watershed assessment, trans-boundary analysis country report Egypt*. Prepared by Hydrosult Inc., Tecsult, DHV, and their Associates Nile Consult, Comatex Nilotica; and T and A Consulting. Report for Nile Basin Initiative-Eastern Nile Technical Regional Organization (NBI-ENTRO).

Elshamy, M.E., Seierstad, I.A., and Sorteberg, A., 2008. Impacts of climate change on Blue Nile flows using bias-corrected GCM scenarios. *Hydrology and Earth System Sciences Discussions*, 5, 1407–1439.

Erkossa, T., Awulachew, S.B., and Denekew, A., 2009. *Agricultural productivity of the Upper Blue Nile Basin farming systems*. Colombo: CGIAR CPWF.

European Space Agency, 2010. *GlobCover Land Cover v2 2008 database*. Available from: http://ionial.esrin.esa.int/index.asp. [Accessed December 2008].

Farmer, G., 1988. Seasonal forecasting of the Kenya coast short rains 1901–1984. *Journal of Climatology*, 8 (5), 489–497.

FAO, 2007. *State of world fisheries and aquaculture*. Rome: Food and Agriculture Organization of the United Nations. Available from: http://www.fao.org/figis/servlet [Accessed January 2008].

Gamal, K.A.E.M., 2009. *Impact of policy and institutional changes on livelihood of farmers in Gezira scheme of Sudan*. Thesis (MSc). University of Gezira, Sudan.

Gerrits, A.M.J., *et al*., 2007. New technique to measure forest floor interception: an application in a beech forest in Luxembourg. *Hydrology and Earth System Sciences*, 11, 695–701.

Gordon, L.J., Finlayson, C.M., and Falkenmark, M., 2009. Managing water in agriculture for food production and other ecosystem services. *Agricultural Water Management*, 97, 512–519.

Guvele, C.A. and Featherstone, A.M., 2001. Dynamics of irrigation water use in Sudan Gezira scheme, *Water Policy*, 3 (5), 363–386.

Hagos, F., *et al.*, 2009. *Determinants of successful adoption of agricultural water management technologies: case of Ethiopia*. Colombo: CGIAR CPWF. Unpublished data.

Hefny, M. and Amer, S.E.-D., 2005. Egypt and the Nile Basin. *Aquatic Sciences*, 67 (1), 42–50.

Howell, P.P., Lock, M., and Cobb, S., 1988. *The Jonglei canal: impact and opportunity*. Cambridge: Cambridge University Press.

Hughes, R.H. and Hughes, J.S., 1992. *A directory of African wetlands*. Gland, Switzerland, Nairobi, Kenya, and Cambridge, UK: International Union for the Conservation of Nature, United Nations Environment Programme, and World Conservation Monitoring Centre.

Hurst, H.E., 1957. *The Nile*. 2nd. ed. London: Constable.

IPCC, 2007. *Climate change 2007. The IPCC fourth assessment report* [online]. Geneva: Intergovernmental Panel on Climate Change. Available from: http://www.ipcc.ch/publications_and_data/publications_and_data.htm [Accessed January 2010].

ILRI, 2002. *Livestock: a pathway out of poverty. ILRI's strategy to 2010* [online]. Nairobi: International Livestock Research Institute. Available from: http://mahider.ilri.org/bitstream/10568/565/1/Strategy2010.pdf [Accessed 25 July 2010].

IMF, 2009. *World economic outlook*. Washington, DC: International Monetary Fund. Available from: http://www.imf.org/external/pubs/ft/weo/2009/02/pdf/text.pdf

IWMI, 2009. *IWMI climate atlas*, in IWMI Integrated Database Information System (IDIS). Available from: http://dw.iwmi.org/idis_DP/Basin_Kits.aspx [Accessed March 2009].

Kinyangi, J., *et al.*, 2009. *Water and poverty analysis in the Nile basin*. Colombo and Nairobi: IWMI and ILRI. Available from: http://www.slideshare.net/ILRI/improved-agricultural-water-management-in-the-nile-basin-vulnerability-and-poverty-in-agricultural-systems-of-the-nile-basin [Accessed on 25 July 2010].

Kirby, M., Eastham, J., and Mainuddin, M., 2010. *Water-use accounts in CPWF basins: simple water-use accounting of the Nile Basin* [online]. CPWF working paper: Basin Focal Project series, BFP03. Colombo: CGIAR CPWF. Available from: http://www.waterandfood.org/fileadmin/CPWF_Documents/Documents/CPWF_Publications/CP_WP_3_Nile_Basin.pdf [Accessed 22 July 2010].

Kull, D., 2006. *Connection between recent water level drop in Lake Victoria, dam operation and drought* [online]. Available from: http://www.irn.org/programs/nile/pdf/060208vic.pdf [Accessed 22 July 2010].

MoARF, 2006. *Statistical bulletin for animal resources*, 15–16. Khartoum: Sudan Ministry of Animal Resources and Fisheries.

LVFO, 2006. Estimated production of Lake Victoria, 1960s–2006 (unpublished data). Lake Victoria Fisheries Organization. Available from: http://www.lvfo.org

Mohamed, Y.A., Molden, D., and Bastiaanssen, W., 2009. Water accounting at a river basin scale: the Nile basin case. Unpublished data.

Mohamed, Y.A., Savenije, H.H.G., Bastiaanssen, W.G.M. and van den Hurk, B.J.J.M., 2006. New lessons on the Sudd hydrology learned from remote sensing and climate modeling. *Hydrology and Earth System Sciences*, 10, 507–518.

Molden, D., 1997. *Accounting for water use and productivity*. SWIM Paper 1. Colombo: IWMI.

Molden, D., *et al.*, forthcoming. Nile Basin Focal Project, synthesis report. Colombo: CGIAR CPWF.

Mubiru, P., 2006. *Causes of the decline of Lake Victoria levels during 2004 to 2005*. Kampala, Uganda: Commissioner for Energy Resources Department, Government of Uganda.

NEDECO, 1998. *Tekeze River basin integrated development master plan project, sectorial reports – Water resources*. Volumes VI, VII and X: Climatology, Hydrology, and Dams, Reservoirs, Hydropower and Irrigation Development. Addis Ababa: Ministry of Water Resources, Government of Sudan.

Nicholson, S.E., 1996. A review of climate dynamics and climate variability in eastern Africa. *In*: T.C. Johnson and E. Odada, eds. *The limnology, climatology and paleoclimatology of the East African lakes*. Amsterdam: Gordon and Breach, 25–56.

Notenbaert, A., 2009. The role of spatial analysis in livestock research for development. *GIScience and Remote Sensing*, 46, 1–11. DOI: 10.2747/1548-1603.46.1.1.

Okeny, A., 2007. *Southern Sudan: launch of a project to improve animal and fish production*. News release 2007/269/AFR. Washington D.C.: World Bank.

Oosterbaan, R.J., 1999. *Hydrological and environmental impacts of the irrigation improvement projects in Egypt, Nile Valley and Delta* [online]. Report of a consultancy assignment to the Egyptian-Dutch Water management Panel. Wageningen: Waterlog. Available from: http://www.waterlog.info/pdf/irrimpr.pdf [Accessd 22 July 2010].

Otte, M. and Chilonda, P., 2002. *Cattle and small ruminant production systems in sub-Saharan Africa. A systematic review*. Rome: Food and Agriculture Organization of the United Nations.
Payne, W.J.A. (1990). *An introduction to animal husbandry in the tropics*. Harlow: Longman.
Peden, D., *et al.*, 2007. *Investment options for integrated water-livestock-crop production in sub-Saharan Africa*. Nairobi: ILRI.
Peden, D., Taddesse, G. and Haileslassie, A., 2009. Livestock water productivity: implications for sub-Saharan Africa. *The Rangeland Journal*, 31, 187–193.
PEAP, 2004. *Poverty eradication action plan 2004/5-2007/8* [online]. Kampala: Ministry of Finance, Planning and Economic Development, Government of Uganda. Available from: http://www.finance.go.ug/docs/PEAP%202005%20Apr.pdf [Accessed 25 July 2010].
RIS, 2006. *Ramsar information sheet 2006*. Gland, Switzerland: Ramsar Convention. Available from: www.wetlands.org/reports/ris [Accessed May 2009].
Scheren, P.A.G.M., Zanting, H.A., and Lemmens, A.M.C., 2000. Estimation of water pollution sources in Lake Victoria, East Africa: application and elaboration of the rapid assessment methodology. *Journal of Environmental Management*, 58 (4), 235–248.
Seré, C. and Steinfeld, H., 1996. *World livestock production systems: current status, issues and trends*. Animal production and health paper 127. Rome: Food and Agriculture Organization of the United Nations.
Shahin, M., 1985. *Hydrology of the Nile Basin: developments in water science*. Amsterdam: Elsevier.
Strzepek, K., *et al.*, 2001. Constructing not implausible climate and economic scenarios for Egypt. *Integrated Assessment*, 2 (3), 139–157.
Sullivan, C.A., 2002. Calculating a water poverty index. *World Development*, 30 (7), 1195–1210.
Sutcliffe, J.V., 2009. The hydrology of the Nile Basin. *In*: H.J. Dumont, ed. *The Nile: origin, environments, limnology and human use*. Monographiae Biologicae 89. Dordrecht: Springer Science, 335–364.
Sutcliffe, J.V. and Parks, Y.P., 1999. *The hydrology of the Nile*. IAHS special publication 5. Wallingford: International Association of Hydrological Sciences.
Swenson, S. and Wahr, J., 2009. Monitoring the water balance of Lake Victoria, East Africa, from space. *Journal of Hydrology*, 370 (1–4), 163–176.
Thornton, P.K., *et al.*, 2006. *Mapping climate vulnerability and poverty in Africa* [online]. Report to the Department for International Development. Nairobi: ILRI. Available from: http://www.waterandclimateinformationcentre.org/resources/8012007_ILRI2006_mapping-climatevulnpovafrica.pdf [Accessed 22 July 2010].
UBOS, 2007. *The 2002 Uganda population and housing census, economic characteristics* [online]. Kampala: Uganda Bureau of Statistics. Available from: http://www.ubos.org/onlinefiles/uploads/ubos/pdf%20documents/2002%20CensusEconomicXteristicsAnalyticalReport.pdf [Accessed 25 July 2010].
UNDP, 1983. *Sudd fisheries development programme, phase 1*. Khartoum: United Nations Development Programme.
van Breugel *et al.*, 2010. Livestock water use and productivity in the Nile Basin. *Ecosystems*, 13 (2), 205–221.
WEMA Consultants, 2009. *Study on institutional arrangement for the management of Lake Victoria*. Consultancy report submitted to International Water Management Institute as a part of Nile BFP. Unpublished data.
Woldeamlak, B., 2003. *Towards integrated watershed management in highlands of Ethiopia: the Chemoga watershed case study*. Tropical resource management papers 44. Wageningen, Netherlands: Wageningen Agricultural University.
WorldFish, 2009. *The development of aquaculture in Egypt and its impacts on livelihoods and poverty at both local and national levels*. Unpublished report.

Appendix

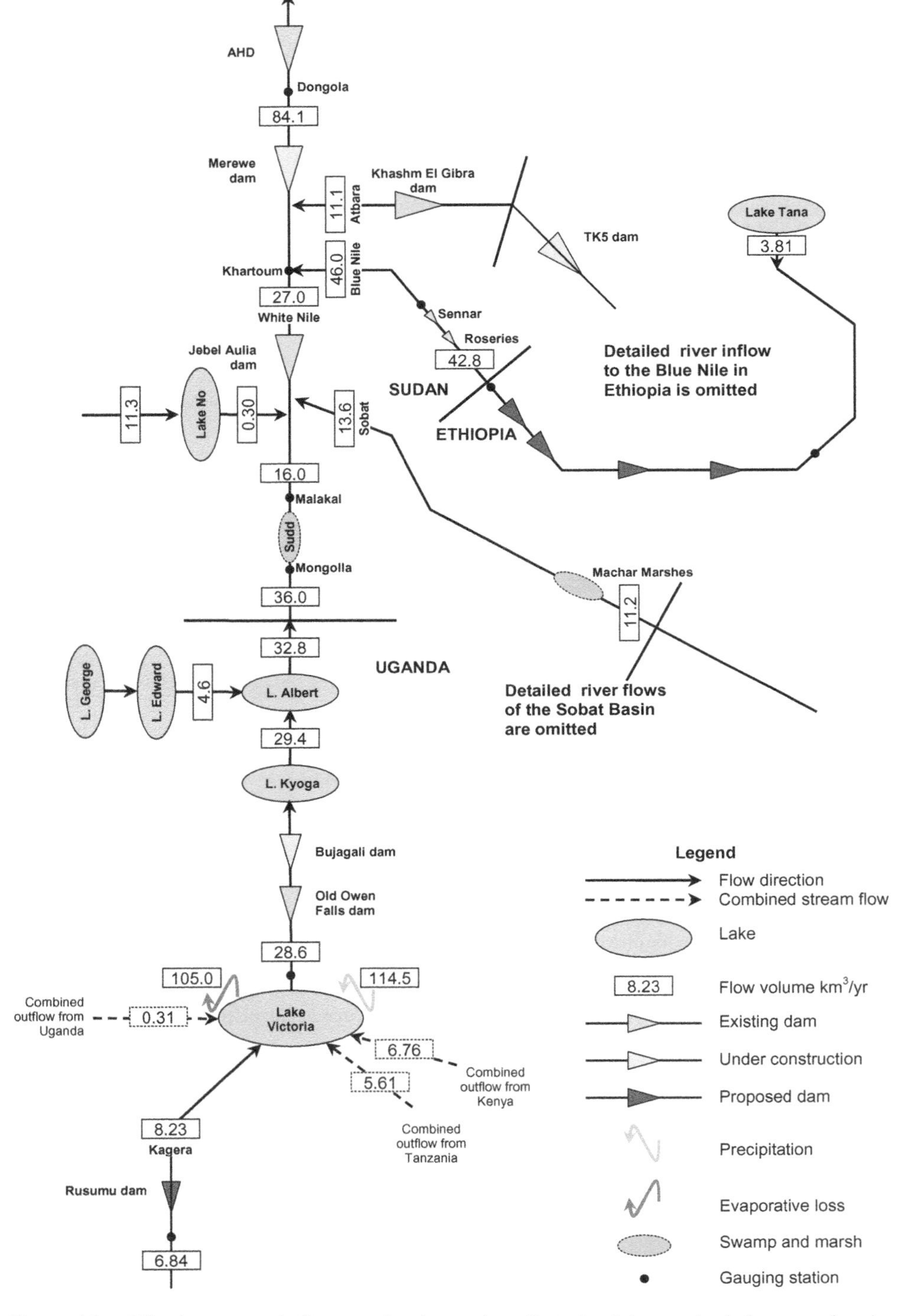

Figure A1. Nile river network diagram showing major tributaries, lakes, water balance, and major dams constructed or planned by 2008.

Table A1. Existing major water control structures in the Nile Basin.

Dam	Country	Live storage km^3	Built	Purpose
Abobo	Ethiopia	0.06	1992	Irrigation; not yet used
Fincha	Ethiopia	1.05	1971	Irrigation, hydropower
High Aswan	Egypt	105.90	1970	Irrigation, hydropower
Jebel El Aulia	Sudan	3.35	1937	Irrigation, hydropower
Khashm El Gibra	Sudan	0.84	1964	Irrigation, hydropower
Koga	Ethiopia	0.08	2008	Irrigation
Chara Chara	Ethiopia	9.13	2000	Hydropower
Owen Falls	Uganda	215.59	1954	Irrigation, hydropower
Roseries	Sudan	2.32	1966	Irrigation, hydropower
Sennar	Sudan	0.75	1925	Irrigation, hydropower

Farming systems and food production in the Volta Basin

Jacques Lemoalle and Devaraj de Condappa

The predominantly rural population in the Volta Basin depends on rainfed crops, a practice which becomes increasingly risky as the rainfall decreases from south to north and rural poverty increases. Yields are low throughout because of drought and dry spells within the growing season, infertile soils, low inputs, poor infrastructure, and low labour productivity. Water-related diseases are widespread and half of all rural households depend on low-quality water. Fertilizer and small-scale irrigation can improve the yields of rainfed crops and alleviate poverty, but the duality between the legal state and the traditional hierarchy complicates land tenure by hindering investment. Small-scale irrigation has little effect on hydropower dams downstream.

Introduction

In this paper, we provide information on water availability in the Volta Basin, including rainfall quantity and distribution, and water use or water accounting. Drought is important in parts of the Volta, and demands particular attention. In parallel, we describe estimates of water productivity in relation to farming practices, and use the combined insight of water and agricultural factors to help explain the condition of rural poverty, mostly in the Ghanaian and Burkinabe parts of the basin.

Policies and institutions in the basin, and the duality between both formal and informal authorities, are important conditioning factors that influence the link between water, agriculture and poverty through land tenure, access to water and access to markets. Finally, we identify opportunities for using water-related interventions to address water and food problems in the Volta, and analyse the consequences of major changes for different water users.

The Volta Basin introduced

The Volta Basin, located in West Africa, covers an area of around 395,000 km^2 across six countries: Benin, Burkina Faso, Côte-d'Ivoire, Ghana, Mali and Togo (Figure 1). Among them, Burkina Faso and Ghana account for 83% of the basin area, while 47% of Togo lies within the basin.

About 20 million people live in the basin. Most (13.5 million) are rural. While inroads have been made to reduce poverty in some areas, rural poverty remains high, especially in Burkina Faso. Population density is around 48 persons/km^2.

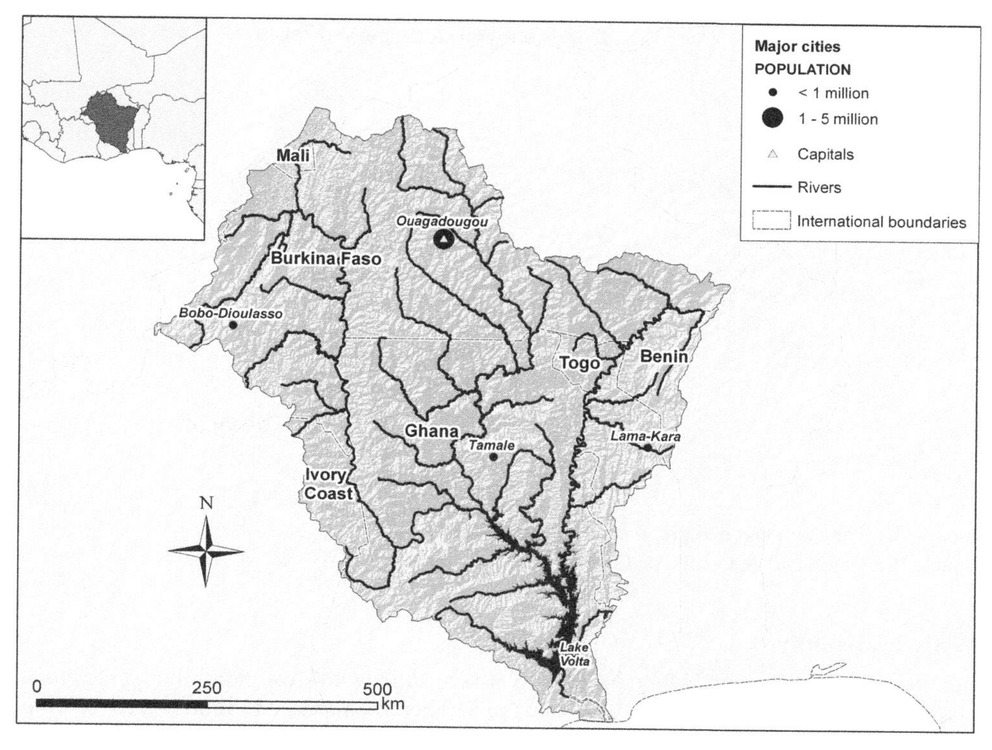

Figure 1. The Volta Basin and its political boundaries.
Source: BFP Volta with the river systems from Dieulin (2007).

About 30–60% of the total land resources in the basin are considered arable. A large proportion of this, however, is left uncultivated (Table 1 and Figure 2). Just outside the basin in southwestern Ghana the population density is much higher and a higher proportion of the land is cultivated.

Expansion of cultivated area over the past 30 years has been related to population growth: there has been little increase in cultivated area per capita (Serpantié 2003). It appears that scarcity of productive assets and water, not land, limits production.

Table 1. Land use in the countries of the Volta Basin.

Country	Arable of total country	Cultivated	Cultivated area	Irrigable	Actually irrigated
		% of arable	*000 ha*	*000 ha*	*% of irrigable*
Benin	62	40	2815	322	1.3
Burkina Faso	33	49	4700	165	n.a.
Côte-d'Ivoire	65	33	7100	475	n.a.
Ghana	42	63	6331	1900	1.5
Mali	35	11	4840	566	n.a.
Togo	60	77	2450	180	3.4

Note: n.a. = not available.
Source: Data for 2003–2007, modified from FAO (2005).

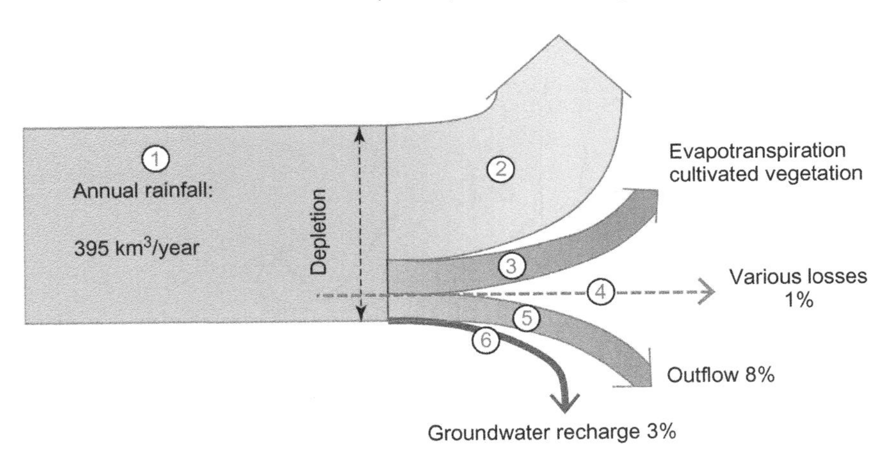

Figure 2. The main flows of green and blue water in the Volta Basin.
Source: Lemoalle and de Condappa (2009).

Water availability

For the Volta Basin as a whole, rainfall is relatively abundant. Mean rainfall over the basin is about 400 km^3/year or about 1040 mm/year. There is a marked gradient between the drier north and the wetter south, however. Annual rainfall varies from about 500 mm in the north of the basin to 1200 mm in the southern, downstream part of the basin. Average basin discharge is about 8–9% of total rainfall (Figure 2).

With more than 2000 m^3 of renewable water resource (surface and groundwater) per capita per year the Volta Basin population is a little above Falkenmark's (1997) per capita water scarcity threshold of 1700 m^3/yr. At the sub-basin scale, however, most of the northern part of the basin suffers physical water scarcity. The per capita renewable water resource in Burkina Faso is about 900 m^3/yr, and many riverbeds are dry for several months during the year (Lemoalle and de Condappa 2009). Moreover, the groundwater resource is distributed unevenly, exacerbating the problem in some areas.

Economic water scarcity occurs when water resources are abundant relative to water use, but insufficient infrastructure or financial capacity prevent people from accessing the water they need. This dilemma plagues predominantly the smallholder farmers in the Volta Basin, most of whom rely on low-yield, rainfed agriculture, and whose livelihoods are constantly threatened by unreliable precipitation, dry spells, droughts and lack of access to good quality household water. For this reason, additional investment in the water sector, especially for small-scale irrigation and dry-season crops, could play a transformative role in poverty alleviation (WRI 2007).

Rainfall, surface water and groundwater

The temporal and spatial distribution of rainfall influences agriculture more than total rainfall. Towards the north, rainfall has a unimodal distribution from May to September and the rainy season becomes decreasingly short. In the south, rainfall is distributed bimodally with long rains from April to July and short rains from September to November (Figure 3). The risk of within-season dry spells influences cropping choices.

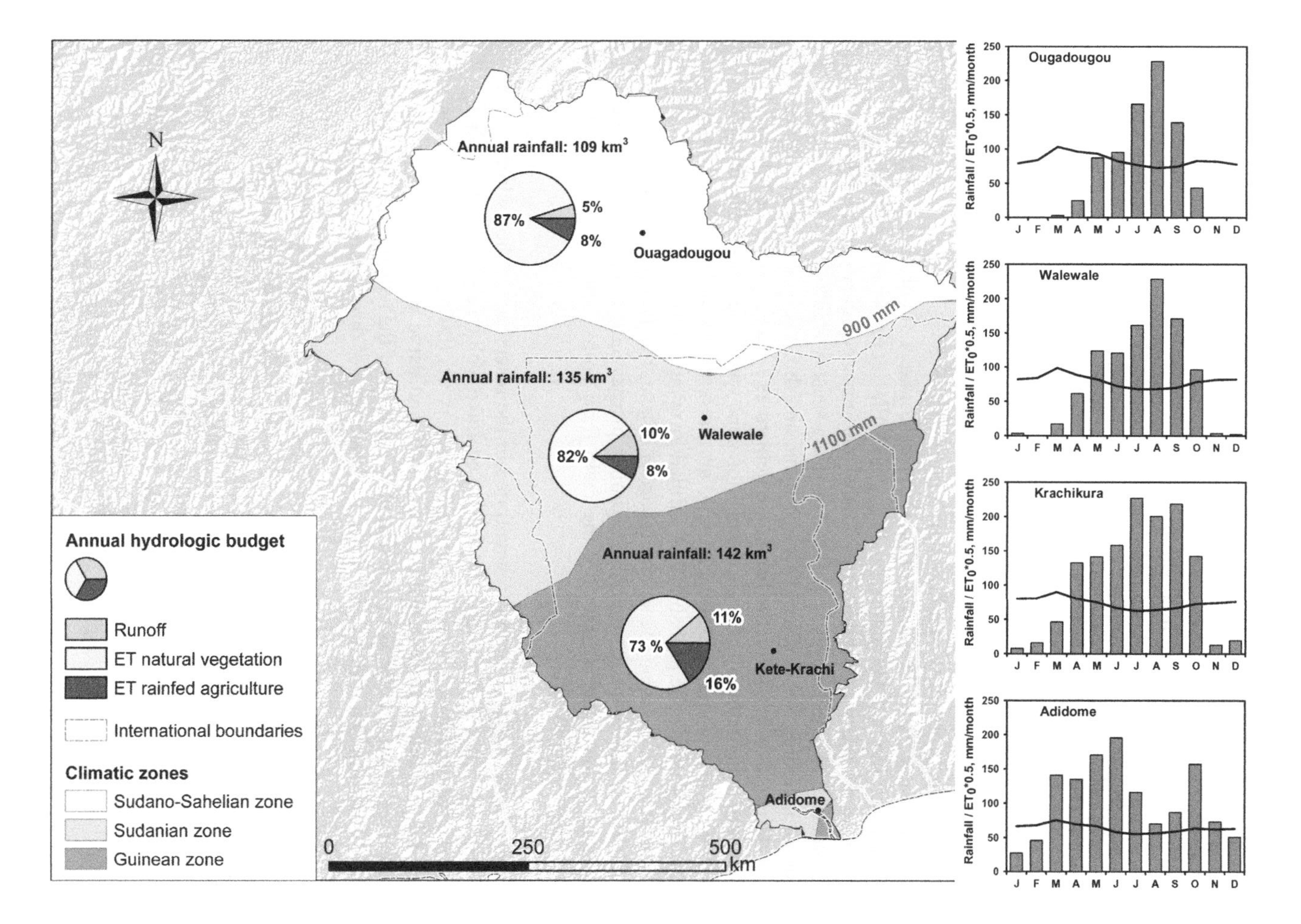

Figure 3. The agro-ecological zones with a summary of the water account in the Volta Basin, and the distribution of rainfall in four stations along the south–north gradient.

Note: ET0 is the reference (or potential) evapotranspiration. The intersection of the 0.5 ET0 line with the rainfall distribution indicates the length of the growing season.

Source: Map by BFP Volta with data from Climate Research Unit, University of East Anglia (Lemoalle and de Condappa 2009).

Although the farmers have developed a variety of strategies to avoid the risk linked to the irregular distribution of rainfall in time and space, failed crops are still quite common. For each of the main cropping systems in sub-Saharan Africa, Hyman *et al.* (2008) have developed a method to assess and map drought risk by estimating the probability of a failed growing season. This probability has been added to the description of the different systems, in order to underline the effects of rainfall variability.

The north to south rainfall gradient has been used as the basis for delineating the following agro-ecological zones based on the Food and Agriculture Organization (FAO) classification (FAO 1996) (Figure 3):

- The Sahel, located in the northernmost part of the basin, is defined as receiving less than 500 mm annual precipitation. The Sahel is a zone of rangeland where livestock herding is the primary activity, complemented with the drought-resistant crops millet and cowpea. The probability of a failed growing season is 53%.
- The Sahelo-Sudan, covering most of Burkina Faso and a small part of Mali, is defined as receiving 500–900 mm annual rainfall. Millet, sorghum and maize are the main cultivated food crops. Cotton, groundnuts and some sedentary cattle contribute to cash income. The probability of a failed growing season is 24%.
- The Sudan, including the northern half of Ghana, and those parts of Côte-d'Ivoire, Benin and Togo that lie within the basin, are defined as receiving between 900 mm and 1100 mm annual rainfall. This is a transition zone with production of both cereals and root crops. Some transhumant cattle are present seasonally and sedentary cattle are widespread. The probability of a failed growing season is 17%.
- The Guinean zone, covering the southern part of Ghana, receives in excess of 1100 mm annual rainfall with distribution becoming increasingly bimodal toward the south. Yam, cassava and plantain are the main food crops. The risk of a failed growing season is only 8%.

Surface water in the Volta is largely concentrated in the reservoirs of the hydropower schemes (Bagré 1.70 km^3; Kompienga 2.03 km^3; and Lake Volta 148 km^3). A smaller amount of surface water is stored in about 1100 small reservoirs (90% located in Burkina Faso) that are used for small-scale irrigation, household and cattle watering. Their total storage capacity is 0.23 km^3. Lake Volta is a major component of the aquatic system, with a surface area of 8500 km^2, in which water has a residence time of three years (Volta River Authority 2010).

Major water management issues focus on surface water, on the impact of the development of small reservoirs, and on the possible effects of climate change on the operation of the Lake Volta hydropower system and on the livelihoods of more than 70,000 fishers.

Although some efforts are being made, groundwater resources are poorly understood but appear to be under-exploited. Groundwater recharge is estimated to be 12.6 km^3/yr or 3.7% of rainfall. Most boreholes are used for rural domestic supply and give access to good quality water for about 44% of the population. The conservative estimate of groundwater use of 0.088 km^3/y corresponds to about 70% of average annual recharge, which leaves ample opportunities for development (Martin and van de Giesen 2005).

The fluctuation of the rainfall regime in West Africa has major implications for water resource management. The decades of the 1950s and 1960s were relatively wet, while the 1970s and 1980s were drier (Figure 4). Mean stream flows were substantially reduced in the latter time period: by 32% at Saboba on the Oti River and by 23% at Nawuni on the White Volta near Lake Volta (Gyau-Boakye and Tumbulto 2000). The change in runoff

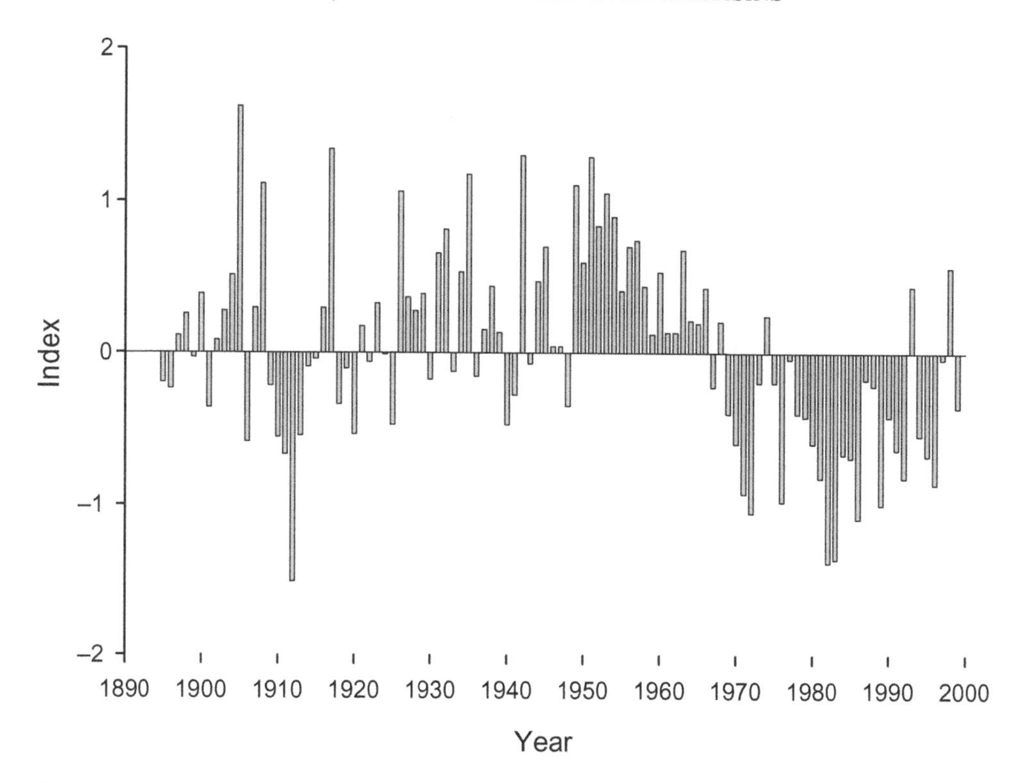

Figure 4. Normalized rainfall anomaly index for the West African Sahel (1896–2000) with a mean value computed for 1921–2000. The normalized anomaly index for a given year is the departure from the mean in mm, divided by the standard deviation (in mm) of the rainfall distribution for the whole period.
Source: L'Hôte *et al.* (2002).

from the wet to the dry period was not uniform across the basin. The runoff coefficient increased during the dry period in the upper basin where annual rainfall is <750 mm/yr, resulting in sometimes higher discharge, while it did not change in the lower basins (>750 mm/yr), where the coefficient was already higher than in the north.

Water use

A detailed water account for the different agro-ecological zones and the sub-basins is presented in Figure 3. Natural vegetation, mostly savannah grasslands, uses the major part of the rainfall (around 80%) throughout the basin. Runoff increases from less than 5% in the northern part of the basin to >10% in the south. The total available stream flow in the basin was about 32 km^3 per year during the period 1990–2000.

National priorities

Most rivers in Burkina Faso dry up during the dry season, except where the two hydropower reservoirs Bagré and Kompienga maintain a continuous flow. With Lake Volta and the Akosombo-Kpong hydropower scheme, Ghana aims primarily at producing energy. There is sufficient rainfall over most of Ghana for a satisfactory production of rainfed agriculture, except in the two northern regions where some small reservoirs and dugouts have been constructed, and in the south near the coast where supplemental irrigation from the reservoirs or Lake Volta is available (van de Giesen *et al.* 2001)

Domestic use

The domestic water demand was estimated as 0.156 km^3/yr in 2005. Large cities in the south, and Ouagadougou, are largely supplied by surface water. Underground water from boreholes supplies most of the other cities in the north and part of the rural population of the basin. Although the situation is improving rapidly, many households in the basin do not have access to good quality water.

Even with increased urban populations in the near future, the demand for domestic water remains a small fraction of the total resource. The same applies to the watering needs for cattle and livestock, with a total annual demand of 0.070 km^3/yr (Clanet and Ogilvie 2009).

Rainfed agriculture

Rainfed agriculture uses only 14% of the total rainfall on the basin. A combination of factors contributes to the gap between potential and observed water productivity. Some of these are directly related to water availability. The inter-annual rainfall variability (droughts) and the intra-seasonal variability (dry spells) lead to low yields, with the combination of high evaporation and poor water-holding capacity of the soil making an important contribution. The soils of the basin come from ancient weathering of the parent rock. Subsequent leaching in the tropical climate made them infertile (Bationo *et al.* 2008), which also contributes to low yields. In the future, gains in the yields and water productivity of the rainfed agriculture appear to be the main prospect for the food self-sufficiency of the basin (Rockström and Barron 2007).

Irrigation

As in most other regions of Africa, the irrigated area in the Volta Basin is only a small fraction (<0.5%) of the total cultivated area, and is also only a small fraction of the irrigation potential (2%). Even the areas for which irrigation water is available from the Bagré and Akosombo/Kpong dams have not been fully exploited. Nevertheless, many small farmers in the drought-prone areas of the basin consider access to dry season irrigation important to alleviate their vulnerability and poverty and to increase food production (Ducommun *et al.* 2005).

The formal, state-owned irrigation schemes amount to 13,000 ha and 6500 ha respectively in Burkina Faso and Ghana. The associated total water demand is 0.509 km^3/yr. Poor management as well as social, cultural and economic reasons contribute to the present underdeveloped state of large-scale irrigation. The policy documents for the development of African agriculture issued by the African Union/NEPAD (2003) and the European Commission (2007) focus mainly on the technical feasibility and on the investments required, but do not take into account properly the social aspects of the implementation of irrigation in West Africa, which remain poorly developed compared with other regions in the tropical world.

About 1400 small reservoirs have been developed in Burkina Faso and northern Ghana. Approximately 1100 are within the basin, with a storage volume estimated as 0.232 km^3. They contribute to dry season vegetable production on about 10,000 ha, and complement rainfed production. The active informal urban and peri-urban irrigation, mostly devoted to vegetable production, remains poorly documented but is rapidly expanding.

Other uses

A main feature of the basin is Lake Volta behind Akasombo Dam. The dam closure in early 1965 severely impacted the downstream environment and its population. It has been complemented since by the smaller Kpong Reservoir (0.19 km^3), a few kilometres downstream. The water released through the turbines amounted to 37 km^3 per year at the beginning of the 2000s. Since 1970, the decrease in rainfall over the basin and probably an increase in electricity production have contributed to a reduction in the water stored in Lake Volta. The energy produced, around 8 billion kWh/year, is an important contribution to the national income, so this water use is a priority for Ghana. A major threat to the lake and hydropower is the presence of the water hyacinth (*Eichhornia crassipes*) in the lower Oti River, one of the main tributaries to the lake. The plant is known to damage the turbines if taken up in the water intakes.

In Burkina Faso, the reservoirs Bagré, for both hydropower and irrigation, and Kompienga only for hydropower generation, contribute also to some inland fisheries. Other important issues are the inland fisheries, livestock access to watering, and the resources associated with the numerous inland valleys and wetlands, which will be discussed below.

Farming systems and food production

Rainfed production of food and cash crops is largely dominant in the basin where only a very small proportion of the cultivated area is irrigated. Meat production from cattle, small ruminants and poultry may be considered as partly derived from rainfed cultivated or wild biomass. Fish catch must also to be taken into account, both as a food production and as an important social and economic activity, especially from Lake Volta.

As food production is especially vulnerable in the north and central parts of the basin, due to the rain variability, we have focused on the rainfed cereal production in these parts of the basin that are prone to dry spells and droughts. Rainfed agriculture is the activity of the majority of the poor rural populations and thus the main tool for improvement in food security. The challenge is to improve the use of the available rainwater. The main systems that contribute to food production are discussed below.

Rainfed farming systems

Variation of the distribution of the main crops across the agro-climatic regions of the basin is mainly a function of rainfall and soil characteristics, with a successive north–south dominance of millet, sorghum and maize, which are replaced by cassava, yam and plantain in the south where the rainfall exceeds 1000 mm/yr and where there is a lower impact of dry spells (Figures 5 and 6). The overall low production of rainfed crops in the basin results from a combination both of a limited use of the arable land and low yields.

The use of arable land

As elsewhere in West Africa, about 50% of the basin area is arable, and only half of this area is cultivated in the countries of the basin (Table 1). The relative increase in cultivated area has been closely related with the changed demography in southwestern Burkina Faso of the

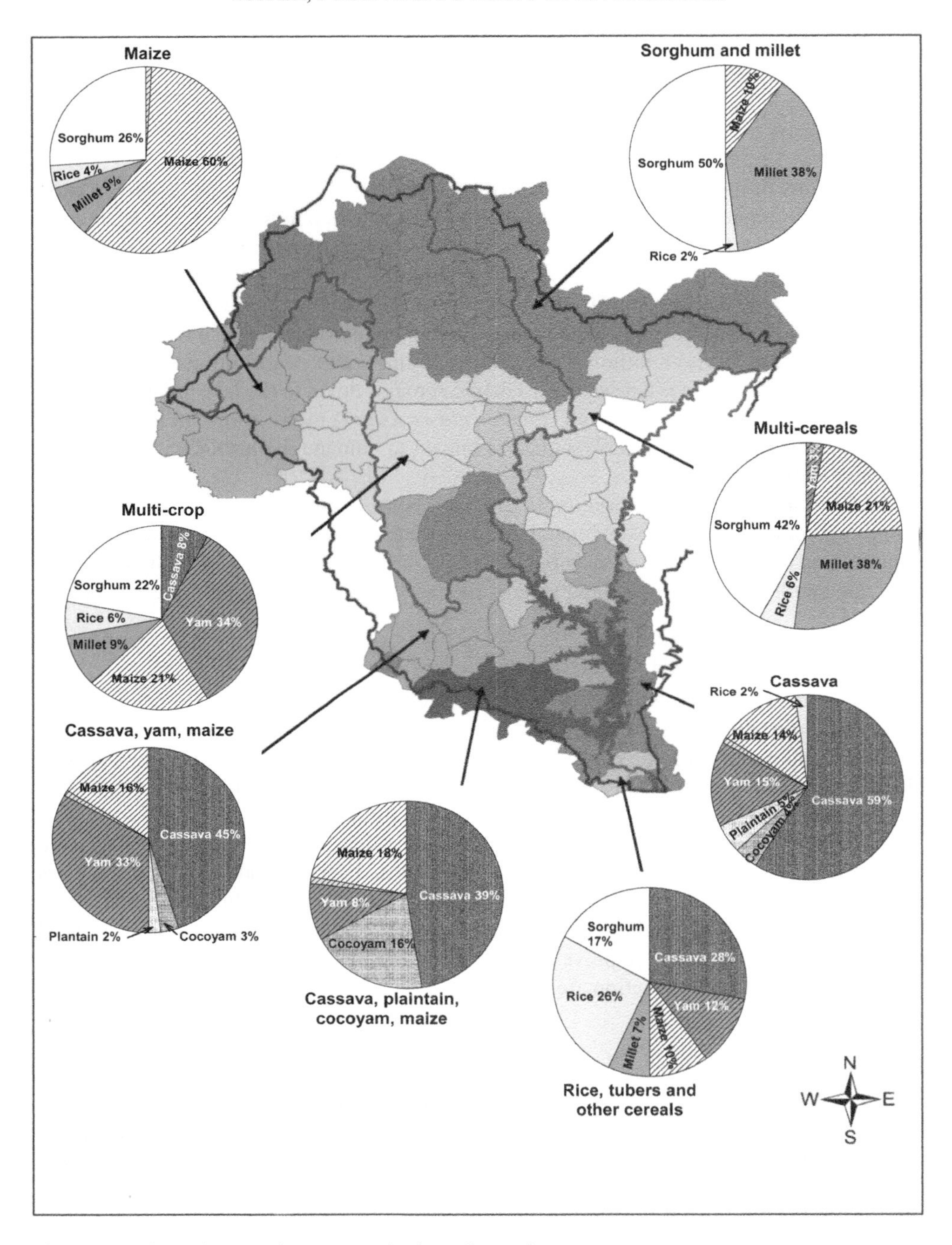

Figure 5. The main cropping systems in the Volta Basin.
Source: Map by BFP Volta (Lemoalle and de Condappa 2009) with production data from MAHRH (2006) and MOFA (2006) data for 1992–2000.

past 30 years, with no net increase of cultivated area per rural inhabitant (Serpantié 2003). More recently, in the Volta Basin parts of Burkina Faso and Ghana, the rate of increase of

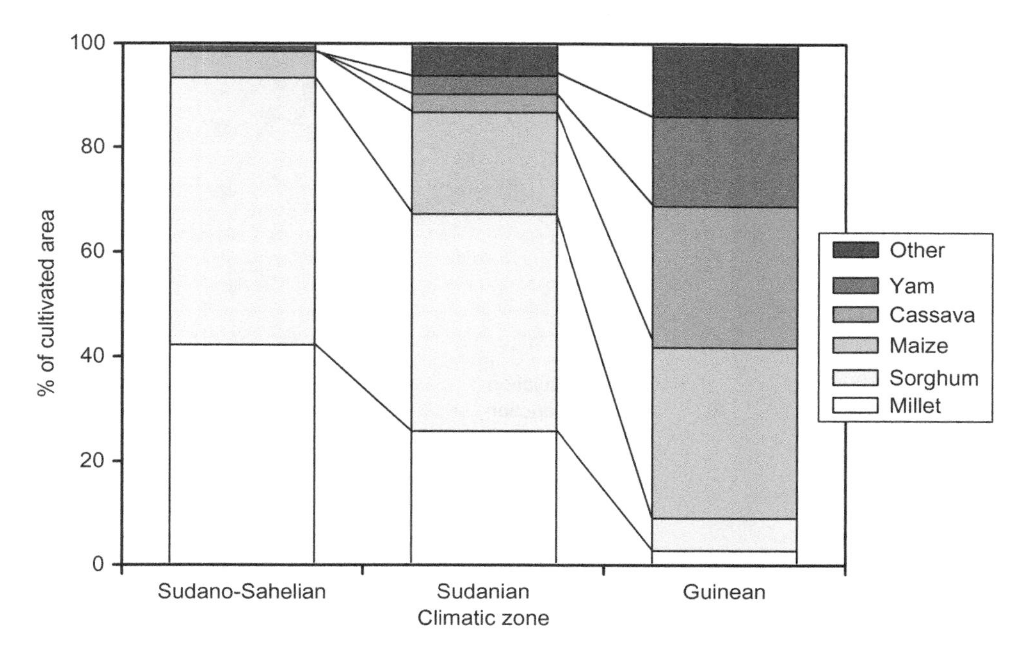

Figure 6. The North-South gradient in agro ecological zones and agricultural systems (as percentage of the cultivated area).
Source: Lemoalle and de Condappa (2009).

the cultivated area has been very similar to the increase rate of the total population between 1992 and 2003 (Figure 7), although there is an indication of a slight increase of cultivated area per rural inhabitant.

There is therefore arable land available at the basin scale for further development of cultivation. This availability leads to some migrations and new settlements within the countries of the basin, but also to conflicts between herders and farmers.

Low yields

Total production is determined by harvested area and yield per unit area. As noted earlier, harvested area in the Volta Basin is relatively low. Crop yields are also low, often less than 1 t/ha for maize. The average cereal productivity was 1.4 t/ha in Ghana for the period 2002–2005, but only 1.0 t/ha in Burkina Faso (Word Bank 2007).

The yields and water productivity figures for the main food crops in the Ghana and Burkina Faso parts of the Volta Basin are below those attained by commercial farmers or experimental stations elsewhere in semi-arid savannahs in Africa (Table 2). The yield gap is mainly manifested in the inability to cope with the dry spells during the rainy season, low crop water-uptake capacity, and poor soil fertility. In general, the soils hold too little available water in the root zone to allow crops to continue growth during within-season dry spells (Rockström 2001).

There are a number of reasons for low crop yields:

- Variability of rainfall in time and space;
- Poor and degraded soils, with little or no input of fertilizers; and
- Lack of manpower and productive assets, such as draught animals and tools.

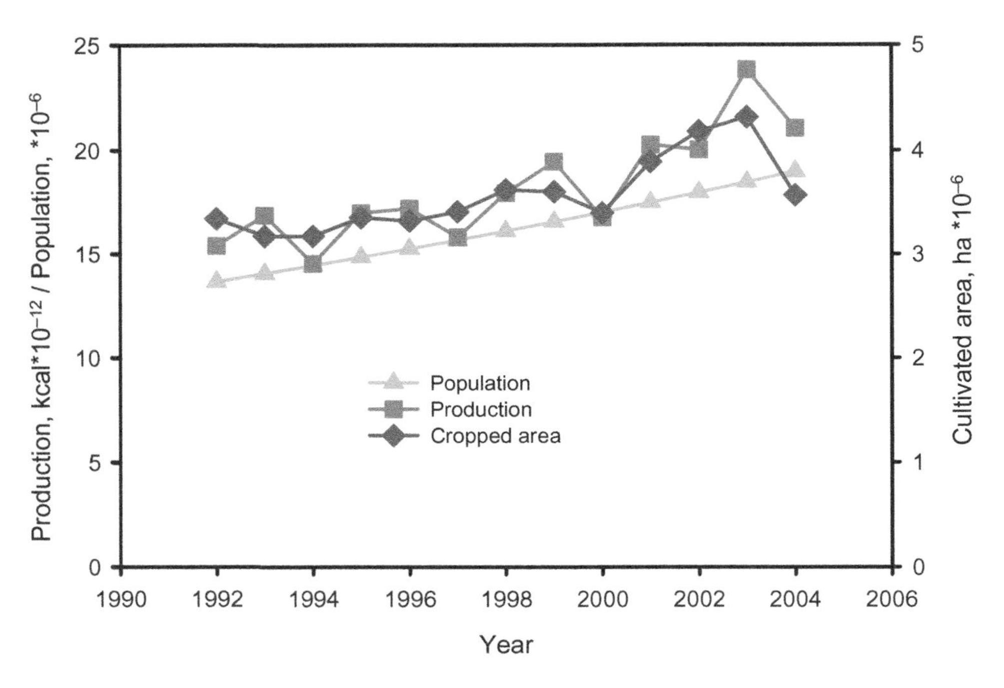

Figure 7. The time trends of cultivated area, total production and increase in population indicate that the increase in production results mainly from increased cropped area.
Source: Data from MAHRH (2006) and MOFA (2006) for cultivated area and production, with an estimated demographic increase of 2.8%/yr in the basin.

Table 2. Mean values for yields and water productivity for the period 1992–2000 in the Volta Basin.

Crop	Yield	Water productivity
	tons/ha	*kg/m³*
Sorghum	0.91 ± 0.26	0.09 ± 0.02
Millet	0.75 ± 0.22	0.07 ± 0.02
Maize	1.25 ± 0.58	0.12 ± 0.03
Yam	9.8 ± 3.6	0.76 ± 0.30
Cassava	11.7 ± 4.4	0.88 ± 0.35
Plantain	7.3 ± 3.2	0.57 ± 0.31

Source: Original data from MOFA (2006) and MAHRH (2006), computed by BFP Volta. Rainfall data from Climate Research Unit (2008).

Water availability has been discussed above. Water harvesting and soil and water conservation techniques can, to some degree, improve the effective use of rain water by crops, especially by increasing the water-holding capacity of the soil by using mulches and incorporating crop residues. Rainfall is spatially heterogeneous, so farmers scatter their fields around the village to minimize risk (Sivakumar and Hatfield 1990).

As described above, most African soils are poor and moreover their nutrient balances are often negative so that they are progressively mined (Smaling 1995, Smaling *et al.*

1999). In contrast to the developed world where excess fertilizer and manure have damaged the environment, insufficient use of fertilizer causes environmental degradation in Africa (Bationo and Mokwunye 1991, Bationo *et al.* 2006).

Simulation modelling showed that fertilizer (N40, P13 kg/ha) can increase yield potential of most soils of the basin (Terrasson *et al.* 2009). The increase ranged from twofold for the better soils to almost sixfold for the poorer soils. The results also point to a distinction between the southern region, where yields vary little with latitude because rainfall is abundant; and the central and northern parts of the basin above latitude 11° N as rainfall becomes more limiting than nutrients. Field observations in the Sahel and farm trials in northern Ghana have also indicated the high potential of fertilizer use (Bationo *et al.* 2008).

Smallholder farmers are often unwilling to use fertilizer technologies because of the social and cultural environment, which include risk avoidance, insecurity of land tenure, and lack of access to credit and to markets. A simulation of farmers' behaviour and needs in the central plateau of Burkina Faso points to three variables that may help towards a more sustainable agriculture: credit, subsidized fertilizers and land tenure (Ouédraogo 2005).

Other cropping systems

Peri- and intra- urban cultivation is an important component of vegetable production, with an estimated 45,000 ha in and around the main cities in Ghana. Peri-urban cultivation is the main user of irrigation in Ghana. Urban production is mostly informal irrigation using wastewater, which will increase as urbanization increases in the future. Research is already under way on the impact of irrigation by wastewater on food safety, which is particularly relevant to improve the health security (Drechsel *et al.* 2008).

Wild cereals and other sources of food or income, such as wild rice, fonio (*Digitaria exilis*) and materials for basketry may be very important in some periods of the year for the poorest rural people. Farmers often grow trees in conjunction with agricultural crops or livestock. These parkland trees generate short-term income, provide environmental benefits (e.g. erosion control), edible leaves, and fruits for household livelihoods or even for the national economy (e.g. shea nuts).

Farmers growing cash crops, cotton and groundnut in both Burkina Faso and Ghana, and cocoa, oil palm and coffee in Ghana, are better off than subsistence farmers. Governments have improved infrastructure to help farmers who produce cash crops because of their importance for export earnings, which has led to improved quantity and quality of cotton (Conley and Udry 2004, World Bank 2007) and cocoa. If similar policies were implemented for food crops, increased production would improve livelihoods and could help avoid food crises.

Livestock systems

Livestock are often a farmer's largest non-land asset. In Burkina Faso they account for more than half of rural households' wealth (World Bank 2007). The patterns of livestock ownership give important insights into rural poverty.

There are three main complementary livestock systems in Burkina Faso (IEPC 2007).

- In the pastoral system (PS), itinerant herders migrate their cattle to find the best food during all seasons and avoid environmental hazards, principally drought. PS occurs in all countries of the basin between latitudes 6°N and 14°N, limited by low rangeland productivity in the north and by trypanosomiasis in the south. Herders constitute 4% of the basin's rural population and 23% of them are highly vulnerable to droughts (Clanet and Ogilvie 2009). There is a trend for PS herders to become sedentary. The PS contributes 65% of the meat export of Burkina Faso (the third-ranked export income of the country) and 585,000 tons of milk. PS is under threat as sedentary farmers increasingly deny herders access to pasture, crop residues and water.

- In the crop-livestock system (CLS), sedentary farmers mix cultivation, arboriculture and herding around their village. The livestock feed on crop residues and the surrounding rangeland. CLS covers 80–90% of all livestock keepers and livestock provide the main cash income to 86% of rural households (IEPC 2007). Sheep and goats are common, but only the wealthiest own cattle.
- In the industrial system (IS) sedentary, market-oriented producers are mostly in the vicinity of cities, or railways. IS is less dependent on natural rangeland and accounts for only about 2% of the total cattle in the basin.

Rangeland provides 90% of the fodder for ruminants in the basin with crop residues and agro-industrial by-products providing the remainder (IEPC 2007). Forage crops provide less than 0.5%. Livestock consume only about one third of the range forage, the rest being lost to stalling, termites and normal senescence processes. Complementary browse from trees and shrubs that use water from deeper in the soil profile, and hence stay green longer into the dry season, increases the water productivity of cattle systems and could be developed further.

The livestock production is limited by access to water and prophylaxis for endemic diseases (anthrax, brucellosis), and parasites. Proper sanitary measures would greatly increase the livestock productivity (Clanet and Ogilvie 2009).

Fisheries systems

The species richness of the fish (147) in the Volta Basin indicates the overall wealth of the aquatic system (Hugueny and Lévêque 1999). Béné and Russell (2007) distinguish three different types of fishing communities in the basin.

Fishing communities around Lake Volta

The Lake supports about 70,000 active full-time fishers (Braimah 2003), suggesting that around 300,000 people depend on fishing, which contributes over 70% of the income of these communities (Pittaluga *et al.* 2003). Braimah (2000) estimated that there were 24,000 fishing canoes on Lake Volta, 95% of which are plank canoes without motors, each operated by three people. In contrast, winch boats represent only 1.8% of the total fleet on the lake, employ only about 5% of the fishers but take 65–70% of the total catch (FAO 1990).

Actual production data for the fishery sector are not entirely coherent. For Lake Volta, figures ranging between 40,000 and 215,000 tons per year have been reported (Braimah 2000, Barry *et al.* 2005, and statistics from MOFA [2006]). Similarly, the value of fish production in the late 90s was estimated at US$30 million (de Graaf and Ofori-Danson 1997) to as much as $160 million (Pittaluga *et al.* 2003), with annual revenues ranging from US$420 to US$2,250 per fisher. In the last decades, the numbers of fishermen and fishing boats have steeply increased, and fishing practices have become more intensified, especially with the introduction of winch nets in the mid-1980s (Ofori-Danson 2005). The lake fishery is said to be overexploited, but there is no scientific evidence to support this (Barry *et al.* 2005).

Farmer-fisher communities around medium- to large-scale irrigation reservoirs

The main communities in this category live around the reservoirs of Bagré (25,000 ha), and Kompienga (18,000 ha) and the Sourou floodplain (68,000 ha) in Burkina Faso and Mali. About 4% of the population relies wholly on fishing (approximately 500 full-time fishers and 300 fish processors). The majority of the population living around these water bodies belongs rather to farming communities for whom fishing is an important secondary activity: about 70% of the households around Bagré diversify their farming activities with seasonal fishing (Béné and Russell 2007).

Fishery around small water bodies and river-floodplains

The fisheries of some 300,000 ha of rivers, ponds, seasonal floodplains and small reservoirs in the basin are estimated to produce about 9000 t/yr for 100,000 people who rely in part on fishing for their livelihoods (Morand *et al.* 2005, Béné and Russell 2007).

Water productivity

Water productivity (WP) is defined as output per unit of water depleted. Output may be measured in terms of the amount or value of crops or livestock produced or (at least in principle) the value of the product in domestic, urban, industrial, hydropower or environmental uses. Water is depleted when it is unavailable for further use, e.g. when it is evaporated or transpired, or polluted to the point where it can no longer be used.

Basin scale

Water productivity at the basin scale is the amount or value of agricultural production divided by the amount of rainfall received. We computed from Ministère de l'Agriculture, de l'Hydraulique et des Ressources Halieutiques (MAHRH) and the Ministry of Food and Agriculture (MOFA) data extrapolated to the whole basin, that the total production of food in the basin for the year 2000 was 1.6×10^{13} kilocalories. This represents a WP of 466 kcal/m^3 if only the rainfall over the cultivated area is taken into account. The amount of water needed per capita for food production in the basin, with an average human energy requirement of 2500 kcal/day, is presently 2000 m^3 per year.

Field scale

Water productivity at the field scale is yield per hectare divided by the amount of rainfall per hectare. Crop WP is generally higher for maize (up to 0.2 kg/m^3 of water depleted) than for sorghum or millet (rarely exceeding 0.1 kg/m^3 of water) (Table 2). The spatial distribution in the Volta Basin of field-scale water productivity for the main food crops is shown in Figure 8.

We analysed the relationships between maize, millet and sorghum yields in different parts of the basin on the one hand, and annual rainfall on the other. Maize is better able than sorghum and millet to take advantage of wetter conditions to produce higher yields. In contrast, yields of sorghum and millet do not decrease as much as that of maize in drier areas or in drier years.

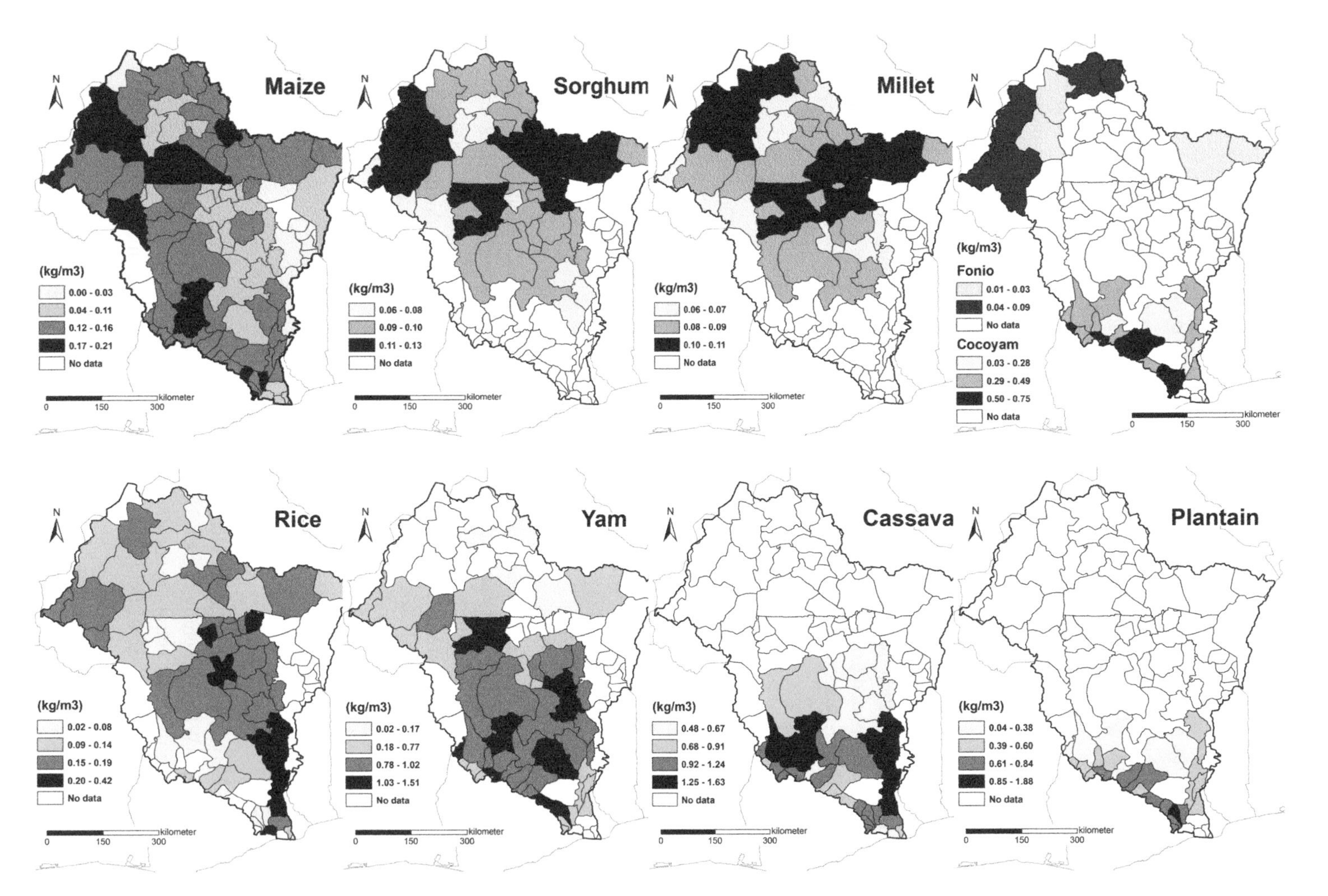

Figure 8. The mean water productivity (1992 to 2000) of the main food crops in the Volta Basin.
Source: Map by BFP Volta with data from MAHRH (2006) and MOFA (2006).

Water and poverty

The Volta Basin's population is 20 million of which 14 million are rural and are mostly very poor (Table 3), amongst the poorest in the world. Many countries in the basin have not yet started the demographic transition of decreased mortality and fecundity rates. There will be development but it will be slow and from a very low base, so that the basin will remain poor for a long time to come (de Fraiture and Wichelns 2007).

The three main causes of water-related poverty in Ghana (Asante and Asenso-Okyere 2003, Asante 2007) are: low productivity of fishing and agriculture; water insecurity (rainfall variability, poor access, health impacts, loss of labour); and water-related diseases (malaria, guinea worm and others).

In Ghana and Burkina Faso and Togo, which make up most of the basin, national surveys show that the incidence of poverty is higher in the drier north than in the wetter south. In Ghana, for example, the proportion of the population below the poverty line is 70% in the rural savannah (north) against 38% in the rural forest (south) (Coulombe and Wodon 2007). Here, the poverty line is set in local currency equivalent to somewhat less than US$100 per year. In Burkina Faso, 52% of the rural population was considered poor in 2003. Because Burkina Faso is predominantly rural, this accounted for around 92% of the poverty found in the whole country. The incidence of poverty was lower among cash crop (groundnut) farmers (46%) than among subsistence farmers (56%) (Lachaud 1998). The poverty line in 2003 was set in local currency equivalent to about US$165.

People are poor because of low agricultural productivity, limited access to markets, unstable prices, and insecure land tenure (Burkina Faso, Ministère de l'économie et du développement 2004). Actions to alleviate poverty are hindered by financial, infrastructure and institutional constraints.

There are more poor and vulnerable households in some categories:

- Subsistence farmers without livestock or draught animals;
- Herders who do not own the cattle under their care; and
- Households that are vulnerable to environmental changes or social restrictions (for example, migrants).

Subsistence farmers who own poultry or small ruminants are typically better off.

Table 3. Information on poverty in Volta Basin countries.

Country	HDI	Rank	GDP per capita	%<national poverty line	<US$1 /day	<US$2 /day
			US$	*%*	*%*	*%*
Benin	0.428	163	1091	29.0	30.9	73.5
Burkina Faso	0.342	174	1169	46.4	27.2	71.8
Côte-d'Ivoire	0.421	164	1551	–	14.8	48.8
Ghana	0.532	136	2240	39.5	44.8	78.5
Mali	0.338	175	998	63.8	72.3	90.6
Togo	0.495	147	1536	32.3	–	–

Notes: HDI = Human development index; rank in the 177 countries for HDI; gross domestic product (GDP) per capita, USD (2004); % of population below national poverty line; % of population living with less than US$1 and US$2 per day.
Source: UNDP Human Development Report (2006).

Poverty and land

Poor families typically lack draught animals and tools such as a plough or a cart so that they can only cultivate as much area as their human work power can manage. The cereal-growing zone of the basin has a Gini coefficient of 0.4, with 10% of farmers cultivating 30% of the land compared with 30% of farmers cultivating only 10% (Ducommun *et al.* 2005). The median farm size in Burkina Faso is 3.1 ha, and the lower quintile is less than 1.4 ha. (MAHRH 2006). Only larger farmers using fertilizer have a food surplus to sell (Ducommun *et al.* 2005).

Water quality and access

Access to domestic water depends both on the time (or distance) to fetch it and its quality. Typically, poor-quality water comes from traditional wells or surface water, while good-quality water comes from modern boreholes. Nearly 50% of households in the basin lack access to good-quality water, especially in the northwest. In Burkina Faso, 82% of the urban population has potable domestic water, but only 44% of the rural population does (INSD 2003).

There are no formal data on the relation between poverty and access to good-quality domestic water. In the rural villages, both rich and poor households use the same water source. In Ghana, households use borehole water when it is available.

Water-related diseases

The water-related diseases malaria, schistosomiasis (bilharzia), trypanosomiasis and diarrhoea from unsafe domestic water (Poda 2007) are a component of poverty in the basin. Malaria and schistosomiasis have effects at the micro (household) and macro (state) scale by reducing labour availability and productivity. Schistosomiasis is directly related with the development of irrigation and small reservoirs, with up to 70% incidence in Burkina Faso. Malaria is widespread especially in the central basin (Figure 9). It is a principal constraint to development and the main cause of mortality in children under five.

The cost of malaria care hits the poor hardest, costing them 34% of income compared with 1% for the rich. At the national level, malaria costs Ghana US$50.1 million in direct and indirect costs (Asante and Asenso-Okyere 2003). Insecticide-treated bed-nets are a cost-effective preventive measure.

Control of water-related diseases in the basin requires proper management of the water and the environment, changes in social and cultural behaviour, and improvements in health services. Onchocercosis was successfully eradicated first through control of the vector (blackflies of the genus *Simulium)* and later of the parasite by prophylaxis with invermectin. This has led to partial recolonization of good land in valleys where formerly river blindness occurred (OCP 1995).

Institutions

The institutional and policy context of a country controls the way in which water and food problems unfold, and within which they must be solved. Institutional and policy changes are often critically important to promote adoption of desirable agricultural and water management practices. "Institutions" are here understood to include informal norms and customs as well as more formal organizations and structures.

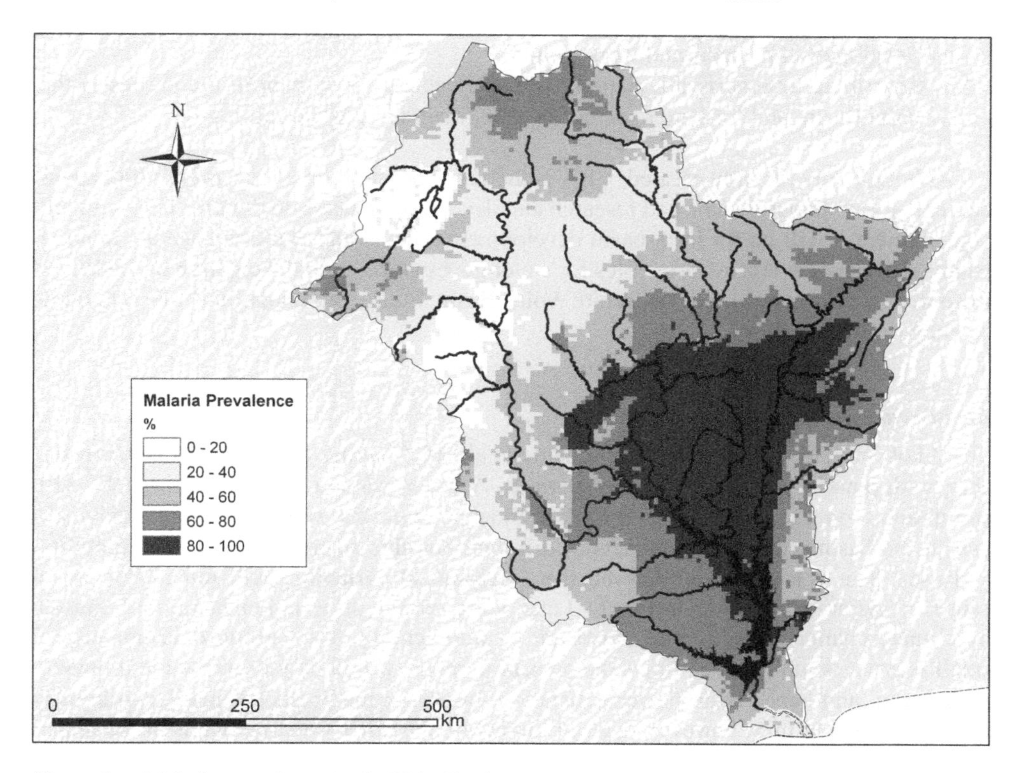

Figure 9. Malaria prevalence in the Volta Basin.
Source: Mapping Alaria Risk in Africa, MARA/ARMA Project (http://www.mara.org.za).

The institutions in the Volta Basin were described by Lautze *et al.* (2006) and by Opoku-Ankomah *et al.* (2006). The main links between rural poverty and institutions were identified as access to credit, to market, to land and to water. Other characteristics of the institutional environment, although not directly related with water, are also important.

The basin context: the Volta Basin Authority

The priorities for water use are not the same in Burkina Faso and Togo (irrigation) as in Ghana (hydropower). In order to implement an international co-operation for the sustainable management of the Volta Basin water resources, the six basin countries, with the support of the Economic Community of West African States (ECOWAS), France, the African Water Facility, and the International Union for the Conservation of Nature (IUCN), created the Volta Basin Technical Committee (VBTC) in 2004, which evolved into the Volta Basin Authority (VBA) in 2009.

The Volta Basin Authority (VBA) has a mandate to:

- Promote permanent consultation tools among the parties for the development of the basin;
- Promote the implementation of integrated water resource management (IWRM) and the equitable distribution of the benefits resulting its utilization;
- Authorize the development of infrastructure and other projects planned by the stakeholders, which could have substantial impact on the water resources of the basin;

- Develop joint projects and works; and
- Contribute to poverty alleviation, the sustainable development of the parties in the Volta Basin, and for better socioeconomic integration in the sub-region.

Although only recently created, VBA is the focus of attention of several funding agencies and acts to transfer identified research and development needs to donors. Time will tell how efficient it will be for intra-basin development. Ghana and Burkina Faso are making attempts to develop IWRM principles for water governance, with the creation of the Water Resources Commission and the White Volta Basin Board in Ghana, and the Nakambe Agency in Burkina Faso.

Basin-scale water management tools

The Akosombo/Kpong hydropower scheme is already making use of all the water available, with power cuts at some critical periods. A water evaluation and planning (WEAP) simulation model of water allocation (de Condappa *et al.* 2009) provides a detailed analysis at the basin scale of the impacts of different possible scenarios of water availability and uses. The WEAP model showed that inter-annual variability will remain the main cause of variation in water inflow to Lake Volta, although variation in rainfall caused by climate change may become important by mid-century. Very strong development of small reservoirs (up to seven times the present number) would only decrease the inflow to Lake Volta, and hence hydropower generation, by 3% in the present climatic conditions. This development may, however, have an impact on the other water demands in the upper sub-basins.

Although the various techniques available for improving water use in rainfed agriculture may have some impact at the basin scale, via modification of runoff and evapotranspiration, their development is more directly related to outcomes at the sub-basin and villages scale.

The national context of dual governance systems

In much of the Volta Basin, the formal, national institutions coexist with the traditional structures of hierarchy. Although governments have attempted decentralization, they have limited ability to enforce policies at the local level, so that policy implementation takes place in the context of multiple power foci and multiple institutions (Lautze *et al.* 2006). The duality impacts everyday life. Decisions regarding land tenure and access to land and water, for example, are often made by the traditional authority, which reduces security of tenure, and can actively discourage investment in technologies that improve land and conserve resources (Lautze *et al.* 2006). Customary land tenure is perceived by community members (if not by outsiders) to be well adapted to local conditions, however other reports state that duality in governance creates uncertainties in land tenure and discourages development (de Zeeuw 1997, Lautze *et al.* 2006, Bugri 2008). Land tenure is in transition in the basin countries, but its progress varies according to the local traditional rules, which vary with the ethnic group concerned.

Competition between local authorities and formal institutions, and a lack of local legitimacy and poor enforceability of official rules, make resource management prone to conflict and renegotiation, for example, in the allocation of land in irrigated schemes in Ghana. This compromised the efficiency, sustainability and equity of the irrigation sector, as well as the prospects for the implementation of water reform (Laube 2005).

Access to markets for food crops

Ghana and Burkina Faso have invested in the development of cash crops, which benefit both the farmers and the nations' hard currency income. In Burkina Faso, a strong political will has accompanied the development of cotton production, with several incentives offered to farmers, including subsidized fertilizer. The result has been a spectacular increase in cotton production, the country ranking as a main producer in Africa (and the first in 2006). This example shows that when the political will exists, important development can be attained in agricultural production.

However, competitors on the world market often subsidize their own production (cereals in Europe, cereals and cotton in the United States where support to national farmers largely exceeds development assistance to developing countries) so that the prices paid to the developing countries remain largely erratic and undervalued. As long as true trade liberalization has not been established, developing countries have little option but to protect their agriculture, and especially smallholder producers of food crops.

Burkina Faso has a strong possibility for the development of national cereal markets, which would alleviate part of the cost of more than 200,000 tons per year of rice and wheat that are currently imported (Ducommun *et al.* 2005). At present, the total sales of food crops in Burkina Faso are estimated to be between US$320 million and US$440 million, that is two to three times the cotton exports. The import of cereals amounts to about US$80 million.

The urgency of the need to develop national food crop markets in Africa is amply demonstrated by the recent world food crisis, and is strongly encouraged by the United Nations (UN) Secretariat (Fleshman 2008). It must involve all the stakeholders, from small producers associations to the higher-level state authorities. In the context of burgeoning urbanization, better organization of national and regional markets is essential and would help alleviate rural poverty. An innovative and pro-poor approach to develop national food-crop markets requires several inputs (Ducommun *et al.* 2005):

- Credit supported by insurance to smaller farms, for equipment;
- Enhanced infrastructures for physical access to local markets;
- Development of small agro-industries to transform the local production for the national market in the first phase, and for export in the second phase;
- Organization of the national market with floor prices; and
- Customs barriers (import tariffs or quotas) to protect local food production from international dumping when needed.

The drivers for change: risks and opportunities

The Volta Basin is affected by global, regional and national drivers of change. These include economic growth, population growth, technical change, land degradation, climate change, globalization, urban development, political change and trade liberalization. In principle, these can profoundly influence future progress by increasing the resilience of the rural poor and of ecosystems at the basin level.

Two drivers of change are discussed in more detail: population growth and its consequences for food demand, and climate change and its consequences for food production (Table 4).

Table 4. Global trends in the Volta Basin and their consequences on rural activities.

Drivers	State variables	Secondary variables	Needed changes
Population change	Demand for food, Ratio urban/rural pop	Cultivated land area Peri-urban production	Increase yields Small-scale irrigation Increase meat production Increase fish production
Climate change	Rainfall distribution Temperature	Water allocation Water availability Length of growing season	

Population growth and food demand

While the limits of the basin and its surface waters are exactly defined by the topography, there are no clear boundaries when social and economical aspects are considered within the context of any given component of national policy. In these instances, the basin limits are not fully operational. Using national estimates of population increase (United Nations Population Division 2008) we have estimated that the total basin population will increase from 17.2 million inhabitants in 2000 to 32 million in 2025. The rate of increase in population may fall in the future, however, and a population of 50 to 60 million in 2050 in the basin seems a reasonable figure.

Together with this large increase, the ratio of rural to urban populations will change dramatically. If 2025 is compared with 2005, the rural population only increases slightly (x 1.2) but that rural population must provide food for a much-increased urban population (16 million instead of 5.8 million in 2000). After 2025, the urban population will become much larger than the rural population with almost two urban dwellers for each rural dweller in 2050.

If the climate remains constant, the greater production of food required cannot be achieved only by an increase in the area of cultivation; large increases in productivity will be necessary as well. Moreover, the nature of the production will also have to change according to the changes in urban diet and the demand for rice, maize and animal proteins.

The main global scenarios, the Millennium Ecosystem Assessment and the Special Report on Emission Scenarios of the Intergovernmental Panel on Climate Change (Nakicenovic 2001) are pessimistic on the possibility for Africa south of the Sahara to be self-sufficient in food production. The identified risk is an increase in the population whose food needs would not be met by a sufficient increase in local food production.

The steep increase in food demand in all scenarios impacts the amount of both arable land and pastureland. Care will be needed to manage properly the competition for land and water, and to protect the natural pastures and water access to livestock the demand for which will increase strongly.

Climate change and food supply

The governments of the states in the basin are well aware of the likely effect of climate change. They are all signatories of both the UN Framework Convention on Climate Change (UNFCCC) and its Kyoto Protocol and all have ratified both. Their First National

Communications to the UNFCCC contain information on national circumstances and vulnerability assessment.

Overall, although no major change in total rainfall appears in climatic models, increased variability and unreliability are common features. Policies and water management projects must take these changes into account, despite some major uncertainties on the exact nature of the changes.

In the Volta Basin, the distribution of the main food crops and the livelihood vulnerability of rural households reflects the rainfall distribution and its north–south gradient. Terrasson *et al.* (2009) using a modelling approach showed a steep sensitivity of cereal yields to rainfall change in the range 500 mm to 1000 mm rainfall per year. Changes in both the nature and quantity of the main staple foods, as well as rangeland productivity, may thus occur with climate change. This will be especially acute in the northern half of the basin, mainly in Burkina Faso, where rainfall is already a limiting factor. One possibility in this region would be to increase the number of small reservoirs, and thus to develop a complement to the farmers' income by dry-season cultivation.

Conclusions: opportunities for implementation and research

A major goal of the Basin Focal Project Volta study (a component of the Challenge Program on water and Food) was to identify researchable questions and breakthroughs in implementation issues that promise major impact and change in both the near future and in the longer term (2025 and 2050). The results clearly indicate that, although there are knowledge gaps on some issues at the basin scale, there is understanding at the local level in some regions of the basin, which allows for development and implementation where the conditions are favourable. Examples are use of fertilizers in semi-arid areas, and irrigation with groundwater where there are sufficient proven water resources available in sedimentary strata.

Opportunities exist in both advanced and applied research institutes for studies to address knowledge gaps and innovative issues. Most of them involve a multidisciplinary approach with a strong social component. While technical or biophysical solutions already exist for many problems in agriculture, their implementation remains limited because of institutional, social, cultural and economic reasons.

The main opportunities for research and implementation are listed below:

(1) *Better access to good quality water:* The governments of the countries in the basin are well aware of the water quality–poverty nexus, and are drilling boreholes and modern wells to provide good-quality water for rural populations. Surveys indicate, however, that a specific effort should also be devoted to maintain the existing boreholes.

(2) *Irrigation development:* The World Bank Africa Region irrigation business plan (2008–12) for poverty reduction ranks Ghana, Burkina Faso and Mali in Group 1 countries, where investment in irrigation and rainfed cultivation is recommended. The main development presently underway is the peri-urban irrigation often using wastewater, which needs more analysis of sanitary issues (Amoah *et al.* 2005).

(3) *Groundwater development:* The groundwater resource should be assessed as a possible tool for extensive development of small-scale irrigation.

(4) *Fertilizer use and micro-credit:* An important opportunity is to develop incentives for smallholder farmers in rainfed systems to invest in fertilizers and small-scale irrigation. Research is needed to reduce barriers to fertilizer use, and increase its socio-economic acceptability.

(5) *Livestock and fisheries:* Livestock and fish systems support many of the poor. Fisheries need to be monitored to allow sound management to be implemented. Services provided by the aquatic ecosystems are not well understood and need socio-economic appraisal. A better use of forage browse for livestock would improve water productivity and animal production. Increasing conflict between nomadic herders and sedentary farmers for access to drinking water for the nomads' livestock requires resolution at the national and transboundary institutional levels.

(6) *Institutional and market development:* Institutional development is key to resolving the ubiquitous traditional/legal duality. The national or regional food market has to be organized in the same way as the cash crop market. Pro-poor activities should involve combined improvements in the access to land, to agricultural water, and to affordable micro-credit for this part of the population.

(7) *Small reservoir development and management:* The storage of renewable water is a requisite for agriculture development (World Water Assessment Programme 2009). Analysis in the central part of the basin indicates that the possibility of cultivating during the dry season would avoid the need for seasonal migration. The construction of small reservoirs provides the possibility of producing out-of-season vegetables as cash crops, even for those who do not possess draught animals. These reservoirs would complement the present use of inland valleys. Small reservoirs require careful management to prevent exclusion of the poorest and careful monitoring to reduce the threat to flows on which Akosombo, Bagré and other large hydroelectric schemes depend.

(8) *Water governance:* Strong political will is needed to implement successful water reforms that mirror the principles and objectives of IWRM. These reforms will create mechanisms to involve local water users and to mediate conflicting interests effectively under the current institutional, administrative and political conditions. The water allocation simulation tool developed by de Condappa *et al.* (2009) for analysis of transboundary water sharing may foster dialogue on IWRM at the basin scale.

The bio-physical constraints on the agriculture production in the basin can be overcome. Answers to the research questions identified above would improve the possibility of enhancing the overall agricultural production, especially food production. Political decisions are needed to put them in practice.

Acknowledgements

This work is part of the Consultative Group for International Agricultural Research (CGIAR) Challenge Program on Water and Food (CPWF)/BFP Volta (http://www.waterandfood.org). The authors gratefully acknowledge that support as well as the co-funding by Institut de Recherche pour le Développement, France.

Note

1. With contributions from Winston Andah, Christophe Béné, Philippe Cecchi, Anne Chaponnière, Jean-Charles Clanet, Myles Fisher, Valérie Hauchart, Edmond Hien, Marie Mojaisky, Madiodio Niasse, Jean-Noël Poda, Jorge Rubiano, Aaron Russel and Isabelle Terrasson.

References

African Union/NEPAD, 2003. *Comprehensive Africa agriculture development programme*. Midrand, South Africa: New Partnership for Africa's Development (NEPAD).

Amoah, P., Drechsel P., and Abaidoo, R.C., 2005. Irrigated urban vegetable production in Ghana: sources of pathogen contamination and health risk elimination. *Irrigation and Drainage*, 54, 49–61.

Asante, F.A., 2007. *Analysis of water-related poverty in the Volta Basin of Ghana*. Accra: Institute of Statistical Social and Economic Research (ISSER), University of Ghana.

Asante, F. and Asenso-Okyere, K., 2003. *Economic burden of malaria in Ghana: Final report*. Accra: ISSER, University of Ghana.

Barry, B., *et al*., 2005. *The Volta River Basin. Comprehensive assessment of water management in agriculture. A comparative study of river basin development and management*. Colombo: International Water Management Institute (IWMI).

Bationo, A., *et al*., 2006. African soils: their productivity and profitability of fertilizer use. *African Fertilizer Summit*, 9–13 June 2006, Abuja, Nigeria. International Fertilizer Development Center.

Bationo, A. and Mokwunye, A.U., 1991. Alleviating soil fertility constraints to increased crop production in West Africa: the experience in the Sahel. *Fertilizer Research*, 29(1), 95–115.

Bationo, A., *et al*., eds, 2008. *Soil, water and nutrient management research in the Volta Basin*. Nairobi: Ecomedia.

Béné, C. and Russell, A.J.M., 2007. *Diagnostic study of the Volta Basin fisheries. Part 1. Livelihoods and poverty analysis, current trends and projections*. Volta Basin Focal Project report 7. Cairo: WorldFish and Colombo: CPWF.

Braimah, L.I., 2000. *Full frame survey at Volta Lake, Ghana 1998. Fisheries subsector capacity building project*. Yeji, Ghana: FAO-Integrated Developement of Artisanal Fisheries in Africa (IDAF) Project.

Braimah, L.I., 2003. *Fisheries management plan for the Volta Lake*. Report. Accra: Directorate of Fisheries, Ministry of Food and Agriculture.

Bugri, J.T., 2008. The dynamics of tenure security, agricultural production and environmental degradation in Africa: evidence from stakeholders in north-east Ghana. *Land Use Policy*, 25(2), 271–285.

Burkina Faso, Ministère de l'économie et du développement, 2004. *Cadre stratégique de lutte contre la pauvreté*. Ouagadougou: Ministère de l'économic et du développement.

Clanet, J.-C. and Ogilvie, A., 2009. Farmer-herder conflicts and water governance in a semi-arid region of Africa. *Water International*, 34(1), 30–46.

Climate Research Unit, 2008. Data set TS 2.19. Available from: http://www.cru.uea.ac.uk/cru/data/hrg/cru_ts_2.10 [Accessed August 2008].

Conley, T.G. and Udry, C., 2004. *Learning about a new technology: pineapple in Ghana*. Working paper series 817. New Haven: Economic Growth Center, Yale University.

Coulombe, H. and Wodon, Q., 2007. *Poverty, livelihoods, and access to basic services in Ghana: an overview*. Washington, D.C.: World Bank.

de Condappa, D., Chaponnière, A., and Lemoalle, J., 2009. A decision-support tool for water allocation in the Volta Basin. *Water International*, 34(1), 71–87.

Dieulin, C., 2007. *Afrique de l'ouest et centrale. Bassins hydrographiques*. Paris: IRD Editions.

Drechsel, P., *et al*., 2008. Reducing health risks from wastewater use in urban and peri-urban sub-Saharan Africa: applying the 2006 WHO guidelines. *Water Science and Technology*, 57(9), 1461–1466.

Ducommun, G., *et al*., 2005. *Commercialisation vivrière paysanne, marchés urbains et options politiques au Burkina Faso: rapport final de synthèse du projet de recherché*. Ouagadougou, Burkina Faso: TASIM-AO, HESA Zollikofen, CEDRES. Available from: http://www.shl.bfh.ch/fef/feprojektef/htm [Accessed August 2008].

European Commission, 2007. *Advancing African agriculture. Proposal for continental and regional cooperation on agricultural development in Africa*. Discussion paper. Brussels: DG Development.

Falkenmark, M., 1997. Society's interaction with the water cycle: a conceptual framework for a more holistic approach. *Hydrological Sciences Journal*, 42(4), 451–66.

FAO, 1990. *Progress report on the FAO/Integrated development of artisanal fisheries project*. Yeji, Ghana: Food and Agriculture Organization of the United Nations (FAO).

FAO, 1996. *Agro-ecological zoning guidelines*. Soil Bulletin 73. Rome: FAO. Available from: http://www.fao.org/docrep/W2962E/W2962E00.htm [Accessed July 2008].

FAO, 2005. *Aquastat country profiles*. Available from: http://www.fao.org/nr/water/aquastat/countries/index.stm. [Accessed July 2008].

Fleshman, M., 2008. Africa struggles with soaring food prices. *Africa Renewal*, 22(2), 12. Available from: http://www.un.org/africa renewal [Accessed December 2008].

de Fraiture, C. and Wichelns, D., 2007. Looking ahead to 2050: scenarios of alternative investment approaches. *In:* D. Molden, ed. *Water for food, water for life*. London: Earthscan and Colombo: IWMI. Available from: http://www.iwmi.cgiar.org/assessment/Publications/books.htm [Accessed 9 August 2010].

van de Giesen, N., *et al.*, 2001. Competition for water resources of the Volta basin. *IAHS Publications*, 268, 199–205.

de Graaf, G.J. and Ofori-Danson, P.K., 1997. *Catch and fish stock assessment in stratum VII of Volta Lake*. IDAF Technical Report 97/I. Rome: FAO.

Gyau-Boakye, P. and Tumbulto, J.W., 2000. The Volta Lake and declining rainfall and stream flows in the Volta River Basin. *Environment Development and Sustainability*, 2(1), 1–10.

Hugueny, B. and Lévêque, C., 1999. Richesse en espèces des peuplements de poissons. *In*: C. Lévêque and D. Paugy, eds. *Les poissons des eaux continentales africaines, diversité, écologie et utilisation par l'homme*. Paris: Orstom, 238–249.

Hyman G., *et al.*, 2008. Strategic approaches to targeting technology generation: assessing the coincidence of poverty and drought-prone crop production. *Agricultural Systems*, 98, 50–61. Available from: http://www.un.org/africa renewal [Accessed December 2008].

IEPC, 2007. *Proposition pour un document national (version provisoire)*. Ouagadougou, Burkina Faso: Initiative Elevage Pauvreté et Croissance.

INSD, 2003. *Enquêtes Burkinabé sur les conditions de vie des ménages (EBCVM)*. Ouagadougou, Burkina Faso: Institut National de la Statistique et de la Démographie.

Lachaud, J.P., 1998. *Modélisation des déterminants de la pauvreté et marché du travail en Afrique: le cas du Burkina Faso*. Documents de travail du CED 32. Bordeaux: Université Bordeaux. Available from: http://ged.u-bordeaux4.fr/ced023.htm [Accessed October 2009].

Laube, W., 2005. *Promise and perils of water reform: perspectives from Northern Ghana*. ZEF working paper series 10. Bonn: University of Bonn.

Lautze, J., Barry, B., and Youkhana, E., 2006. *Changing interfaces in Volta Basin water management: customary, national and transboundary*. ZEF working paper series 16. Bonn: University of Bonn.

Lemoalle, J. and de Condappa, D., 2009. Water atlas of the Volta Basin – Atlas de l'eau dans le bassin de la Volta. Colombo: CPWF and Marseilles: Institut de Recherche pour le Développement.

L'Hôte, Y., *et al.*, 2002. Analysis of a Sahelian annual rainfall index from 1896 to 2000: the drought continues. *Hydrological Sciences Journal – Journal des Sciences Hydrologiques*, 47(4), 563–572.

Martin, N. and van de Giesen, N., 2005. Spatial distribution of groundwater production and development potential in the Volta River Basin of Ghana and Burkina Faso. *Water International*, 30(2), 239–249.

MAHRH, 2006. *Agristat 3.5.1*. Ouagadougou, Burkina Faso: Direction générale des prévisions et des statistiques agricoles. Burkina Faso: Ministère de l'Agriculture de l'Hydraulique et des Ressources Halieutiques.

MOFA, 2006. *Agricultural data for major crops*. Accra: Statistics, Research, Information and Public Relations Directorate, Ministry of Food and Agriculture.

Morand, P., Sy, O.I., and Breuil, C., 2005. Fishing livelihoods: successful diversification, or shrinking into poverty? *In*: B. Wisner, C. Toulmin, and R. Chitiga, eds. *Towards a new map of Africa*. London: Earthscan, 71–96.

Nakicenovic, N., *et al.*, 2001. Special report on emissions scenarios. Arendal, Norway: Global Resources Information Database Arendal.

Ofori-Danson, P., 2005. An assessment of the purse-seine (winch-net) fishery in Volta Lake, Ghana. *Lakes and Reservoirs: Research and Management*, 10(3), 191–197.

OCP, 1995. Twenty years of onchocerciasis control in West Africa. Geneva: Onchocerciasis Control Programme in West Africa, World Health Organization.

Opoku-Ankomah, Y., *et al.*, 2006. *Hydropolitical assessment of water governance from the top-down and review of literature on local level institutions and practices in the Volta Basin*. IWMI Working Paper, 111. Colombo, Sri Lanka: International Water Management Institute.

Ouédraogo, S., 2005. *Intensification de l'agriculture dans le Plateau Central du Burkina Faso: une analyse des possibilités à partir des nouvelles technologies*. Doctoral thesis. Rijksuniversiteit Groningen, The Netherlands.

Pittaluga, F., *et al.*, 2003. *Poverty profile of riverine communities of southern Volta Lake.* SFLP/FR/18. Cotonou, Benin: Sustainable Fisheries Livelihoods Programme, FAO.

Poda, J.N., 2007. *Les maladies liées à l'eau dans le bassin de la Volta: état des lieux et perspectives.* Volta Basin Focal Project report 4. Montpellier, France: IRD and Colombo: CPWF.

Rockström, J., 2001. Green water security for the food makers of tomorrow: windows of opportunity in drought-prone savannahs. *Water Science and Technology*, 43(4), 71–78.

Rockström, J. and Barron, J., 2007. Water productivity in rainfed systems: overview of challenges and analysis of opportunities in water scarcity prone savannahs. *Irrigation Science*, 25(3), 299–311.

Serpantié, G., 2003. *Persistance de la culture temporaire dans les savanes cotonnières d'Afrique de l'Ouest. Etude de cas au Burkina Faso.* Thèse doctorat. Institut National d'Agronomie, Paris-Grignon.

Sivakumar, N.V.K. and Hatfield, J.L., 1990. Spatial variability of rainfall at an experimental station in Niger, West Africa. *Theoretical and Applied Climatology*, 42(1), 33–39.

Smaling, E.M.A., 1995. The balance may look fine when there is nothing you can mine: nutrient stocks and flows in West African soils. *In*: H. Gerner and A.U. Mokwunye, eds. *Use of phosphate rock for sustainable agriculture in West Africa.* Proceedings of a seminar on the use of local mineral resources for sustainable agriculture in West Africa, 21–23 November 1994. Lomé, Togo: International Fertilizer Development Center-Africa.

Smaling, E.M.A., Oenema, O., and Fresco, L.O., eds., 1999. *Nutrient disequilibria in agroecosystems. Concepts and case studies.* Wallingford: CABI Publishing.

Terrasson, I. ,*et al.*, 2009. Yields and water productivity of rainfed grain crops in the Volta basin, West Africa. *Water International*, 34(1), 104–118.

UNDP, 2006. *Human Development Report* [online]. Available from hdr.undp.org/statistics/ [Accessed October 2008].

United Nations Population Division, 2008. *World population prospects: the 2008 revision. Population database.* Available from: http://esa.un.org/unpp/index.asp [Accessed October 2008].

Volta River Authority, 2010. *The story of Akosombo Dam: a historical perspective.* Available at: http://www.vra.com/ [Accessed October 2008].

WRI, 2007. Earth trends. Washington, D.C.: World Resources Institute. Available from: http://www.wri.org/ecosystems [Accessed October 2009].

World Bank, 2007. *World development report 2008. Agriculture for development.* Available from: http://www.worldbank.org/afr/findings/infobfre/infobf83.pdf [Accessed October 2008]

World Water Assessment Programme, 2009. *The United Nations world water development report 3: water in a changing world.* Paris: UNESCO and London: Earthscan.

de Zeeuw, F., 1997. Borrowing of land, security of tenure and sustainable land use in Burkina Faso. *Development and Change*, 28(3), 583–595.

Yellow River basin: living with scarcity

Claudia Ringler, Ximing Cai, Jinxia Wang, Akhter Ahmed, Yunpeng Xue, Zongxue Xu, Ethan Yang, Zhao Jianshi, Tingju Zhu, Lei Cheng, Fu Yongfeng, Fu Xinfeng, Gu Xiaowei and Liangzhi You

The Yellow River basin is a key food production centre of global importance facing rapidly growing water scarcity. Water availability for agriculture in the basin is threatened by rapid growth in the demand for industrial and urban water, the need to flush sediment from the river's lower reaches, environmental demands and growing water pollution. Climate change is already evident in the basin with long-term declines in river runoff, higher temperatures, and increasing frequency and intensity of drought. The Chinese government has exhausted most options for improving water supply. The challenge will be to switch to improved water demand management, which is hampered by existing governance structures, and lack of integrated agriculture and water resource policies.

Introduction

China is facing growing water scarcity in many river basins due to its rapid economic development, an expanding population, growing urbanization and limited scope to develop new supplies. Water overdrafts, both from surface and sub-surface sources, are causing serious environmental problems ranging from the degradation of ecosystems in the deltas of major rivers to aquifer depletion in northern China. The Yellow River basin (YRB) is symptomatic of the challenges facing China's water economy. The YRB, which is the second largest basin in China, is a key agricultural and industrial region in the country and also considered the "cradle of Chinese civilization". However, the basin faces severe water shortages. The particular climatological and hydrologic conditions together with very rapid industrial and urban development are making sustainable water supply for all users and uses a complex and difficult task. Given the extreme water shortages in the basin, how can water resources be managed to continue to support agricultural and economic development while also improving outcomes for the environment?

In this paper we report results of a two-year study on biophysical and socioeconomic aspects of the water and related resources in the YRB; the relation of water development and agricultural and economic growth; and options for enhancing water availability and access for sustained agricultural and economic development, while maintaining environmental sustainability.

Background on the Yellow River basin

Water resources in the Yellow River basin

The Yellow River (or "Huanghe" in Chinese) is the second longest river in China. It rises in the Bayangela Mountains in western China, dropping a total of 4500 m as it loops north into the Gobi Desert before turning south through the Loess Plateau and then east to its mouth in the Bohai Sea (Figure 1). The river flows 5464 km and passes through nine provinces and autonomous regions, with a basin area of 795,000 km^2, which includes 42,000 km^2 of inland river catchments in the northwest of the basin. Rainfall averages 450 mm and annual average natural runoff is 53.5 km^3, which is less than runoff estimates of 58 km^3 during the 1960s to the 1980s. Total annual water resources, including groundwater, are 64.7 km^3 (YRCC 2006).

The basin faces severe pressures on available water resources. With an estimated 150 million people benefiting from Yellow River water resources, both inside and outside the basin area, per capita water availability today is already only 430 m^3, less than half the 1000 m^3 threshold for chronic water scarcity (Falkenmark and Widstrand 1992). Other

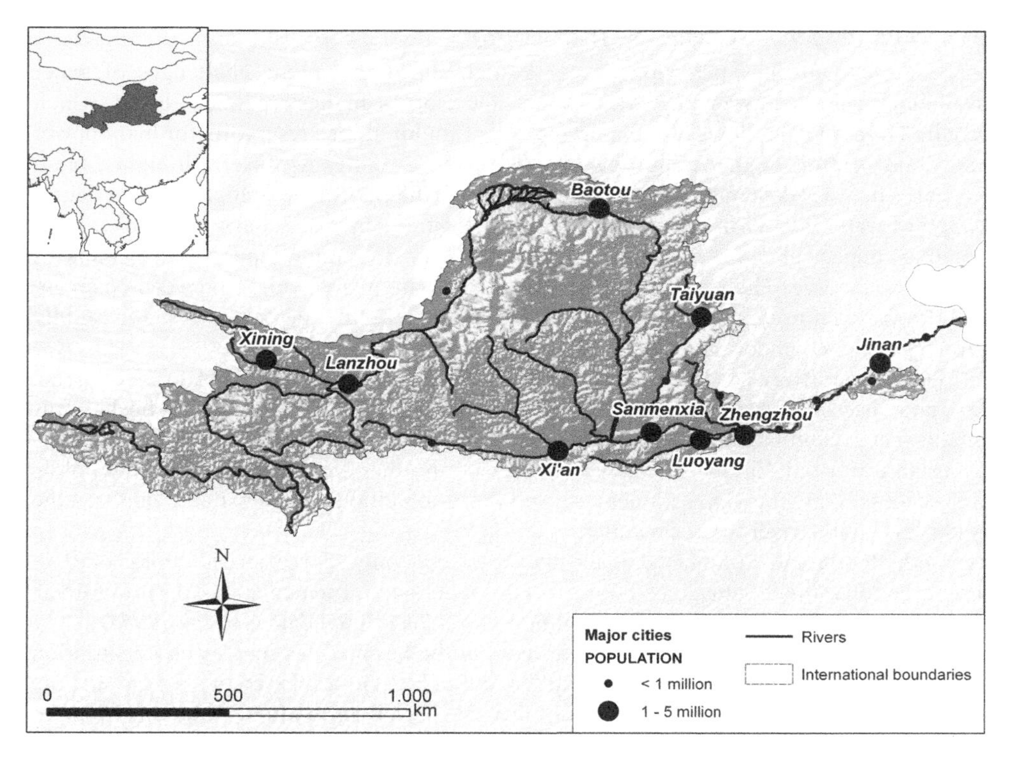

Figure 1. Map of the Yellow River basin.

indicators of water scarcity include the ratio of annual water withdrawal to renewable water resources, which exceeds 75% in the basin (Ministry of Water Resources 2001), and the high fraction of consumptive water use (ratio of consumption to withdrawal), estimated to be 75%, which is far higher than the global average of 43%. This is due to the comparatively high share of water use by agriculture, almost full use of return flow in downstream irrigation districts through conjunctive use of canals and wells, as well as considerable non-beneficial water losses (Rosegrant and Cai 2003, Cai and Rosegrant 2004).

In 2000, the basin produced 14% of Chinese grain harvest and 14% of the country's GDP using only 2% of national water resources. Total agricultural area is approximately 13 million ha of which 7 million ha are irrigated. To address past flooding problems and support agricultural and economic development, the Yellow River has been heavily engineered, with 15 large reservoirs that store 566 km^3 with an installed hydropower capacity of 10,380 MW.

The basin is typically divided into an upstream, midstream and downstream area. The mountainous upstream area generates most of the river flow, has a relatively low population density, limited agricultural and industrial development, and concentrates most of the poverty in the basin. The midstream area includes both the fragile Loess Plateau and semi-arid and arid agricultural areas that heavily depend on irrigation. Over the last decade, there has been important industrial development in several of the mid-stream provinces, competing with irrigation for limited water resources. The downstream area contains most of the urban-industrial development, a combination of ground- and surface-water irrigation, and fragile wetland ecosystems at the river mouth.

Key challenges for YRB water and food security

In addition to the low per capita water availability, other unique challenges of water availability and access in the YRB include the world's highest sediment loads, which require large flushing flows; the important role of multipurpose reservoirs for hydropower and flood control; large flooding events, and, more recently, significant droughts; rapid increases in water demand from industries, cities and the environment; high levels of degradation of water quality in the middle and downstream main channel and tributaries; large potential impact of climate change and variability; and continued poor management of the water resource. These developments have led to sharp competition between upstream and downstream users, between irrigators and industry in the midstream area and rapidly growing water degradation.

The Yellow River has the highest sediment concentration in the world, at 37.6 kg/m^3 (Shi and Shao 2000, Xue *et al.* 2010). Therefore, since 2002 the YRB annually flushes sediments that accumulate in the lower reaches of the river through targeted reservoir releases, using an estimated 15 km^3 of water resources during the rainy season. While the policy was successful at removing sediment, several irrigation intakes are now too high above the water level in the river to access water.

Both floods and droughts damage the YRB economy. For example, from 1950 to 1990, the total direct damage of floods and droughts was estimated at 116.4 billion RMB (1 RMB = US$0.146), with floods accounting for 45% of total damage (Ma 1996).

Agriculture, which is the major water user in the basin, faces increasing competition for water resources as a result of rapid urban and industrial development. In 50 years, the irrigated area in the YRB increased more than 350% and agricultural water use by more than 250% (YRCC 2006). Water demand from industry and domestic use increased even more steeply, but from a very low base. The largest adverse impacts on the availability of

Table 1. Water Allocation Agreement of 1987 and actual withdrawals, 1998 and 2008 (km^3).

		2008			1998		
	UWFR	Total withdrawals	Surface water	Ground Water	Total withdrawals	Surface water	Ground water
Qinghai	1.41	1.86	1.47	0.39	1.92	1.63	0.29
Sichuan	0.04	0.03	0.03	–	0.02	0.01	0.01
Gansu	3.04	4.44	3.80	0.64	4.15	3.53	0.62
Ningxia	4.00	7.63	7.12	0.51	9.68	9.14	0.54
Inner Mongolia	5.86	9.37	6.97	2.40	9.26	7.30	1.96
Shaanxi	3.8	6.27	3.22	3.05	5.51	2.42	3.09
Shanxi	4.31	4.14	1.70	2.44	3.64	1.18	2.46
Henan	5.54	6.63	4.20	2.43	5.77	3.34	2.43
Shandong	7.00	7.99	7.07	0.92	9.77	8.46	1.32
Heibei/Tianjin	2.00	0.73	0.73	n/a	n/a	n/a	n/a
Total	**37.00**	**49.10**	**36.31**	**12.78**	**49.71**	**37.00**	**12.71**

Note: UWFR = Unified Water Flow Regulation (enforced since 1999).
Source: YRCC (2005).

irrigation water came in 1998 from the decision by the government of China to stop the increasing flow cutoff periods to the downstream river reaches, which had attracted international attention. Flow stoppages in the YRB were the most striking evidence of excess water withdrawal and consumption in the basin with increasing cutoff periods from 1972–98. In 1997, there was no discharge from the basin to the sea for 226 days, and the river dried up to Kaifeng, 600 km inland from its mouth (Cai and Rosegrant 2004, Ke and Zhou 2007). Flow cutoffs were eliminated through unified water flow regulation (UWFR), which was implemented by the YRCC in 1999 as enforcement of the 1987 cross-provincial water allocation agreement (Table 1). Implementation of the UWFR contributed to a decline in total irrigation water use in the mid- and downstream areas by 4.8 km^3 from 1988–92 to 2002–2004 (Chen 2002, YRCC 1998–2006), while urban-industrial uses continued to grow (Table 2). The enforcement of the UWFR has not led to any compensation of irrigation water users. Moreover, declines in surface-water use for irrigation directly contributed to increased groundwater withdrawals, particularly in the downstream areas. Despite the maintenance of year-round flows since 1999, flows remain insufficient to prevent seawater intrusion and wetland recession.

Water quality problems have grown in the YRB, both as a result of the reduced capacity of the river to dilute waste and growing domestic, industrial and agricultural effluents. Their combined effect has reduced the Yellow River's service functions for decades to come. According to Li *et al.* (2003), in 1998 water pollution cost the YRB a total of 14.97 billion RMB or 2.6% of its GDP. In early 2010, the government of China, for the first time, released national-level estimates of pollution that also included agriculture. According to the study, agriculture is responsible for 43.7% of the nation's chemical oxygen demand (the main measure of organic compounds in water), 67% of phosphorus and 57% of nitrogen discharges (*The Guardian*, 9 February 2010). This is not surprising as the country consumes more than 30% of the world's nitrogen fertilizer, which is applied to only 7% of the world's land area. While no basin-level figures are available, the data are likely representative for water quality in the YRB. This first official recognition by the government of serious agricultural pollution will likely support a review and revision of the incentives and subsidies provided for agricultural inputs, particularly fertilizer.

Table 2. Irrigation water use, YRB, 1988–92 and 2002–2004 (km³).

Years	Reach	Total	Agricultural	Industrial	Domestic
1988–92[a]	Upper	13.11	12.38	0.51	0.22
	Middle	5.44	4.77	0.38	0.28
	Lower	12.18	11.24	0.55	0.38
	Basin	30.72	28.39	1.45	0.89
2002–2004[b]	Upper	17.54	15.71	1.42	0.41
	Middle	5.71	4.16	0.97	0.58
	Lower	8.44	7.04	0.82	0.58
	Basin	31.69	26.91	3.21	1.57
Difference	Upper	34%	27%	179%	84%
	Middle	5%	−13%	155%	108%
	Lower	−31%	−37%	49%	54%
	Basin	3%	−5%	121%	77%

Sources: [a]Chen (2002); [b]YRCC (2002–2004), as used in Cai (2006).

To investigate the impact of climate change on future stream flows in the Yellow River, we applied the predictions of the HadCM3 global circulation model using the SRES B2 scenario.[1] We applied the statistical downscaling model to the Yellow River's headwater catchment areas using the SWAT-BNU (Soil and Water Assessment Tool developed at Beijing National university) model. According to the downscaled values, maximum air temperatures are predicted to increase by 1.3°C by the 2020s, 2.6°C by the 2050s, and up to 3.9°C by the 2080s in the basin; minimum temperatures are expected to increase by 0.9°C, 1.5°C and 2.3°C, for the same periods. Annual precipitation volumes under this scenario would be 3.5% higher in the 2020s, 6.4% higher in the 2050s, and 8.7% higher in the 2080s. The combined impact from higher temperatures and slightly higher precipitation levels on Yellow streamflows would be declines of 88 m^3/s, 117 m^3/s and 152 m^3/s, for the three periods, respectively. Thus, even the relatively moderate and precipitation-abundant HadCM3 SRES B2 scenario would severely affect future regional water supply and water security in the basin, putting further pressure on food security in the country (Xu *et al.* 2009).

When linking the SWAT-BNU model results with the water simulation model of the YRCC river basin authority, we find that under climate change the annual water budget deficit would rise to 4.2 km^3. The situation would be even worse in dry years: in one out of four years the water shortage would reach 15.1 km^3 and in one out of 20 years, the shortage would reach 21.0 km^3 resulting in basin water deficit ratios of 28% and 37%, respectively.

Food and water in the Yellow River basin

Most of the irrigation water in the YRB, and in China in general, is used for the production of basic staple crops. In 2005, production of irrigated cereals accounted for 85% of total national production, on 82% of the harvested area for cereals, up from 74% of total production on 70% of area in 1995. In comparison, worldwide, irrigated cereal production accounted for 52% of total production on only 38% of the global area harvested for cereals. The YRB accounted for 14% of irrigated harvested cereal area and production in China. While agricultural area in China is expected to continue to contract and irrigated area to barely increase over the next decades, the national share of irrigated area of the YRB is expected to increase up to 18% due to more rapid declines in irrigated rice area in other Chinese river basins (International Food Policy Research Institute [IFPRI] 2009).

Total demand for cereals in China in 1995 was estimated at 375 million metric tons: 69% for direct human consumption, and most of the reminder for animal feed. Ten years later, demand had increased to 400 million metric tons and is projected to further increase to 492 million metric tons by 2050. By then, only 44% will be destined for direct human consumption, given the large increase in the use of maize for animal feed (IFPRI 2009). In 2005, China already accounted for a quarter of the world's total livestock production (FAOSTAT 2010).

In 2007 China was the third largest bioethanol producer in the world after the United States and Brazil with an annual production of 1.35 million tons. As a result of growing concerns for food security at the national level, the government has since prohibited bioethanol production using maize and wheat as feedstocks, except for four plants that were allowed to maintain their output but not expand (Qiu *et al.* 2010).

Concerns about food security have been at the heart of much of the policy on agricultural development in China for decades. China's medium- to long-term policy for grain security 2009–20 sets a target of 95% self-sufficiency in grain production, slightly less than the 98% for the preceding period. To achieve these levels of production, the government focuses chiefly on investments in science and technology, combined with direct support to farmers. Key farm support measures include the abolition of the agricultural land tax in 2006 and continued support and subsidies for crop inputs, particularly fertilizers, fuel and water, many of which have been gradually decoupled and converted to direct transfer payments to farmers.

Similar to other parts of Asia, overall farm support measures have been growing as a result of the food price crisis, which peaked in 2007/08. By 2008, Chinese farmers received US$34.4 per acre, comparable to the per-acre level of subsidy (but not per-household support) in the United States (Huang and Rozelle 2009, Rosegrant *et al.* 2009a, Huang *et al.* 2010). Despite the government's strong efforts to achieve close to food self-sufficiency in key crops, it is likely that net food imports will increase from approximately 18 Mt to 50 Mt by 2050 given the growing land and water shortages (IFPRI 2009, Rosegrant *et al.* 2009a).

The government's goals and supporting policies on food self-sufficiency have had direct negative impacts on water availability and use in the YRB. For example, the abolition of the agricultural land tax, which traditionally was collected together with service fees for irrigation water, has increased the relative cost and difficulty of collecting the latter because the collection costs are now spread over a smaller fee base. Moreover, rates of collection have fallen because some farmers believe that following the demise of the land tax, they should also not have to pay irrigation service charges. Another example is the government support for nitrogen fertilizer, which has contributed to their over-use, resulting in heavy non-point source pollution in the YRB and elsewhere in China.

At the same time, policies in the water sector have harmed agriculture, such as the silt-flushing policy and flow-cutoff implementation discussed above. Various policies implemented to conserve irrigation water have had other adverse effects, such as reduced maintenance of the irrigation systems and reduced salaries of irrigation system managers, who are paid according to volume of water delivered, measured at the off-take level, and not volume of water conserved.

Water legislation and administration

Water legislation

In 2002, the government of China passed a new water law. Key elements include the emphasis on river basin management; a strong focus on water savings and improved water-use efficiency; the implementation of water-use quotas, permits and fees for large withdrawals; and the recognition of water for ecological uses as equal in importance to water used for industry and agriculture. Given the limited water resources of the country, the government has also widened and deepened legislation on water pricing over the last ten years (Ministry of Water Resources 2003, 2005, Wang 2007, Fu *et al.* 2008). Several regulations released since 2004 support a water rights system as well as water rights transfers, particularly for the YRB. These include the *Guidance on water rights transfer demonstration works in Inner Mongolia and Ningxia,* the *Management and implementation measures on water rights transfer in the Yellow River basin,* and *Management regulation on water-saving engineering*. These regulations have provided the legal foundation for water right transfers in the YRB.

However, regulations to implement the national laws and national-level regulations in many cases are still lacking at the provincial level. The slow pace of promulgation of implementing regulations at the provincial level is likely due to provincial officials not seeing the legislation as a priority; a sheer lack of capacity and understanding by provincial officials, and a lack of financial resources to support implementation at lower administrative levels in the provinces (Wang and Zhang 2009).

Water administration in the YRB

In China, water resources are administered through a nested hierarchical administrative system (Wang *et al.* 2007). The Ministry of Water Resources (MWR) is at the highest central level directly under the State Council, with Water Resource Bureaus at the provincial, prefecture and county levels, and water management stations in townships at the lowest level of administration. Water Resource Bureaus at the provincial, prefecture and county levels are controlled jointly by the respective government at the same level and the MWR. Irrigation districts administer water resources that span lower-level administrative boundaries. This system of water administration is supplemented by seven river commissions, including the YRCC, which are administered by the MWR. However, many other agencies have retained direct or indirect responsibilities for water management such as bureaus or agencies of construction, land resources, environmental protection, energy resources, meteorology and finance, key among which are the State Environmental Protection Agency and the Ministry of Energy Resources.

In this environment, local governments tend to focus on maximizing local revenues and economic growth subject to given requirements for grain self-sufficiency, rather than focusing on conserving scarce water resources. Thus, water administration and management generally see their priorities to achieve these local goals. The often contradictory objectives of the various water, agriculture and energy agencies continue to hamper integrated water resources management in China and the YRB. For example, while the YRCC is authorized by the State Council of China to control Yellow River water resources, some provinces have continued to withdraw water in excess of agreed-upon quotas without penalties. The recent Yellow River Water Regulation Act (2006) allows for punishment of those provinces that exceed their water quota, but provides no implementation mechanisms. Furthermore, the YRCC has focused on integrated surface water management on the mainstream, while most tributaries and groundwater remain without integrated management. Since 2006, YRCC has assumed some control over two key tributaries, the Weihe and Qinhe.

A cross-provincial water allocation agreement was developed in 1987 and has been enforced by YRCC since 1999 to counteract the downstream flow cutoffs in the basin as discussed earlier (see Table 1). The Agreement distributes a total of 37 km^3 across the riparian provinces, including 2 km^3 to downstream urban-industrial centres outside the basin area.

What is the role of water development in the YRB for poverty reduction and agricultural and economic development?

To assess the role of water development for poverty alleviation and agricultural development in the YRB, we used income data from the 2001 household income and expenditure survey conducted by the National Bureau of Statistics of China. No later data were available. Because we used the international purchasing power parity (PPP) exchange rate with 2005 as the base year, we adjusted per capita income data from the 2001 survey for inflation using the consumer price index for China, with the base year 2005 = 100. The data set used for the analyses represents the rural communities and households in the YRB and included 5085 households in nine basin provinces (Ahmed *et al.* 2009).

Based on the PPP US$1.25 a day poverty line and current per capita levels of income, 30.5% of the population in the rural regions of the YRB were living in poverty in 2001. The poverty rate was highest in the mountainous areas far from the mainstream and lowest on the plains. The poverty rate in the upstream area (47.5%) is nearly five times higher than that in the downstream area (only 9.9%), while the rate in the midstream area (29.6%) is three times higher. The headcount poverty rate ranged from a high of 52.2% in the upstream province of Gansu in western China to a low of only 3.1% in the downstream province of Shandong.

There is a direct empirical link between irrigation development and poverty reduction and agricultural and economic development in the basin. The US$1.25 a day headcount poverty rate was significantly lower in irrigated areas than in non-irrigated areas of the YRB region: while 19.4% of all households living in irrigated villages are poor, the rate was more than double (41.4%) in villages without irrigation. Figure 2 shows the concentration of non-poor in irrigated villages, such as Shandong, Inner Mongolia and Henan provinces, whereas Gansu and Qinghai are provinces with the lowest share of population in irrigated villages and also the highest levels of poverty among the nine provinces sharing the YRB.

The percentage of households using electric tubewells for irrigation increases considerably from upstream to midstream areas, and increases dramatically from midstream to downstream areas. However, village-level coverage of surface irrigation reveals a rather different pattern with the largest share of cultivated land being irrigated in the midstream area. The patterns of surface irrigation and tubewells suggests that households living in the downstream area rely on groundwater for irrigating their crops, as a result of increased flexibility and reliability of the resource, particularly following the implementation of the UWFR.

A further indicator of the role of water development for agricultural and rural economic growth is the school enrolment rate. The gap in enrolment between the poor and the non-poor is smaller in irrigated villages than in non-irrigated villages. This indicates that the availability of irrigation at the community level is not only associated with increased school enrolment in the community, but the improvement also seems to benefit the poor more than the non-poor.

Irrigation also contributes to improved access to safe drinking water in the YRB. In irrigated villages 78% of households have access to safe water compared to only 47% in non-irrigated communities. While the difference in access to safe water between poor and non-poor is 16 percentage points in non-irrigated villages, it is only three percentage points in irrigated villages. These findings have important policy implications, as access to safe water is critical for improved health and nutrition, particularly for children.

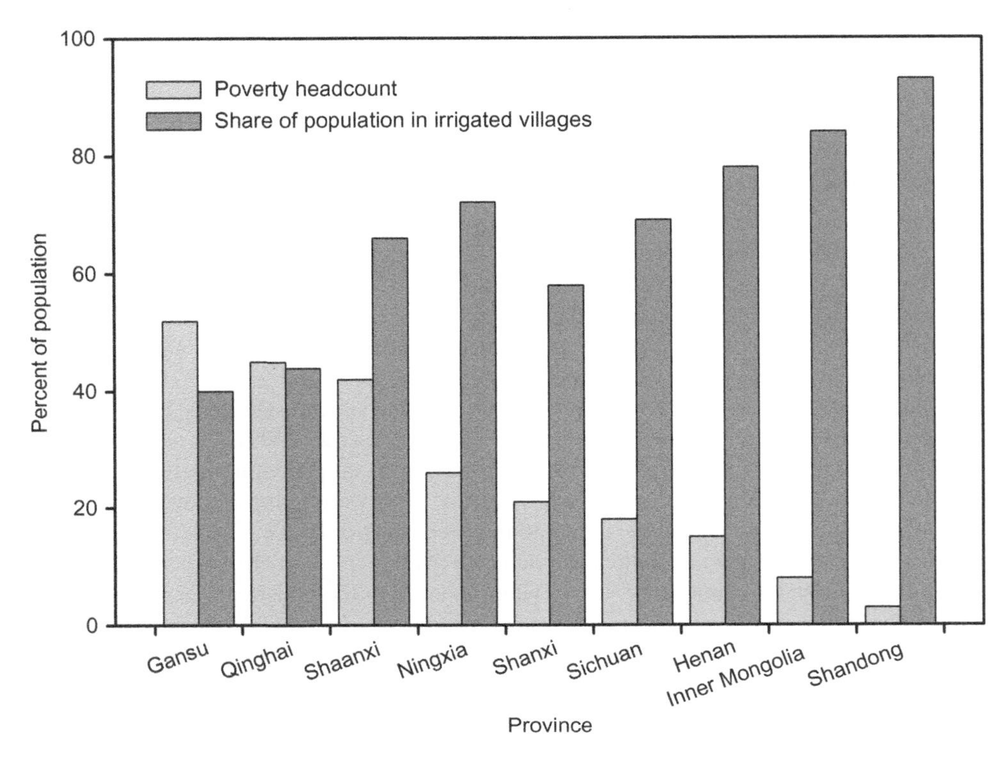

Figure 2. Headcount poverty and share of population living in irrigated villages.
Source: The authors.

The relationship between irrigation coverage and agricultural productivity is direct and very strong. Higher irrigation coverage of cultivated land at the village level is associated with greater crop productivity. Yields of various crops grown in the YRB are substantially higher in irrigated villages than in non-irrigated villages, particularly for rice (although rice only accounts for 2% of total crop area). Yield of rice in irrigated villages reached 7.2 tons/ha compared to 3.3 tons/ha in non-irrigated villages. Furthermore, the shares of land under high-yielding varieties (HYVs) of wheat and maize, the two main crops grown in the YRB, increase considerably with higher coverage of irrigated land. Cultivation of HYVs of crops is more capital-intensive than traditional varieties, but irrigation takes much of the risk out of crop production as the dependence on rainfall is reduced and encourages farmers to invest more in seed and other agricultural inputs.

Coverage of surface water for irrigation is a statistically significant determinant of per capita household income. Results suggest that a 10% increase in village-level coverage of surface irrigation increases the per capita income of households living in that village by 1.7%, on average. Furthermore, increasing irrigation coverage by 10% reduces the incidence of poverty by 5.1%. As expected, irrigation gives a higher marginal return in communities where rainfall is relatively low.

While expanding irrigation has a large impact on reducing poverty, we also found large and positive effects caused by increasing the opportunities to earn off-farm income in the rural YRB. Our econometric analyses show that the headcount poverty rate declines by 4.5% if the share of non-farm income in total per capita household income increases by 10%.

Addressing water scarcity in the YRB: options and investment needs

Improved water use efficiency, particularly in irrigated agriculture, but also for domestic and industrial uses, is key to meeting future growth in the demand for water for sustained agricultural and economic development. Water-use efficiency can be increased through engineering, agronomic, institutional and economic measures. More recently, economic and institutional means have become more important. When implemented appropriately, economic incentives for water management (prices, taxes, subsidies, quotas and use or ownership rights) can affect the decisions made by water users and motivate them to conserve and use water more efficiently. Efficiency pricing works well in the domestic and industrial sectors, but it is much more challenging for irrigation as price increases are often punitive to farmers because water is a large input to generally low-value production. This is particularly so in China where much of the irrigated area produces basic grains.

In the past, increasing the supply of water through new water development has been a common strategy to address water shortages. However, in maturing water economies, which are characterized both by increasing scarcity of water (Randall 1981), and by increasing transfers of water both in scale and amount, managing the demand for water becomes more important. The task of demand management is to generate both physical savings of water and economic savings by increasing the output per unit of evaporative loss of water, by reducing water pollution, and reducing non-beneficial water uses. This can be supported through a variety of policy measures, including economic incentives to conserve water, for example, through pricing reform and reduced subsidies. Other demand-side measures include regulations on the rights to use water, education campaigns, leak detection, retrofitting, recycling and other technical improvements, enhanced pollution monitoring, and quota and licence systems. While many measures of demand management have targeted irrigation as the largest water user, municipal and industrial water use cannot be allowed to grow unchecked. Regulation and economic incentives are needed to reduce the negative ecological, economic, and social impacts of these uses, especially on water quality.

Given the size of the YRB, no single intervention could possibly do justice to the extreme diversity of water-related challenges found in the basin, which ranges from the upstream mountainous areas dominated by livestock herders, to the hilly/mountainous Loess Plateau with severe erosion challenges, the semi-arid to arid irrigated plains in Inner Mongolia/Ningxia with rapidly growing industries, to the key urban-industrial centres interspersed with highly productive irrigation downstream.

Many interventions have been implemented in the past to increase water supply and enhance flood control. Key among these are the construction of the Xiaolangdi reservoir, completed in 1999, which has increased the designed flood-control period from 60 years to over 1000 years (Cai and Rosegrant 2004); the construction of several thousand silt-trap dams across the Loess Plateau (Brismar 1999); and two large watershed rehabilitation projects implemented by the government of China and the World Bank (World Bank 2003, 2007).

Additional interventions in recent years to address growing water shortages and the need for food include the conversion of hillside production into terraces (this was also done as part of the watershed rehabilitation project); rainwater harvesting schemes in the western upland areas; the use of plastic sheeting to contain soil moisture and reduce evaporation in the arid parts of Inner Mongolia and Ningxia; and the resettlement of people out of extremely dry areas.

Technical solutions: role of engineering measures and enhanced water productivity

The south-to-north water transfer (SNWT) project, if fully implemented, would be the largest engineering feat to date to address water challenges in northern China and the YRB. The SNWT was planned in the 1950s and officially launched in 2002. Once completed, it could transfer up to 50 km^3 (comparable to total Yellow River runoff) a distance of more than 1000 km, from the Yangtze River in southern China, to the North China Plain. A western, middle and eastern route have been planned and work is progressing on the technically and economically more feasible middle and eastern routes.

The general objective of the project is to sustain economic growth in northern China (Yang and Zehnder 2005, Pietz and Giordano 2009). The objective of the middle and eastern routes is to provide water for water-short regions in the Haihe and Huaihe River basins, particularly Beijing, Tianjin, Hebei, Henan and Shandong, with limited impact or benefit for the YRB. Due to the high (and increasing) construction cost, the price of water delivered through the SNWT could easily surpass the estimated "affordable" price of US\$0.70/m^3. The western route, on the other hand, could transfer 20 km^3 to irrigate an additional 1.3 million ha and provide water for economic development in Qinghai, Gansu, Shanxi and Shanxi provinces, as well as Ningxia and Inner Mongolia, all in the YRB. However, economic, engineering and ecological side effects prevent this route from development in the foreseeable future.

Even without the SNWT, engineers at YRCC still see some potential for water savings in the basin, amounting to 5.7 km^3 by 2020 and 7.6 km^3 by 2030, mostly in the agriculture sector. According to their calculations, water savings in agriculture of 4.0 km^3 by 2020 and 5.4 km^3 by 2030 can be achieved through adjustments in planting dates and crop species, crop yield improvements and lining of canals. These calculations take into account continued agricultural and economic growth and allow for small increases in irrigated area. Furthermore, the industrial sector is expected to reduce its water use by 1.5 km^3 by 2020 and by 2.1 km^3 by 2030 through increased water reuse and recycling. In the domestic sector, potential water savings have been estimated at 0.12 km^3 by 2020 and 0.17 km^3 by 2030 for the YRB, chiefly through increased leak detection and other efficiency-enhancing programmes and disconnection of illegal users.

As the simulations for climate change presented earlier show, these savings will not be sufficient to turn around trends of growing water deficits in the basin, particularly in dry years. Thus, even more investment in agricultural research and development will be needed to achieve even more rapid improvements in crop yields without use of more irrigation water; this is the current focus of the Chinese government as we discussed above.

To assess water productivity (WP) further across the YRB for key rainfed (WPR) and irrigated (WPI) crops, we used data from 60 counties from the upstream, midstream and downstream basin areas (including downstream areas irrigated outside the hydrologic boundaries), and extrapolated the results to the entire basin. We then assessed the spatial variability of water productivity as well as associated water and energy factors with regard to climate, land cover and agricultural practices (Cai *et al.* 2010).

All crops of rice and wheat receive some form of irrigation in the YRB.[2] Wheat grows during the winter–spring season, during which precipitation is less than 30% of the crop water requirement. In contrast, about 11% of maize and 17% of soybean area are rainfed. Table 3 presents average values of irrigated and rainfed area and yield by basin area. While irrigated maize yields are, on average, 77% higher than rainfed yields, basin-wide average soybean yields are similar for both rainfed and irrigated areas.

Table 3. Irrigated and rainfed area and yield of key crops by sub-basin in the YRB.

	Crops	Basinwide	Midstream	Downstream
Irrigated area (000 ha)	Rice	25.3	13.0	12.3
	Maize	540.2	254.3	284.9
	Wheat	1141.0	536.4	597.7
	Soybean	149.6	80.6	69.0
Rainfed area (000 ha)	Rice	0.0	0.0	0.0
	Maize	68.8	30.3	37.9
	Wheat	0.0	0.0	0.0
	Soybean	30.1	14.3	15.8
Irrigated yields (ton/ha)	Rice	5.4	5.5	5.3
	Maize	5.3	5.0	5.7
	Wheat	3.7	2.8	4.4
	Soybean	1.4	1.2	1.7
Rainfed yields (ton/ha)	Rice	n/a	n/a	n/a
	Maize	3.0	1.9	4.0
	Wheat	n/a	n/a	n/a
	Soybean	1.4	1.0	1.9

Source: Authors.

Table 4. Area-weighted WPI and WPR for different regions in YRB.

	WPI (kg/m^3)				WPR (kg/m^3)			
Region/Crops	Rice	Maize	Wheat	Soybean	Rice	Maize	Wheat	Soybean
Basin-wide average	0.50	0.97	1.39	0.26	–	1.09	–	0.41
Standard deviation	0.25	0.32	0.51	0.13	–	0.36	–	0.16
Midstream	0.49	0.94	1.16	0.26	–	0.68	–	0.28
Standard deviation	0.22	0.33	0.49	0.13	–	0.35	–	0.15
Downstream	0.51	0.99	1.57	0.27	–	1.41	–	0.52
Standard deviation	0.26	0.30	0.34	0.12	–	0.33	–	0.12

Source: Authors.

Using the cropped area as a weighting factor, we interpolated irrigated and rainfed water productivity to the entire YRB. Table 4 presents the results for upstream, midstream and downstream areas. Results fit the range of values previously published by Zwart and Bastiaanssen (2004), who reported water productivity values of 0.6–1.6 kg/m^3 for rice, 1.1–2.7 kg/m^3 for maize, and 0.6–1.7 kg/m^3 for wheat. While values of water productivity for rainfed and irrigated crops are quite different in the midstream basin, they are similarly high in the downstream area.

It is interesting to note that WPR for maize and soybean is slightly higher than WPI in the downstream area and also for soybean in the midstream basin. This implies that in parts of the basin, irrigated maize and soybean may not be as water-efficient as rainfed crops. This is likely a result of inefficient water use (i.e. the divisor in the equation is higher than it should be). The standard deviation of the WPR data is higher than that of the WPI data. Thus, irrigation stabilizes crop yield and production, which is important under increasing climate variability and climate change.

While there is still scope for increased water-use efficiency in irrigated agriculture in the YRB (National Bureau of Statistics of China 2003, Yang *et al.* 2003), the scope is limited and further declines in allocation of water to irrigation will eventually result in reduced

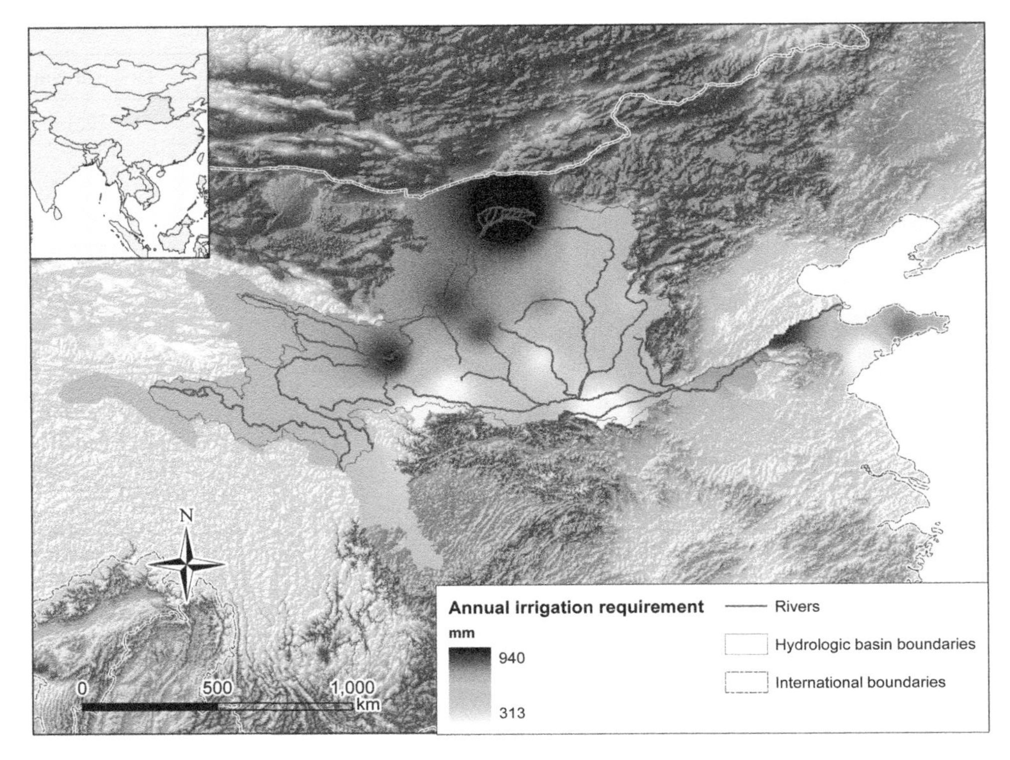

Figure 3. Irrigation requirement in the Yellow River basin.
Source: The authors.

food production with serious implications for local food security and farmer incomes, as well as potential impacts on global food prices and trade.

Figure 3 is a map of annual requirements for irrigation water for the YRB, computed as reference evapotranspiration minus precipitation, summed over the crop season. The spatial pattern of irrigation water requirements mirrors the location of irrigated areas in the basin. The highest requirement for irrigation water in the northwestern part of the midstream basin is close to one metre. The map clearly demonstrates that irrigation is and will continue to remain a major factor for agricultural production if the goals of food production are to be achieved.

Institutional solutions: irrigation management reform

Despite high levels of water scarcity in the country and in the YRB, integrated water management in China remains elusive as a result of fragmented management and conflicts among water users at the national, provincial and local levels. Key challenges in Chinese water legislation and administration relate to the lack of regulations supporting implementation of the 2002 water law, and poor incentives for water conservation at the level of the irrigation system. To address growing water scarcity, in addition to water supply/engineering measures, the government has started to support reform of irrigation management.

At the level of the irrigation system, reform since the early 1990s has successively established water-user associations (WUAs) and contractors (hired technical experts) in place of collective management (water allocation through village leadership) to enhance irrigation management. A survey of irrigation districts in Ningxia and Henan provinces in 2001 and 2005 showed that by 2004, 30% of villages managed their water under contract and 21% through WUAs. However, 85% of WUAs still used the village leadership as governing board, at least in their initial set-up.

The key difference for water conservation outcomes was not the type of administration but the type of incentives. Our econometric analysis showed that regardless of the water management institution, managers who faced positive incentives – typically receiving direct compensation for reducing water applications below estimated targets – were able to reduce water use per hectare of wheat by nearly 1,000 m^3, or 20%, in the sampled irrigation districts, but wheat yields also declined by approximately 4%. Results were statistically inconclusive for maize and rice. While changes in institutions and incentives have successfully reduced water applications in the YRB, the sustainability of these measures remains doubtful. This is because the savings provide limited benefits to local governments and farmers when the water is transferred to other provinces without compensation. One way to address compensation is through a system of transfer of water rights as discussed below (Wang and Zhang 2009).

Economic solutions: role of water pricing and water markets

Water pricing

China has gradually moved toward efficiency-oriented policies of water pricing as a method to help rationalize water allocation and alleviate water scarcity for both the urban and irrigation sectors. Generally, water pricing is instituted to (1) create incentives for efficient water use; (2) recover costs of water service provision; and (3) ensure financial sustainability for water supply systems and irrigation, including the ability to raise capital for expansion of services to meet future demand. In water-scarce economies, such as the YRB, the efficient allocation of water across sectors is also an important consideration.

Although water prices have been steadily raised over the past several decades in China, and particularly since the latest round of water-pricing reforms started in 1997, agricultural water is still thought to be much under-priced. As a result, water charges remain a limited instrument to increase water use efficiency and productivity further (Wang and Zhang 2009). Moreover, given the growing rural–urban income divide, it is unlikely that the government will raise irrigation fees to levels high enough to reduce irrigation water use seriously (Rosegrant and Cai 2003). Furthermore, in the YRB, the UWFR has led to income shortfalls of districts in parts of the basin where irrigation water supplies were cut considerably, particularly in midstream provinces. To make up for water shortfalls, these provinces were allowed to increase irrigation service fees. For example, Ningxia doubled the price of the service charge for irrigation to 0.012 RMB/m^3 (US$0.002/$m^3$) in 2000 (Wang *et al.* 2003). Downstream provinces, on the other hand, have generally maintained lower and simpler area-based fee structures.

Water rights, markets and transfers

Clearly defined and legally enforceable water rights and responsibilities for water operators and users in an irrigation system are the foundation underlying the incentives for conserving water and improving irrigation efficiency (Bruns and Meinzen-Dick 2000, Yang *et al.* 2003). The establishment of systems of water-use rights could empower water users in all sectors, as it establishes both rights and responsibilities to specified water use. If water is allocated to other sectors, typically urban and industrial, irrigators and other users would need to be compensated. Moreover, establishing water rights can serve as an incentive to invest in productive water uses, as they convey security to use water for a prolonged period. Furthermore, water-use rights provide incentives for all sectors to invest in water-saving technologies, as water outside of the existing water use right would have to be bought and paid for (Rosegrant and Binswanger 1994, Rosegrant *et al.* 2009b).

China does not currently have formal water markets that are supported by transparent and universal water property rights. There are, however, non-market mechanisms for assigning water-use rights and allocating water in China. Usufructuary rights to water use have evolved either explicitly through laws and regulations or implicitly through conventions (Ma *et al.* 2007). These rights are generally assigned based on one of three systems: first-come first-served allocation (prior appropriation rights); allocation based on proximity to water bodies ("riparian" rights); and public allocation (Heaney *et al.* 2006). Moreover, as discussed above, the government of China has released several regulations supporting water trading.

Since 2000, YRCC has promoted the establishment of water-right systems by conducting demonstration projects aimed at reducing water competition among sectors. The purpose of these demonstration sites is to reallocate water from agriculture to industry through increasing irrigation efficiency, generally through engineering measures, such as canal lining (Wang *et al.* 2006, Chen *et al.* 2007, Li 2007, Liu *et al.* 2007, Wang 2007). One transfer pilot project operates in Ningxia Province and 16 projects signed transfer contracts in Inner Mongolia, with a value of US$100 million. Under these projects, irrigation districts transfer part of their water use rights to industrial enterprises for a period of 25 years. However, analyses showed that water users in the irrigation districts are generally not aware of the water rights transfer; transfers are determined by the administration, not markets, and there are no adjustments based on market signals or economic measures. Thus, major challenges remain until a true market for water rights can be established.

The intra-provincial irrigation-to-agriculture transfers in the YRB provide important inputs for the potential development of inter-provincial water trading, which has been discussed by both policy makers and water allocation managers at the MWR and YRCC for several years. Such a reallocation could increase the water allocation efficiency of the 1987 cross-provincial water allocation agreement. Upstream provinces have a strong interest in maintaining the *status quo* in water allocation, however, and thus avoid the political costs of changing the current allocation. Moreover, given the large share of return flows in the YRB, changes in provincial permits from upstream to downstream might be inconsequential. It is therefore important to assess the full costs and benefits of changing the current system of water quotas.

Heaney *et al.* (2006) assess the benefits of water reallocation across YRB water resource regions using a production-function approach without accounting for the river hydrology (flow routing or return flows). They estimate economic benefits through increased value of agricultural production at 1 billion RMB per year, with reallocation chiefly occurring from the midstream to the downstream area. The authors caution, however, that for the benefits to be reaped, in addition to administrative challenges, new agricultural areas and labour would need to be made available downstream.

The most successful administrative water transfer to date in the YRB was the enforcement of the UWFR that ensured that flow to the Yellow River mouth was not cut off after 1999. This policy was in line with the refocus, over the last decade, on sustainable water use and keeping the Yellow River "healthy" promoted by the government of China. However, as we pointed out above, no compensation was paid to those provinces and water users that had to give up water as a result of the enforcement of the 1987 agreement.

Given the key importance of compensating irrigators for giving up water for both flows at the river mouth and rapid urban–industrial development downstream, we analyse the potential impact on basin GDP of water rights trading using a multi-agent system (MAS) modelling framework developed for the YRB (Yang *et al.* 2009). The model is populated with aggregated data from the YRCC water simulation model. A total of 52 water-use agents are defined, nine for the provinces sharing Yellow River flows, three to reflect downstream ecological needs, five to represent key reservoirs, and the reminder to represent key tributaries and inflows. The model is calibrated to 2000 data. Using the MAS model, we compared two scenarios to evaluate the consequences of changes from the current scheme of water allocation (business-as-usual of the current UWFR based on the 1987 allocation agreement): (1) water allocation across provinces without quotas; and (2) a market-based approach of water allocation for irrigation.

Under the UWFR, YRCC determines targets of monthly water releases for each of the major reservoirs on the main channel, based on the current reservoir storage, the future weather forecast, and the downstream water demand. The scenario without regulation assumes no administrative allocation mechanisms; agents are free to maximize water use subject to available resources. Thus, upstream water users will maximize off-takes, leaving less water available for downstream users; similarly, reservoir agents will maximize hydropower generation. The water rights trading scenario uses the UWFR as an initial water entitlement, based on which water can be traded among agents. To avoid adverse impacts on the downstream ecosystem, minimum downstream flows achieved under the UWFR scenario are set as constraints.

Figure 4 compares business-as-usual with the second scenario without any allocation rules for both water consumption and gross domestic product (GDP). Overall annual water consumption under the scenario without regulation is 38.3 km^3, 11% higher than the 34.5 km^3 under the UWFR scenario. The system-wide GDP is 1123.26 billion RMB under the scenario without regulation, 10% less than the 1246.68 billion RMB from the UWFR scenario. As expected, impacts on downstream ecosystem agents from unmanaged flows are considerable. For the most downstream ecosystem agent, flow stoppages start in February and continue through December, reflecting reality from 1972 to 1998 before the UWFR was enforced (Zhao *et al.* 2009). On the other hand, water consumption declines and GDP increases under the UWFR. For example, upstream GDP declines by 2.5 billion RMB annually, without compensation.

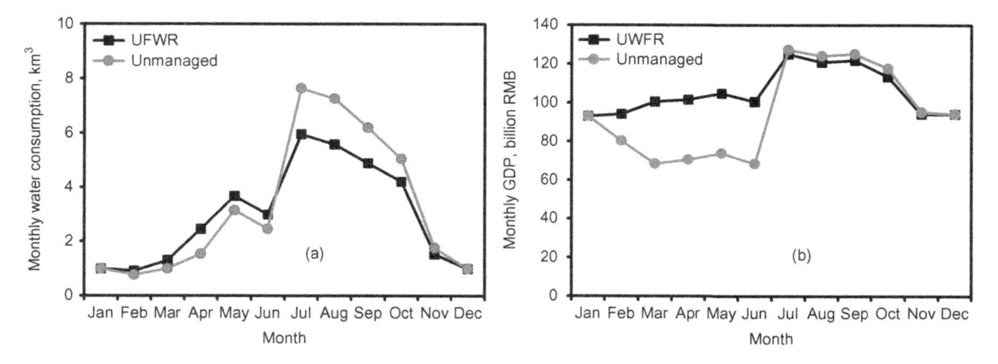

Figure 4. System-wide comparisons between UWFR scenario (baseline) and unmanaged scenario in (a) monthly water consumption; (b) monthly GDP.
Source: The authors.

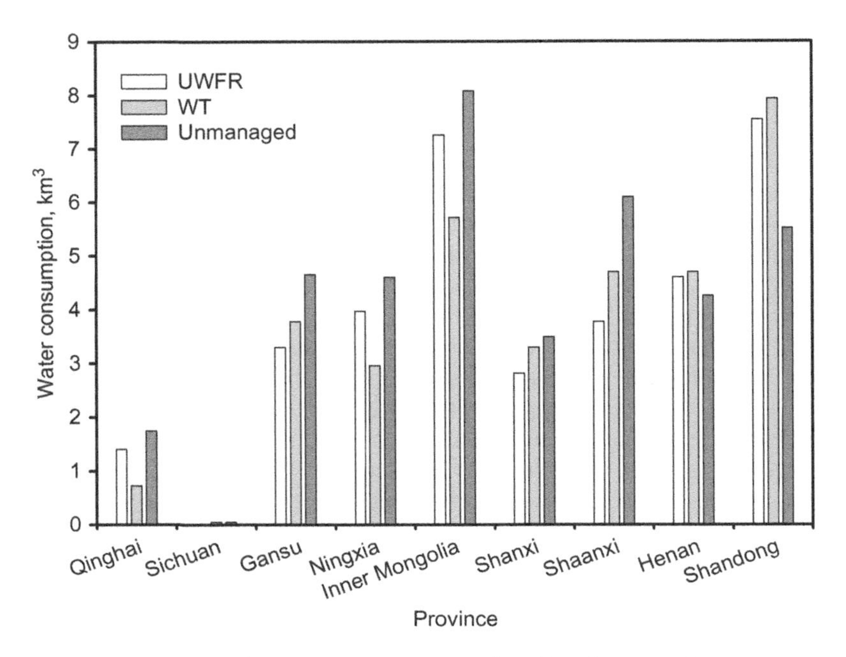

Figure 5. Water consumption in the YRB under alternative allocation scenarios.
Note: Provinces are shown from upstream to downstream.
Source: The authors.

The water trading scenario allocates priority for water use to the manufacturing and industrial (M&I) sector, after which trading can occur among irrigation sites across provinces. To ensure that the UWFR allocation to the downstream ecosystem is maintained, minimum downstream flows are included as hard constraints. Figure 5 presents total water consumption across provinces for the UWFR baseline as well as for the unmanaged and the water trading scenarios. Compared to the UWFR, water depletion increases upstream under the unmanaged scenario that basically supports maximization of withdrawals upstream. Under UWFR, on the other hand, water flows toward the downstream areas increase significantly, supporting much higher withdrawals in Shandong province, for example. Thus, the water-trading scenario follows prescribed withdrawals downstream to support the ecosystem agents, but also reallocates water from Inner Mongolia and Ningxia to Shanxi, Shaanxi and Gansu provinces.

Generally, water-trading prices are higher in low-flow months and highest in tributaries, as water transactions from the mainstream into tributaries is not feasible. Furthermore, the monthly water consumption with agents under the water price scenario is lower than that under UWFR, saving a total of 0.73 km^3. The annual GDP from the trading scenario is 1270.1 billion RMB, compared to 1246.7 billion RMB under UWFR, an increase by 23.5 billion RMB. GDP for individual agents is either the same or higher than that under UWFR. The GDP increase is 5.64 billion RMB for upstream agents and 17.8 billion RMB for midstream and downstream agents. Overall more water is sold than is bought in all months. Water without an agent buyer is purchased by the government to support ecosystem flow requirements. The annual total transaction costs (the sum of the water price multiplied by the amount of actual water transactions) is 2.95 billion RMB. This amount could be interpreted as a maintenance cost that the government has to pay to prevent flow stoppages otherwise occurring at the river mouth.

In summary, compared to the baseline scenario with UWFR, the scenario without regulation results in higher water consumption and lower GDP, and significant flow cutoffs at the river mouth. The water-trading scenario, on the other hand, results in a small decline in water consumption, combined with a significant increase in GDP. GDP, basically for agriculture, increases by 1.9%, which compares well with results of Heaney *et al.* (2006), even though they did not model the basin hydrology. GDP increases would be much higher if M&I would not receive first priority and thus would become an active water-trading sector.

Conclusions

The government of China has recognized the severe water constraints in the YRB. To address growing water scarcity, the government has started to change its approaches from management of water supply toward improved management of water demand. Signs of the new approaches are the 2002 water law, the increased number of regulations on water prices and a series of water trading pilots implemented in the YRB. However, most organizations concerned with water management are still headed and staffed by engineers, and traditional water-engineering measures, as exemplified by the SNWT, still dominate interventions in terms of funding.

There is no panacea for addressing the severe water scarcity challenges in the YRB. Based on an assessment of options available, we believe that the government should continue to reform the institutions responsible for irrigation management and water pricing across all water-using sectors, but current users need to be compensated for ceding water resources to users with higher-valued uses. Projects that transfer enhanced water rights that follow market mechanisms, and include the establishment of water rights and related responsibilities would be a first step in that direction. Reform is also required at all the administrative levels from the central to the local government to support fully integrated land and water management at the basin level and to avoid large inefficiencies caused by conflicting objectives of the various agencies involved in water supply and food production in the YRB.

Expanding irrigation in the YRB will help boost crop yields, which in turn will increase incomes of the poor and reduce poverty. However, the potential for expanding irrigation is limited, and labour productivity is known to be lowest in agriculture. Therefore, accelerating a shift of the rural labour force out of agriculture by creating off-farm employment opportunities in higher-productivity sectors in rural areas is arguably even more important for future rural economic development.

Other ancillary measures that need to be continued include further adoption of water-saving technologies, and continued support to agricultural research and development to increase crop productivity for both irrigated and rainfed crops. Continued productive investment is needed, rather than subsidies, in the rural non-farm sector to ensure that the urban-rural poverty gap does not widen even further. There is still scope for savings of agricultural water through improved water productivity. But continued transfers of water out of agriculture will eventually result in declining production, with implications for national food production as well as global food prices and trade.

Acknowledgement

This study forms part of the Consultative Group on International Agricultural Research Challenge Program on Water and Food.

Notes

1. HadCM3 stands for United Kingdom Meteorological Office Hadley Centre's Coupled Model, version 3. The B2 storyline and scenario family characterizes a world in which the emphasis is on local solutions to economic, social, and environmental sustainability, with slowly increasing population and intermediate economic development. It is considered a very moderate scenario.
2. While in government statistics some wheat areas are shown as rainfed, these areas generally have access to water harvesting facilities, ponds or groundwater. Based on our methodology, we classified them as irrigated.

References

Ahmed, A., *et al.*, 2009. *Water and poverty in China's Yellow River Basin*. Mimeo. Report prepared for the Challenge Program on Water and Food Yellow River Basin Focal Project. Available from: http://www.ccapwater.com/river/Ahmed%20et%20al%20_Water%20and%20Poverty%20in%20China_FINAL_Dec%2016,%202009.pdf

Brismar, A., 1999. *Environmental challenges and impacts of land use conversion in the Yellow River basin*. Interim Report IR-99-016. Laxenburg, Austria: International Institute for Applied Systems Analysis.

Bruns, B.R. and Meinzen-Dick, R.S., eds. 2000. *Negotiating water rights*. London: ITDG Publishing.

Cai, X., 2006. *Water stress, water transfer and social equity in Northern China: implications for policy reforms*. Human Development Report Office. Occasional Paper. New York: United Nations Development Programme.

Cai, X. and Rosegrant, M., 2004. Optional water development strategies for the Yellow River basin: Balancing agricultural and ecological water demands. *Water Resources Research*, 40, W08S04, doi:10.1029/2003WR002488.

Cai, X., *et al.*, 2010. *Water productivity assessment for the Yellow River Basin*. Mimeo. Report prepared for the Challenge Program on Water and Food Yellow River Basin Focal Project. Available from: http://www.ccapwater.com/river/YellowBasinWaterProductivityAssessment.pdf

Chen, J., Zhang W., and He, H., 2007. Implementation effects of water rights transfer in the Yellow River basin. *China Water Resources*, 19, 49–50 [in Chinese].

Chen, Z.K., 2002. *Impact of the successive drought on the North China Plain ecological environment*. Working paper of the China Institute of Water and Hydropower Research [in Chinese].

Falkenmark, M. and Widstrand, C., 1992. *Population and water resources: a delicate balance*. Population Bulletin. Washington: Population Reference Bureau.

FAOSTAT 2010. Food and Agriculture Organization statistical databases. Available from: http://faostat.fao.org/ [Accessed February 2010].

Fu, T., Chang, M., and Zhong, L., 2008. *Reform of China's urban water sector*. London: IWA Publishing.

Heaney, A., *et al.*, 2006. Water reallocation in northern China: towards more formal markets for water. *In*: I.R. Willett and Z. Gao, eds. *Agricultural water management in China*. Canberra: Australian Center for International Agricultural Research, 130–141.

Huang, J. and Rozelle, S., 2009. *Agriculture, food security, and poverty in China: past performance, future prospects, and implications for agricultural R&D policy*. International Food Policy Research Institute – Australian Centre for International Agricultural Research Policy Brief. Washington DC: IFPRI.

Huang, J., *et al.*, 2010. Subsidies and distortions in China's agriculture: evidence from producer-level data. Mimeo. Beijing: Center for Chinese Agricultural Policy.

IFPRI, 2009. IFPRI IMPACT Simulations. Mimeo.

Ke, S. and Zhou, K., 2007. Integrated water resources management in the Yellow River basin. *People's Yellow River*, 29 (1), 5–7 [in Chinese].

Li, G., 2007. Exploring and practice of water rights transfer in the Yellow River Basin. *China Water Resources*, 19, 30–31 [in Chinese].

Li, J., *et al.*, 2003. Estimation of the economic loss of water pollution in China. *China Water Resources*, 11, 63–66 [in Chinese].

Liu, X., Wu, L., and Wan, Z., 2007. Research on water rights transfer in the Inner Mongolia in the Yellow River basin. *People's Yellow River*, 29 (10), 16–17 [in Chinese].

Ma, X., 1996. *Flood and drought disasters of the Yellow River basin*. Zhengzhou: Yellow River Press.

Ma, X., Han, J., and Chang, Y., 2007. Research on evolution of water rights transfer in the Yellow River Basin. *Research of China's Economic History*, 1, 41–47.
Ministry of Water Resources, 2001. *Water resource bulletin of China 2000* [online]. Beijing: Ministry of Water Resources. Available from http://www.chinawater.net.cn [Accessed 2002].
Ministry of Water Resources, 2003. *Water resources yearbook*. Beijing: Ministry of Water Resources.
Ministry of Water Resources, 2005. *Water resources yearbook*. Beijing: Ministry of Water Resources.
National Bureau of Statistics of China, 2003. Available from http://www.stats.gov.cn/tjsj/qtsj/hjtjzl/hjtjsj2003/t20050706_402261014.htm. [Accessed 2002].
Pietz, D. and Giordano, M., 2009. Managing the Yellow River: continuity and change. *In*: F. Molle and P. Wester, eds. *River basin trajectories: societies, environments and development*. Wallingford: CAB International, 99–122.
Qiu, H., *et al.*, 2010. Bioethanol development in China and the potential impacts on its agricultural economy. *Applied Energy*, 87 (1), 76–83.
Randall, A., 1981. Property entitlements and pricing policies for a maturing water economy. *Australian Journal of Agricultural Economics*, 25 (3), 195–212.
Rosegrant, M.W. and Cai, X., 2003. Rice and water: an examination from China to the world. *In*: T.W. Mew, *et al.*, eds. *Rice science: innovations and impact for livelihoods*. Manila: International Rice Research Institute, 847–867.
Rosegrant, M.W. and Binswanger, H.P., 1994. Markets in tradable water rights: potential for efficiency gains in developing country water resource allocation. *World Development*, 22 (11), 1613–25.
Rosegrant, M.W., Ringler, C., and Zhu, T., 2009a. Water for agriculture: maintaining food security under growing scarcity. *Annual Review of Environment and Resources*, 34, 205–223. doi: 10.1146/annurev.environ.030308.090351.
Rosegrant, M.W., Fernandez, M., and Sinha, A., 2009b. Looking into the future for agriculture and AKST. *In*: B.D. McIntyre, *et al.*, eds. *International assessment of agricultural knowledge, science and technology for development (IAASTD): Global report*. Washington DC: Island Press, 307–376.
Shi H. and Shao M., 2000. Soil and water loss from the Loess Plateau in China. *Journal of Arid Environments*, 45 (1), 9–20.
Wang, J. and Zhang, L., 2009. *Water policy, management and institutions: role in pro-poor water allocation in the Yellow River Basin*. Mimeo. Report prepared for the Challenge Program on Water and Food Yellow River Basin Focal Project. Available from: http://www.ccapwater.com/river/Report/Policiesandinstitutionsfinal.pdf
Wang, J., *et al.*, 2003. *Pro-poor intervention strategies in irrigated agriculture in China*. Report. Colombo: International Water Management Institute and Manila: Asian Development Bank. Available from: http://www.ccapwater.com/river/index.html
Wang, J., *et al.*, 2007. Agriculture and groundwater development in Northern China: trends, institutional responses, and policy options. *Water Policy*, 9 (S1), 61–74.
Wang, S., *et al.*, 2006. Transfer water rights to optimize water allocation in the southern irrigation districts of Inner Mongolia in the Yellow River. *Water Conservancy in Inner Mongolia*, 2, 56–59.
Wang, Y., 2007. Comments on the reform of water price, water rights and water markets in China. *China's population, resources and environment*, 5: 153–158 [in Chinese].
World Bank, 2003. *Implementation completion report for loess plateau watershed rehabilitation project*. Report No. 25701. Washington, D.C.: The World Bank.
World Bank, 2007. *Project performance assessment report for second loess plateau watershed rehabilitation project and Xiaolangdi multipurpose project I & II and Tarim Basin II project*. Report No. 41122. Washington, D.C.: The World Bank.
Xu, Z.X., Zhao, F.F., and Li, J.Y., 2009. Response of streamflow to climate change in the headwater catchment of the Yellow River basin. *Quaternary International*, 208 (1–2), 62–75.
Xue, Y., Sun, Y., and Ringler, C., 2010. *A review of governance, laws and water interventions in the Yellow River Basin over the last 60 years*. Report prepared for the Challenge Program on Water and Food Yellow River Basin Focal Project. Mimeo. Available from: http://www.ccapwater.com/river/Report/WP5a_final.pdf
Yang, H., Zhang, X., and Zehnder, A.J.B., 2003. Water scarcity, pricing mechanism and institutional reform in northern China irrigated agriculture. *Agricultural Water Management*, 61 (2), 143–161.
Yang, H. and Zehnder, A.J.B., 2005. The South-North Water Transfer Project in China. Water International, 30 (3), 339–349.

Yang, Y.-C.E., Cai, X., and Stipanović, D.M., 2009. A decentralized optimization algorithm for multiagent system–based watershed management. *Water Resources Research*, 45, W08430, doi:10.1029/2008WR007634.

YRCC, 1998–2006. *Yellow River water resources bulletins* [online]. Yellow River Conservancy Commission. Available from: http://www.yrcc.gov.cn/ [in Chinese]. [Accessed 2006].

Zhao, J., *et al*., 2009. Evaluation of economic and hydrologic impact of unified water flow regulation in the Yellow River Basin. *Water Resources Management*, 23 (7), 1381–1401.

Zwart, S.J. and Bastiaanssen, G.M., 2004. Review of measured crop water productivity values for irrigated wheat, rice cotton and maize. *Agricultural Water Management*, 69 (2), 115–133.

Water, food and poverty: global- and basin-scale analysis

Simon Cook, Myles Fisher, Tassilo Tiemann and Alain Vidal

Global population growth exerts stresses on river basins that provide food, water, energy and other ecosystem services. In some basins, evidence is emerging of failures to satisfy these demands. This paper assembles data from nine river basins in a framework that relates water and food systems to development. The framework provides a consistent basis for analysis of the water and food problem globally, while providing insight into specific conditions within basins. The authors find that successes occur when demand is met by increased productivity, while failure occurs when factors conspire to prevent development of land and water resources.

Rationale: water and food systems support development within a global environment

The global environment currently supports nearly seven billion people through a range of ecosystem services that include food production, water supply and sanitation. By 2050, the global population is projected to increase to over nine billion (UN 2009) with concomitant increase in the demands on the natural environment. There is evidence that, in reacting to meet some of these demands, human societies are damaging the environment's capacity to satisfy other demands. In river basins, this is manifested through the inequitable sharing of finite water and land resources.

The Food and Agriculture Organization of the United Nations (FAO) predicts that 70% more food will be required by the year 2050 (Bruinsma 2009). Due to evolving diets, especially for growing urban populations, demand for animal products is estimated to increase by 74%. FAO estimates that over 900 million people currently go hungry. Domestic and industrial demand for electric power will increase by about 50%, of which hydropower is expected to supply about one third (EIA 2010). Most of the increased food production is forecast to come from intensification of production systems, but about 15% is expected from extension of the agricultural area. Urban populations will expand from a current estimate of 3.5 billion (50%) to 6.3 billion (69%) by 2050 (UN 2010). The impacts of these changes will be compounded by other factors, in particular global climate change, which imposes major uncertainties on future water availability, environments of crop production, and disease (IPCC 2007).

We choose river basins as the environmental entity with which to study this problem since this is the only way to understand flows and exchanges of water. The global picture translates into very different outcomes within individual river basins. Ten river basins were chosen for study in the Basin Focal Projects (BFPs) of the Challenge Program on Water and Food (CPWF) of the Consultative Group for International Agricultural Research (CGIAR). The 10 basins chosen are in developing countries where the disjunct between poverty, water and food is particularly acute. According to UN projections (UN 2009) almost all population growth in the next 40 years will be in developing countries. Increasing populations will increasingly stress the environment as countries attempt to reduce poverty and improve food security. While we can learn some lessons from river basins in high-income countries, we do not believe that they can offer us universal insights that we could apply to basins in developing countries.

Introducing papers reporting the outcomes of the BFPs, Fisher and Cook (2010) present evidence of these conditions, showing that in some areas, average annual water availability is falling below 1700 m^3/capita, an arbitrary level some consider necessary for food security (Falkenmark *et al.* 2009). Per capita land holding is falling, even while the number of people in urban areas increases. Environmental flows are strongly affected by fragmentation and altered flows in 37% of 227 large river basins assessed globally and moderately affected in 23% of them (WRI 2003). Flows have reduced below levels regarded as safe in many rivers (Pearce 2007) and at times have ceased entirely in some major rivers such as the Yellow and the Amu Darya.

The loss of ecosystem function impacts economics and social development. Livelihoods are affected in different and often subtle ways that can be difficult to quantify. It is difficult to persuade people, and the politicians who need to respond on their behalf, to react to problems that are not clear-cut. The situation is particularly complicated when the cause is a small group pursuing their own self-interest, which affects the collective common good. For this, we need compelling and unambiguous evidence on the conditions, causes, and solutions to the problem, to support what Cohen (1989) calls deliberative democracy. Research is therefore required to provide unambiguous evidence of the condition of water and food systems as they support development within river basins. A clear analysis of the constraints that the conditions of water and food systems impose upon development is also required.

The questions that need to be answered are therefore: (1) can global food and water systems support nine billion people in 2050 without seriously compromising the functionality of ecosystems and their services?; (2) what is the condition of water and food systems within basins that explains the current and future constraints to development?; and (3) what kinds of interventions within river basins seem most promising to enable sustainable development?

Components of the problem: organizing information to help explain conditions in river basins

The problem is made up of aspects of poverty and development: water resource management, agriculture and institutions. Poverty is described within basins according to measures of income, consumption or livelihood assets. Of course, poverty and food insecurity are related and we understand from Byerlee *et al.* (2008) that food security is a necessary if not sufficient basis for poverty alleviation. We consider poverty to be a dependent variable, which represents the degree to which people are not supported by the development of water and food systems. A range of metrics are used to identify the number of people who are not supported, and from these we determine the relationships between poverty and data that describe water and food systems. This process tends to accentuate the negative aspect of development, but does not describe the positive aspects by which development supports populations.

We describe water in river basins[1] according to its receipts and balances, including estimates of evapotranspiration (ET), which we use to estimate agricultural water productivity. Water-use accounts indicate how the water resource is distributed amongst major users at sub-basin level, focussing on agriculture's use of blue and green water. Further analysis indicates how water availability varies. Agriculture is described according to land use and productivity of the land. We measure productivities either as yield or monetary value, and compare them with estimates of water use to determine their water productivity. Institutional analysis is required to explain the social processes that assist or obstruct "improved" use of water and agricultural resources, which occur at scales that range from local to global, but whose form and function are often ambiguous. At a local scale, such processes are exemplified by reform of land and water rights as in the Limpopo and Volta basins; at the basin scale, by trading of ecosystem services as being developed in the Andes; or internationally through processes that influence trade of "virtual water" between countries.

A framework to analyse conditions in basins

We now come to the problem of formulating the analysis. Conventionally, analysis of development in river basins has approached the problem from the hydrologic perspective, with scant reference to the activities of the agricultural systems that operate within it. In this approach, water flow is analysed, using water-use accounts, or "finger diagrams" to identify where water flows within basins, and to which uses. In contrast, agricultural research has focussed strongly on aspects of the farming systems, with little reference to their interaction with water systems. Land productivity is the normal focus of agricultural research, with the individual aspects of agricultural systems usually studied separately. Food-systems approaches analyse its different components without accounting for water use. There is therefore a clear disconnect between the three approaches.

From the development perspective, neither the water nor the food-systems approach is sufficient to explain how either system interacts *with* the other to produce livelihood outcomes. Focusing only on the food system provides no insight into the implications of variations in use of a shared water resource. Focus on the water productivity of food systems takes no account that livelihood systems gain support from a wide range of support mechanisms. Water may flow through to a final benefit by many different processes, which operate in parallel or serially. Moreover, benefits may substitute one for the other, for example, people may be supported by food from irrigation, by livestock feed from rainfed grassland, by fish that live in the aquatic environment that the irrigation water might otherwise support, or by the benefits of non-farm employment enabled by hydropower. A focus only on agricultural production can therefore omit important off-farm contributions.

Molden (2007) provides the *Comprehensive assessment of water management in agriculture*, which was a major programme of research describing a wide range of aspects of the water and food systems. The *Assessment* provides valuable general advice to policy makers but it does not, however, attempt to assemble these components within specific basins. The international model for policy analysis of agricultural commodities and trade (IMPACT)-Water model of Rosegrant *et al.* (2002) and the policy dialogue model (PODIUM, de

Fraiture *et al.* 2001) both assemble selected components of food systems within river basins, but these models do so at a relatively broad level of generalization at a country scale.

This paper draws upon the logical structure of nine projects in river basins to provide observations according to a single analytical framework (Fisher and Cook 2010) (Figure 1). Starting with a background briefing of prevailing economic and demographic conditions, each project analyses the basin's basic hydrology, using data both from within the basin and available globally. Each also identifies the performance of the agricultural system, using the key concept of water productivity, that is, the measured gain from water consumed by agriculture. The analysis also describes the strong influence of institutional factors on both water availability and water use.

Both water and food influence livelihoods in diverse ways. We analyse these with the intention of identifying what will happen to people if conditions vary substantially. For example, if the number of small reservoirs in the Volta increases towards its potential, how will people downstream be affected, especially those dependent on electricity generated at the Akasombo Dam on the lower Volta? How will flow in the Mekong be affected by climate change? What are the consequences of water allocation policies to smallholder and commercial farmers in the Limpopo Basin?

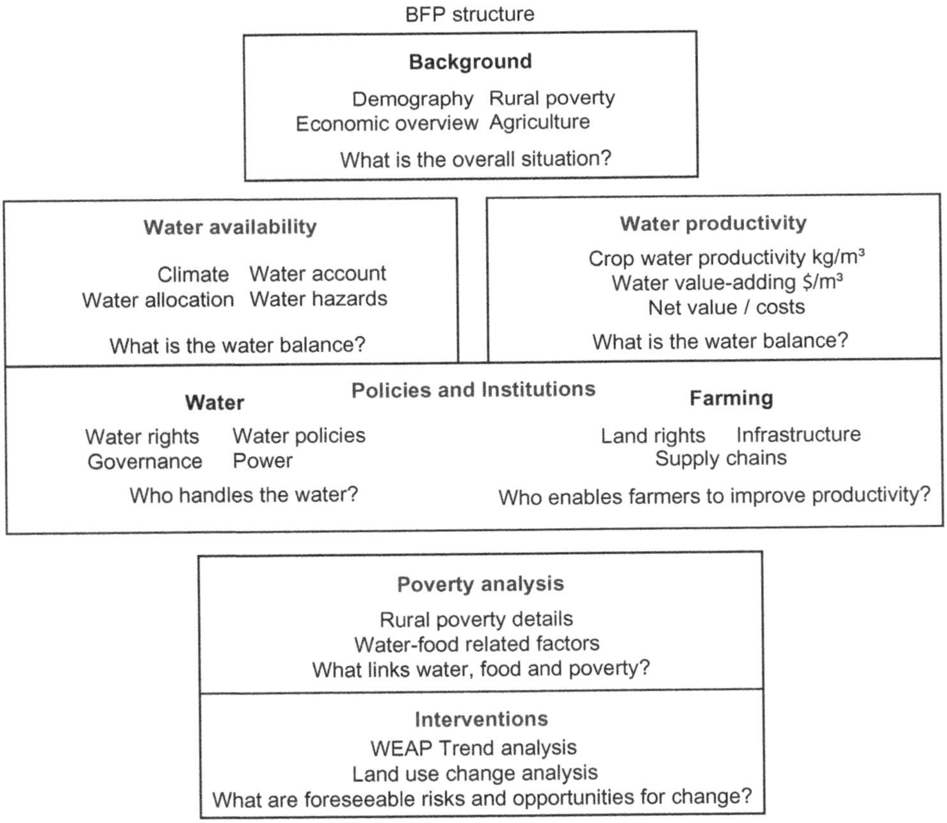

BFP structure

Background
Demography Rural poverty
Economic overview Agriculture
What is the overall situation?

Water availability
Climate Water account
Water allocation Water hazards
What is the water balance?

Water productivity
Crop water productivity kg/m³
Water value-adding $/m³
Net value / costs
What is the water balance?

Policies and Institutions

Water
Water rights Water policies
Governance Power
Who handles the water?

Farming
Land rights Infrastructure
Supply chains
Who enables farmers to improve productivity?

Poverty analysis
Rural poverty details
Water-food related factors
What links water, food and poverty?

Interventions
WEAP Trend analysis
Land use change analysis
What are foreseeable risks and opportunities for change?

Figure 1. Analytical framework of a logical structure of observations in nine projects in river basins (WEAP = water evaluation and planning).

Insights from basins

The observations from basins are detailed in the September 2010 special issue of *Water International* (Fisher and Cook 2010). The analyses showed that while there are clear examples where increasing water scarcity reduces livelihood support, in most places a more nuanced approach is required to explain how water and food systems interact, and how this interaction influences development.

Water influences development in many different ways. It does so indirectly through its impact on irrigated or rainfed agriculture, through its support of aquatic systems, and also through the provision of urban water, power, sanitation and transport. In some basins, economic systems have responded to increasing water scarcity with no discernible impact on rate of development. In other areas, water-related factors have a clear impact on rural development, which can be felt concurrently through food, income or environmental security. This can be difficult to analyse, since impacts can be interchangeable. Moreover, development of one may threaten another, such that it is the total picture that needs to be considered.

For example, the range of observations from the studies in basins include:

- Over-development of irrigation in the Yellow River led to basin closure and subsequent attempts to manage demand (Ringler *et al.* 2010);
- Tension is increasing in the lower Mekong between proponents of hydropower development and those who rely on the aquatic system, which this is likely to modify (Kirby *et al.* 2010);
- The sustainability of groundwater use in the western Ganges is under threat as a result of expanded use of tube wells (Sharma *et al.* 2010);
- Increasing pressure on the Andean basins for urban water supply is increasing the opportunities for supply of ecosystem services in the form of avoiding contamination by effluent from mines and from urban sources (Mulligan *et al.* 2010); and
- Development of upstream agriculture in the Nile raises concerns in Egypt, which depends totally on in-flow (Awulachew *et al.* 2010).

Four factors that couple water to development

From observations in 10 basins we conclude that there are four factors that link water to development.

Water scarcity

Water scarcity has been described as either physical or economic (Seckler *et al.* 1998). Here we focus on physical water scarcity, as reported in the Yellow River, Karkheh, Indus and upper Limpopo basins. While the Nile as a whole is not considered water-scarce, political tensions occur because Egypt and Sudan rely totally on inflows from the less-developed upstream countries. Population density is low in the Limpopo where less than 1% of available water is used for irrigation. Conversely, population density in the Yellow River is extremely high and irrigation consumes 14% of average basin flow.

On average, the Niger, Nile, São Francisco and Volta basins are moderately water-scarce. Data of average water availability hides spatial variations, such that less populated parts of all are water-scarce. The Ganges is, in general, only moderately water-scarce but contains areas of extreme or increasing scarcity.

Areas of low scarcity include the Andean system and the Mekong. Nevertheless, the Andes basins are extremely diverse and contain some of the driest places on earth; average annual rainfall in Lima is less than 100 mm and the city is dependent on outside water supply. Areas of northeast Thailand, southern Laos and Cambodia in the Mekong are frequently affected by drought and seasonal water scarcity due to prolonged dry seasons of up to six months.

Evidence suggests that variation of water availability is not strongly correlated with poverty. An illustration of this apparent paradox is the comparison between the Democratic Republic of Congo (DRC) and Israel: the water-rich DRC is amongst the poorest in the world, while water-poor Israel ranks among the richest (Molle and Mollinga 2003). Analysis from basins suggests that water scarcity is one factor controlling development. It is undoubtedly a major concern for those agricultural economies such as Pakistan, Egypt, India and China that have developed their agriculture through intensive use of irrigation. The poorest people, however, live in areas where such development has not occurred, and where other factors therefore influence development more strongly. It is important to understand this in the light of geographical variation of development processes.

Dry areas do not support large populations without irrigation but, in some basins, irrigation has supported the development of intensified agriculture using rates of abstraction that now seem unsustainable. In some cases further growth may be possible by improving water productivity. The situation becomes problematic when areas (for example, Indus and Yellow River basins) that already face moderate to severe water scarcity are squeezed by demands from non-agricultural sectors. Parts of these basins seem to have reached maximum water productivity so that the options for further growth of low value agriculture are limited. In such cases, further development depends on a move away from dependence on basic agriculture towards higher value crops or non-agricultural activities.

Lack of access to water

Lack of access to water resources, sometimes referred to as economic water scarcity (Seckler *et al.* 1998), was reported as a widespread problem in basins and occurs even in relatively well-watered areas such as central Ghana. It occurs because either (1) there is a lack of infrastructure development; or (2) there are institutional constraints, which may grant access to some, but deny access to others, usually to the poorest, most disadvantaged people.

The Asian basins Karkheh, Mekong, Indus, Ganges and Yellow generally have well-developed infrastructure and high levels of access to water. Gini coefficients for income are generally moderate or low, suggesting that economic benefits are relatively widely distributed. Available groundwater resources are generally exploited, with over-exploitation is common in some regions.

The Andes and Limpopo have high Gini coefficients (inequitable income distribution). Both have experienced political tensions over inequitable access to limited water resources. In the Niger, Nile, and Volta basins, infrastructure is very underdeveloped. Access to sanitation in rural areas is very poor.

In Asia, the variability in degree of development of water resources between countries is quite high, while in Africa the degree of development is low nearly everywhere. Excluding Egypt and South Africa from the data, access to water in African countries is lower than Asian countries (index of 7.3 vs. 10.8), but the standard deviation is only half of that of the Asian selection (2.38 vs. 4.38), which shows that access is uniformly low in Africa.

The correlation between access and capacity for the 147 countries analysed worldwide by Lawrence *et al.* (2002) is relatively high ($r^2 = 0.68$). Access to water and the level of development are strongly linked.

Exposure to water related hazards of drought, flood and disease

Floods and droughts occur sporadically, but they have a disproportionately negative impact on the poor. This is because they push them into survival conditions in critical years, and overall deter critical investment that may allow them to escape from poverty. Although droughts tend to occur sporadically, in recent years serious droughts have led to food insecurity over large parts of sub-Saharan Africa and South Asia. Severe drought frequently affects the Limpopo, Nile and Niger basins, with several major events in the past decade. Though less frequent, drought has recently damaged food security in the Ganges.

Floods are generally of more limited geographic extent but – as demonstrated by the major 2010 floods in the Indus, Niger and Volta – they cause intense disruption and loss of life and property. Floods are a serious problem in the Limpopo Basin, where there are extreme year-to-year variations in flows with major floods in Mozambique in 2000 and 2008. In the Andes and upper Ganges, floods are of small magnitude but associated landslides disrupt transport infrastructure. Floods in the lower Ganges pose a serious hazard that appears to be increasing in magnitude. Devastating historic floods in the lower Yellow River are a reason for strict control of flow. In the Mekong, 90% of flow occurs in three months, leading to widespread flooding in the lower basin. Contrary to being considered a hazard, however, the inundation is regarded as a vital to the aquatic resource on which 65% of the basin's population depend.

Water-related diseases such as malaria, schistosomiasis, or onchocerciasis (river blindness), impose serious constraints on land use over large parts of sub-Saharan Africa. The central Volta is a hotspot for malaria, which is also an important hazard in other African basins. Together with the widespread but now controlled incidence of onchocerciasis, it is one reason for low agricultural activity of the central Volta.

Water productivity

Following Hoff and Rockström (2009), we divide water productivity into productivity of green and blue water, and restrict our comments here to water productivity of crops. Blue water productivity describes the conversion of water abstracted from rivers for irrigation. Green water productivity describes the conversion of precipitation in rainfed agricultural systems. Cai *et al.* (2011) provide more details on the water productivity of crops, livestock and fish. This and the basin reports (*Water International,* September 2010) suggest that green water use is far greater in most basins than blue water use and also that green water productivity is substantially lower than blue water productivity. This supports the conclusion of Molden *et al.* (2007) that improvement of rainfed agriculture presents a major opportunity to meet the demand for more food without increasing agricultural water use.

Blue water productivity

Productivity of blue water is very high for irrigated areas in the Yellow River basin. Estimates for some areas approach the likely maximum for wheat of approximately 1kg/m^3. Slightly lower estimates are recorded for irrigated areas in the upper Ganges, although the lower Ganges and the Indus are much lower. Values from the Nile Delta have been boosted in recent years through production of high value crops, in addition to the integration of aquaculture and agriculture.

Irrigation is less widespread in the Mekong, but values of water productivity from the Delta region are also very high, and have increased over time more than keeping pace with the demand for food. In the Karkheh, water productivity of irrigated crops is much lower.

With the exception of the Delta and Gezira region of the Nile, irrigation consumes less than 1% of water in the Limpopo, Niger, Nile and Volta basins and contributes little to economic activity of the basins.

Green water productivity

Overall, green water accounts for over 70% of the water flux in the 10 basins included in the BFP studies of the CPWF.

Water productivities in parts of the Yellow and Ganges basins are high (>0.5 kg/m^3), although less than that of irrigated areas. Elsewhere, water productivity varies between 0.1–0.5 kg/m^3, suggesting widespread low activity of the agricultural system. Water productivity of rainfed agriculture in the African basins is generally extremely low (water productivity <0.1 kg/m^3).

The economies of the Andean countries are mostly classified as in the transition to industrialization. Agricultural production remains important, but less so than in basins more completely dependent on agriculture. Preserving the quality of water destined for urban use and environmental uses is more important than any economic services it might provide, with the exception of hydropower, which is a non-consumptive use. The São Francisco contains a contrasting dualistic agricultural economy of smallholder agriculture and commercial agriculture. Commercial agriculture is expanding while smallholder farmers are migrating to urban centres, which converts rural poverty to urban poverty.

Low water productivity is a consequence of a wide range of limitations that collectively constrain agricultural production, and cause the widespread limit on the contribution of water and agriculture to economic development. Estimates of water productivity indicate that activity is well below potential, and taken together, low blue and green water productivity represents a systemic failure of agriculture to the convert water into food or income. This is by far the most important water-related constraint to improved food, income and environmental security by consumption of water by agriculture.

Crossing from local to global scales

Coherence between different scales: global, basin and local

The linkages between the conditions that are discernible at a global scale and what happens in catchments and farming or fishing communities are complex and need clarification. Data at broad global scales indicate the emerging tension between food and water systems. Population increase and changing dietary habits are expected to double the demand for food and animal feed. Present and projected conditions are expressed very clearly in the *Comprehensive assessment of water management in agriculture* (Molden 2007). At a local scale, the situation can appear very different, reflecting the strong influence of local conditions on the way people manage water and food to support their livelihoods. The systems are connected within river basins through transfers of water, food or other products of agriculture (Figure 2). But how, exactly? What are the particular pressures and opportunities that occur within individual basins? How can these complex behaviours be described and analysed, and what are their impacts on poverty locally? By adopting the basin as the

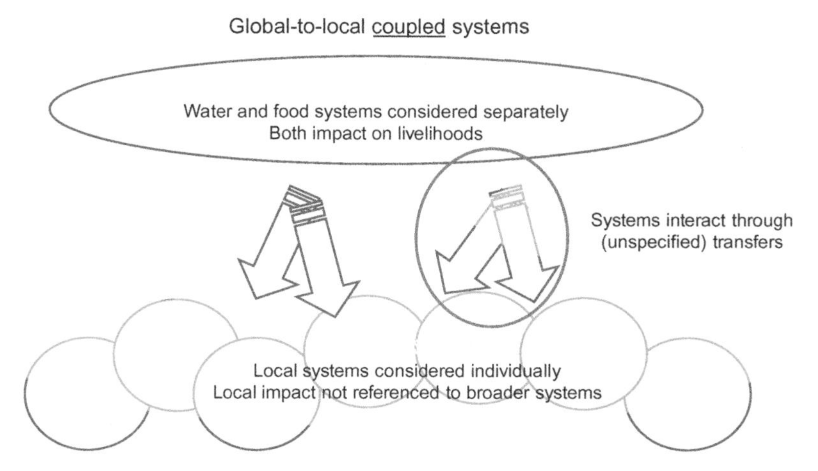

Figure 2. Food and water systems interact strongly at basin scale but linkages are difficult to see.

prime object of analysis the BFPs connect what is happening within basins to trends and pressures that are evident globally.

To move beyond analysis within individual basins towards a global view, we organize observations from basins according to the themes of the papers in this issue: water availability (Mulligan *et al.* 2011); water productivity (Cai *et al.* 2011); and poverty (Kemp-Benedict *et al.* 2011). The overall condition of economic development in river basins can be understood using the scheme of the *World development report* (Byerlee *et al.* 2009). Figure 3 shows basins arranged according to two variables: rural poverty and agriculture as a percent of GDP. The arrow tracks a generalized "development trajectory" in which agriculture is seen as a necessary, but not sufficient basis for development. The trajectory passes from strongly agricultural economies in the Niger, Volta, Nile and Limpopo,

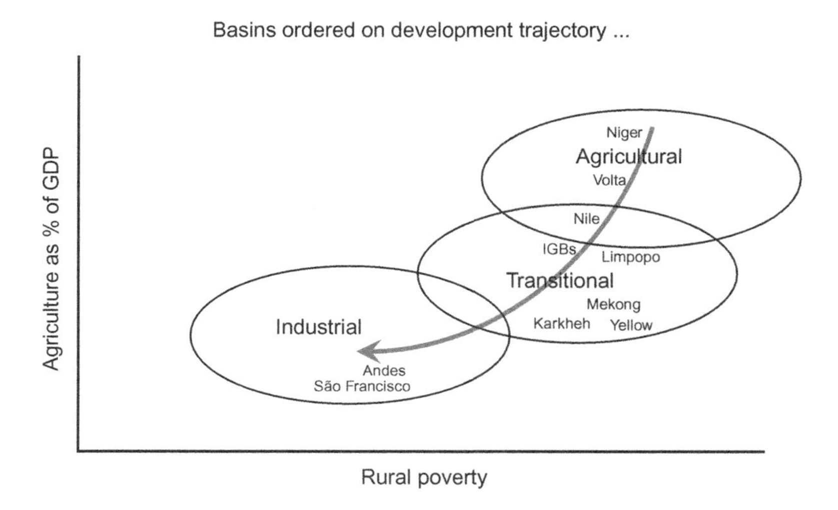

Figure 3. BFP basins ordered according to rural poverty and agricultural contribution to gross domestic product (GDP).

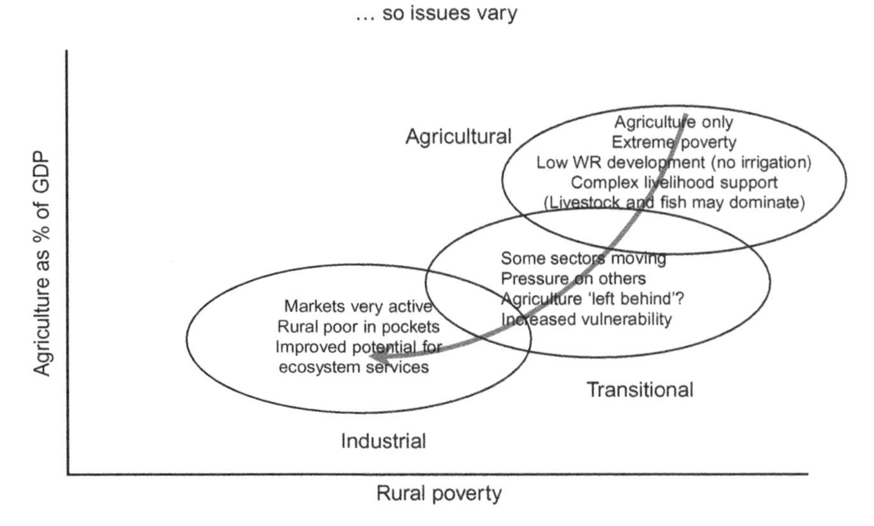

Figure 4. Prevailing conditions in three classes of basins.

through those of transitional economies (Indus, Ganges, Mekong, Yellow); to basins containing industrial economies such as the Andes. It should be noted that most basins contain large variations: for example, the Mekong is transitional, but contains areas of Laos that are strongly agricultural and others in Thailand and Vietnam that are industrial.

We suggest this as a "first cut" of generalizing conditions throughout the developing world, since it is the development drivers, *in conjunction* with resource constraints that determine the broad variation of constraints and opportunities (Figure 4). The prevailing conditions and opportunities for each type of situation can be summarized in the following paragraphs.

Agricultural basins: Niger, Nile, Volta

Agriculture dominates activities in these basins but agricultural productivity is very low and rural poverty widespread. With the exception of mining industries in some countries, non-agricultural activities contribute little to economic activity. Water infrastructure is poorly developed, normally at 1% or less of its potential. "Non-engineered" agriculture is relatively more important than in transitional or industrial economies. In drier basins, livestock systems are very important for the poorest. In wetter areas, fish and wetlands provide vital livelihood support, on which the poorest and landless depend heavily. Demands on water resources are not fully expressed as population densities remain low, though increasing. People are exposed to water-related hazards (Figure 5).

Byerlee *et al.* (2009) show clearly that improving agricultural productivity is a necessary step to move economies along their development pathway. A prime development objective in these basins is to focus on the provision of basic needs of sanitation, healthcare, education and transport. Market development and infrastructure are key issues that are poorly developed. The major opportunities for improvement of agriculture are to support food security through rainfed agriculture without compromising livelihoods of those dependent on marginal livestock or aquatic systems. Development of water resources for irrigation may deliver local benefits, but from such a low base, this seems likely to have limited impact on rural poverty.

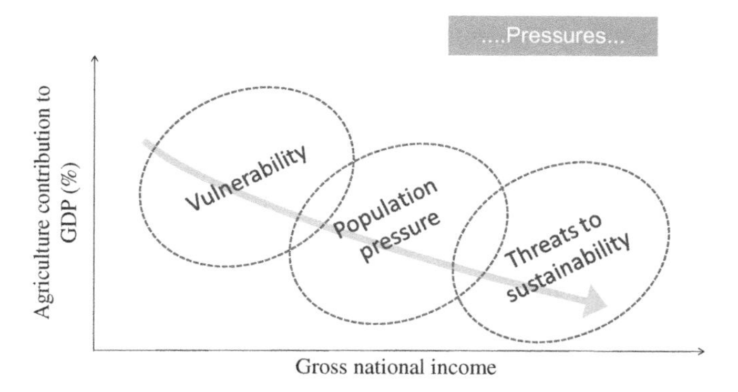

Figure 5. Pressures change with development status.

Transitional basins: Indus-Ganges, Limpopo, Karkheh, Mekong and Yellow

In transitional economies, activities that are non-agricultural and that add higher value make increasing contributions to gross domestic product (GDP) and attract people out of agriculture. These are the areas that are expected to experience most rapid growth of economic activity and population, although the population within the Yellow River basin is declining. It is in these regions that demand for food is expected to intensify and where water resources are generally well developed. Non-farm activities expand, and may be inter-woven with agriculture to support development. Overall, however, the process seems to be one of uneven and localized development in which many gain but some are left behind. Agriculture remains important at a national level, and agricultural productivity increases in response to market demands and requirements for food security of an increasingly urban population. Agriculture may also wield considerable political power. Greatly increased agricultural activity may exert pressure on water resources and compete with expanding non-agricultural demands. Issues of water quality emerge but lack the institutional capacity for ecosystem servicing. There is only partial protection against hazards of flood, drought and water-related diseases.

The main opportunity seems to be institutional development to enable transparent, informed and broadly based processes of change, which can distribute benefits and capacity without constraining development. This can occur under a range of political environments.

Industrial basins: Andes, São Francisco

Agriculture is no longer a major economic contributor in industrial economies though rural poverty remains in localized areas. In such conditions agriculture retains its importance as a means of reducing the risk of social unrest caused by depopulation to urban areas. Markets are highly active. Direct food security may decrease in importance as income security increases through exploitation of higher value agricultural activities. Water resources may be highly stressed, but resources may be managed intensively. Ecosystem services and benefit sharing become increasingly recognized as a means of ensuring environmental security. Greater levels of economic activity afford a high degree of protection from water-related hazards.

Food security is reasonably assured in these conditions. Income security is relatively insensitive to agricultural activity. Therefore, while agricultural activity remains supported, often for political reasons, increasing opportunities are sought for development of ecosystem services. These are required to maintain the environmental security of water supplies to urban and industrial consumers, hydropower and high-value agricultural activities. Aesthetic and cultural factors play increasingly important roles at steering development through regulation and political norms.

Developing insight to support change

Change will occur in food and water systems according to prevailing drivers in each and how institutions respond to the drivers. Groups of people will be affected by these changes as they strive to feed themselves and maintain economic activity. In a situation of unequal power, some are likely to be left behind, or lose their livelihood support from water and other resources as they are commandeered by others.

Change can occur in diverse ways but intervention can assist either by increasing the capacity of existing resources to improve productivity or by reducing the likelihood of loss of livelihood. This can only be achieved by negotiation, through the process of deliberative water politics (Dore 2007) in which people agree to adopt or accept actions based on informed and transparent debate and deliberative consideration of the options. An example is the change from the self-interest of the riparian states (provinces) in the Murray-Darling Basin in Australia to rational allocation within the limits of the resource under the aegis of the federal government, a political solution forced on the states by the devastating droughts of the last decade (NWC 2009). Examples from the basins include the emergence of payment for ecosystem services in the Andes, (Mulligan *et al.* 2010), and intensifying dialogue over issues of hydropower development in the Mekong (Molle *et al.* 2009).

Change is a political process that is mediated by people who represent interested groups. But how much do such people know about their basins? What do people and their political representatives need to be informed and with what outcome in sight?

(1) How much water? It is difficult to see how any political process of agreement about sharing river water resources can proceed without basic information about volumes of water moving through river basin systems. Yet major uncertainties exist in many river basins. How much water flows through different parts of the river basin? What extreme flows occur? Where are groundwater reserves? – and many other basic questions. Uncertainty persists even in major basins such as the Nile and Ganges for which vast amounts of information exist.

(2) Who uses the water? As water flows through basins it is partitioned through different sub-systems: irrigation, rainfed crops, livestock systems, non-agricultural land. The balance supports aquatic systems, on which fishing, sanitation and hydropower depend. Water balances are generally unknown. They vary widely within basins; some areas are essentially consumers of water, other areas are providers. Differentiation according to such hydronomic zones, as they are sometimes called, characterizes these broad demands and supplies. Where are these zones? How sensitive are they to changes in agricultural or urban activity? Green water dominates in most basins, yet it is rarely quantified. How can discussion proceed without this basic information?

(3) How effectively is water used by agriculture? Agriculture is the dominant user of water, but how efficiently is this use converted into food, income or other livelihood support? What are the primary and secondary gains of water used by agriculture? What are the opportunities for increasing agricultural output without increasing

water use? By what process can downstream users of agricultural flows, such as in the Nile, negotiate to compensate upstream providers for guaranteed supply rather than insist upon it as a right?[2]

(4) What are the consequences to livelihoods of tensions? Water is just one of many factors that determine livelihood support from systems that process it. A basic understanding of how the livelihoods of different people are affected by water availability, access and use is necessary to understand how change will affect them. How else can the impact of changes in basins be agreed?

(5) Who has the power to change? Institutional analysis is the most poorly developed of these methods, yet it is clear that institutions hold the key to explain the points of influence. A key requirement is not to focus solely on traditional institutions that control water but to understand how institutions can work together to enhance or constrain the ability of the agricultural system to change. This is our focus in this paper.

(6) How sensitive is the system to change? The uncertainty that is most difficult to remove is that caused by non-linear dynamics of river basin systems. For example, how much land-use change can Andean basins tolerate before suffering serious consequences to flow regimes? The unsustainable use of groundwater in the upper Ganges (Sharma *et al.* 2010) cannot continue, but entrenched political support of the status quo will make change difficult. How much further development of hydropower can the Mekong accommodate before there is serious disruption of aquatic systems such as the Tonle Sap in Cambodia on which the livelihoods of more than a million people depend? Such questions can only be addressed by analysis of linked social and ecological systems.

Conclusions

While it is convenient to visualize a global water and food crisis in which increasing demand for food and water results in increasing poverty, food insecurity and political conflict, detailed analyses from the BFPs show a far more nuanced reality. Analysis of conditions in basins shows a complex dynamic between development processes and the natural resources they consume. This dynamic can push river basins, or parts of them, beyond the level at which ecosystem services of water provision, food production, energy and other services can be delivered in a sustainable manner. This raises problems of potential conflict over limited resources between different communities within river basins. An alternative situation occurs when resources are effectively under-developed. In such cases, poverty is associated with low productivity of land and water.

The relation between water and food systems and the development that they support is bi-directional. Water and food systems influence development and development influences the use of water and food resources. Societies use a range of ecosystem services as they develop, but conversely, the way these are used depends strongly on the development status of those societies, their power, their capacity to govern themselves and their capital. Consequently, while development in the Yellow River has allocated virtually all the water resources in the basin, it is also worked to increase the productivity of the system by assembling all components into a highly productive system.

The global environment supports people through the provision of ecosystem services such as food production, water supply, sanitation and hydropower. People appropriate services individually or communally, through institutions that govern sharing, production and investment. The optimist would observe that total possible livelihood support from all services globally exceeds the demands from present and future populations. However, problems occur locally that obstruct development and push people into poverty. These problems tend to be of two broad types.

In some basins, development by some sectors of society has appropriated available land or water resources to the point at which it constrains development for others. In the Limpopo, for example, appropriation of water and land is the consequence of its colonial history, which gives disproportionate access to better land and limited reliable irrigation water to some commercial farmers. In the Nile, entitlements claimed by downstream Egypt and Sudan impose constraints on water uses by upstream Ethiopia. In the Mekong, hydropower development to satisfy demand from industrial economies threatens the livelihoods of fishers who depend on existing flow patterns. In all these cases, it is clear that the livelihoods of some are compromised by the use of water resources by others. Use of water resources interacts physically with the activity of the food system but institutionally, the two are largely independent. As we noted in point (5) above, institutions involved with water have traditionally focused on the hydrologic aspects, but this unfortunately gives them a narrow view. We emphasize again that institutions need to evolve a holistic approach to address issues of unequal development that leads to unequal sharing of resources and benefits. In many cases this requires a complete rethink of how departments of water resources, agriculture, mining, and health can be restructured to avoid the compartmentalized, independent institutions of the past that have proved so inadequate to confront the issues of water, food and livelihoods.

In many of the river basins studied in the BFPs, a serious problem is the underdevelopment of land and water resources. The underdevelopment is indicated by low water productivity, which is widespread in rainfed cropping systems. Low water productivity of rainfed systems is a general feature in the African basins, with only the Egyptian Nile demonstrating high productivity. While intensively managed areas within the Ganges, Yellow and Mekong basins are very highly productive, even these basins contain substantial areas with low productivity. Lack of development is related to many factors that can be summarized collectively as a lack of coherence within farming systems, in which lack of access to resources, to finance, or markets prevent farmers developing land to its potential productivity. In the poorest areas, we attribute lack of development to water-related hazards such as drought, floods, or disease, which have a known negative impact on the investments that are essential to escape poverty.

Our analysis above of conditions in basins shows the need for a detailed synthesis across all the BFP basins of water availability, water productivity, institutions that underpin how people use water and food systems, and the specific consequences of these factors to livelihoods and poverty. The papers herein address these issues and show, in response to the questions originally posed, that there are problems of water and land resource scarcity, especially in basins characterized by transitional economies. They also show that a more widespread condition is low water productivity, particularly of green water. A general observation, explored in more detail in the basin reports, is that while serious problems exist at national or sub-national scale, at a global scale, the capacity exists to meet, in theory at least, future global demand for water and food. The basin reports indicate the problems of exploiting this capacity in a sustainable and equitable manner. They also point to the overriding need for institutions that will balance the demands of different groups of people within basins, in addition to balancing the pressures for development and environmental protection within the environment it uses.

Notes

1. Basin teams produced reports for the Andes, Indus-Ganges, Limpopo, Karkheh, Mekong, Niger, Nile, São Francisco, Volta and Yellow river basins. All except the São Francisco are published in the September, 2010 issue of *Water International*: Ahmad and Giordano (2010), Awulachew *et al.* (2010), Kirby *et al.* (2010), Lemoalle and de Condappa (2010), Mulligan *et al.* (2010), Ogilvie *et al.* (2010), Ringler *et al.* (2010), Sharma *et al.* (2010), and Sullivan and Sibanda (2010).
2. The issue here is that valuation of different water uses cannot proceed in a vacuum about the per-unit consumed. Water resources are rarely valued, but their products generally are.

References

Ahmad, M.D. and Giordano, M., 2010. The Karkheh River basin: the food basket of Iran under pressure. *Water International*, 35 (5), 522–544.

Awulachew, A., *et al.*, 2010. The Nile Basin: tapping the unmet agricultural potential of Nile waters. *Water International*, 35 (5), 623–654.

Bruinsma, J., 2009. By how much do land, water and crop yields need to increase by 2050? FAO Expert meeting on how to feed the world in 2050, 24–26 June, 2009 Rome [online]. Available from: ftp://ftp.fao.org/docrep/fao/012/ak971e/ak971e00.pdf [Accessed 13 November 2010].

Byerlee, D., de Janvry, A., and Sadoulet, E., 2009. Agriculture for development: toward a new paradigm. *Annual Review of Resource Economics*, 1 (1), 15–31.

Cai, X., *et al.*, 2011. Producing more food with less water in a changing world: water productivity assessment in ten major river basins. *Water International*, 36 (1), 42–62.

Cohen, J., 1989. Deliberative democracy and democratic legitimacy. *In*: A. Hamlin and P. Pettit, eds. *The good polity*. Oxford: Blackwell, 17–34.

Dore, J., 2007. Multi-stakeholder platforms (MSPS): unfulfilled potential. *In:* L. Lebel, *et al.*, eds. *Democratizing water governance in the Mekong region*. Chiang Mai: Mekong Press, 197–226.

de Fraiture, C., *et al.*, 2001. PODIUM: Projecting water supply and demand for food production in 2025. *Physics and Chemistry of the Earth, Part B: Hydrology, Oceans and Atmosphere*, 26 (11–12), 869–876.

EIA, 2010. *International energy outlook 2010* [online]. Washington, DC: US Energy Information Administration (EIA). Available from: http://www.eia.doe.gov/oiaf/ieo/index.html [Accessed 22 September 2010].

Falkenmark, M., Rockström, J., and Karlberg, L., 2009. Present and future requirements for feeding humanity. *Food Security*, 1 (1), 59–69.

Fisher, M. and Cook, S., 2010. Introduction. *Water International*, 35 (5), 465–471.

Hoff, H. and Rockström, J., 2009. Green and blue water in the global water system – a model synthesis [online]. *Global Water News*, 8, 10–12. Available from http://www.gwsp.org/fileadmin/downloads/GWSP_Newsletter_no8_web.pdf [Accessed 22 September 2010].

IPCC, 2007. *Climate change 2007: impacts, adaptation, and vulnerability*. *In*: M.L. Parry, *et al.*, eds. Contribution of Working Group II to the Fourth Assessment Report of the Intergovernmental Panel on Climate Change (IPCC). Cambridge: Cambridge University Press.

Kemp-Benedict, E., *et al.*, 2011. Connections between poverty, water, and agriculture: evidence from ten river basins. *Water International*, 36 (1), 125–140.

Kirby, M., *et al.*, 2010. The Mekong: a diverse basin facing the tensions of development. *Water International*, 35 (5), 573–593.

Lawrence, P., Meigh, J., and Sullivan, C., 2002. *The water poverty index: an international comparison* [online]. Keele Economic Research Papers. Available from: http://citeseerx.ist.psu.edu/viewdoc/download?doi=10.1.1.13.2349&rep=rep1&type=pdf [Accessed 22 September 2010].

Lemoalle, J. and de Condappa, D., 2010. Farming systems and food production in the Volta Basin. *Water International*, 35 (5), 655–680.

Molden, D., ed., 2007. *Water for food, water for life: a comprehensive assessment of water management in agriculture*. London: Earthscan and Colombo: International Water Management Institute (IWMI).

Molle, F. and Mollinga, P., 2003. Water poverty indicators: conceptual problems and policy issues. *Water Policy*, 5 (5), 529–544.

Molle, F., Foran, T., and Käkönen, M., eds, 2009. *Contested waterscapes in the Mekong Region: hydropower, livelihoods and governance*. London: Earthscan.

Mulligan, M., *et al.*, 2010. The Andes basins: biophysical and developmental diversity in a climate of change. *Water International*, 35 (5), 472–492.

Mulligan, M., *et al.*, 2011.Water availability and use across the Challenge Program on Water and Food (CPWF) basins. *Water International*, 36 (1), 17–41.

NWC, 2009. *Australian water markets report 2008–2009*. Canberra: National Water Commission (NWC).

Ogilvie, A., *et al.*, 2010. Water agriculture and poverty in the Niger River basin. *Water International*, 35 (5), 594–622.

Pearce, F., 2007. *When the rivers run dry: what happens when our water runs out?* Toronto: Key Porter Books.

Ringler, C., *et al.*, 2010. Yellow River basin: living with scarcity. *Water International*, 35 (5), 681–701.

Rosegrant, M., Cai, X., and Cline, S., 2002. *World water and food to 2025. Dealing with scarcity*. Washington, DC: International Food Policy Research Institute.

Seckler, D., *et al.*, 1998. World water demand and supply, 1990–2025: scenarios and issues. International Water Management Institute Report 19. Colombo: IWMI.

Sharma, B., *et al.*, 2010. The Indus and the Ganges: river basins under extreme pressure. *Water International*, 35 (5), 493–521.

Sullivan, A. and Sibanda, M.L., 2010. Vulnerable populations, unreliable water and low water productivity: a role for institutions in the Limpopo Basin. *Water International*, 35 (5), 545–572.

UN, 2009. *World population prospects: the 2008 revision* [online]. Population Newsletter 87, June 2009. New York: Population Division of the Department of Economic and Social Affairs of the United Nations (UN) Secretariat. Available from: http://www.un.org/esa/population/publications/popnews/Newsltr_87.pdf [Accessed 22 September 2010].

UN, 2010. *World urbanization prospects: the 2009 revision* [online]. New York: Population Division of the Department of Economic and Social Affairs of the United Nations Secretariat. Available from: http://esa.un.org/unpd/wup/Documents/WUP2009_Highlights_Final.pdf [Accessed 22 September 2010].

WRI, 2003. *Water Resources eAtlas. Watersheds of the world: global maps, 14. Degree of river fragmentation and flow regulation* [online]. World Resources Institute (WRI), World Conservation Union (IUCN), IWMI and Ramsar Convention Bureau. Available from: http://earthtrends.wri.org/pdf_library/maps/watersheds/gm14.pdf and http://www.wri.org/publication/watersheds-of-the-world [Accessed 22 September 2010].

Water availability and use across the Challenge Program on Water and Food (CPWF) basins

Mark Mulligan, L.L. Saenz Cruz, J. Pena-Arancibia, B. Pandey, Gil Mahé and Myles Fisher

This paper analyses water availability and use *within and between* the Challenge Program on Water and Food basins. It describes the main features of water demand and supply in the basins and indicates where there are deficits and opportunities for development of water resources. A typology of basin water resources status uses a range of global spatial datasets. The main outcomes of basin activities on water availability are identified. Interbasin assessment of water availability is very challenging for such very large basins, due in large part to difficulties in collecting and integrating local data sets.

The context of the Challenge Program on Water and Food basins

The aim of this paper is to provide consistent analysis of water availability and use *within and between* the 10 basins of the Challenge Program on Water and Food (CPWF) Basin Focal Projects (BFP). We begin by providing the context for the basins using the best available global datasets, re-sampled to a common 1-km pixel resolution, cut to the boundaries as defined by the CPWF and summarized as mean and range of each basin. While some of these data are at coarse scale (the coarsest being the actual evapotranspiration data at 45-km resolution), these are also large basins. There is therefore no other way to provide consistent measures between the basins than to use global datasets that use a common methodology. The 10 BFP basins are a collection of small basins in the Andes, and the São Francisco in South America; the Volta, Niger, Limpopo, and Nile in Africa; the Karkheh, Indus-Ganges, Mekong and Yellow in Asia. Together they account for some 9,686,000 km^2, representing 6.5% of the Earth's land surface. The countries that occupy these basins are given in Table 4. Data for the Indus were not available so the analysis for the Indus-Ganges is confined to the Ganges. We indicate in the text where information refers to the Ganges alone and to the combined Indus-Ganges.

The largest of the BFP basins, as defined by them, is the Nile (2.59 Mkm^2), followed by the Andes (2.32 Mkm^2), Niger (1.09 Mkm^2), Yellow (0.86 Mkm^2), Ganges (0.81 Mkm^2), São Francisco (0.61 Mkm^2), Mekong (0.54 Mkm^2), Volta (0.36 Mkm^2), Limpopo

(0.36 Mkm^2) and the smallest is the Karkheh (0.11 Mkm^2). The basins cover a wide range of geographic settings on three continents with many covering more than 10 degrees of latitude so that a single basin may have a number of climate zones.

The Andes basins are highly mountainous, cool and very wet in the north, and colder and drier in the south. Permanent glaciers in the Andes are generally small and confined to single mountain peaks. Conflicts between land uses and their impact on downstream water supply are common. The São Francisco is highly variable from semi-arid to wet and humid, with low seasonality of temperature. It has important hydropower development and irrigation for agriculture.

The Niger rises in the mountainous areas of Guinea and flows northeast to the inner delta in Mali where seepage and evaporation reduce flows considerably. The basin is warm and wet and largely confined to lowlands. The Indus-Ganges is very mountainous and wet in the north, with large glacial zones, but low, flat and dry in the populous south. It includes part or all of the territories of Pakistan, India, Nepal, Bhutan and Bangladesh. The plains are made up of highly productive soil and are thus intensely cultivated by hundreds of millions of rural people across the region. There are distinct (monsoonal) wet and dry seasons with flooding in the wet seasons but water shortages in the dry season leading to heavy use of irrigation.

The Limpopo is a lowland, dry and relatively flat basin with high temperatures and a short and intense rainfall season of unreliable rainfall, leading to frequent droughts. Flows are thus intermittent and crop production insecure. Soil erosion and degradation are issues as is pollution from upstream mines and urban areas. The Nile is the longest river in the world and carries water from the humid and mountainous south to the dry and flat north. It is considered a "lifeline" providing most of Egypt's water, soil fertility and food security.

The Yellow (Huang) River is the second longest river in China, carries high sediment loads from the loess plateau in the northwest and is prone to frequent flooding in its lower course. The basin is highly seasonal with cold winters and very low annual rainfall supporting intensive agriculture and significant populations through exploitation of groundwater reserves. River water is highly contaminated in parts because of industrial effluent, with water in some parts of the river unfit even for agricultural use. The Karkheh is mountainous in parts and very dry with strong seasonality for temperature. It has areas of significant overgrazing. Poor farming practices and heavy use of agricultural inputs is leading to land degradation and most of the available water resources are utilized.

The trans-national Mekong Basin is highly mountainous and snowfed in the north where the basin is narrow and the water clear. The lower Mekong in the south the basin is wide and lowland systems, which supply significant sediment loads, dominate. Flow varies strongly on a seasonal basis and human populations are highly dependent on important fisheries. The river is increasingly being dammed for hydropower. The Volta is hot and highly seasonal, especially for rainfall, and has a strong N–S gradient in rainfall and population. The Akosombo Dam on the Volta holds back the largest manmade lake in the world storing some 150 km^3 and producing more than 80% of Ghana's electricity (Barry *et al.* 2005).

Population and water use

Population density varies significantly within and between the basins. Mean densities for 2010 (CIESIN/CIAT 2005) vary from 486 persons/km^2 (Ganges) down to 26 persons/km^2 (São Francisco) giving a total population loading varying from 395 million (Ganges) to nine million (Karkheh) (Table 1). The Karkheh's nine million out of 72 million for all

Table 1. Human and agricultural characteristics of the CPWF basins.

						GlobCover						
	Population density	Total population	Urban area	Maximum population density	Total crops	Dry crops	Grassland	Irrigated crops	FAO Irrigation	Mean livestock	Mean barnfed	Mean GDP (1990)
	/km²	*millions*	*%*	*persons/km²*	*%*	*%*	*%*	*%*	*%*	*head/km²*	*head/km²*	*MUSD/yr*
Andes	33	80	0.08	9,961	2	2.1	3.8	0.0	3.2	20	102	1.4
Ganges	486	400	0.6	43,464	67	15.4	5.3	51.6	30.5	203	144	4.2
Karkheh	82	9	0.1	267	5	4.5	0.01	0.0	16.1	92	369	4.0
Limpopo	39	14	0.7	13,273	0	0.1	17.7	0.0	6.0	19	118	2.8
Mekong	109	60	0.01	156,208	37	28.2	1.1	9.2	9.1	17	277	2.0
Niger	70	77	0.07	7,327	10	10.2	9.5	0.0	3.1	58	110	0.4
Nile	66	170	0.2	105,363	3	1.6	5.6	1.1	21.9	65	48	0.6
São Francisco	26	16	0.09	10,118	10	10.0	0.0	0.0	0.5	33	90	2.2
Volta	59	21	0.04	1,994	12	11.9	1.7	0.0	0.6	64	104	0.3
Yellow River	174	150	1.0	5,312	33	28.0	12.1	5.5	22.4	121	690	1.6

of Iran is much greater than the 3.7 million for the basin defined by Marjanizadeha *et al.* (2010) but the BFP definition of the basin includes parts of Iraq and Kuwait, which changes the population loading. The BFP basins (excluding the Indus) together are home to 995.4 million people, or 14.5% of the current global population. Built land varies between basins with some 0.96% of the Yellow River being classified as urban land cover (GlobCover 2008) compared with only 0.01 % for the Mekong. The *intensity* of the urban condition also varies between the basins with maximum population density of 156,000 persons/km^2 for the Mekong, 105,000 persons/km^2 for the Nile, 43,000 persons/km^2 for the Ganges, but only 266 persons/km^2 for the Karkheh.

There are few spatial datasets on gross domestic product (GDP) or poverty and as yet no accessible global poverty or development map at sufficient spatial resolution for this analysis. While the individual BFPs produced much more sophisticated analyses of poverty and water poverty than presented here, those analyses are not readily comparable between basins because of differences in scope and method. As an indicator of near-current basin-level economic development we use the special report on emissions scenarios (SRES) B2 GDP estimate (CIESIN 2002) for 1990 on a unit area basis (that is, GDP/km^2). The basin mean values of this index hide significant variability within basins between urban centres (with values in the millions of USD/km^2 over small footprints) and the rural hinterland (with values close to zero). The basin mean represents the balance between this urban wealth and rural poverty and is clearly affected by population dispersion since areas with dispersed populations will have a higher mean GDP than those in which populations are highly concentrated with the remaining depopulated rural areas contributing very little. Basin mean GDP is therefore highest for the Ganges at 4.21 MUSD/km^2/yr (range 0–82 MUSD/km^2/yr), followed by the Karkheh (4.00, range 2–16 MUSD/km^2/yr). The lowest basin GDP is for the Volta at 0.34 MUSD/km^2/yr (range 0–6 MUSD/km^2/yr). Whilst the Andes shows a low overall GDP of 1.3 MUSD/ km^2/yr (because of low rural GDP especially in the near-deserted southern deserts), the range varies from 0 MUSD/km^2/yr to 182 MUSD/km^2/yr.

Agriculture and land use varies significantly between basins with total cropland according to GlobCover (2008) varying from 67% of the basin area (Ganges) down to near zero (Limpopo) and grasslands varying from 18% of the basin area (Limpopo) to close to zero (São Francisco). Though these figures are very likely subject to localized misclassification – especially given the range of vegetation characteristics for, for example, grasslands between humid and dry areas, the relative trends between basins should be correct. Irrigation data are difficult to validate at these spatial scales. Here we use the GlobCover (GlobCover 2008) irrigated cropland land cover and the Food and Agriculture Organization (FAO) irrigated area analysis (Siebert *et al.* 2007) as independent indicators of the degree of irrigation in the basins. Differences between the two irrigation assessments are sometimes significant, especially for the Ganges, Karkheh, Nile and Yellow River. The basin with the greatest area apparently under irrigation is the Ganges at 30% (FAO) and 52% (GlobCover), followed by the Yellow River (22% FAO, 5% GlobCover) and the Nile (22% FAO, 1% GlobCover). The magnitude of difference warrants caution in the use of irrigation estimates like these.

A number of the catchments have significant livestock densities and distributions. Using the Wint and Robinson (2007) analysis extracted for these catchments we find that total (wildland) livestock headcount varies from 203 head/km^2 (Ganges) through 121 head/km^2 (Yellow River), down to 17 head/km^2 (Mekong). Densities of barnfed animals (poultry and pigs) are much greater than (wildland) livestock for all basins except the Nile and the Ganges, with the greatest densities recorded for the Yellow River (690 head/km^2) and the Karkheh (369 head/km^2). There is no global database for aquaculture, but the individual BFPs indicate basins where aquaculture is important, notably the Mekong and the Nile delta.

Climatic context

Climate varies within as well as between the basins, as a function of latitude, altitude and continentality (Table 2). Many of the basins cover significant climate gradients with changing latitude or altitude, for example from hyper-humid to desert environments in the Andes. Thus, in setting the context for the basins, we provide mean and variance information for each. Like many such datasets, regional climate data are at best an estimate, and at worst a "guesstimate", of the distribution of climates. Where possible, we use multiple datasets to capture the diversity of opinions on these variables. We demonstrate how little we know about spatial variability at the mega-basin scale that we cannot even say with certainty what the rainfall inputs are to many of the basins. Without a good knowledge of water inputs, calculation of water balances and water productivity is difficult and, whilst we can make water balances "close" in relation to observed stream flows, the uncertainty associated with measurements of stream flow for mega basins can be very high. Moreover, water budgets can close because more than one variable (rainfall, evapo-transpiration, stream flow and change in basin storage) are in error. Closure is easier for much smaller basins where the density of climate stations can be higher and the measurement of stream flow is more accurate.

Precipitation

We used independent datasets of rainfall to analyse the distribution of precipitation in the basins. The WorldClim (WC) dataset (Hijmans *et al.* 2005) is interpolated from rainfall stations (for 1950–2000) at 1-km spatial resolution using a thin plate spline with elevation as a co-variable. The tropical rainfall measuring mission (TRMM) satellite rainfall climatology was aggregated from 50,000 satellite swaths (from 1998–2006) and re-sampled from 5-km to 1-km spatial resolution by Mulligan (2006) from the National Aeronautical and Space Administration TRMM 2b31 rainfall product. The means of rainfall for whole catchments are broadly similar between the two datasets. They indicate that the wettest catchment is the Mekong (1710 mm/yr [WC] and 1610 mm/yr [TRMM]), followed by the Ganges, the Niger and the Volta. The driest catchment is the Karkheh (350 mm/yr [WC] and 540 mm/yr [TRMM]). The most spatially variable catchment for rainfall is the Andes with values from 0 to 7700 mm/yr (WC) and 0–19,350 mm/yr (TRMM). The figure of 19,350 mm/yr, which occurs in the Choco of the Colombian Pacific coast is undoubtedly an overestimate. It occurs for a single 1-km pixel with the next highest value 9000 mm/yr and most other values less than 5000 mm/yr. Other areas of important spatial variability are the Ganges (basin mean of 4140 mm/yr for WC and 8820 mm/yr for TRMM), the Niger (basin mean of 3450 mm/yr for WC and 6490 mm/yr for TRMM) and the Nile (basin mean of 2280 mm/yr for WC and 7000 mm/yr for TRMM).

The WorldClim dataset is an interpolation from sparsely distributed rainfall stations (with the data for more than 99% of pixels being interpolated values) while the TRMM gives a spatially detailed distribution based on space-borne rainfall radar. But there are only 10 years with observations of each pixel about each 10 days. Although there are significant differences between the two climatologies, it is not clear which of them is closer to the

Table 2. Climate characteristics of the basins.

	Annual total rainfall (WorldClim)	Annual total rainfall (TRMM)	Catchment max–min rainfall (WorldClim)	Catchment max–min rainfall (TRMM)	Mean temperature (WorldClim)	Temperature seasonality (WorldClim)	Precipitation seasonality (WorldClim)
	mm/yr	*mm/yr*	*mm/yr*	*mm/yr*	°*C*	*SD*	*CoV %*
Andes	784	832	7,697	19,349	12	2.3	78
Ganges	1,073	1,025	4,137	8,816	21	5.5	125
Karkheh	348	541	449	3,786	16	8.6	89
Limpopo	547	558	1,430	2,868	19	3.7	84
Mekong	1,713	1,606	2,797	8,925	24	2.0	86
Niger	1,017	964	3,453	6,487	25	2.1	108
Nile	618	716	2,279	6,997	24	2.6	103
São Francisco	975	823	1,608	3,418	21	1.4	84
Volta	973	869	1,298	4,191	26	1.8	96
Yellow River	438	n/a [a]	976	3,490	7	9.8	93

Note: [a] The TRMM data only covers the southern half of the basin.

actual, unmeasured precipitation for a given pixel. It is important also to note that for some of these basins precipitation will be in excess of rainfall because of significant inputs from snow (Ganges), fog (Andes) or other forms of precipitation that are not measured by standard rain gauges.

Temperature

Temperature is hydrologically important as a major driver of evapotranspiration but is also important as an independent control of crop growth. The catchment means of annual mean temperature derived from WorldClim is highest for the Volta (26°C), followed by the Niger, Nile, Mekong and São Francisco. The Andes (12°C) and Yellow River (7°C) have the lowest annual mean temperatures reflecting altitudinal (Andes) or latitudinal (Yellow River) effects. Within-catchment spatial variability in temperature is highest for the Ganges and the Andes which both have some permanently ice-capped mountain peaks with annual mean temperature of −17°C, while some of the desert regions of the Andes or lowlands of the Ganges have annual mean temperatures of 31°C, giving within-basin ranges of 40°C (Andes) and 46°C (Ganges). The Nile and the Karkheh are also highly spatially variable with respect to mean annual temperature.

Variability and seasonality

Inter-annual and long-term variability of rainfall are an important component of water availability in some of the basins, but there are few datasets that would allow us to compare them across basins (Mulligan *et al.* 2011). We defer here to the BFP studies to indicate in which basins this variability is important (see the later section on water resources in basins). Mean annual values are important but seasonality is also a strong driver of crop growth and water resources in many basins. Seasonality can be measured in a number of ways. Rainfall seasonality can be expressed as the coefficient of variation (CoV, which is the standard deviation divided by mean expressed as a percentage) for monthly rainfall totals. It is particularly high for the WorldClim data for the Ganges (125%), the Niger (108%) and the Nile (103%) and least for the Andes (78%). There is significant variation in the strength of seasonality within basins such that the Andes has a CoV for rainfall varying from 0% to 238 % and the Niger from 0% to 235%. The Karkheh is least spatially variable in rainfall seasonality (varying from 74% to 125%). Areas that have significant dry seasons can, of course, receive water from neighbouring upstream areas that are less seasonal, but only where the less seasonal areas lie upstream, as with the Himalayas for the Indus-Ganges, and in the Nile and the Andes basins, for example.

Temperature seasonality is important in setting the limits to crop-growing seasons and interacting with rainfall seasonality to determine the seasonal progression of the water balance and productivity. Temperature seasonality can be expressed as the standard deviation (SD) of monthly mean temperatures and is highest for the higher latitude Yellow River (SD = 98) and Karkheh (SD = 86) and least for the tropical São Francisco Basin (SD = 14). Spatial variability in temperature seasonality (not shown) is greatest for those basins that cross a range of latitudes with the Nile at SD = 67, the Andes at SD = 63 and the Yellow River at SD = 6.

Landscape

In addition to their climatic differences, the CPWF basins fall in very different landscape settings. These settings determine the climate but also the landscape conditions for

Table 3. Terrain characteristics of the CPWF basins.

	Elevation	Slope gradient	Mountainous
	masl	*degrees*	*%*
Andes	2,453	5.7	83
Ganges	859	2.7	23
Karkheh	1,335	4.7	62
Limpopo	796	0.9	7
Mekong	364	2	28
Niger	359	0.7	5
Nile	718	0.9	10
São Francisco	653	1.1	8
Volta	254	0.5	1
Yellow River	1,773	2.2	41

crop growth as well as the potential for geomorphic and hydrological hazards affecting agriculture, infrastructure and human life.

Terrain

According to the shuttle radar topographical mission (SRTM) digital elevation model (DEM) (Farr and Kobrick 2000), re-sampled to 1-km resolution (Table 3), the mean basin elevation is greatest for the Andes Basin (at 2450 masl) followed by the Yellow River (1770 masl), then the Karkheh (1340 masl) and the Ganges (860 masl). The lowest basin is the Mekong at 360 masl. Slope gradient is greatest for the Andes (5.7° at 1-km resolution), the Karkheh (4.7°), the Ganges (2.7°) and the Yellow River (2.2°) and is least for the Niger (0.7°). Defining mountains using the SRTM terrain data and according to the definition of Kapos (Kapos *et al.* 2000), 83% of the Andes is considered mountainous, 62% of the Karkheh, 40% of the Yellow River and 28% of the Mekong. The least mountainous (most lowland) basin is the Volta at 0.9%. It is almost counterintuitive that the Karkheh is the second most mountainous basin after the Andes, but it is caused by the rugged Zagros Mountains where the river rises. This is important in understanding its water resource characteristics. In contrast, while the Ganges includes the Himalayas in its headwaters, they make up a only a small proportion of its otherwise lowland-dominated basin.

Water resources infrastructure

Rivers and lakes

The water resources of a basin are not solely the outcome of rainfall minus evapotranspiration, but also include water storage, which may be infrastructure such as dams as well as natural stores such as soils, lakes, snow and ice, and groundwater. There are, unfortunately, no global datasets available for soil water storage, aquifers or groundwater resources, so we will examine these on the basis of published analyses for the basins. There are global datasets for dams and water bodies, however. Some of the CPWF basins also have important open-water reserves. The surface area of lakes, according to the water bodies dataset of – which includes natural and artificial lakes (Lehner and Döll 2004), is greatest for the Volta Basin (0.9% of the basin surface area), followed by the Nile Basin (0.58%), and the São Francisco (0.52%). The Karkheh (0.06%), Limpopo (0.05%) and Niger (0.08%) have the least areas of open water. Depths of the lakes may differ significantly, and there is no global dataset available to indicate capacity, but nevertheless surface areas do give an indication of the size of the resource.

Snow and ice

Most of the basins have no permanent snow and ice cover (GlobCover 2008) but 1.5% of the surface area of the Ganges is permanent snow and ice, 0.49% of the Andes and 0.21% of the Yellow River. Seasonally, snow and ice cover may be greater than this but seasonal variation represents a delayed flux of precipitation through the basin rather than a significant permanent store of water resource and we do not consider it here.

Dams

There are no global databases accessible on the volume of dam storage, which would allow us to compare storage volumes between basins. We therefore use the number of dams as an indicator of (realized) demand for surface water storage. According to the tropical database of dams, digitized from satellite imagery for the tropics and beyond (Mulligan *et al.* 2009), the CPWF basins incorporate at least 2212 large dams with the Indus-Ganges having both the most large dams (785 with 84 in the Indus and 701 in the Ganges) and the greatest density of dams per unit area (1037/Mkm^2), followed by the Limpopo (418 dams, 532/Mkm^2), the Mekong (344 dams, 419/Mkm^2), the Andes (174 dams, 219/Mkm^2), the São Francisco (165 dams, 199/Mkm^2) and the Yellow River (125 dams, 184/Mkm^2). The River Niger has few dams, but a series of large dams are planned for the coming years in the upper watershed. These will strongly impact the river regime and especially the inland delta (Lienou *et al.* 2010).

In all basins' dams make important contributions to hydropower and water resources management, especially for irrigation. The current dams, and a number of proposed dams, in some cases present important impediments to benefit sharing, trans-national water management, and the sustainability of fisheries, for example in the Mekong.

Groundwater

There are no global databases for accessible groundwater, although Struckmeier *et al.* (2010) have produced an analysis of global groundwater resources. Each CPWF basin is very different in its groundwater characteristics, which we summarize below.

Andes. Aquifers in the eastern and western slopes of the Andes are mostly formed by alluvial fans occurring in the narrow valleys or of lacustrine deposits (Reboucas 1999). Recharge in the western Andes averages <100 mm/yr (Struckmeier *et al.* 2006). Lowland rainfall is markedly seasonal, so that recharge in the eastern Andean foothills, together with long groundwater residence times, may provide an important source of water during the dry season (Peña-Arancibia *et al.* 2010). Groundwater is used extensively in the arid coastal zone for domestic supply and irrigation, providing as much as 80% of the total water demand in Lima and other Peruvian cities (Reboucas 1999) The eastern Andean foothills have important aquifers also in alluvial fans, with thicknesses of more than 500 m. Recharge occurs from rainfall and floods, and is >100 mm/yr and exceeds 300 mm/yr at higher elevations (Struckmeier *et al.* 2006). The aquifers in this area are mainly used for urban supply (Cochabamba, Bolivia and Santiago and Copiapo, Chile) and for irrigation (around Cordoba, Mendoza and Tucuman in Argentina) (Reboucas 1999).

São Francisco. A large groundwater basin underlies the south of the São Francisco Basin. Sandstone deposits with thickness of >8000 m have a groundwater store estimated at 400 km^3, sufficient to supply irrigation, domestic and industrial requirements (Reboucas 1999). Recharge is 100–300 mm/yr in parts of the basin (Struckmeier *et al.* 2006). Rainfall in the basin is seasonal and groundwater discharge to rivers may sustain river flows and diversion for irrigation particularly in the semi-arid northern parts of the basin (Peña-Arancibia *et al.* 2010).

Nile. Two large groundwater basins (the Nubian Sandstone aquifer system and the Upper Nile Basin system), together with important alluvial deposits along the river flood plain, are used for domestic water supply and irrigation (Struckmeier *et al.* 2006). The Nubian Sandstone is a large trans-boundary aquifer shared by Libya, Sudan and Chad. It is up to 500 m thick, close to the surface, and a total storage of about 500 km^3 (Abu-Zeid 1995). The Upper Nile aquifer is in the southern part of the basin. The mountains of the Ethiopian Plateau and the equatorial zone provide recharge >100 mm/yr, which then discharges downstream in the Nile's lower reaches (Abu-Zeid 1995, Struckmeier *et al.* 2006). Groundwater helps maintain the river's flows to the arid north (Hassan *et al.* 2004).

Limpopo. The most important aquifers in the Limpopo are located in shallow alluvial deposits along the main river stem and fractured rocks in the tributaries (Busari 2008). Most productive activities use surface water, but in the northwestern areas irrigation is by groundwater pumped from the alluvial deposits (Busari 2008). The headwaters in the east have a tropical climate with high rainfall so that recharge is 100–300 mm/yr (Struckmeier *et al.* 2006). Recharge in the semi-arid parts of the basin is <50 mm/yr.

Niger. Fractured rocks and shallow aquifers with limited groundwater characterize the upper Niger Basin. Recharge from the headwaters is generally low (Fontes *et al.* 1991). Recharge increases in the lower reaches and alluvial aquifers such as the l'Air crystalline aquifer are an important source of groundwater. Recharge rates are 100–300 mm/yr in the alluvial deposits alongside the main river channel. In the inland delta, the Continental Terminal aquifer, with a thickness >100 m, is widely used, particularly in Niger (Andersen *et al.* 2005). The Continental Shale Band aquifer borders the Niger River just north of Benin but also runs under the semi-arid areas of Mali and Niger (Andersen *et al.* 2005). Recharge is >300 mm/yr in some areas (Struckmeier *et al.* 2006). There are also large, thick sedimentary and alluvial deposits in the lower Niger, which provide good-quality groundwater compared with surface waters (Andersen *et al.* 2005). The water level in the crystalline aquifers has fallen markedly since the drought in 1970 (Mahé 2009), leading to dry wells. On the other hand, the level of near-surface aquifers in Niger and northern Mali has been increasing, supplying water longer into the dry season, despite reduced rainfall over the period (Leduc *et al.* 2001, Mahé and Paturel 2009).

Volta. The Voltaian system and the Basement Complex dominate the hydrogeology of the Volta Basin. The rocks in these systems are primarily impermeable, but there is groundwater in fractured geologies (Barry *et al.* 2005), which is mainly used for domestic supply. Weathered aquifers in this area are up to 100 m thick in the forested southwest but are thinner in the arid northeast (Kortatsi 1994). Recharge, mainly from rainfall rather than stream flow, is >100 mm/yr in the southwest, decreasing towards the more arid interior (Struckmeier *et al.* 2006).

Indus-Ganges. Aquifers in the Indus-Ganges basins are dominated by large and deep alluvia. The Indus-Ganges-Bramaputhra basins are one of the largest groundwater reservoirs in the world, being over of 700 m thick (Mahajan 2008). The Indus-Ganges is one of the world's greatest concentrations of groundwater use to grow vast areas of irrigated crops, and for urban and rural water supply, as well as industrial use. Recharge is generally higher in the Ganges at >100 mm/yr, whilst in the Indus it is generally <100 mm/yr (Struckmeier *et al*. 2006).

Mekong. There are shallow alluvial aquifers in the deep valleys of the upper Mekong. Recharge is generally >100 mm/yr. There are also large groundwater aquifers in the Mekong Delta, with recharge up to 300 mm/yr, but they are not much used for water supply.

Yellow River. Two separate groundwater units are recognized in the Yellow River basin: mountainous areas and plains. Han (2003) estimated that groundwater resources of the Yellow River in 1997 were 40.6 km^3/yr, two thirds of which originates in the mountainous areas unit. Recharge was estimated at 40.4 km^3/yr. Groundwater contributes about 30% of total water use in the basin, 14.0 km^3/yr, which represents some 76% of the total exploitable groundwater resource of 18.2 km^3/yr at the basin level so the resource is being used close to capacity. Locally, there can be significant overexploitation.

Water access

There are no consistent data between basins on access to water via water treatment plants, reservoirs and distributed pipeline networks either for irrigation or for domestic use. Most analyses therefore only focus on water *availability* whereas a better measure of the influence of water on productivity and poverty is a person's actual physical and economic *access* to water, irrespective of its hydrological availability in the landscape. We can get an indication of the difficulty of accessing water in the basins by comparing water costs for typical domestic use as a proportion of per capita GDP across basins. Table 4 indicates the countries that make up each basin and the mean International Water Association (IWA) data on the cost of water for those years for which data are available 1995–2008 for each country (normalized to 2007 US dollars [USD]). The costs, of course, vary with the prevailing economic conditions of the countries concerned, so we also calculated cost as a proportion of per capita GDP at purchasing power parity (PPP) (World Bank 2010). Total cost is calculated assuming a per capita daily use of 200L (the upper limit of Gleick [1999]). There are significant differences in cost and percentage of per capita GDP between countries, even countries in the same basin. Clearly, many factors other than hydrological availability of water affect price since prices can be higher in countries with high water availability (for example, São Francisco, Brazil) compared with dry countries (for example, Chile). Access to water is thus defined in part by the water availability analysed here, but this is superimposed on complex infrastructural, socio-economic and political considerations.

Water supply, demand and use across the basins

Water supply (precipitation-evapotranspiration)

The *water balance* at a point is a measure of its water supply from incoming rainfall minus losses from outgoing evapo-transpiration (ET) (from vegetation, soil and free water surfaces). In some areas, there are important inputs from upstream runoff and losses to

Table 4. Estimates of water cost for the basins based on International Water Association (IWA) figures.

Basin	Countries		Water cost from IWA (www.iwahq.org)	Water cost
	In basin	*Costed*	*USD/m³ (2007)*	*% of per capita GDP (Purchasing Power Parity [PPP]) @200L/day use*
Andes	Argentina, Bolivia, Chile, Colombia, Ecuador, Peru	Chile	0.41	0.21
		Argentina	0.20	0.10
		Colombia	0.04	0.03
Indus-Ganges	Afghanistan, Bangladesh, Bhutan, China, India, Nepal, Pakistan	China	0.45	0.48
		India	0.12	0.27
		Nepal	1.39	8.78
Karkheh	Iran		n/a	
Limpopo	Botswana, Mozambique, South Africa, Zimbabwe	South Africa	0.36	0.26
		Zimbabwe	0.24	17.53
Mekong	Cambodia, China, Lao PDR, Myanmar, Thailand, Vietnam	China	0.45	0.48
		Thailand	0.11	0.10
Niger	Burkina Faso, Cameroon, Chad, Guinea, Mali, Niger, Nigeria	Burkina Faso	0.44	2.70
Nile	Burundi, Democratic Republic of Congo, Egypt, Eritrea, Ethiopia, Kenya, Rwanda, Sudan, Tanzania, Uganda	Ethiopia	0.11	1.72
		Uganda	0.57	3.42
São Francisco	Brazil	Brazil	0.86	0.60
Volta	Benin, Burkina Faso, Ghana, Togo	Burkina Faso	0.44	2.70
		Benin	0.28	1.35
Yellow River	China	China	0.45	0.48

downstream discharge, which must be considered. The balance between rainfall and ET represents the water available for infiltration, runoff and groundwater recharge and storage. The proportion of ET that leaves the stomata of food crops as transpiration is productive water (or green water). Where pastures are important to maintain livestock, then the transpiration of pastures also contributes to food production.

There are various measures of ET although few are available at global scales and at high spatial resolutions. Trabucco and Zomer (2009) provide a 1-km resolution grid for potential evapotranspiration (PET) based on the Hargreaves (1985) model as adapted by Droogers and Allen (2002). Using this dataset, the mean PET for the CPWF basins varies from 1980 mm/yr (Nile), 1940 mm/yr (Niger), through 1910 mm/yr (Volta). The catchments with the lowest PET are the Andes (1330 mm/yr) and the Yellow River (960 mm/yr). PET is a measure of atmospheric demand for water rather than the ability of the land surface to meet that demand, which is usually referred to as actual ET (AET). The FAO provides global crop reference ET at 10-arc-minute resolution (about 18 km) (Allen *et al.* 1998). Reference ET (RET) is defined as ET from a reference surface (usually short grass), which is not short of water. It is thus not affected by soil water, crop or phenological characteristics, only by climate and is closer to PET than AET. Observed patterns between catchments are similar to those of PET with the highest values estimated for the Nile (1700 mm/yr), Niger (1870 mm/yr) and Volta (1700 mm/yr) and the lowest for the Andes (1310 mm/yr) and Yellow River (960 mm/yr).

Although Mulligan and Burke (2005) provide data at 1-km resolution of AET estimates for the tropics, some of the CPWF catchments are not covered so we do not use them here. Ahn and Tateishi (1994) provide 30-arc-minute-resolution (about 45 km) AET and water balance data globally (Ahn and Tateishi 1994). These data indicate much lower values of AET than PET for the CPWF basins, as is to be expected. The greatest AET is for the Mekong (1050 mm/yr), followed by São Francisco (930 mm/yr), Volta (910 mm/yr), Niger (800 mm/yr). The basins with the lowest AET are Karkheh (290 mm/yr), Yellow River (460 mm/yr), Nile (610 mm/yr) and Andes (630 mm/yr). Spatial variability in AET is high within all catchments, often varying from 0 (for example, over bare rock) to more than 1000 mm/yr, with the exception of the Karkheh (380 mm/yr maximum) and Yellow River (770 mm/yr maximum).

Ahn and Tateishi (1994) also provide estimates of water balance (rainfall-AET). These show an annual water deficit (excess of AET over rainfall) for all catchments except the Mekong (340 mm/yr). The catchment with the most negative water balance is the Limpopo (−1170 mm/yr), followed by Karkheh (−850 mm/yr), Nile (−770 mm/yr) and Andes (−490 mm/yr). The least negative water balances are given by the Ganges (−40 mm/yr) and São Francisco (−70 mm/yr). Since we have summarized these data over entire watersheds, we would expect rainfall to be greater than AET by the volume which leaves the catchment as runoff over the long term (with any short term deviations from this budget "closure" resulting from changes in storage within the basin such as groundwater recharge). Since most basins are showing a negative balance with these data, we suggest that at the basin level either Ahn and Tateishi (1994) underestimate rainfall or they overestimate AET. Since the AET values are reasonable it is likely that the much more spatially variable rainfall is overestimated at the coarse resolution used for the basins.

Since rainfall is the dominant variable in determining water balance and is highly spatially variable and difficult to represent accurately over large areas, here we calculate water balance on the basis of other, higher resolution rainfall datasets for comparison (Table 5). Using the WorldClim rainfall data and the Ahn and Tateishi (1994) data of AET, we have the same patterns of water balance between catchments (the highest water balance for the Mekong, the lowest for the Limpopo), and the water balances are generally positive with only the Limpopo and the Yellow River negative. Using the TRMM data for rainfall, most basins remain with a positive water balance, the Yellow River becomes the most negative (−140 mm/yr) followed by São Francisco (−100 mm/yr) and Limpopo (−80 mm/yr). This analysis suggests that the higher resolution rainfall datasets from WC and TRMM provide more realistic water balance results than the coarse resolution rainfall datasets used by Ahn and Tateishi (1994) but also that there are still significant difficulties in calculating a simple water balance for basins of this scale.

To test the balances for each basin we provide the Global Runoff Data Centre (GRDC) estimates of long-term river discharge (GRDC 2010), using data for the gauging station closest to the basin outlet, and catchment sizes as calculated by the CPWF (Table 5). There is some doubt whether the GRDC discharge is for the basin as defined by the CPWF; for

Table 5. Various water balance estimates compared with estimated discharge for the basins.

Basin	Mean PET [a]	Mean ET REF [b]	Mean AET [c]	Water balance Mean	Water balance Mean (TRMM-AET)	Water balance Mean (wc-AET) (rank)	Long-term mean GRDC discharge (rank)	Difference in catchment area
	mm/yr	*mm/yr*	*mm/yr*	*mm/yr*	*mm/yr*	*mm/yr*	*km^3/yr (mm equivalent)*	*Mkm^2*
Andes	1,334	1,307	632	−490	203	152 [4]	n/a	n/a
Ganges	1,534	1,399	746	−39	279	327 [2]	351.7 (434)[2]	0.04
Karkheh	1,562	1,558	291	−852	250	58 [6]	24.4 (219)[3]	−0.05
Limpopo	1,691	1,496	640	−1,173	−82	−93 [10]	2.0 (5.4)[9]	−0.16
Mekong	1,597	1,413	1,049	337	555	662 [1]	418.2 (771)[1]	0.12
Niger	1,945	1,872	804	−413	160	212 [3]	159.6 (145)[4]	n/a
Nile	1,980	1,969	606	−768	111	12 [8]	39.5 (15)[8]	0.30
São Francisco	1,731	1,558	928	−66	−104	48 [7]	89.2 (144)[5]	0.01
Volta	1,915	1,700	910	−445	−42	63 [5]	34.3 (94)[6]	0.03
Yellow River	960	961	458	−266	n/a	−20 [9]	45.0 (51)[7]	−0.14

Notes: [a]Trabucco and Zomer (2009); [b]Allen *et al.* (1998); [c]Ahn and Tateishi (1994); [d]The TRMM data only covers the southern half of the basin. The discharge stations used for each basin were as follows: Ganges (Hardinge Bridge), Karkeh (Karun–Ahvaz), Limpopo (Beitbrug U/S), Mekong (Phnom Penh Chroui Changvar), Niger (Lokoja), Nile (El Ekhsase), São Francisco (Traipu), Volta (Senchi–Halcrow), Yellow (Huayuankou).

example in some basins there is no gauging station close to the basin outlet. Where the difference in area is negative the GRDC area is lower than the CPWF area. We also converted the data of long-term mean discharge values to mm equivalent based on the GRDC reported basin area. Where this mm equivalence is close to the estimated water balance figure, we can be more confident of the water balance and discharge data. The discharge values are generally closest to the WorldClim-AET-derived water balance, especially for basins in which the GRDC and CPWF areas are close. In general, the rank order of the water balances and GRDC discharge estimates are the same. The differences that exist between the water balance estimates and the discharge estimates indicate the errors in both and the difficulty of closing, even long-term, water balances for mega-basins like these.

Water demands

Agriculture

By calculating the proportion of AET occurring on pastures estimated by GlobCover (2008) that are used for livestock (that is, with Wint and Robinson [2007] livestock counts greater than 0) or on GlobCover (2008) cropland, we can calculate the amount of AET that contributes to production and thus agricultural water use, at least in a crude sense (Table 6). We emphasize that these figures do not include irrigation inputs except insofar that irrigation permits agriculture in areas where it be impossible climatically. The area of GlobCover cropland is therefore greater in irrigated drylands than they would otherwise be. The proportion of total AET used in agriculture varies from 67% for the Ganges, 50% for the Yellow River, and 38% for the Mekong at the higher end, down to only 7% for the Andes, 6% for the Nile and 4% for the Karkheh. Although the Karkheh falls in the middle of the range for livestock densities, little of the catchment is considered grassland by GlobCover because the pastures are very sparsely vegetated. Also the area of cropland is only about 5% and sparsely vegetated so that not all the water is counted as transpiring from the crops, which gives lower estimates of crop AET for the Karkheh than those of Harrington *et al.* (2009) or Qureshi *et al.* (2010). As a proportion of the WorldClim rainfall, water use is 52% of the Yellow River's rainfall, 47% of the Ganges rainfall and 23%

Table 6. Measures of agricultural water use in the basins.

Basin	Mean AET for productive pastures and agricultural areas[a]	Mean AET	Agricultural use of AET	Annual total rainfall (WorldClim)	Agricultural use of mean water balance (wc-AET)
	mm/yr	*mm/yr*	*%*	*mm/yr*	*%*
Andes	43	632	7	784	6
Ganges	499	746	67	1,073	47
Karkheh	13	291	4	348	4
Limpopo	103	640	16	547	19
Mekong	393	1,049	38	1,713	23
Niger	116	804	14	1,017	11
Nile	36	606	6	618	6
São Francisco	94	928	10	975	10
Volta	98	910	11	973	10
Yellow River	229	458	50	438	52

Note: [a]Ahn and Tateishi (1994).

of the Mekong rainfall but only 4% of the Karkheh rainfall. It is important to emphasize that AET is not lost or consumed water. We are dealing with a hydrological cycle so that AET returns water to the atmosphere from which, at the scale of the CPWF basins, it may well fall again as rain in the basin. If AET was not used to grow crops, cropland would be replaced by other plant covers so that AET would likely still be the same as for dryland agriculture. In other words, a reduction of dryland agriculture does not necessary provide more water for downstream use, although a reduction in irrigation generally does.

Domestic

Assuming per capita domestic (non-agricultural) water demands are similar across basins, a reasonable estimate of domestic water demand is the population of the basin (Table 7). According to Gleick (1999), however, there is significant variation in per capita demands for domestic water ranging from 20–200 L/person/day, but there are no data on the distribution of per capita demands within or between the CPWF basins. Demands of 200 L/person/day would produce basin-wide domestic demands from 28.92 km^3/yr for the Ganges through 12.47 km^3/yr (Nile), 11.19 km^3/yr (Yellow River) to a minimum of 0.67 km^3/yr (Karkheh). It is important to note that this water is not used *per se* but rather its properties are changed (it may become polluted for example) and it is returned to the hydrological stores and fluxes in the basin. Moreover, although the numbers seem large, if averaged over the area of the basin and expressed as a depth of water, as is commonly done for rainfall and ET, domestic (non-agricultural) use accounts for only a small fraction of a millimetre for each basin. Moreover as a percentage of rainfall, no basin requires more than 0.0034% of its rainfall input for domestic use. Domestic use may be locally important and may require higher quality of water than other uses, but at a catchment scale the volumes of water "used" are insignificant.

Industrial and energy

Industrial and energy uses are highly specific to the basins and often localized. Where water is used by industry it may become contaminated, where it is used for the purposes of hydropower generation, the timing of its flows may be changed and thus its availability for

Table 7. Measures of water demand for domestic use in the basins.

		At water use of 200 litres/person/day		
Basin	Total population	Water demand	Catchment depth of water equivalent to demand	Demand as a percentage of rainfall
	millions	*Mm^3 /yr*	*mm*	*%*
Andes	79.57	5,817	0.0025	0.000319
Ganges	395.61	28,919	0.0357	0.003328
Karkheh	9.15	669	0.0060	0.001723
Limpopo	13.86	1,013	0.0028	0.000514
Mekong	59.63	4,359	0.0080	0.00047
Niger	76.90	5,621	0.0051	0.000504
Nile	170.55	12,467	0.0025	0.000778
São Francisco	15.90	1,162	0.0357	0.000193
Volta	21.12	1,544	0.0060	0.000439
Yellow River	153.13	11,194	0.0028	0.002943

other uses. Generally industrial and energy uses, through important, are a small proportion of total water use at the mega-basin scale and are very localized in nature. We refer the reader to individual basin papers (Fisher and Cook 2010) and Table 8 for insights into specific issues about water resources in the CPWF basins.

Water quality

Water availability is in part an issue of water quantity (water balance), but in many basins there is also an issue of water quality, especially for domestic users, which need low quantities of high quality water. It is impossible to monitor water quality spatially at the mega-basin scale because water quality is not amenable to remote sensing. Water quality is, however, determined by a few variables that can be assessed globally. Deterioration in water quality can result from inputs from point sources (mining, oil and gas, roads, urban areas) and non-point sources (pasture and cropland). Contaminated water can be diluted when water flowing from upstream areas such as wildlands and protected areas that are little impacted by human activity mixes with water flowing from human-impacted areas. By calculating the rainfall-weighted inputs at each point in the hydrological flow network from areas of point source, non-point source and no-human-impact zones upstream, Mulligan (2009) calculated the human footprint on water quality, globally at 1-km resolution. By summarizing this human footprint over the CPWF basins, we can understand current and emerging water quality issues, which may have impacts on domestic and industrial water uses and may require significant investments in water treatment in the basins. The human footprint on water is measured as the percentage of water in each 1-km pixel that fell as rain on a human-impacted area and whether the human impact is point or non-point source. It thus can be indicator of potential contamination, not a measure of actual contamination, which will vary with agricultural, industrial and urban contamination loads, regulatory frameworks and water treatment infrastructure.

The most human-impacted water as a basin average is in the Karkheh (73%), Yellow River (62%), Ganges (58%) and Limpopo (52%), because of the dominance of cropland and pasture in the landscape. Since these values do not consider the intensity of agriculture and grazing (only its presence or absence) they likely overestimate the impact for basins with extensive but sparse (low intensity) grazing or croplands, such as the Karkheh. The least human-impacted waters are in the Andes (25%) and Mekong (26%), because of the preponderance of natural environments in those basins and their impact on diluting the human influence on water flows. The basin averages mask considerable within-basin variability in the human footprint on water and the precise impacts on people will depend upon the distribution of downstream populations versus upstream natural and human-dominated landscapes (Mulligan 2009).

Drivers of change, expected crisis points and strategies for increasing availability and sustainability

We used global datasets to examine some future scenarios for the basins and went on to examine the potential implications of the changes for water resources. There are few spatial datasets available for how population and economics might change in the future. Spatial projections for population and GDP are publicly available only for the Intergovernmental Panel on Climate Change (IPCC) Special Report on Emissions Scenarios (SRES) B2 scenario (CIESIN 2002). This scenario emphasizes local solutions to sustainability, with increasing population and intermediate levels of economic development (Nakicenovic *et al.* 2000). We re-sampled the grids and extracted data for the mega-basins to calculate mean changes for each basin. Under this scenario, mean basin GDP by 2025 compared with 1990 increases in all basins. The greatest increase is for the Yellow River basin, with an increase in annual GDP of 28 MUSD/km^2, followed by the Mekong (19 MUSD/km^2), and the Ganges (17 MUSD/km^2). The Karkheh and Limpopo increased by 9 MUSD/km^2. All other basins had modest increases in GDP. These GDP changes are the result of both growth in per capita GDP and of population growth in the basins.

Compared with 1990, some 250 million more people are expected to occupy the CPWF basins by 2025 in a SRES B2 world with a total population of 8.039 billion people. The basins would need to support 9% of the global population increase of 2.757 billion in 6.5% of the Earth's land surface. The Ganges is expected to have the greatest growth with an extra 82 million people, followed by the Nile (60 million) and the Niger (40 million) with other basins showing modest increases. The least population increases are in the Karkheh (3.5 million) and the São Francisco (2.8 million). Total population change is, of course, proportional to basin area so that a better measure of population pressure is population density. The greatest increases in population density (persons/km^2) are expected in the Ganges (77), Niger (30), the Volta (23) and the Nile (18). The least increase in population density is expected for the Andes (6.1) and the São Francisco (3.8).

Within-basin variability

For the sake of brevity and simplicity we have focused on basin mean values in this paper. It is important, however, to understand that the spatial variability of many of these properties *within* basins is significant and in some cases as great as the variability *between* basins. This is exemplified in Figure 1 in which we present WorldClim rainfall for the basins as a set of frequency distributions in common frequency classes between the basins. Some basins, like the Karkheh, Yellow River and Limpopo, have very low rainfall and thus low spatial variability within the basin, others like the Andes and Mekong vary over a very wide range. Both the form of the frequency distribution and the range over which variability occurs differs between basins. Some of this variability is effectively reduced by upstream–downstream flows so that even areas with very little rainfall, such as parts of Egypt, can have considerable water availability through flows from upstream. Nevertheless away from river and groundwater reserves, the accessibility of these flows for use in agricultural production is more limited and expensive than for those that are rainfed, so this spatial variability of rainfall can be particularly difficult for the poorest farmers without access to water resources infrastructure.

Water resources syndromes represented in the BFP basins

We have seen a range of water availability characteristics for each of the CPWF basins and will now attempt to assign the basins into a set of water resource syndromes on the basis of their present situation and likely future changes. Figure 2 shows seven key characteristics of the basins and their likely future. These characteristics are classified on the basis of population, change, development and water as the symptoms that comprise the syndrome.

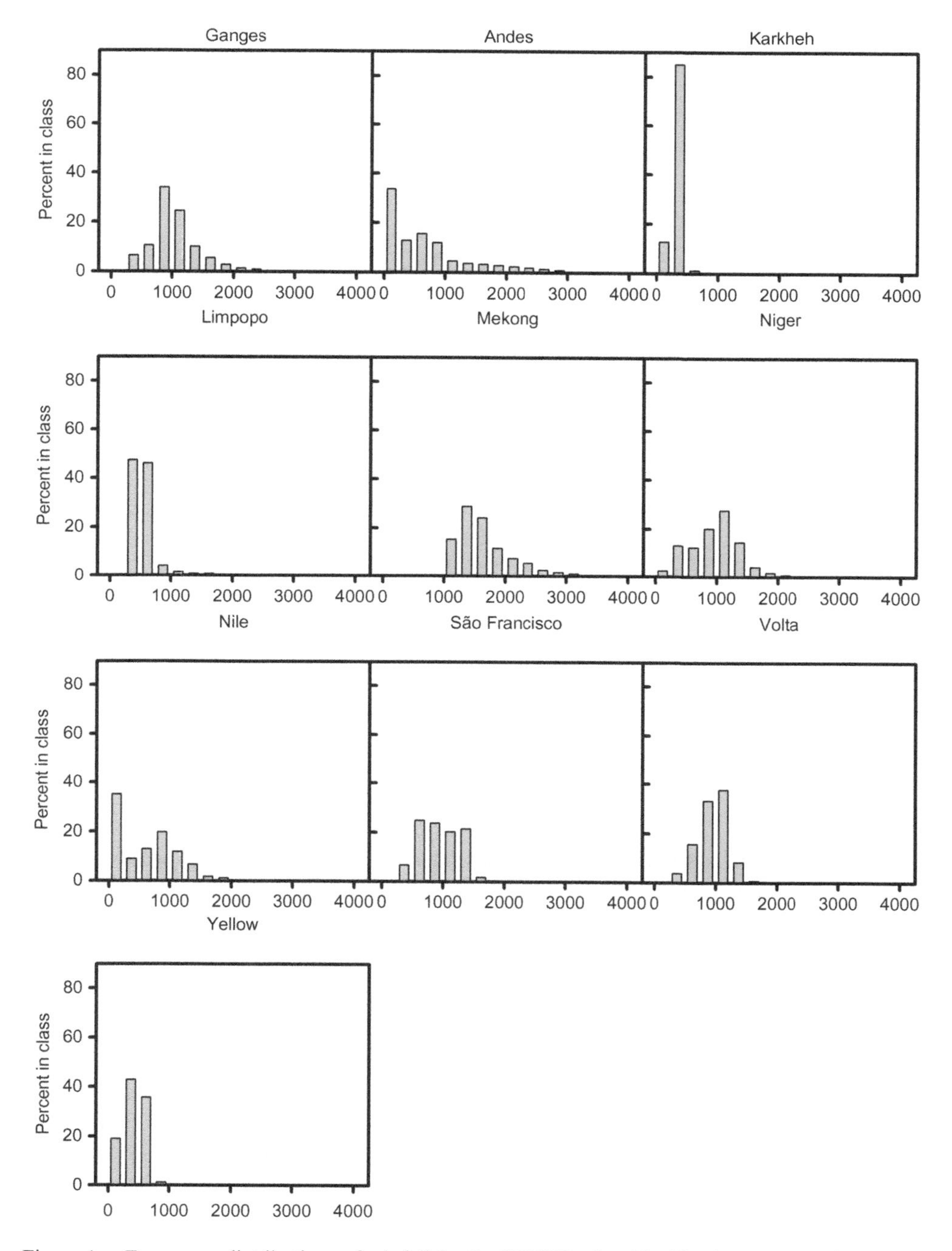

Figure 1. Frequency distributions of rainfall for the CPWF basins. The Y-axis represents the percent of 1-km pixels of the basin that have each class of rainfall.

For each basin the characteristics are expressed in relation to the maximum across all basins for that characteristic (and drawn as a percentage of that maximum). All values are mean values for the basin and expressed as densities where necessary to remove the confounding effect of basin size.

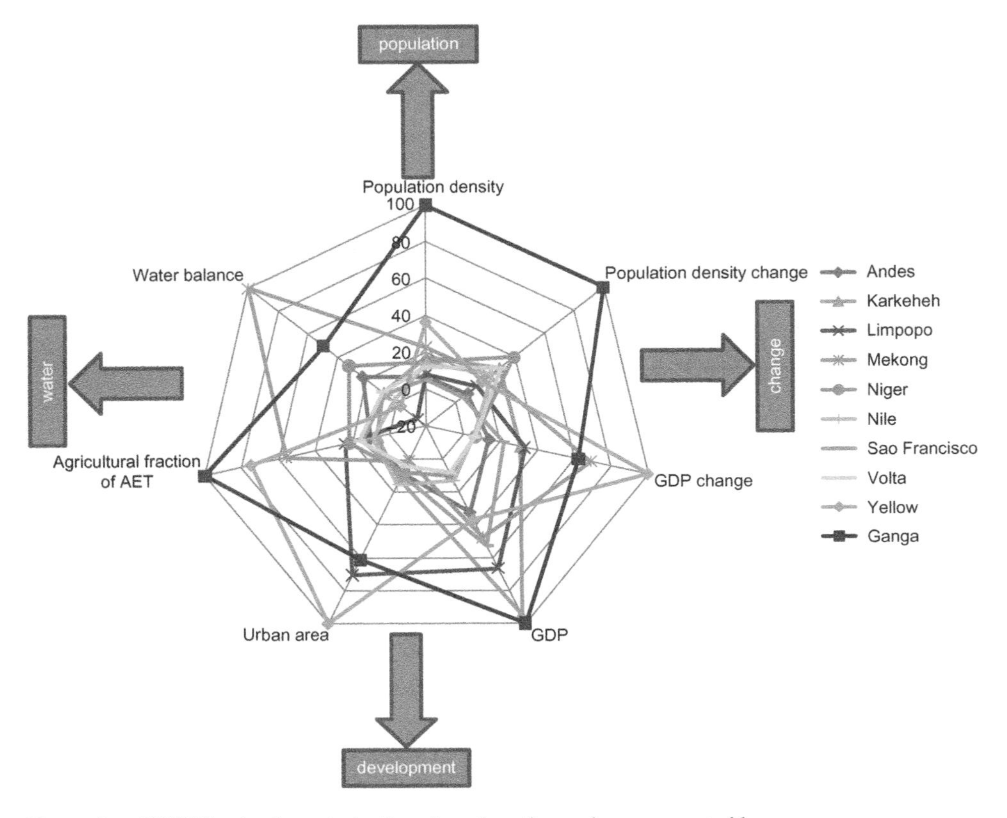

Figure 2. CPWF basin characterizations based on the analyses presented here.

Clearly the Ganges Basin has the highest values for all variables except water balance, urban area and GDP change. It can be characterized as the basin with the highest population pressure and the greatest potential for population change, but also significant GDP and potential for GDP growth, which is important given its already extensive use of available water resources in agricultural production. The Ganges may be characterized as a basin *nearing its capacity*. Other basins can be classified as *highly urban* (that is, with population concentration in urban areas): Yellow River, Limpopo; *with high agricultural water demands*: Ganges, Yellow River, Mekong; and *highly populous*: Ganges, Yellow River, Mekong. The Yellow River, Mekong and the Ganges can also be considered *highly agriculturally developed* (much of the AET being used by agriculture) compared with the other basins. The remaining basins can be considered the *very poor* (Volta, Nile, Niger). In terms of trajectories for change, the Nile, Mekong, Karkheh, Andes, São Francisco, Limpopo can be considered as *rapidly developing*.

Key water resource issues in basins

According to the CPWF BFPs (see papers in Fisher and Cook [2010]), each basin has a particular set of water resource issues. These are identified for each basin, based on the outcome of the BFPs, in Table 8. There are some common water resource issues and issues of assessment of water resource across basins. All basins lack sufficient high-quality, high-resolution and long-term hydrological data to access fully the detail of water resources in

Table 8. Key water resource issues in the CPWF basins as identified by the BFPs.

Basin	Key water resource related issues identified from basin focal projects
Andes	Aridity in parts, excess water (and resulting hazards) in others. Strong influences on downstream users of upstream activities, especially mining, land use, dams. Inequality of benefit sharing. Water quality and conservation of nature/ecosystem services important. Lack of data and knowledge sharing.
Indus-Ganges	Most of the available water resources utilized. High and expanding populations requiring some 30%–50% growth in cereal production and thus higher water demands (Nellemann and Kaltenborn 2009). Potential impacts of climate change in Himalaya. Drawdown of water table and over-dependence on "fossil" groundwater. Lack of data and knowledge sharing.
Karkheh	Water-poor basin with significant use. Rapidly growing population with requirement for food self-sufficiency. High dependence on irrigation. Competition for water from industry, extractives (oil) and the environment. High degree of rainfall seasonality. Lack of data and knowledge sharing.
Limpopo	Physical water scarcity and issues of poor water access. Highly variable and unpredictable rainfall renders dryland crop productivity vulnerable. Competition for water with mining and tourism. Importance of livestock. Inequality of benefit sharing around water. Lack of data and knowledge sharing.
Mekong	Issues of trans-boundary water management. Dependence on aquatic production (fisheries) and potential impact of dams. Growing population, increasing development and resource use. Competition between upstream and downstream users. Inequality of benefit sharing. Lack of data and knowledge sharing.
Niger	Lack of water and severe droughts leading to crop failure. Potential impacts of climate change. Importance of RAMSAR wetlands. Importance of fishing and livestock. Competition for water and lack of benefit sharing especially around dams. Poor agricultural water management. Projections for significantly increasing populations. Lack of data and knowledge sharing.
Nile	High rainfall variability leading to food insecurity. High dependence on subsistence agriculture. Significant population growth rate. Poor land management and resulting resource degradation. Little potential for expansion of irrigation. Lack of data and knowledge sharing.
São Francisco	Competition between commercial farming operations and subsistence farming. Strong rainfall gradient. Some important water-dependent ecosystems. Expansion of irrigation. Lack of data and knowledge sharing.
Volta	Dominated by rainfed production. Poor soil fertility. High degree of rurality. Increasing population. Strong rainfall gradient. High rainfall seasonality and inter-annual variability. Little use of groundwater. Competition between agriculture and water supply for hydropower. Poor rainwater use and cultivation practices, lack of irrigation. Water quality issues. Lack of data and knowledge sharing.
Yellow River	Low and highly seasonal precipitation, heavy reliance of irrigation. Important food production region. Competition with industrial and urban uses and resulting pollution. Impacts of climate change. Increasing water productivity. High levels of water loss through ineffective management practices. Rural-urban inequality. Upstream–downstream conflict. Lack of data and knowledge sharing.

the basin. In particular, good rainfall, land use, water use and population data are often lacking at the mega-basin scale, notwithstanding the significant efforts of the BFPs and basin authorities and the globally available resources that we used here. Flow data can be poor, especially for the largest rivers or those rivers that do not form a single basin as in the Andes. There is also sometimes inefficiency in sharing knowledge in the basins such that where data do exist at the mega-basin scale, they are not available for use, or in forms that are usable. This fundamental lack of even biophysical data limits our ability to quantify and understand the water resources of the basins. If we cannot close a hydrological budget at the mega-basin scale, the uncertainties associated with more sophisticated empirical analyses at this scale become important. Detailed local studies can improve closure for individual sub-basins of the CPWF mega-basins (for example, Muthuwatta *et al.* [2010] for the Karkheh), though it is difficult to carry out such detailed studies at the mega-basin scale.

In terms of water, water, productivity and poverty, some of the basins are limited by rainfall (Karkheh, Limpopo) or by rainfall seasonality and uncertainty (Niger, Nile). Some are limited by hydrological hazards such as soil erosion and flooding (Andes, Volta). Some are limited by over-population (Indus-Ganges) or having reached cropping or irrigation capacity (Yellow River). Some basins are limited by competition for water and lack of benefit sharing between water users (Andes, Mekong, Karkheh, Limpopo). Many of the basins are also limited by factors including water infrastructure (Mekong), water access (Limpopo), economic poverty (Nile, Niger), poor land management and soil fertility (Limpopo) and institutions (Andes, Mekong) as are discussed in other papers in this issue.

Policy implications

There are no simple solutions to these problems. First of all the problems are not always fully understood at the basin level through lack of appropriate data at the scales required, which requires more research and better knowledge sharing with and between relevant local, national and international institutions. Moreover, many of these basins are transnational, which can create challenges for managing water at the basin level. Many basins face growing water shortages as population and water demands grow, unless policies and investments are undertaken successfully to change the growing scarcity or unless climate change increases rainfall inputs. Water investments will need to be set within the context of globalization, trade policy, and agricultural water policy, including benefit-sharing agreements. Many basins are subject to climate variability on a range of scales meaning that populations and agricultural systems, even if developed over decades, may be rendered inappropriate during periods of very different rainfall. This means that all basins need capacity to adapt and need to consider long-term sustainability in agricultural and water planning. Moreover, humans are not the only users of water in basins. Ecosystems also need water to continue to provide the ecosystem services upon which other ecosystems and human populations depend. Maintaining these ecosystem services is an important element of sustainable water management in many of the basins: agricultural (and aquacultural) investments should consider their impacts on environmental services so as not to create local benefits with excessively deleterious costs downstream.

Finally, there are particular scale issues associated with working in mega-basins like the CPWF basins. The advantages many of the BFPs found of working on water availability at this scale include an increased policy relevance, the ability to take a basin-wide and thus more holistic view, the focus on upstream–downstream relationships and the ability to understand better the role of within-basin variability. The challenges of working on water availability at this scale centre on gathering and managing data and the representation of fine-scale hydrological processes such as rainfall at sufficiently detailed resolution over such large areas. The distinction between blue (runoff) and green (ET) water is clear at the field and the landscape scale, but less so at the mega-basin scale since hydrology is cyclic. Blue water may run off upstream but be diverted for irrigation and ET downstream, green water may evaporate in one part of the basin but generate clouds, reducing solar radiation loads and thus ET and then for as rain in another part of the basin. The spatially explicit mega-basin approach can provide insights that are necessary to develop policies for management of water resources that are targeted to conditions at the local scale, but are also adapted so as not to be counter-productive at the basin scale.

Acknowledgements

The authors would like to thank the CPWF and its donors for funding the Basin Focal Projects. This paper benefited greatly from reviews by the CPWF BFP team, the volume editors and anonymous reviewers. Valuable contributions were received from reports authored by BFP water availability teams including contributions from: Yasir Mohammed (Nile), Kevin Scott (Limpopo), Jianshi Zhao (Yellow River), Luna Bharati and Bharat Sharma (Indus-Ganges), Jacques Lemoalle (Volta), Mac Kirby (Mekong), Steve Vosti (São Francisco), Mark Giordano and Mobin Ahmad (Karkheh).

References

Abu-Zeid, M., 1995. Major policies and programs for irrigation drainage and water resources development in Egypt. *Options Méditerranéennes,* Series B, No 9, *Egyptian agriculture profile.*

Ahn, C.-H. and Tateishi, R., 1994. Development of a global 30-minute grid potential evapotranspiration data set. *Journal of the Japan Society of Photogrammetry and Remote Sensing*, 33 (2), 12–21.

Allen, R.G., *et al.*, 1998. Crop evapotranspiration: guidelines for computing crop requirements. Irrigation and Drainage Paper No. 56. Rome: Food and Agriculture Organization of the United Nations (FAO).

Andersen, I., *et al.*, 2005. *The Niger River basin: a vision for sustainable management* [online]. Washington, DC: The World Bank. Available from: http://siteresources.worldbank.org/INTWAT/Resources [Accessed 14 May 2010].

Barry, B., *et al.*, 2005. *The Volta River basin: comprehensive assessment of water management in agriculture* [online]. Colombo: IWMI. Available from: http://www.iwmi.cgiar.org/Assessment [Accessed 14 May 2010].

Busari, O., 2008. Groundwater in the Limpopo Basin: occurrence, use and impact. *Environment, Development and Sustainability*, 10 (6), 943–957.

CIESIN, 2002. Country-level population and downscaled projections based on the SRES B2 scenario 1990–2100 [digital version]. Palisades, NY: Center for International Earth Science Information Network (CIESIN), Columbia University. Available from: http://www.ciesin.columbia.edu/datasets/downscaled [Accessed 19 November 2010].

CIESIN/CIAT, 2005. Gridded population of the world version 3 (GPWv3) [online]. Palisades, NY: Socioeconomic Data and Applications Center (SEDAC), Center for International Earth Science Information Network (CIESIN), Columbia University and Cali, Colombia: Centro Internacional de Agricultura Tropical (CIAT). Available from: http://sedac.ciesin.columbia.edu/gpw/ [Accessed 19 November 2010].

Droogers, P. and Allen, R.G., 2002. Estimating reference evapotranspiration under inaccurate data conditions. *Irrigation and Drainage Systems*, 16 (1), 33–45.

Farr, T.G. and Kobrick, M., 2000. Shuttle radar topography mission produces a wealth of data. *Eos, Transactions, American Geophysical Union*, 81 (48), 583–585.

Fisher, M. and Cook, S., 2010. Introduction. *Water International*, 35 (5), 465–471.

Fontes, J.C., *et al.*, 1991. Paleorecharge by the Niger River (Mali) deduced from groundwater geochemistry. *Water Resources Research*, 27 (2), 199–214.

Gleick, P., 1999. Basic water requirements for human activities: meeting basic needs. *Water International*, 21 (2), 83–92.

GlobCover, 2008. *GlobCover land cover v2 2008 database*. European Space Agency GlobCover Project led by MEDIAS-France. 2008. Available from: http://ionia1.esrin.esa.int/index.asp [Accessed 19 November 2010].

GRDC, 2010. *Global Runoff Data Centre station database* [online]. Available from: http://www.bafg.de/cln_007/nn_294146/GRDC/EN/02__Services/01__RiverDischarge/GoogleEarth/google__earth__node.html?__nnn=true [Accessed November 2010].

Han, Z., 2003. Groundwater resources protection and aquifer recovery in China. *Environmental Geology*, 4 (1), 106–111.

Hargreaves, G.L., Hargreaves, G.H., and Riley, J.P., 1985. Irrigation water requirements for Senegal River basin. *Journal of Irrigation and Drainage Engineering ASCE*, 111 (3), 265–275.

Harrington, L., *et al.*, 2009. Cross-basin comparisons of water use, water scarcity and their impact on livelihoods: present and future. *Water International*, 34 (1), 144–154.

Hassan, T.M., Attia, F.A., and El-attfy, H.A., 2004. Groundwater potentiality map of the Nile Basins countries. A step towards integrated water management. *International Conference and Exhibitions on Groundwater in Ethiopia: Providing Water for Millions*. Addis Ababa, Ethiopia, 25–27 May 2004.

Hijmans, R.J., *et al.*, 2005. Very high resolution interpolated climate surfaces for global land areas. *International Journal of Climatology*, 25 (15), 1965–1978.

Kapos, V., *et al.*, 2000. Developing a map of the world's mountain forests. *In*: M.F. Price and N. Butts, ed. *Forests in sustainable mountain development: a state of knowledge report for 2000*. Wallingford, UK: CABI, 4–9.

Kortatsi, B.K., 1994. Groundwater utilization in Ghana. *In*: J. Soveri and T. Suokko, eds. *Future groundwater resources at risk*. Proceedings of the international conference *Future groundwater resources at risk (FGR 94)*. Helsinski, June 1994. IAHS Publication 222, 149–156. Wallingford, UK: International Association of Hydrological Sciences.

Leduc, C., Favreau, G., and Schroeter, P., 2001. Long-term rise in a Sahelian water-table: the continental terminal in south-west Niger. *Journal of Hydrology* 243 (1), 43–54.

Lehner, B. and Döll, P., 2004. Development and validation of a global database of lakes reservoirs and wetlands. *Journal of Hydrology*, 296 (1–4), 1–22.

Lienou, G., *et al.*, 2010. The river Niger water availability: facing future needs and climate change. *In:* E. Servat *et al.*, eds. *Global change: facing risks and threats to water resources*. Proceedings of the sixth world FRIEND conference. Fez, Morocco, 25–29 October 2010. IAHS Publication 340, 637–645. Wallingford, UK: International Association of Hydrological Sciences.

Mahajan, G., 2008. *Evaluation and development of groundwater.* New Delhi, India: APH Publishing.

Mahé, G., 2009. Surface/groundwater relationships in two great river basins in West Africa, Niger and Volta. *Hydrological Sciences Journal*, 54 (4), 704–712.

Mahé, G. and Paturel, J.E., 2009. 1896–2006 Sahelian rainfall variability and runoff increase of Sahelian rivers. *Comptes Rendus Geosciences*, 341 (7), 538–546.

Marjanizadeha, S., de Fraiture, C., and Loiskandl, W., 2010. Food and water scenarios for the Karkheh River basin, Iran. *Water International*, 35 (4), 409–424.

Mulligan, M., 2006. TRMM 2b31-based rainfall climatology version 1.0 [online]. London: Kings College. Available from: http://www.ambiotek.com/1kmrainfall [Accessed 19 November 2010].

Mulligan, M., 2009. The human water quality footprint: agricultural, industrial, and urban impacts on the quality of available water globally and in the Andean region. *In*: *Proceedings of the international conference on integrated water resource management and climate change*. Cali, Colombia, 10–12 November 2009. Available from: http://www.ambiotek.com/publications/CINARA_Industry_and_mining.pdf

Mulligan, M. and Burke, S.M., 2005. DFID FRP Project ZF0216 Global cloud forests and environmental change in a hydrological context. Final Report [online]. December 2005 Available from: http://www.ambiotek.com/cloudforests/cloudforest_finalrep.pdf [Accessed 24 November 2010].

Mulligan, M., *et al.*, 2009. Global dams database and geowiki. Version 1. Available from: http://www.ambiotek.com/dams [Accessed 26 November 2010].

Mulligan, M., *et al.*, 2011. The nature and impact of climate change in the CPWF basins. *Water International*, 36 (1), 96–124.

Muthuwatta, L.P., *et al.*, 2010. Assessment of water availability and consumption in the Karkheh River basin, Iran - using remote sensing and geo-statistics. *Water Resource Management*, 24 (3), 459–484.

Nakicenovic, N. *et al.*, 2000. Special report on emissions scenarios: a special report of working group III of the Intergovernmental Panel on Climate Change [online]. Cambridge: Cambridge University Press. Available from: http://www.grida.no/climate/ipcc/emission/index.htm [Accessed 19 November 2010].

Nellemann, C. and Kaltenborn, B.P., 2009. The environmental food crisis in Asia: a "blue revolution" in water efficiency is needed to adapt to Asia's looming water crisis. *In*: *Water storage: a strategy for climate change adaptation in the Himalayas*. Sustainable mountain development No. 56. Kathmandu, Nepal: International Centre for Integrated Mountain Development (ICIMOD), 6–9.

Peña-Arancibia, J.L., *et al.*, 2010. The role of climatic and terrain attributes in estimating baseflow recession in tropical catchments. *Hydrology and Earth System Sciences*, 14 (11), 2193–2205.

Qureshi, A.S., *et al.*, 2010. Water productivity of irrigated wheat and maize in the Karkheh River basin of Iran. *Irrigation and Drainage*, 59 (3), 264–276.

Reboucas, A., 1999. Groundwater resources in South America. *Episodes*, 22 (3), 232–237.

Siebert, S., *et al.*, 2007. *Global map of irrigation areas version 4.0.1*. Frankfurt (Main): University of Frankfurt and Rome: Food and Agriculture Organization of the United Nations (FAO).

Struckmeier, W.F., *et al.*, 2006. *Groundwater resources of the world. Transboundary aquifer systems 1:50 000 000. Special edition for the 4th World Water Forum, Mexico City* [online]. Available from: http://www.whymap.org [Accessed 14 May 2010].

Trabucco, A. and Zomer, R.J., 2009. *Global aridity index (global-aridity) and global potential evapo-transpiration (global-pet) geospatial database* [online]. CGIAR Consortium for Spatial Information. Available from: http://www.csi.cgiar.org/ [Accessed 19 November 2010].

Wint, G.R.W. and Robinson, T.P., 2007. *Gridded livestock of the world*. Rome: FAO. Available from: ftp://ftp.fao.org/docrep/fao/010/a1259e/a1259e01.pdf [Accessed 19 November 2010].

World Bank, 2010. *World development indicators database*. Washington, DC: The World Bank. Available from: http://data.worldbank.org/data-catalog/GNI-per-capita-Atlas-and-PPP-table [Accessed 24 November 2010].

Producing more food with less water in a changing world: assessment of water productivity in 10 major river basins

Xueliang Cai, David Molden, Mohammed Mainuddin, Bharat Sharma, Mobin-ud-Din Ahmad and Poolad Karimi

This article summarizes the results of water productivity assessment in 10 river basins across Asia, Africa and South America, representing a range of agro-climatic and socio-economic conditions. Intensive farming in the Asian basins gives much greater agricultural outputs and higher water productivity. Largely subsistence agriculture in Africa has significantly lower water productivity. There is very high intra-basin variability, which is attributed mainly to lack of inputs, and poor water and crop management. Closing gaps between "bright spots" and the poorly performing areas are the major tasks for better food security and improved livelihoods, which have to be balanced with environmental sustainability.

Introduction

The concept of water productivity

The world is under great pressure to feed nine billion people by 2050. Total evapotranspiration (ET) from global agricultural land could double in the next 50 years if trends in food consumption and current practices of production continue (de Fraiture *et al.* 2007). With increasing demand from non-agricultural sectors and the uncertainties in water management brought about by climate change, the agricultural sector in many areas will get less water in the future (Bakkes *et al.* 2009). Together, the increasing demand for water for food production and the limits of the availability of water resources suggest that agriculture must produce more food with less water, that is, make more productive use of water resources (Cai and Sharma 2010).

Water productivity (WP) measures how the system converts water, together with other resources, into goods and services. It is defined as the ratio of net benefits from crop, forestry, fishery, livestock and other mixed agricultural systems to the amount of water used in the production process (Molden *et al.* 2010). The benefits can be measured with various terms including physical mass (kilogram), economic value (monetary) and nutritional value (calorie). The water input, denominator in the WP equation, also has a set of choices depending on the purpose of the examination and the availability of data. For example, irrigation diversion, gross/net inflow, evapotranspiration and precipitation can all be used to calculate WP indicators. These variations give WP assessment flexibility and robustness as a tool to measure efficiency of water use.

WP indicators enable performance assessment and comparison for better water management, which is driven by water scarcity and increasing food demand (Cook *et al.* 2006, Ahmad *et al.* 2009a). Different WP indicators enable different stakeholders to examine water-use performance according to their specific interests. For example, WP as expressed in kg/m^3 of irrigation diversion might be of interest to an irrigation manger, while WP in US$/m^3 of ET is more interesting to a basin development agency that is concerned with overall water consumption and the outputs generated. A water resources planner might look at the difference between kg/m^3 irrigation diversion and kg/m^3 ET to assess how well irrigation water is managed (Molden *et al.* 2007). With increasing problems of water scarcity, the food and other manufacturing industries have also started measuring the water required to produce a unit of output, which is termed "water footprint", for example, water footprints of milk, coffee or palm oil production (Hoekstra *et al.* 2009).

The WP indicators need to be used with caution, however, especially when making cross-system comparisons. The WP concept is more relevant in water-scarce regions where increased production is constrained by water. WP values are affected by many factors including natural and management conditions. They are better understood in conjunction with specific system settings, for example, whether it is a water-abundant or water-scarce area; whether WP is constrained by yield or water use; whether it is an irrigated or a rain-fed system. WP is scale dependent, which is related to specific geographic extents as well as the types of farming systems involved. The interpretations are thus restricted by the boundary conditions. As WP has variable forms, the user needs to make sure to use the same form when making intra- (or inter-) system comparisons.

Assessment of basin water productivity: the implications for livelihood, food security and sustainable development

Holistic understanding of water management needs to examine issues at large scales such as a whole river basin. Agricultural water management has been traditionally organized around farming and irrigation systems. Hydrological processes are often more complex because of extensive interactions between water, ecosystems and people. Control and use of water upstream often impacts downstream users. In water-rich basins the impact might be in terms of water quality alone, while in water-scarce basins both quality and quantity might be affected. Full water accounting is required for integrated water management within the confines of the natural basin boundary (Turner *et al.* 2009).

Basin WP has important implications for regional food security and livelihoods, especially in places where there is poverty and crop production is low. Improved WP means more food, more income, less vulnerability to risks, and possibly less water used. Improving WP may leave more water for other sectors to produce more, which in turn contributes to regional development. Africa withdraws less than 4% of its renewable water resources for agriculture (Hanjra and Gichuki 2008). Lack of inputs such as land preparation (energy), seeds, fertilizer, pesticides and irrigation water limit the opportunity to increase productivity. Increasing water productivity through higher yields and improved water use helps to enhance crop production, generate and stabilize income, boost employment, reduce consumer prices and reduce costs. Mixed agricultural systems that include crops, livestock and fisheries help to distribute risks and maximize benefits from limited water resources (Ahmad *et al.* 2009b, Mainuddin and Kirby 2009). Cash crops are efficient in reducing poor communities' vulnerability to natural disasters such as droughts (Mertz *et al.* 2005).

The importance of livestock and fisheries is often underestimated. At the global level, livestock is responsible for around 20% of agricultural water ET (de Fraiture *et al.* 2007). With the changing diet of a wealthier population, consumption of milk, meat and fish is projected to double by 2050. Both livestock and fish can be part of integrated water management systems, with less additional water required for fish and more land management for livestock.

Gains in WP need to be balanced with environmental sustainability. Environmental degradation has emerged as a severe problem in several of the world's largest river basins such as the Yellow and the Indus-Ganges (Singh and Sontakke 2002, Economy 2004). Non-point pollution from agriculture is a major threat to the health of rivers (Carpenter *et al.* 1998), and over-exploitation of limited water resources, especially groundwater, is another (Rodell *et al.* 2009). Wetlands have been diminishing at a fast pace, and so have forests. Sustainable development needs to take account of the environmental consequences of the exploitation of basins' water resources. In cases where the volume of water is obviously insufficient, actions need to be taken to balance agricultural gains and environmental sustainability (Molden *et al.* 2010).

The water productivity assessment across river basins

Assessment of WP is a key procedure to link basin social and physical settings, water management and agricultural production. As part of the Basin Focal Projects (BFP) of the Challenge Program on Water and Food (CPWF), assessments of WP were carried out in 10 river basins with different physical and social settings: the Yellow River in China, the Mekong in Southeast Asia, the Indus-Ganges in South Asia, the Karkheh in Iran, the Nile in eastern Africa, the Limpopo in southern Africa, the Niger and the Volta in western Africa, the São Francisco and a collection of basins in the Andes in South America.

The 10 river basins are at different development stages and have distinct differences in their social and physical settings. They are located on three continents in regions ranging from humid to semi-arid. The contribution of agriculture to gross domestic product (GDP) varies from 6.5% in the Limpopo to more than 50% in the Mekong, Nile, and Niger (Figure 1), and involves 23% to more than 90% of the basin population. The percentage rural poverty also varies widely, from 11% in the Yellow River to more than 70% in the Niger. The three Asian basins have the highest total and percentage rural population, closely followed by the African basins, but both are much lower in the Andes and the São Francisco. The contributions of agriculture to GDP and the incidence of poverty in the African basins are significantly higher than in other basins. Although the rural population in South American basins is relatively low, rural poverty is high. It is noteworthy that there is high variability in some transboundary basins where the characteristics of one country may skew the basin average.

Irrigation development is diverse in the 10 basins. The Indus-Ganges, with mean rainfall of 1250 mm, has by far the highest percentage of irrigated area (78%), followed by the Yellow with 45%. The São Francisco and the Volta have the lowest ratios of about 6% and 1% respectively. Maize is the dominant crop in commercial cultivation in most basins except the Indus-Ganges, the Mekong and the Niger. Wheat and rice are the dominant crops in the Yellow, Mekong, Indus-Ganges, Nile and Karkheh. The Niger, however, is dominated by millet and sorghum and the São Francisco by maize, beans and rice. Among these major crops rice usually requires heavy irrigation. Wheat is also intensively irrigated in the Yellow and the Indus-Ganges. Maize in Africa and South America is mostly rainfed.

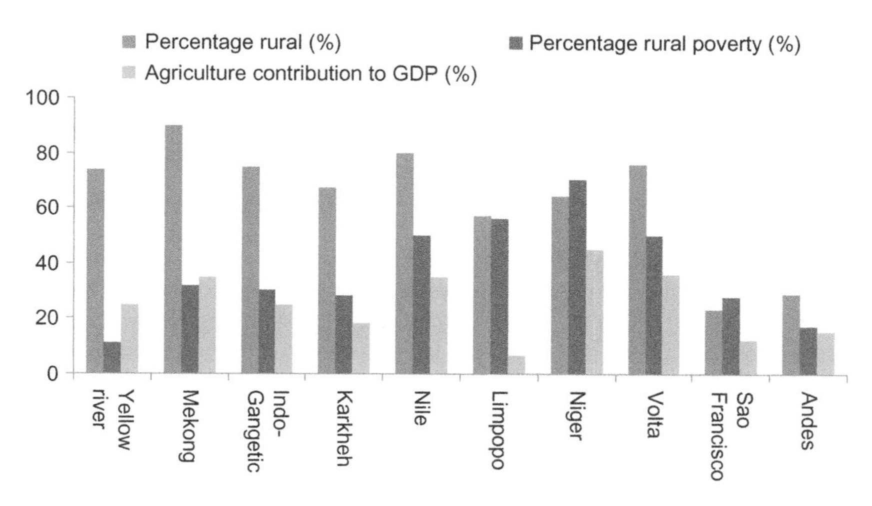

Figure 1. Rural population, poverty, and contribution of agriculture to GDP in the 10 BFP basins.

The objectives of this paper are to examine the methodologies used in the assessments; to give an indication of water productivity across different locations of the world, and to draw inferences about the underlying reasons.

Methodology overview

There were various approaches, combining a range of datasets, used to assess WP, because the 10 basins have diverse social and physical settings with different levels of data availability. Crops dominate agricultural production and are the biggest water consumer, so we gave them more attention in WP assessment. In some basins, where livestock and fisheries are important contributors, we also included them. The basic equation to calculate WP is (Molden 1997):

$$WP = \frac{Output\ derived\ from\ water\ use}{Water\ input} \tag{1}$$

The numerator, and particularly the denominator, were determined using various methodologies such as field experiments, (agro-) hydrological modelling, spreadsheet calculations and remote sensing. Datasets come from sources such as field monitoring, household surveys, official statistics, weather stations, remote-sensing imagery and literature review. We combined the datasets to make assessments of WP at various scales from the field, through sub-catchments, to the basin.

Crop water productivity

An important prerequisite to assess crop WP at the basin scale is to understand the cropping pattern/distribution, which is usually complex and dynamic. Land-use/land-cover (LULC) maps based on remote sensing provide pixel-level information on the distribution of agriculture. They help identify the distribution of crops and subsequently estimate crop production and water consumption (ET). Global-, regional-, and national-level LULC products were used to assess WP. These include GlobCover from the European Space

Agency (http://ionia1.esrin.esa.int/) in the Nile, moderate resolution imaging spectro-radiometer (MODIS) continuous vegetation fields (www.landcover.org/data/vcf/) in the Andes, the global irrigated area map (GIAM) (http://www.iwmigiam.org) in the Indus-Ganges, and the South Africa national farmland boundaries in the Limpopo. Crop types are rarely distinguishable in large-scale LULC maps. However, a crop-dominance map, based on GIAM products and other LULC maps, was produced for crop-specific analysis (Cai and Sharma 2010).

There are a number of ways to assess crop production, the numerator in the WP equation. Gross value production (GVP) of crops was frequently used in the BFP basins to calculate the outputs of all crops. GVP was calculated by summing local market value of the different crops and converting them into US dollars for a constant year, which enabled comparisons across basins (Molden *et al.* 2003). Crop production was also converted into energy (calories) in the Volta Basin. Statistical data of cultivation and production for major crops, which differentiate well the spatial variation within basins, are usually available at the district level. Remote sensing is an effective technique to monitor crop condition and yield accumulation. Vegetation dry matter production was mapped in the Andes basins based on a time series from the Système Probatoire d'Observation de la Terre vegetation sensor (SPOT-VGT) data. The spatial and temporal distribution was then analysed against land-use maps to examine crop and pasture production. A method combining remote sensing and census data was developed to estimate yields of major crops at pixel level in the Indus-Ganges basins (Cai and Sharma 2010). This method maximizes the utilization of existing data from different sources, eases the often-seen gap between census and remote sensing, and produces pixel-based yield maps to depict the natural distribution of crop yields regardless of administrative boundaries.

There was a range of denominators used to assess WP in the BFP basins: potential ET, actual ET, rainfall, evapotranspirable water, irrigation diversion and gross inflow. Potential ET is the maximum water requirement, which can be calculated from the weather data that are available in every basin. Calculation of actual ET across large scales is more challenging but enables estimates of real agricultural water use. Water consumption was estimated using hydrological models: for example, water balance for the whole of the Nile and the Indus-Ganges was computed with water evaluation and planning (WEAP) system. Estimates of ET were provided by two-dimensional modelling, for example, the decision-support system for agrotechnology transfer (DSSAT) crop simulation in the Volta and a soil water balance model in the Mekong.

Remote sensing provides spatially explicit maps of ET, which avoids the complex hydrological processes on the ground. Several basins adopted remote sensing approaches to estimate ET including the surface-energy balance system (SEBS) in the Karkheh (Muthuwatta *et al.* 2010), the surface-energy balance algorithm for land (SEBAL) in the Nile (Mohamed *et al.* 2004) and the upper Indus basin (Ahmad *et al.* 2009a), and a simplified surface-energy balance (SSEB) model in the Indus-Ganges (Cai and Sharma 2010) and the Limpopo. While all three models are based on the principle of land-surface energy balance, SEBS and SEBAL derive most of the parameters from satellite imagery and require few ground data inputs. In contrast, SSEB combines ground-measured weather data and land-surface temperature measured by satellite. The grid-based maps allow analysis of water consumption across different spatial scales and patterns of land use.

Livestock

WP of livestock is the market value of meat, milk, eggs and other items, divided by water used to raise and maintain the animals. Previous estimates have shown that 1 kg of animal products requires 1 to 20 m^3 water (Molden *et al.* 2007). The water directly consumed by animals is tiny compared with the water required to produce their feed, which varies according to management practices and types of feed (for example, grassland or crop residues). Current assessments in the 10 basins focused more on production analysis using statistical data. In the Karkheh Basin, however, ET estimated from remote sensing for grazing land, including pasture and crop residues, were used to estimate water use (Ahmad *et al.* 2009b). A more rigorous approach was developed in the Nile Basin where animal production was linked to the quantity of feed and then to water required to produce it. While there were still a number of assumptions in the parameterization of these two approaches that need to be validated, they present new methods to examine livestock water consumption and productivity.

Fisheries

Fisheries, including capture fish and aquaculture, are also mostly reported based on analysis of production. In the Nile Basin, water balance of aquaculture ponds was monitored in the field and the impacts of dams and water quality on capture fish production assessed (Alemayehu *et al.* 2008). It was assumed that evaporation losses estimate the water used to produce fish, which, however, reflects only the climatic conditions but does not reflect the fish-farming practices. In the Niger Basin, the catch of capture fish is directly related to the amount of water available in river channels. In the inland delta area, fish production increases by 27.8 t for each increase of 1 m^3/s in river flow. Case studies in the Gorai subbasin (Bangladesh) of the Ganges collected primary data on water productivity in terms of weight, economic value, and drivers for different capture and aquaculture systems (Mustafa *et al.* 2010).

Discussions on methodology

The various techniques described above were used in the 10 river basins, giving WP with different numerators and divisors. Estimates of WP with same terms are directly comparable, for example, kg/m^3 for major crops or GVP/m^3 for all crops. Obviously, results are not comparable when the terms used are different. For example, kg/ET will give different results than kg/applied water. The term ET as the denominator has inherent strength at the basin scale because the hydrological processes are especially complex at large scale. The ability to assess ET (crop water consumption) directly allows the assessment of actual water use versus water inputs in different conditions of water availability.

Remote sensing has proved to be a promising approach in assessing WP. It provides information of actual water consumption while avoiding the complex ground hydrological processes. It enables analysis at sub-catchment level and down to the level of individual pixels, which is often more informative than arbitrary administrative levels. Maps based on this level of analysis enables assessment of the magnitude and variations in WP to pinpoint hot and bright spots, which can then be linked to factor analysis and assessment of potential performance. A major challenge is to produce maps of dynamic crop rotation, which are essential for explicit estimates of production and water use.

Remote sensing can readily be integrated with hydrological modelling, which simulates the processes of water flows on the ground, to assess agricultural WP. This process-based analysis provides better opportunities to study the nexus between water, interventions and the responses of the production system (for example, Roost *et al.* 2008, Kim *et al.*

2009). The approach differs from monitoring by remote sensing alone, which only measures results (for example, yield and ET). Spatially explicit maps from remote sensing, when combined with the results of hydrological modelling at fine temporal resolution, strengthen the analysis. A further step is to use hydro-economic modelling, which allows comprehensive analysis of food security and income generation to support strategic decisions.

Assessment of WP for crops is relatively well developed compared with WP of livestock and fisheries, for which both the concept and methods need further development and validation. Livestock and fisheries are important contributors to food security and livelihood in rural communities (Béné and Friend 2009, Descheemaeker *et al.* 2009), yet there are still large gaps in identifying the water they consume. The difficulty comes from accounting the water required to produce feed, which comes from grazing land, crop residues and commercial grain-based products. Commercial feed also requires consideration of the trade in virtual water, which might go beyond the basin boundary. Water consumption by fisheries is even more complicated. The yield of capture fisheries is related to the volume of water in the watercourses (see above), but losses to evaporation occur regardless of the fish catch. Evaporation from single-purpose aquaculture pond can be treated as water consumed, as for crops. Water lost to seepage and percolation is likely to be available for further use downstream, so that evaporation is the appropriate measure of water use. In addition, however, and analogous to livestock, the water used to produce the fish feed also needs to be considered. Multiple-purpose fish ponds (for example, a rice–fish system) are more difficult since their operation is very flexible and their water use is highly dynamic. They are, however, important to the livelihoods of local communities hence should be included in overall agricultural WP assessment.

This discussion of methodology demonstrates the need to develop a standard framework to assess WP at the basin scale. In the absence of such a framework, it is difficult to compare results across sites, although agro-hydrological modelling together with remote sensing can help. The extension of agricultural WP to include livestock, fisheries, probably also agro-forestry, as well as other relevant outputs needs further development.

Water productivity and its importance in the BFP basins

Agricultural productivity, water use and water productivity

Yields of the major crops (maize, wheat and rice) vary both across and within basins (Figure 2). All three crops in the Yellow River basin have relatively high yields, although not the highest for maize and rice. The yields for the Nile in Egypt are very high for all three crops, although low yields elsewhere reduce the Nile's overall figures. The Indus-Ganges basins are the most populous and have the most intensive cultivation, but have relatively low yields overall for both rice and wheat, which are the major sources of food and income.

There is large intra-basin variability in all the basins. The average yield of maize in the Limpopo is 3.6 t/ha. While the irrigated commercial farms yield as high as 9 t/ha, the large area of subsistence farms, which are threatened by frequent droughts and crop failure, yields less than 2 t/ha. The Indian states of Punjab and Haryana, the "bright spots" in the Indus-Ganges basins, yield more than double elsewhere. The highest yields of rice among all BFP basins are in the Nile Basin, because of the very high yields in the delta in Egypt. Elsewhere in the basin, including most of Ethiopia and Sudan, yields are very low. There is similar variability in the Mekong Basin, where yields are high in the delta, but low in Cambodia and northeast Thailand.

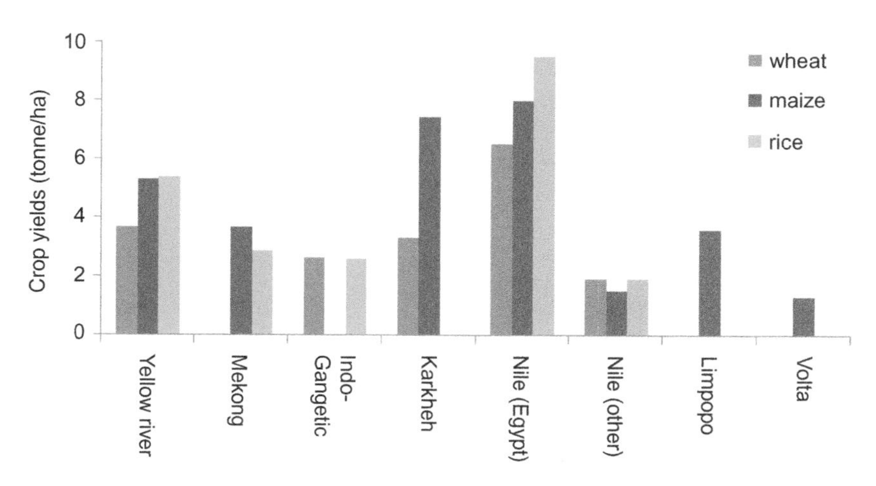

Figure 2. Average yield values of major crops in BFP basins.

WP values in terms of GVP divided by ET are available for some of the basins, and vary widely between them (Table 1). For each m^3 water consumed, irrigated crops in the Karkheh generate US$0.22, the Indus-Ganges US$0.13, while the Mekong only US$0.012–0.059. Crop yields in the Karkheh are high, which accounts for the high crop WP. The Indus-Ganges basins also stand out with relatively high WP in spite of moderate yields, which contrasts with the Mekong Basin with higher yields but much lower WP. Both basins have more than 1200 mm/year rainfall, slightly higher in the Mekong. However, the cropping intensity in the Indus-Ganges is much higher so that there is a high demand for water for irrigation. Farmers are therefore under pressure to increase water-use efficiency. Crop diversification in the Indus-Ganges also contributes significantly to high WP. Cash crops such as millet, sugarcane and pulses greatly increase the economic returns to irrigation, and reduce the risks imposed by climate extremes such as floods and droughts (Sharma *et al.* 2010).

Several basins calculated WP in terms of ET for maize, rice and wheat separately so that we can compare the performance of the production systems of these major food staples. WP of maize is highest in the Yellow River (0.97 kg/m^3), followed by the Mekong (0.58 kg/m^3). Maize is the single dominant crop in the Limpopo but WP is very low at 0.14 kg/m^3. For wheat, ET used in the Limpopo Basin is the yearly total while for other basins it is only for the wheat-growing season. However, even when we correct ET for the crop growth period in the Limpopo, which has pronounced seasonality, WP is still considerably lower than in the Yellow River. The difference is explained by the difference in yields of the mostly irrigated, high-yielding Yellow River compared with the rainfed, low-yielding Limpopo. The WP of wheat in the Yellow Basin is 48% higher than that of the Indus-Ganges, although wheat is a major crop in both basins. WP of rice showed less variation. The Indus-Ganges lead all the basins with 0.74 kg/m^3, closely followed by the Yellow River (0.50 kg/m^3), the Mekong (0.43 kg/m^3) and the Nile (0.14–0.67 kg/m^3). Rice is an important component of regional food security in the Indus-Ganges and the Mekong and the areas of rice cultivation are much bigger than in any other basins.

Yields and WP of irrigated crops were compared with rainfed crops in some basins. As expected, irrigated crops yield much higher than rainfed crops in all cases, although WP showed different trends. In the Yellow River yields of maize were 3.7 and 3.0 t/ha

Table 1. Water consumption and water productivity of the 10 BFP basins. Combined data for the latest years.

Basin	Water source	Cropland area	Precipitation	ETa[a]	ETp	Crop productivity: Crop types	Crop productivity: Yields	Crop productivity: As ET		Crop WP
		million ha	mm	mm	mm		t/ha	kg/m³	US$/m³	CV of kg/m³
Yellow River	irrigated	7.5	452			wheat/maize/soybean/rice	3.7/5.3/ 1.4/5.4	1.39/0.97/ 0.26/0.5	–	0.51/0.32/ 0.13/0.25
	rainfed	8.8				maize/soybean	3.0/1.4	1.09/0.41	–	0.36/0.16
Mekong	irrigated	3.28	1516	657	1457	rice/sugarcane/maize	2.87/64.5/ 3.79	0.431/9.81/0.576	0.059	0.16
	rainfed	11.22								
Indus-Ganges	irrigated	62.1	1254	416/299	610	rice/wheat	2.6/2.65	0.74/0.94	0.131	0.44/0.7
	rainfed	17.5								
Karkheh	irrigated	0.45	358	323		wheat/barley/maize	3.3/2.6/7.4	–	0.22	0.45
	rainfed	1				wheat/barley	1.5/1.4	–	0.051	0.2
Nile[c]	irrigated	5.5	563	554		maize/rice/wheat	8/9.5/6.5		0.042	
	rainfed	20.5				maize/rice/wheat	1.5/1.9/1.9	–		–
Limpopo	irrigated	0.244	530	779[b]	1676	maize	3.6	0.14	0.012	–
	rainfed	2.06				maize				
Niger	irrigated	0.075				rice		0.14–0.67		
	rainfed	0.425				millet, sorghum	1.0	0.1	–	
Volta	irrigated	0.036								
	rainfed	2.9				maize/sorghum/millet	1.3/1/0.9	0.15/0.1/0.08	–	
São Francisco	irrigated	0.355				banana/mango/coconut/grape		–	–	
	rainfed	5.15				maize/beans		–	–	
Andes	irrigated	3.09	1835					–	–	
	rainfed	10.88				potato/maize/pasture		–	–	

Notes: [a]Actual ET of growth periods of major crops unless specified; [b]Actual ET of a calendar year; [c]It is assumed the three major crops in the Egypt are fully irrigated and elsewhere are rainfed.

irrigated and rainfed respectively. In contrast, WP of irrigated maize is 0.97 kg/m^3 compared to rainfed 1.09 kg/m^3. Patterns in the Karkheh were different. Both the yields and WP of rainfed crops are lower than irrigated crops, which suggest that there may be several factors contributing to water productivity. In assessing WP performance, WP values have to be considered in relation to the specific basin setting and comparisons can only be made between the same settings.

There was wide spatial variability of crop WP within each of the basins. The coefficient of variation (CV) of WP for maize, rice and wheat in the four Asian basins varied mostly from 0.3 to 0.5, with the extreme high of 0.7 for wheat in the Indus-Ganges and extreme low of 0.16 for rice in the Mekong. Higher CVs indicate higher levels of heterogeneity in the basin, suggesting greater chances to close the gap between the good and the poor performers. Figure 3 illustrates the magnitude and variations of crop WP in the Nile and Indus-Ganges basins. WP in the Indus-Ganges is greater than that of the Nile whose narrower range of WP lies within the broader variation of the Indus-Ganges. The intra-basin variability is also clearly demonstrated in the two pixel-based WP maps. There are bright spots in the upstream of the Indus and the delta of the Nile with average high WP of US$/m^3 0.19 and 0.20 respectively, which are 1.5 and 4.8 times of basin averages respectively. Understanding the reasons for these differences would both assess the potential for improvement and identify priority interventions in low-performing areas. We discuss this aspect in later in the paper.

Analysis of time-series data on crop production, water consumption and WP shows the temporal evolution of agricultural systems. Where time-series data were available, physical WP generally increased, though at different rates. Figure 4 shows the trend of ET, yield and WP of rice in the lower Mekong Basin 1995–2003. Yields have gradually increased while ET has decreased. As a result, WP for ET has increased 20% from 0.344 kg/m^3 to 0.431 kg/m^3. WP of rice in US$/m^3, however, has declined significantly over this period. Markets have played an important role here. The farm gate price of rice has decreased from US$166/t in 1995 to US$126/t in 2003 (FAOSTATS,

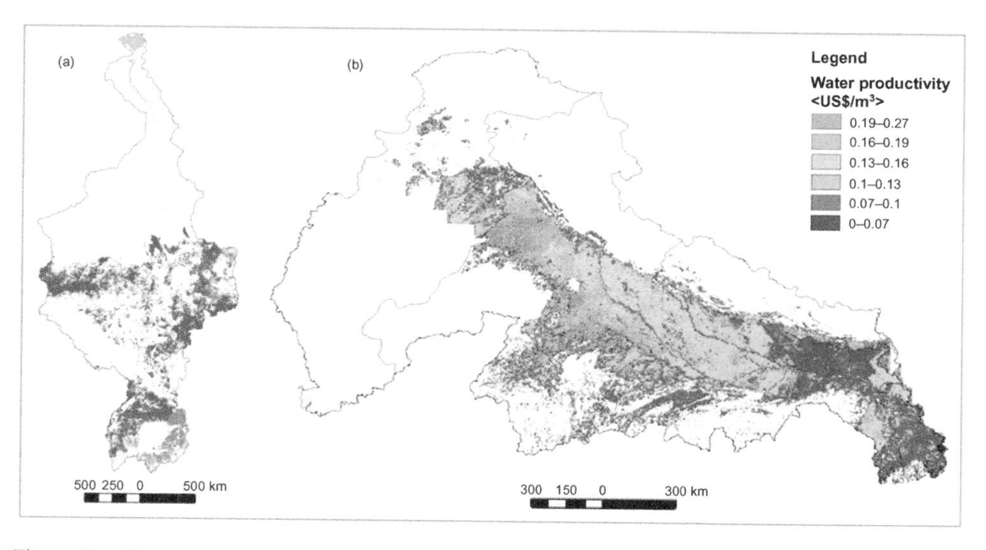

Figure 3. Crop water productivity and the variations in the Nile (left) and the Indus-Ganges basins (right).

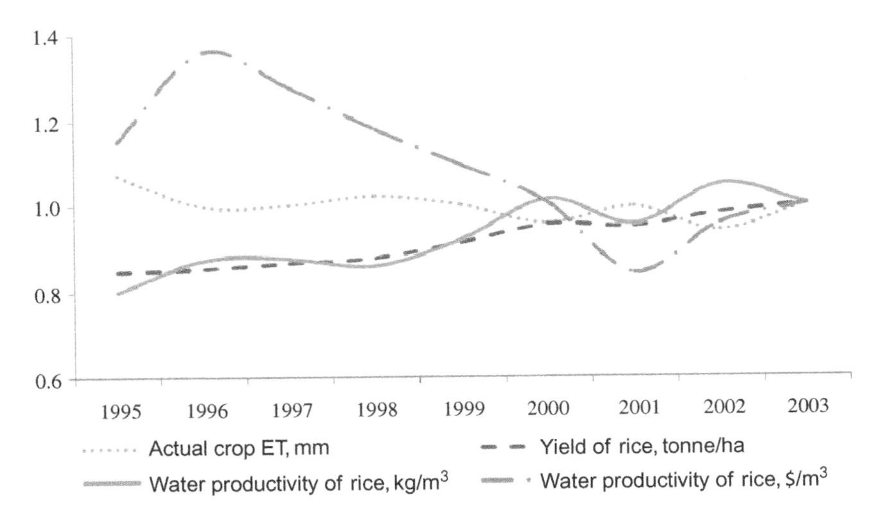

Figure 4. ET, yield and water productivity of rice in the lower Mekong Basin. The data were normalized to 2003 values.

http://faostat.fao.org/site/570/default.aspx), with even greater inter-season fluctuations within this period. This means that the farmers' gain from improved crop and water management has been offset by the market. In Bangladesh, WP of *kharif* (wet season) rice has always been lower than that of the *rabi* (dry season) rice. This is possibly due to the faster pace of adoption and deeper penetration of the high-yielding variety (HYV) technology and groundwater irrigation during the *rabi* season.

The importance of livestock and fishery production systems

Livestock and fisheries are important sources of farmers' food production and income in the BFP basins. When we add data for livestock and fisheries to crops, the agricultural production changes dramatically. Figure 5 shows the time trend 1999–2004 of GVP of crops, livestock and aquaculture in Cambodia and Vietnam in the Mekong (Kirby *et al.* 2009). There are no reliable figures for capture fish except in 2000. Even without capture fish, livestock and aquaculture fisheries contributed 24–34% to the overall agricultural GVP in the basin in 2003/04. The rapid development of aquaculture ensured substantial improvement

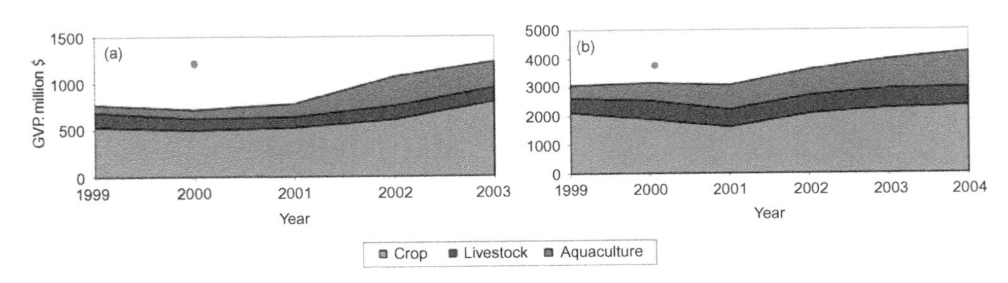

Figure 5. The gross value production of crops, livestock, and aquaculture in: (a) Cambodia; and (b) Vietnam in the Mekong Basin. The dot indicates the addition of GVP of capture fish. Data were only available for 2000. Source: Kirby *et al.* (2009).

in agricultural production, especially in Vietnam where crop production remained static. Results were similar in the Karkheh Basin, where livestock made a major contribution to WP (Ahmad *et al.* 2009b).

Contribution of WP to food security and livelihood

The contribution of agriculture to GDP ranges from 7% to 55% across the 10 basins. This contribution supports one billion people living in rural areas, which account for 72% of the total population of the 10 basins. More than 300 million people live below poverty lines as defined by the United Nations or the respective countries. The poor are overwhelmingly dependent on agriculture for their livelihoods.

Improved agricultural water productivity could significantly contribute to food security and livelihoods (Cook *et al.* 2009) of rural and city dwellers. Improved agricultural WP is possible through either increased production or reduced water use. When production is increased, either through higher yields or diversified crops, food security increases, and most likely the increment will be converted into higher income. Figure 6 illustrates the relation between revenue from land and crop WP of lime fruit at Buriti Vermelho, in the São Francisco Basin in Brazil. Higher WP is positively linked to higher income generated by the crop. When water is saved, the cost of its diversion, that is, labour, electricity and water fee, are reduced. The saved water will also contribute to downstream users for more agricultural production, recreation, or city uses, which overall contributes to regional development and livelihoods of all people in the basin.

Irrigation helps assure higher yields and water productivity. The Yellow, the Indus-Ganges, and the Karkheh basins are better equipped with irrigation and drainage facilities, farm machinery, plant breeding, pesticide control and other technologies than most other BFP basins. Consequently, production of agriculture is reliably sustained at higher levels. In contrast, the predominantly rainfed agriculture of the African basins is most vulnerable to droughts. With the exception of the large commercial farms in the South African part of the Limpopo, crop failure due to drought is common. Small-scale supplementary irrigation has been successful in the Indus-Ganges, Nile, Volta, Niger and Limpopo basins. The technology uses minimum water to help crops survive short dry spells and hence increase both crop production and water productivity (Sharma *et al.* 2010). Poor farmers with access to the technology are better protected from the hazard of drought.

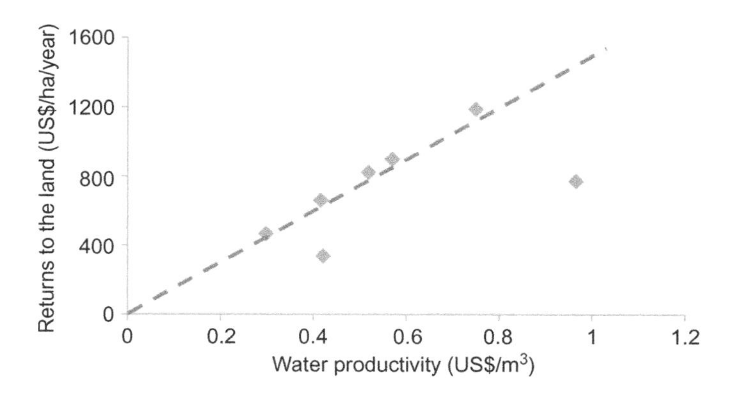

Figure 6. The relation between land revenue and WP of lime fruit in Buriti Vermelho, São Francisco Basin.

Livestock and fisheries are also important components of food security and livelihoods. We have described their contribution to agricultural production in the Mekong Basin above. In the riparian countries of the Limpopo (South Africa, Mozambique, Botswana and Zimbabwe) livestock contributes 45% of agricultural production. In the lower reaches of the Ganges, Bangladesh produces 2.6 Mt of fish annually, which contributes 21% of the country's total agricultural production. It provides more than 60% of the population's intake of animal protein and is the second most important sector in export earnings. It employs 12.5 million people, most of whom are rural poor, and thereby makes a substantial contribution to poverty reduction. Livestock and fisheries offer more livelihood opportunities for the poorest farmers who have limited land or poor access to water and other inputs.

Towards improved water productivity for people's livelihoods

Causes of variations and threats to water productivity

Land and water productivity varies widely both across and within basins. The Asian basins, Yellow, Mekong, Indus-Ganges and Karkheh have more productive farming systems than those in Africa and South America. There are many causes for the variation, some of which we discuss in this section, together with some of the threats to sustainable agriculture.

The level of socio-economic development has a big impact on agriculture. Hanjra and Gichuki (2008) suggested that, in most cases, the higher the contribution of agriculture to GDP the higher the incidence of poverty. In turn, this limits farmers' capacity to increase inputs to agriculture, improve WP, and cope with climate extremes such as droughts and floods. The contribution of agriculture to GDP is slightly lower in Asia and highest in the Nile, Volta and Niger. The African basins mostly rely on rainfed agriculture with poor infrastructure, low inputs of fertilizer and irrigation, and consequently low crop yields and low crop WP. But the Limpopo and the Andes basins, which have the lowest contribution of agriculture to GDP, also have low agricultural and water productivity. The reason is that South Africa skews the mean because it is a relatively well-developed country with severe inequity of highly developed commercial agriculture juxtaposed with a large number of poor subsistence farmers. The latter, together with poor smallholders from other Limpopo countries are most vulnerable to the droughts that occur frequently.

Water stress is a determining factor for all basins. Water stress occurs either due to physical scarcity, poor access, floods or waterlogging. Water for crop production is a concern in most areas including the extremely water-scarce basins of the Yellow, the Limpopo and the Indus. Water scarcity has worsened over years and the trend will continue due to the competitive demand from the rapidly expanding cities. Access to water is another constraint. The basins in Africa and South America are not the driest, but overall they do have the poorest access to water. Lack of appropriate storage and diversion infrastructure exposes farmland to droughts. At the other end of the scale, waterlogging affects the Mekong, and downstream parts of the Ganges, Limpopo and the Andes basins. Waterlogging is largely due to shallow groundwater or excess rainfall, whose effects can be reduced by suitable drainage infrastructure and appropriate management of irrigation water.

Improved seed varieties, traction energy, fertilizers, pesticides and soil management are critical inputs for large areas of low productivity. Land degradation is often another serious problem. Tanner and Sinclair (1983) concluded that, where actual yields are less than 40–50% of the potential, the effects of non-water factors are more pronounced. Combined management of soil, water, plants and pests is required to overcome these limits and give yield increases (de Wit 1992, Deborah *et al.* 2010).

Often site-specific technologies are required to achieve better outputs from land and water. For example, maize in the downstream parts of the Limpopo often suffers from waterlogging, indicating a need for improved drainage. It might also be practical however, to change cropping patterns to include more tolerant crops. Rice is an important crop in Ningxia Province, upstream in the Yellow Basin. However, the increasing demand on water reallocation from downstream provinces has put pressure on farmers to reduce rice cultivation by 30% or more. In the Indian state of Punjab, rice irrigated with groundwater cannot by law be transplanted until the monsoon starts, avoiding high evaporation losses in the hot weather that precedes the monsoon. Monocrops are popular in the Indus-Ganges Plain for the convenience of supplying irrigation water and operation of farm machinery. In most other areas, however, crop diversification is important to improve overall land and water productivity and reduce risks. For example, the smallholder farmers in the Lao PDR have increased production of various upland crops while production of rice has remained static.

Access to well-functioning markets is central to determining the overall value of agricultural production and net returns to farmers. Although agriculture is often subsidized, markets often are not accessible to many farmers. Subsistence farmers in the Limpopo Basin are often obliged to sell their produce to big farmers, who have the resources and bargaining power to send it to distant markets (Rosemary and Johann 2009). In the Gazeira in Sudan, cotton is a mandatory crop financed and marketed by the government. It is increasingly grown at a loss as production falls because of the lack of incentives to the tenant farmers (Abdel Karim and Kirschke 2009). Price fluctuations have strong impacts on the value of agricultural crops regardless level of crop yields. For example, maize yields are similar in the riparian countries of the Limpopo Basin, but local market prices are widely different in each causing huge differences in GVP and consequently the returns to land and water. Figure 4, discussed above, shows that increases in crop yield and WP in the Mekong basin were totally offset by falls in the market price of rice. On the other hand, the minimum support price (MSP) of wheat and rice by the government of India provides good remuneration to the farmers but discourages them from diversifying into crops with lower water requirement (Joshi 2005).

Although there have been continuous efforts to enhance the efficiency of agricultural systems, new threats are emerging, among which environmental degradation and climate change are the two major concerns. As agriculture develops it almost certainly has negative impacts on the environment (Bakkes *et al.* 2009). But the externalities can be managed to different degrees. Agriculture withdraws and consumes water that could be otherwise beneficial to other users including the environment (Molden *et al.* 2010). In closed basins, where there is competitive demand for water, the environment is often the loser. Environmental flows, which are the minimum flows required to maintain the health of rivers, are often ignored (Smakhtin and Anputhas 2006). The Yellow River ceased to reach the sea in the 1990s. This is no longer the case, but the pressure of high demand for water keeps increasing. Although the national south-to-north water transfer project will deliver a good amount of water to the basin, it does not consider environment flows due to the very high cost of diversion. The Indus is another closed basin where both surface water and groundwater are over exploited, causing drastic declines in groundwater table, which threatens sustainability of the agricultural systems that depend on it. In these cases the broadly defined WP, including industry outputs, are increasing but at the expense of agricultural sustainability, which supports multi-billion rural populations.

For the limited quantity of water left in river channels and aquifers, water quality often becomes a major concern. A survey in the Yellow River in 2007 found that 33.8% of the river system registered worse than level V for water quality, which is classified as not suitable for drinking, aquaculture, industry, or even agriculture (YRCC 2007). The Andes basins, due to their steep slopes, suffer from soil erosion in the upstream areas, which imposes another kind of water scarcity downstream due to loss of quality. In the Mekong Basin, water quality issues are more often confined to water-poor areas, for example, water pollution in the Tonle Sap, quality of groundwater in Northeast Thailand, and water salinization and acidification in the delta area of Vietnam (Kirby *et al.* 2009). In the lower parts of the Ganges Basin (West Bengal, Bangladesh), arsenic contamination of groundwater is widespread and is linked to its over-exploitation. Degradation of water quality is mainly caused by effluent from cities and rural households, but also from agriculture itself. Non-point-source pollution from agriculture is a major threat to water quality in areas of intensive irrigation, where it is often accompanied by high fertilizer inputs (FAO 1996). The severely degraded water quality that results threatens water supplies and consequently, water productivity.

Climate change is projected to have various impacts on agricultural production systems. In spite of forecasts that total rainfall may increase in some regions, the water available to agriculture will likely fall if water storage, diversion infrastructure, and management remain at their current levels (Backlund *et al.* 2008). More extreme climatic events, such as shorter and more intense rainy seasons and longer and more intense dry seasons will make agriculture, especially rainfed agriculture, more vulnerable, and hence reduce agricultural WP. The faster snow melt caused by increased temperature will increase flows in the river channels in the short term but in the longer term flows will change their seasonality as rain replaces snow in the headwaters. The glaciers of the Tibetan plateau and surrounding mountains that border South and East Asia will be significantly affected by global warming. The changes in this region could affect billions of people in the East, Southeast, South and Central Asia, many of whom depend on river diversion for irrigated agriculture.

Sea level rises will adversely affect the deltas, causing greater risks of floods and saltwater intrusion. It is estimated in the Mekong Basin more than one million people are expected to be affected by 2050 (Nicholls *et al.* 2007). Climate change will also have negative impacts on crop yields on a global scale as a result of shorter growth periods and lower soil moisture caused by increased temperatures and more uneven distribution of rainfall (Wheeler *et al.* 2000, Lobell and Field 2007). Crop water-use efficiency will decrease both because of lower yields as well as increasing evaporation from bare soil (Xiao *et al.* 2009). Figure 7 illustrates the sensitivity of runoff to climate change together with land-use change in the Andes basins, with the most sensitive regions indicated by the red circles. Although it is forecast that climate change will have an impact on agriculture and water management, more detailed studies on the effects of changed rainfall and particularly the impacts on runoff are required. Assessments of the impact of climate change on agriculture and agricultural water productivity is especially needed.

The rapid expansion of cities is also a potential threat to agriculture by diverting resources away from the agricultural sector. Cities are much more powerful in negotiating water reallocation (Molle and Berkoff 2009). They convert surrounding farmlands, which are usually favourable for cultivation with high agricultural productivity, into factories, settlements and roads. Cities also attract large numbers of rural youth. A survey in the Yellow River basin revealed that the average age of household heads increased from 42.7 years in 2004 to 50 years old in 2008, and women in the labour force increased from 62% to 78% respectively (Song *et al.* 2009). The rural areas are likely to face shortage of experienced male labour in the near future, which will potentially lead to poorer land and water management.

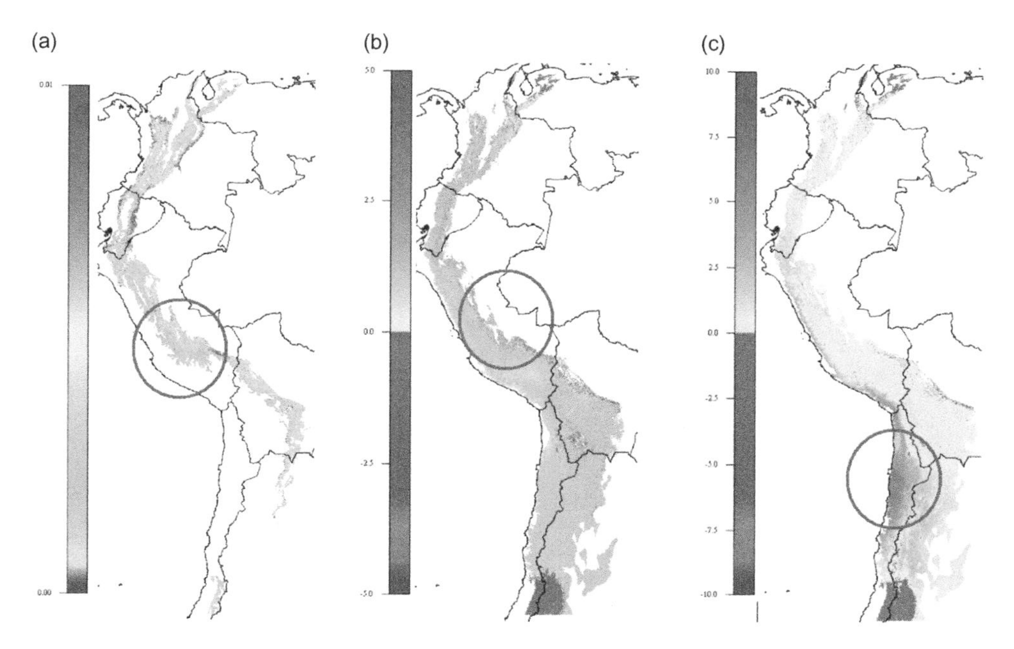

Figure 7. Sensitivity of the Andes to land use and climate change: What will happen by 2050 as we continue to push? (a) runoff sensitivity to tree cover change (% change in runoff per % change in tree cover); (b) runoff sensitivity to precipitation change (% change in runoff per % change in precipitation); (c) runoff sensitivity to temperature change (% change in runoff per % change in precipitation). Red circles indicate areas of particularly high sensitivity. Source: Mulligan *et al.* (2010).

Potential for improvement

We can look at the potential for improvement of water productivity in two steps: firstly through the comparison of "bright spots" and "hot spots" to identify visible yield gaps, and secondly through site-specific assessment of the biophysical potential. The second step involves local analysis based on solar radiation and soil of the region to explore water-fertilizer applications in conjunction with crop genetic innovations. This approach was successful in achieving the green revolution and still remains a major strategy to achieve the world's long-term goal of food security. While agriculture in most of the developing world is still at a low level of management, however, the first approach provides a greater chance for improved agriculture performance in near future.

As described above, there is large variation in WP across basins as well as within basins. Crop WP from remote sensing at the pixel level provides explicit description of both the magnitude and the variation. The levels of WP for rice and wheat in the Indus-Ganges Basin are plotted in Figure 8. The CVs, 0.44 for rice and 0.70 for wheat, are much higher than those for the world (Zwart and Bastiaanssen 2004). This is because remote sensing captures every pixel in a basin, including extreme spots. The CV of wheat is much higher than that of rice, meaning water management of wheat is more diverse compared with rice. This is because there is little rainfall during the wheat-growing season and hence the crop depends on irrigation. Improving the irrigation management of wheat is probably the most urgent and easiest way to improve basin WP.

The potentials of rice and wheat are different, both in terms of magnitude and areas of focus. Water productivity generally increases with increasing yield (Figure 9). The "bright

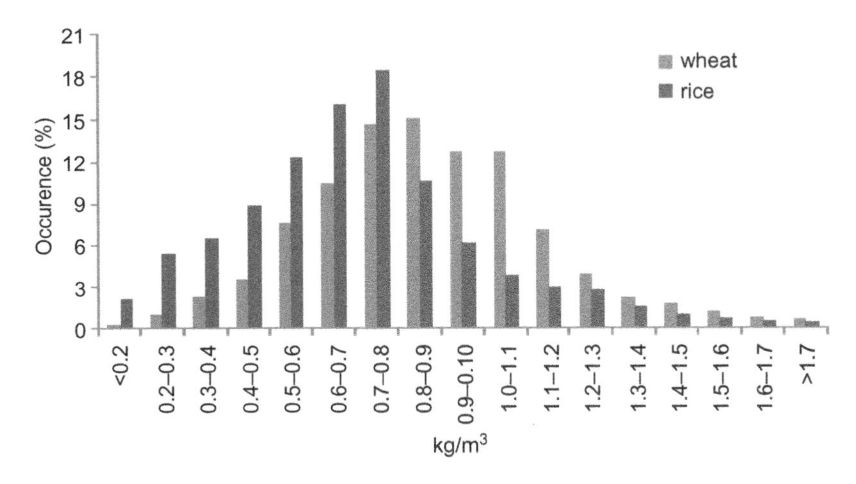

Figure 8. The histogram distribution of WP values for rice and wheat.

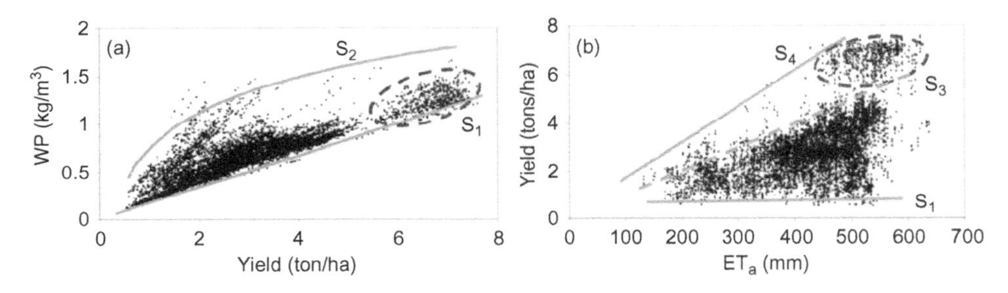

Figure 9. Relations between (a) water productivity and yield and (b) yield and evapotranspiration of rice in the Indus-Ganges basins. Adapted from Cai and Sharma (2010).

spot" of the Indian Punjab clearly stands out in both plots (circled). The yields of the Punjab are so high that the slope of yield against ET changes from S_3 to S_4. The largest scope for improvement of rice in the region will be firstly to improve the low yield and low WP areas to match those that provide the S_2 trend. That is, to maintain yields with less consumption of water. The final target would be to increase the yield levels of all areas to "bright spot" values, during which the water consumption too will increase.

Such well-performing "bright spots" with higher crop yields, higher gross value of production and higher water productivity exist in every basin, for example, the Delta in the Nile and the centre-pivot irrigated commercial farms in the Limpopo. There are local constraints to prevent large areas from achieving the same, but clearly there is large scope for improvement.

Implications for interventions

Increasing agricultural WP can contribute to better food security and enhanced livelihoods, but requires addressing a number of institutional and technical issues amongst which is the need to avoid environmental degradation (Molden *et al.* 2010).

Bright spots identify sites where farm yields most nearly match the biophysical potential through a range of interventions, which may not be directly replicable but are indicative of success. There are sites in every basin that have high yields, and often high WP. Molden *et al.* (2010) and Cai and Sharma (2010) conclude that further increasing yields and WP in these areas will be a long and gradual process, including innovations in plant breeding. A more pressing priority will be to increase the yields and WP of the large under-performing areas through tailored interventions.

Improving WP through better water management is central to the solutions for improved productivity. In rainfed systems, it is often the poor distribution of rainfall rather than the total precipitation that limits productivity (Sharma *et al.* 2010). On-farm conservative farming practices can reduce non-beneficial evaporation and increase soil water content. In irrigated systems, Molden *et al.* (2010) suggested that the general perception about the potential to save water is commonly overstated. This is especially true at the basin scale, where irrigation return flows are reused (Ahmad *et al.* 2007). Reliable, low-cost irrigation would enable more poor farmers to improve their productivity. Appropriate drainage systems can have major impacts on downstream lowland areas. Enhancing the level of water management together with management of soil, pests and diseases together could significantly improve WP of many agricultural systems.

Strategies to manage water better, together with other production resources, are different in different regions. In water-scarce situations, improving WP is critical to produce more food with less water. In water-rich areas, improvement of water productivity is not necessarily the target. Rather, it will be to produce as much as possible from limited land resources.

Agricultural sustainability is a key element to be maintained in the process of improving agricultural and water productivity. Over-exploitation of surface and ground water, as often seen in the Asian basins, leaves little water or low quality water for environment flows. These issues are often ignored by local stakeholders and few basins have the institutional capacity to deal with it (Giordano 2003). Localized institutions for sub-watersheds and basins covering broader social issues have to be established, which introduce tradeoffs between agriculture and the environment and include incentives and compensations.

Conclusions

Agricultural WP is a key indicator to link basin water resources and agricultural outputs, which sets a useful baseline for efficient agricultural water management. A holistic overview of agricultural production systems at the basin level can help to identify the issues that are relevant for informed policy making. Agricultural WP is important to regional food security as well as to farmers' livelihoods. Increasing WP needs to consider the economic costs of doing so. While bright spots illustrate the potential for improvement, it might not be feasible to achieve same level of WP elsewhere in a cost-efficient way. Small-scale interventions, such as supplemental irrigation, are relatively easy to adopt with less investment, and could significantly improve agricultural WP.

The potential to save water is not as big as many think, but there is significant scope to improve WP. Different farming systems have different priorities in different locations. Markets play a key role in converting agricultural production to income. We note, however, the need to balance WP and environment.

The concept of WP has gone through various stages of development, and recently has been expanded to the basin context and to cover broad systems of agricultural production. However, methodology development is still a major issue for large-scale assessment of whole basins. Current assessment relies overwhelmingly on statistical data, which are variable in quality, the spatial scale is often poor and does not correspondent to hydrological boundaries. Remote sensing is a powerful tool to estimate both crop production and consumptive water use. Basin-scale hydrological modelling also helps to study the processes of water cycling and to examine existing interventions. Combining the two provides greater opportunities to capture images of WP as well as understanding processes on the ground. WP for livestock and fisheries needs major development of both concepts and methodology. The key issue is to measure the actual water use, which can then guide potential interventions.

References

Abdel Karim, I.E.E.A. and Kirschke, D., 2009. World cotton markets, liberalisation and cotton trade in Sudan. *World Review of Entrepreneurship, Management and Sustainable Development*, 5 (3), 299–310.

Ahmad, M.D., *et al.* 2007. At what scale does water saving really save water? *Journal of Soil and Water Conservation*, 62 (2), 29A–35A.

Ahmad, M.D., Turral, H., and Nazeer, A., 2009a. Diagnosing irrigation performance and water productivity through satellite remote sensing and secondary data in a large irrigation system of Pakistan. *Agricultural Water Management*, 96 (4), 551–564.

Ahmad, M.D., *et al.*, 2009b. Mapping basin-level water productivity using remote sensing and secondary data in the Karkheh River Basin, Iran. *Water International*, 34 (1), 119–133.

Alemayehu, M., *et al.*, 2008. Livestock water productivity in relation to natural resource management in mixed crop, livestock production systems of the Blue Nile River basin, Ethiopia. *In:* E. Humphreys, *et al.*, eds. *Fighting poverty through sustainable water use: volume II. Proceedings of the CGIAR Challenge Program on Water and Food 2nd International Forum on Water and Food*, 10–14 November 2008, Addis Ababa, Ethiopia. Colombo: CPWF.

Bakkes, J., *et al.*, 2009. *Getting into the right lane for 2050: a primer for EU debate.* Bilthoven, Netherlands: Netherlands Environmental Assessment Agency.

Backlund, P., *et al.*, 2008. *The effects of climate change on agriculture, land resources, water resources, and biodiversity in the United States.* Synthesis and Assessment Product 4.3. Washington, DC: Climate Change Science Program, US Environmental Protection Agency.

Béné, C. and Friend, R.M., 2009. Water, poverty and inland fisheries: lessons from Africa and Asia. *Water International*, 34 (1), 47–61.

Cai, X.L. and Sharma, B.R., 2010. Integrating remote sensing, census and weather data for an assessment of rice yield, water consumption and water productivity in the Indo-Gangetic river basin. *Agricultural Water Management*, 97 (2), 309–316.

Carpenter, S.R., *et al.*, 1998. Nonpoint pollution of surface waters with phosphorus and nitrogen. *Ecological Applications*, 8 (3), 559–568.

Cook, S., Gichuki, F., and Turral, H., 2006. *Water productivity: estimation at plot, farm, and basin scale.* Basin Focal Project Working Paper No. 2. Colombo: Challenge Program on Water and Food.

Cook, S.E., *et al.*, 2009. Water, food and livelihoods in river basins. *Water International*, 34 (1), 13–29.

Deborah, B., Kim, G., and William, C. 2010. Managing water by managing land: Addressing land degradation to improve water productivity and rural livelihoods. *Agricultural Water Management*, 97 (4), 536–542.

Descheemaeker, K., Amede, T., and Haileslassie, A., 2009. *Livestock and water interactions in mixed crop-livestock farming systems of sub-Saharan Africa: interventions for improved productivity.* IWMI Working Paper 133. Colombo: International Water Management Institute (IWMI).

de Fraiture, C., *et al.*, 2007. Looking ahead to 2050: scenarios of alternative investment approaches. *In*: D. Molden, ed. *Water for food, water for life: a comprehensive assessment of water management in agriculture.* London: Earthscan and Colombo: IWMI, 91–145.

de Wit, C.T., 1992. Resource use efficiency in agriculture. *Agricultural Systems*, 40 (1–3), 125–151.

Economy, E.C., 2004. *The river runs black: the environmental challenge to China's future*. Ithaca, NY: Cornell University Press.

FAO, 1996. Control of water pollution from agriculture. Irrigation and Drainage Paper 55. Rome: Food and Agriculture Organization of the United Nations (FAO).

Giordano, M.A., 2003. Managing water quality in international rivers: global principles and basin practice. *Natural Resources Journal*, 43 (1), 111–136.

Hanjra, M.A. and Gichuki, F., 2008. Investments in agricultural water management for poverty reduction in Africa: case studies of Limpopo, Nile, and Volta river basins. *Natural Resources Forum*, 32 (3), 185–202.

Hoekstra, A.Y., *et al.*, 2009. *Water footprint manual: state of the art 2009*. Enschede, Netherlands: Water Footprint Network.

Joshi, P.K., 2005. Crop diversification in India: nature, pattern and drivers [online]. New Delhi: National Centre for Agricultural Economics and Policy Research. Available from: http://www.adb.org/Documents/Reports/Consultant/TAR-IND-4066/Agriculture/joshi.pdf [Accessed 18 November 2010].

Kirby, M., *et al.*, 2009. *Mekong Basin focal project: final report*. Colombo: CPWF.

Kim, H.K., *et al.*, 2009. Estimation of irrigation return flow from paddy fields considering the soil moisture. *Agricultural Water Management*, 96 (5), 875–882.

Lobell, D.B. and Field, C.B., 2007. Global scale climate-crop yield relationships and the impacts of recent warming. *Environmental Research Letters*, 2, 014002.

Mainuddin, M. and Kirby, M., 2009. Spatial and temporal trends of water productivity in the lower Mekong river basin. *Agricultural Water Management*, 96 (11), 1567–1578.

Mertz, O., Wadley, R.L., and Christensen, A.E., 2005. Local land use strategies in a globalizing world: subsistence farming, cash crops and income diversification. *Agricultural Systems*, 85 (3), 209–215.

Mohamed, Y.A., Bastiannssen, W.G.M., and Savenije, H.H.G., 2004. Spatial variability of evaporation and moisture storage in the swamps of the upper Nile studied by remote sensing techniques. *Journal of Hydrology*, 289 (1–4), 145–164.

Molden, D., 1997. *Accounting for water use and productivity*. System-wide Initiative on Water Management (SWIM) Paper No.1. Colombo: IWMI.

Molden, D., *et al.*, 2003. A water-productivity framework for understanding and action. *In*: W. Kijne, R. Barker, and D. Molden, eds. *Water productivity in agriculture: limits and opportunities for improvement*. Colombo: IWMI and Wallingford, UK: CABI, 1–18.

Molden, D., *et al.*, 2007. Pathways for increasing agricultural water productivity. *In*: D. Molden, ed. *Water for food, water for life: a comprehensive assessment of water management in agriculture*. London: Earthscan and Colombo: IWMI, 279–310.

Molden, D., *et al.*, 2010. Improving agricultural water productivity: between optimism and caution. *Agricultural Water Management*, 97 (4), 528–535.

Molle, F. and Berkoff, J., 2009. Cities vs. agriculture: a review of intersectoral water transfers. *Natural Resources Forum*, 32 (1), 6–18.

Mulligan, M., *et al.*, 2010. *Andes basin focal project: final report*. Colombo: CPWF.

Mustafa, M.G., *et al.*, 2010. Assessing potential interventions to maximize fisheries - water productivity in the Eastern Gangetic Basin (EGB), evaluation of constraints and opportunities for improvement: context Gorai-Madhumati (GM) sub-basin. WorldFish-IWMI Research Report. Penang: WorldFish and Colombo: IWMI.

Muthuwatta, L.P., *et al.*, 2010. Assessment of water availability and consumption in the Karkheh River basin, Iran using remote sensing and geo-statistics. *Water Resources Management*, 24 (3), 459–484.

Nicholls, R.J., *et al.*, 2007. Coastal systems and low-lying areas. Climate change 2007: impacts, adaptation and vulnerability. *In:* M.L. Parry, *et al.*, eds. *Contribution of working group II to the Fourth Assessment Report of the Intergovernmental Panel on Climate Change*. Cambridge: Cambridge University Press, 315–356.

Rodell, M., Velicogna, I., and Famiglietti, J.S., 2009. Satellite-based estimates of groundwater depletion in India. *Nature*, 460 (7258), 999–1002.

Rosemary, A.E. and Johann, K. 2009. Supermarket expansion in developing countries and their role in development: experiences from the Southern African Development Community (SADC). Paper presented at the *International Association of Agricultural Economists Conference*, Beijing, China, 16–22 August 2009.

Roost, N., *et al.*, 2008. An assessment of distributed, small-scale storage in the Zhanghe irrigation system, China. Part II: impacts on the system water balance and productivity. *Agricultural Water Management*, 95 (6), 685–697.

Smakhtin, V. and Anputhas, M., 2006. *An assessment of environmental flow requirements of Indian river basins*. IWMI Research Report 107. Colombo: IWMI.

Sharma, B.R., *et al.*, 2010. Estimating the potential of rainfed agriculture in India: Prospects for water productivity improvement. *Agricultural Water Management*, 97 (1), 23–30.

Singh, N. and Sontakke, N.A., 2002. On climatic fluctuations and environmental changes of the Indo-Gangetic plains, India. *Climatic Change*, 52 (3), 287–313.

Song, Y.C., *et al.*, 2009. Feminization of agriculture in rapid changing rural China: policy implication and alternatives for an equitable growth and sustainable development. Paper presented at *the FAO-IFAD-ILO Workshop on Gaps, trends and current research in gender dimensions of agricultural and rural employment: differentiated pathways out of poverty*. Rome, 31 March–2 April 2009.

Tanner, C.B. and Sinclair, T.R., 1983. Efficient water use in crop production: research or re-search? *In*: H.M. Taylor, W.A. Jordan, and T.R. Sinclair, eds. *Limitations to efficient water use in crop production*. Madison, WI: American Society of Agronomy, 1–27.

Turner, G., Baynes, T., and McInnis, B., 2009. A water accounting system for strategic water management. *Water Resources Management*, 24 (3), 513–545.

Wheeler, T.R., *et al.*, 2000. Temperature variability and the yield of annual crops. *Agriculture, Ecosystems and Environment*, 82 (1–3), 159–167.

Xiao, G.J., *et al.*, 2009. Effects of temperature increase on pea production in a semiarid region of China. *Air, Soil and Water Research*, 2 (1), 31–39.

YRCC, 2007. Yellow River water resources bulletin (online). Yellow River Conservancy Commission (YRCC). Available from: http://www.yellowriver.gov.cn/other/hhgb/2007.htm [Accessed March 2010]. (In Chinese).

Zwart, S.J. and Bastiaanssen, W.G.M., 2004. Review of measured crop water productivity values for irrigated wheat, rice, cotton and maize. *Agricultural Water Management*, 69 (2), 115–133.

The resilience of big river basins

Graeme S. Cumming

Big river basins are complex systems of people and nature. This article explores the resilience of nine case studies of big river basins. A system description and generic conceptual model suggests that resilience to changes in water quantity is critical. When water becomes limiting, the social-ecological system must adapt rapidly if key elements (for example, communities, biodiversity) are to be maintained. Water limitation imposes a water economy and alters political and institutional links between actors. Pro-active management for resilience demands politically acceptable participatory processes that use the best possible science and incorporate social, ecological and economic elements in problem definition and solution.

Introduction

This paper explores the central issues relating to ecosystem services and human well-being in a set of 10 big river basins from around the world. Each basin has important social, economic and ecological elements; each can be considered a social–ecological system (SES), a linked system of people and nature. As the preceding papers make abundantly clear, the big river basins in this set of case studies are similar in some important ways and different in other, equally important ways. They all face big challenges in the future, including problems that are global in nature (such as climate change and human population expansion) and problems that are more localized (such as regional conflict or inequities). From both analytical and management perspectives, it is important to understand the ways in which what has happened in one basin can inform our understanding of other basins. Can we learn from comparisons between these basins whether, and how, they will be able to cope effectively with what the future holds for them? And more specifically, can analysis of their current strengths and weaknesses identify possible pitfalls or pro-active strategies through which additional coping ability could be built?

As with any comparative analysis, answering these questions requires a level of generalization. Generalization in turn requires reference to a set of underlying concepts that can be used to identify commonalities between systems that are superficially quite different. There are many different ways in which such an analysis could be framed, and no single guide (aside from eventual utility) as to which framing is best. Traditional approaches to the problems of big river basins have tended to approach specific issues from a disciplinary perspective. For example, the allocation and efficiency of water use have been major themes in assessing the economics of river basin management (for example, Rosegrant *et al.* [2000]). While traditional approaches often provide essential information about individual aspects of a broader problem, they also make a large number of assumptions about the constancy of causal relationships, the prevalence of linear relationships, and the structure of the study system. The accuracy of standard scientific predictions is heavily contingent on things remaining as they are (Clark *et al.* 2001). In the context of river management, which includes high levels of uncertainty in many social and ecological parameters, conventional approaches to minimizing risks are quite likely to be sub-optimal (Clark 2002). Furthermore, the real world includes "messy" details like pushy stakeholders, political processes, technological change and feedbacks between science and human behaviour (Ludwig 2001, Stirling 2003, Waltner-Toews *et al.* 2003).

Big river basins are similar to many other systems of people and nature in that they involve a set of complex, "wicked" interactions[1] and non-linear feedbacks. They are united by a shared biophysical component – water – which flows directionally along a well-defined gradient. Human societies have in many cases organized themselves around and along this gradient, with predictable variation in human-dominated components of the system (for example, income or city size) often occurring from headwaters to estuaries. The inescapable importance of environmental gradients, and the heavy reliance of human communities in a basin on a single resource (water), are probably the unique features of big river basins that distinguish them from other social–ecological systems.

A number of approaches to thinking about "wicked problems" in complex systems exist (e.g., Holling 2001, Norberg and Cumming 2008, Waltner-Toews *et al.* 2008, Ostrom 2009). The approach that I will take focuses on the idea of resilience (Holling 1973, Holling 2001), and particularly on the ability of key elements of the system to persist into the future (Cumming *et al.* 2005). I will first describe some elements of a general framework and then attempt to apply it in an integrative (if somewhat qualitative) analysis of the resilience of big river basins.

Resilience thinking

Resilience is used here as an emergent attribute of a social–ecological system. Resilience thinking (Walker and Salt 2006) has its roots in systems approaches, in which the world is perceived and described as a set of interacting "systems" (Checkland 2009). Systems are a form of model in that they are defined by the observer as a way of making sense of the complexities of the real world; they are intended to capture essential elements of a problem, rather than to describe accurately the full range of complexity that exists "out there". Analysis of resilience focuses on understanding how the number and nature of current system components, and their interactions and broader context, influence the ability of important elements of that system to persist into the future.

Although the concept of resilience has become more widely used in recent years, some confusion still exists about the definition of resilience and some related terms. Some current definitions of resilience-related terminology as used in this paper are offered in Table 1.

Table 1. Definitions of some of the terms used in thinking about resilience.

Term	Definition
System (Cumming and Collier 2005)	A cohesive entity consisting of key elements, interactions and a local environment; must show spatio-temporal continuity. An SES includes interactions between people and ecosystems.
Resilience (Holling 1973; Carpenter *et al.* 2001; Holling 2001; Cumming *et al.* 2005)	There are at least two complementary definitions of resilience that are currently used in the study of SESs. Carpenter *et al.* (2001) define resilience as: (1) the amount of change that a system can undergo and still maintain the same controls on function and structure; (2) the system's ability to self-organize; and (3) the degree to which the system is capable of learning and adaptation. Second, based on Cumming *et al.* (2005), system identity depends on maintaining essential elements through space and time. Resilience can thus be viewed as the ability of the system to maintain its identity in the face of internal change and external perturbations. Note that these definitions offer slightly different perspectives on the same concept; they are not in competition with each other.
Robustness (Anderies *et al.* 2004; Levin and Lubchenco 2008)	Anderies *et al.* (2004) cite an engineering definition (Carlson and Doyle 2002) focusing on "maintenance of system performance either when subjected to external, unpredictable perturbations, or when there is uncertainty about the values of internal design parameters". Levin and Lubchenco (2008) equate robustness and resilience, stating that both "mean much the same thing: the capacity of systems to keep functioning even when disturbed". Maintenance of function is a different criterion from maintenance of *identity* (unless identity is defined purely in terms of function, which is not usually the case) and this functional emphasis appears, as best I can tell, to be the primary distinguishing feature of "robustness".
Vulnerability (Adger 2006)	Vulnerability is a measure of the extent to which a community, structure, service or geographical area is likely to be damaged or disrupted, on account of its nature or location, by the impact of a particular disaster hazard (Organisation for Economic Co-operation and Development [OECD] Glossary) . A more resilience-oriented definition is offered by Adger (2006): "Vulnerability is the state of susceptibility to harm from exposure to stresses associated with environmental and social change and from the absence of capacity to adapt."
Sustainability (Norberg and Cumming 2008)	First introduced to the policy arena by the Bruntland Report (WECD 1987) which defined sustainable development as development that "meets the needs of the present generation without compromising the ability of future generations to meet their needs". Also defined as "the equitable, ethical, and efficient use of natural resources" (Norberg and Cumming 2008). Inclusion of "efficiency" is contentious in some circles.
Regime, regime shift (Carpenter 2003; Scheffer and Carpenter 2003)	A locally stable or self-reinforcing set of conditions that cause a system to vary around a local attractor; the dominant set of drivers and feedbacks that lead to system behaviour; a "basin of attraction". For ecosystems, a regime shift is "a rapid modification of ecosystem organization and dynamics, with prolonged consequences" (Carpenter 2003, chapter 1).

(*Continued*)

Table 1. (Continued)

Term	Definition
Threshold (Scheffer and Carpenter 2003; Scheffer 2009)	In non-linear relationships, the point at which a function (or a rate) changes sign; in state space, the location at which a system tends towards an alternative local equilibrium or new attractor; the combination of variables under which a system enters a new regime.
Feedback (Norberg and Cumming 2008)	Occurs when a change in system element A triggers changes in other system elements (B, C, and so on) that eventually influence element A. Feedbacks may be self-regulating (negative, homeostatic) or self-amplifying (positive). For example, in the human body, sweating is a self-regulating response that cools the body down, pushing it back towards an equilibrium temperature; while hyperthermia and convulsions in response to heat serve only to exacerbate the problem, moving the body further away from its preferred equilibrium point. Self-amplifying feedbacks are particularly important in social–ecological systems as they can push a system over a threshold and into a new regime.

The references in the left-hand column offer a recent example of a peer-reviewed publication in which the term is used and/or defined. Note that in many cases, different research groups have been working with very similar concepts using different terminology; differences here tend to matter less than similarities. Terminological differences also reflect the plurality of complex systems (that is, most complex systems can be viewed from many different but equally "correct" perspectives). The table is modified from Cumming (2011).

In thinking about system resilience, it is necessary first to define the system of interest (with particular emphasis on its most important elements) and then to consider how the system is likely to respond to specified future changes. System definition entails many subjective elements (Checkland 1981, 2009) and should be approached in a systematic manner to avoid excessive bias on the part of the analyst. One approach to system description is to break down the basic elements of a system into structural, spatial and temporal categories. These categories can be considered as axes, in the sense that each system element has a structure or nature, a location in space, and a location in time. A similar conceptual framework has been applied by Agarwal *et al.* (2002) to compare models of land use and land-cover change. As proposed by Cumming *et al.* (2005) and Norberg and Cumming (2008), structural elements in SESs include system components, internal interactions, and other system constituents that contribute to continuity, information processing and adaptation. The system interacts with its environment through input and output functions; and the environment provides a set of "givens" (for example, geographic location) as well as being a potential source of perturbations. The elements that are included in the structural axis should in theory be those that are considered to be most central to the system's identity (see detailed discussion in Cumming *et al.* [2005]). Examples of each element in the system description are given in Table 2.

The spatial axis captures spatial variation, hierarchical system arrangements in space, and spatial interactions and connections at multiple scales. Spatial variation is relevant over at least three scales; internally within the system ("local"), and externally at the scale of the immediate context ("regional"), and the larger ("global") set of social–ecological systems that impact the SES. Outside the boundaries of the focal SES, the primary spatial elements of interest include context (spatial surroundings, defined at the scale of analysis); connectivity; and spatial dynamics that are driven by connections or system inputs, such as spatially driven feedbacks and spatial subsidies (Polis *et al.* 1997, Polis *et al.* 2004). Both internal and external spatial elements of an SES must be considered in relation to the structural aspects of the system, including (as outlined above) the number and nature of components and interactions, the ability of the system to undergo change while maintaining its identity, system memory, and the potential inherent in the system for adaptation and learning. Note that despite its temporal role, memory is treated as a structural aspect of an SES because it requires a structure (for example, a brain, an archive, or a long-lived individual) that records change in time.

The temporal axis captures temporal variation, rates and temporal hierarchies, and the temporal location of a system within a set of processes or along a trajectory. Many of the aspects of temporal variation can be broken down in the same way as spatial elements, by considering local, regional and global dynamics and the relevance of sets of fast and slow variables. In addition, key aspects of this component of the framework include the diagnosis or assessment of the location of the system within a broader state-space, particularly in regard to its proximity to a potential threshold or regime shift; its location within an adaptive cycle (if the adaptive cycle is deemed to fit the system of interest), including whether key elements of the SES are growing, collapsing, or reorganizing; and assessment of the degree to which the system's history has created path dependence in its current and future dynamics (see Martin and Sunley [2006], for example).

Slow variables are considered to be particularly important in SES theory because of their role in creating regime shifts and alternative stable states. For example, soil fertility may be a slowly changing variable in an agricultural system (Issaka *et al.* 1997). If crop rotation times and types are insufficient to allow the restoration of soil nutrients, and if communities lack the means to generate or purchase fertilizer, changes in soil nutrients

Table 2. Summary of elements of a generic big basin SES.

System attribute	Generic elements
1. Structural axis	
SES components	Actors: farmers, fishers, household users, management agencies, local and international non-governmental organizations (NGOs), researchers, different ethnic or racial groups, mining companies, civil society groups, hydroelectric power producers, bridging groups or fora that connect different stakeholders. Biophysical components: water, vegetation, crops, livestock, fish. Built components: impoundments, roads, canals, cities. Institutions: laws, policies, regulations.
SES interactions	Implementation of land tenure and water rights; negotiations over water withdrawals; upstream–downstream user interactions; transboundary politics.
SES continuity and memory	Provided by long-term policies, laws and conventions; long-standing management agencies; and knowledgeable individuals.
SES information processing	Refers to information availability, decision making and response capacity. Although research is typically undertaken by universities and NGOs, information processing is often undertaken by individuals (for example, farmers may respond to predictions of drought by storing excess water) and by organizations.
SES adaptation	Related to research and local innovations; may also occur through the action of selection on diversity, although high social diversity may inhibit collective action.
Environment (context for SES):	
Inputs and outputs to SES	External drivers, such as rainfall and global market demands for local produce; international and national influences such as resettlement schemes, economic incentives, flood compensation and others. Outputs include water, food and information.
Perturbations	Floods, droughts, climate change, economic shocks, politics (for example, governmental changes).
Asymmetries	Gradients exist along each case study from headwaters to foothills. These include biophysical gradients (elevation, river size, water quality, and so on) and socioeconomic gradients in household income and production systems as one moves from highlands to lowlands.
2. Spatial axis	
Connectivity	Human communities in big river basins are connected by the water that flows through them and by networks of roads and waterways. Downstream ecosystems depend on upstream water production. Social networks provide an additional kind of spatial connectivity.

Subsidies	Many large downstream towns are to some extent subsidised by the opportunity costs incurred by upstream communities (for example, salinity outputs from irrigated agriculture are capped in the Goulburn-Broken catchment of the Murray-Darling system in Australia, in part to keep water drinkable for Adelaide (Anderies 2004). Similarly, the inhabitants of a river basin may produce or import their food; dependence on external sources can alter system resilience.
3. Temporal axis	
Slow variables	Groundwater extraction, soil fertility, human population increase, changes in attitudes of local community.
Evolution/selection	As the system changes through time, elements (such as farming systems or organizations) are lost and gained.
Path dependence	In some case studies the role of history is an important determinant of current patterns in such things as tenure, land use, soil fertility and equity.
Phase of adaptive cycle	Is the system increasing in one or more capitals, reaching a ceiling, in a state of collapse, and/or just emerging from a collapse?

This list is not intended to be exhaustive; its focus is on identifying the most important elements that are present in a large catchment. This table provides both a simplified inventory for thinking about any individual case study and a generalised inventory for thinking about between-case comparisons. In this chapter it represented the starting point from which to select elements to include in a simple systems model.

can become a slow variable that ultimately limits the faster dynamics of crop planting and harvesting, with significant consequences for human wellbeing and social dynamics. In the social realm, a similar problem exists in regard to climate change; the supposedly "fast" political and social processes that should be reducing "slow" increases in carbon dioxide emissions to the atmosphere are currently operating at too slow a rate to prevent global climate change.

Achieving a working system description, with a focus on the most important dynamics, is the first step towards assessing system resilience. It is not possible to think constructively about system dynamics without an understanding of what constitutes the system and its primary elements. The next step towards assessing resilience is to pull together selectively the pieces of the system description into a more unified whole, using models. It is often particularly useful to think through some of the ways that a particular kind of system may respond to specified perturbations. This may be done formally, with quantitative models, or in a more qualitative manner as I have adopted here.

Resilience of big river basins

One entry point for thinking about the resilience of big river basins in this case study set is to take the primary elements identified in the case study description and use some of them to construct a simple generic system model (that is, one that captures important elements that are common to nearly all of the case studies). In order to build a generic system model we first need to have a clear idea of what the purpose of the model is; that is, to clarify in our own minds the aim of the analytical exercise and the kinds of outcome that we will consider important. As Carpenter *et al.* (2001) point out, resilience is contextual; in applying these concepts it is always necessary to ask "resilience of what to what?" To assist in determining a suitable focus for the analysis, I asked each case study team to rank a set of potential focal problems on a scale from one to five. I summarized the data in two different ways (Figure 1). First, I designated problems that were rated over 2.5 as being "important" for a given case study to obtain a chart that shows the number of catchments for which a given issue is important. I also calculated the total number of "votes" that each issue received by summing the scores given to each. The second plot gives an indication of the relative importance of each issue.

According to this simple survey, the most important concerns that this set of big river basins shares are water quantity, the demands and consequences of irrigated agriculture and poverty. Institutional issues, population growth, and land-cover change also appear to be important in most of the case study catchments. For the subsequent analysis I have therefore focused on the resilience of these big river basin SESs to changes in water quantity as driven by irrigation, land-cover change and climate change. Of course, these issues are strongly interrelated. Questions of institutional arrangements and equity can be seen both as important influences on the primary drivers and as important outcomes from changes in them.

Table 2 suggests a common set of system elements that occur in most of the case studies. Big river basins share a common set of biophysical elements, such as headwaters, tributaries, river deltas and reliance on rainfall. The human elements of individual systems show more variation, although in many cases these are variations along a relatively small set of common themes (Table 3).

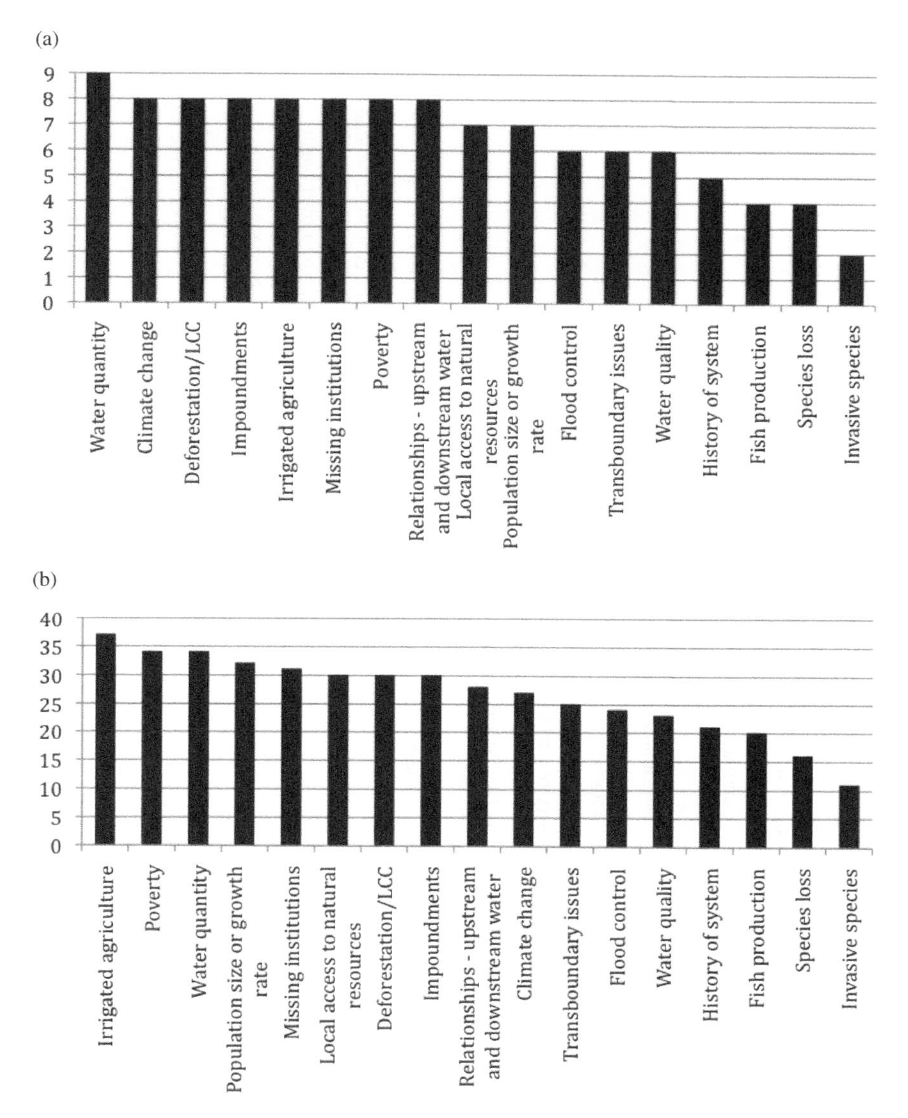

Figure 1. (a) Salience of issues; (b) The total score for each issue, as obtained by the survey. Note: (a) The number of catchments, from a total of nine, for which a given issue was considered "important". Water quantity was the only single issue that was considered important in all basins; (b) Since issues were rated from one to five across nine catchments, the maximum possible score was 45. Note the importance of irrigated agriculture in big river basins. LCC, land cover change.

Table 3 suggests some fundamental differences between different basins. In particular, there is considerable variation between big river basins in the degree to which agriculture is large-scale and commercial versus small-scale and subsistence-oriented. For example, water use in the Limpopo Basin is dominated by mining companies and commercial irrigation farmers, with small-scale farmers constituting a relatively marginalized voice; whereas in the Indus–Ganges basins, the number of small-scale cultivators is enormous and

Table 3. Important human system elements in each of the eight BFP big river basin case studies.

Basin	Local	National	International
Limpopo, Southern Africa	Mining and large-scale agriculture; Rural poor, often small-scale farmers; NGOs such as World Vision, Working for Water, and the Kalahari Conservation Society.	National governments of Botswana, Zimbabwe, South Africa, and Mozambique. Different ministries, often planning in isolation from one another. DWAF (Department of Water Affairs, South Africa) ARA-SuL (Administração Regional de Áquas do Sul, Mozambique) ZINWA (Zimbabwe National Water Authority) Botswana Water Affairs	LIMCOM (Limpopo Basin Commission), an emerging bridging organization
Karkeh, Iran	Provincial governments, farmers; poverty-alleviation organizations such as the Red Crescent Society and the Emam Khomeini Relief Committee.	Power ministry offices, which determine water allocations Ministry of Energy, Ministry of Jihad-e-Agriculture and the Ministry of Jihad-e-Sazandagi (watershed management and rural development)	Not directly relevant since basin is internal to Iran.
Nile, North and East Africa	Commercial agriculture, small-scale farmers, fisheries.	National governments and ministries of bordering nations.	The Nile Basin Initiative (NBI) is the primary organization in the basin that works towards confidence, trust and capacity building among the riparian countries. The ultimate goal of the NBI is to facilitate the formation of a river basin commission and enabling environment for integrated water resources management. The Eastern Nile Technical and Regional Organization (ENTRO) is a subsidiary action programme of the NBI that works on the joint multi-purpose investment projects among Egypt, Sudan and Ethiopia. A similar subsidiary action programme exists for the Nile Equatorial Lakes (NELSAP) region. There are also legal institutions in the Nile Equatorial Lakes region, such as the Lake Victoria Basin Commission for the East African Community.

Volta, West Africa	Agriculture, almost entirely small-scale farmers	Governments of Burkina Faso and Ghana, including relevant ministries. The Volta River Authority, which is in charge of producing electricity.	The Volta BASIN Authority, a multinational bridging organization.
Niger, West Africa	Agriculture, almost entirely small-scale farmers; nomadic herders; Office du Niger in Mali; Sub-basin agencies (Bani, Liptako Gourma); NGOs; Traditional governance and farming systems, including fadama/irrigation projects in Northern Nigeria. The Inner Delta in Mali is an important Ramsar site (3 M ha) and ecological/environmental interests in conserving this resource are also important.	National ministries of water and agriculture.	ECOWAS (West African EU) at the regional level. Niger Basin Authority (which currently seems unable to fulfil its difficult role).
Yellow River, China	Irrigated agriculture, Industry, rural poor; agricultural research centres, stations and policy. Provincial-level governments and water resource bureaux (provinces control tributary flows within their boundaries). Some environmental NGOs are having growing impacts. Water-use rights transfer pilot projects to conserve water without adverse impacts on irrigation benefits, seen by many as a way to address the water crisis.	Ministry of Water Resources Ministry of Agriculture Ministry of Environmental Protection Yellow River Conservancy Commission (YRCC), under the Chinese Ministry of Water Resources, which has the mandate for allocating flows in the downstream part of the basin. The Yellow River Water Allocation Scheme, established and issued by the State Council of China in 1987, set a cap on abstraction at 37 km^3 per year and quotas for each province tied to average runoff of 58 km^3. 2002 Water Law (focus on river basin management, water savings, and many other topics)—highest-level legislation, but	Not directly relevant.

(*Continued*)

Table 3. (Continued)

Basin	Local	National	International
	Regulations on flow and water quality at controlling points along the river.	becomes only effective if supporting regulations are both developed and implemented.	
Indus-Ganges Basins, India and Pakistan	Farmers' organizations – formal and informal – who push to get water services at nominal or free costs. These organizations also impact the energy policies of the state for keeping the energy tariffs low or zero (for groundwater pumping). Non-governmental/ civil society organizations/ donors/ other voice groups who clamour to strike a balance between development and environment, inclusiveness of the poor and marginalized and long-term sustainability. Industries and domestic supplies are regulated to some extent, but irrigation supplies remain largely unregulated. One of the main reasons is the sheer number of small cultivators, which defies any governance mechanism.	Federal governments/ ministries of the water resources of the four major riparian countries: Pakistan, India, Nepal, Bangladesh. State/ provincial governments and irrigation bureaucracies in each of the states in the country (since water is considered a state subject, there is perpetual conflict between the federal and the state governments). Institutional basis under which water resources are not public/national; any person having land rights also has the right over water resources as an adjunct to the land.	Lacking international bridging organizations.

Mekong, Asia	Local inhabitants, especially farmers and fishers.	Governments of the riparian countries: China, Myanmar, Lao PDR, Thailand, Cambodia, Vietnam. Relevant ministries and departments in each country (including agriculture, fisheries, forestry, water resources and hydropower). National Mekong Committees.	Mekong River Commission (MRC); relationship between MRC and China and Myanmar (two upstream countries), particularly relationship between China and MRC, as most of the developments such as construction of dams are in the upper part of the basin in China; bilateral relationship and interactions between the MRC countries such Cambodia–Thailand and Cambodia–Vietnam relationship.
Andes, South America	Farmers, communities, industrialists, dam operators for hydroelectric power and irrigation, mining companies, urban water companies.	National governments and ministries; regional and national policy institutes.	International environment, agriculture and development donors, international conservation NGOs

Table 2 provides a generic listing of case study elements; this table provides specific details of the human elements considered as important by participating case study teams. I have broken them down according to their scale of action. Note that this list is intended to provide supporting information, rather than an exhaustive inventory.

although some powerful organizations exist, there is no single channel through which they can all be negotiated with. This makes small-scale farmers difficult to include in water-use debates and equally hard to govern.

Some kinds of system element are particularly relevant to big river basins. A number of big river basins in this set of case studies have some form of bridging organization that attempts to coordinate and integrate management across different sectors and between upstream and downstream water users. Examples from the case studies in the BFPs include the Limpopo Basin Commission (LIMCOM); the Mekong River Commission (MRC); the Nile Basin Initiative (NBI); the Volta BASIN authority; and the Yellow River Conservancy Commission (YRCC). The commonness of bridging organizations in big river basins, and the importance accorded to such organizations in the resilience literature (Hahn *et al.* 2006, Olsson *et al.* 2004, 2007), suggests that basins without bridging organizations (or where they exist but are dysfunctional) may be less capable of responding effectively to environmental change, and hence less resilient. Bridging organizations have also been proposed by Ernstson *et al.* (in press) as potentially important contributors to the resolution of scale mismatch situations, which occur when the scale of management is not appropriately aligned to the scale at which ecological and social processes occur (Cumming *et al.* 2006).

Drawing on the summaries in Tables 2 and 3 and the issues considered to be of central importance by case study teams, a simplified generic model of a big river basin case study might look something like Figure 2.

As Tables 2 and 3 and Figure 2 suggest, big river basins have the potential for many different kinds of feedback. To make Figure 2 a little more specific, consider a downstream system that is in an economic growth phase and currently experiencing a beneficial climate (Figure 3).

Tracing some possible feedback through Figures 2 and 3, for example, considering what happens if rainfall is reduced or if agricultural demands for water increase, suggests that the potential for both self-regulating and self-amplifying feedbacks increases hugely when water is scarce. Once water becomes limiting, small changes in either water availability or production systems will inevitably influence actors and land cover, which in turn will influence water and production systems. For example, when water demand from irrigated farm land starts to approach the limit that rainfall in an average year can support, water availability for other sectors (for example, fisheries, hydroelectric power companies, or downstream ecosystems) is reduced, depriving the associated actors of resources and leading to conflict. Similarly, excessive land clearing for agriculture (or the planting of water-hungry timber plantations in place of native shrubs) can reduce the total amount of water entering the basin, leading to reductions in crop and livestock production, increasing poverty, additional land clearing, and a downward spiral that includes further environmental degradation and a decrease in human wellbeing.

The point at which the demand for water approaches or exceeds the supply of water is possibly the single most important tipping point for many of the big river basins considered in the BFPs. Once it is exceeded, a number of initially dormant or potential interactions between actors, production systems and land cover start to become increasingly important. Given that the focus here is on the resilience of big river basin SESs to changes in water quantity as driven by irrigation, land-cover change and climate change, water quantity thresholds offer a convenient entry point for thinking about basin resilience.

Water scarcity may occur as a consequence of a number of different factors. There are many routes to crisis and collapse. Some of the most important driving factors in the case studies in BFPs include expansion of domestic demand (Indus–Ganges, Yellow); increasing water demands for irrigation or mining (common to all case study systems); excessive

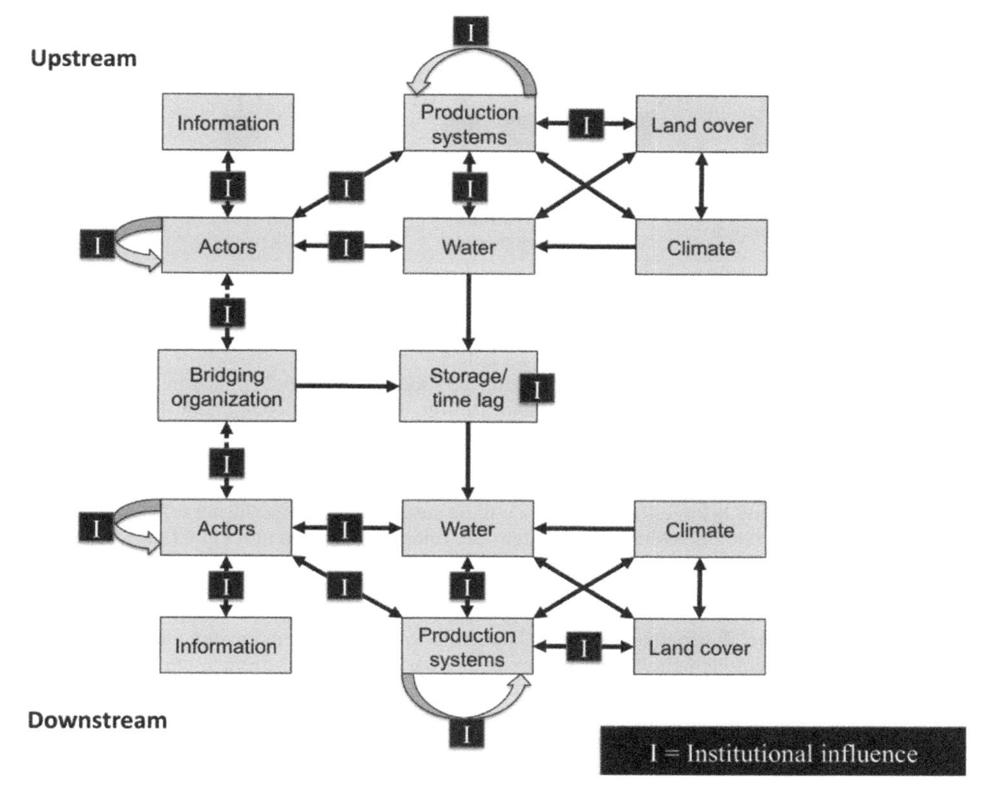

Figure 2. A generic model of a big river basin. Spatial dynamics are represented by a simple upstream–downstream divide; in reality the spatial distinctions between river sections will often include three or more recognisably different use-zones, corresponding to the headwaters, mid-sections and river delta, and some elements of the system may differ between zones. Upstream and downstream zones are connected by water, often with a delay via impoundments. Bridging organizations connect actors across different zones and/or at different scales. In each zone, local actors and production systems use and are influenced by water, and a variety of important interactions and feedbacks occur within both of these boxes. Climate (for example, via snowmelt or evaporation) determines water quantity, while also influencing land cover; and feedbacks from land cover (for example, vegetation via evapotranspiration, urban heat islands, or other albedo-altering effects) influence local rainfall. Production systems include different sectors (for example agriculture, mining, power generation and household users) and may directly influence land cover. Although actors may of course influence land cover directly for non-productive uses, I have chosen to focus on the interaction that has most impact (that is, LCC caused by production systems). Production systems respond to and influence water quantity and quality by such activities as extraction, alterations to runoff, and pollution. Land cover influences infiltration and runoff rates, with effects on water quantity and quality. Actors generate information, which may influence water use. Lastly, institutional interventions or guidelines, denoted in this diagram by "I", can influence some but not all interactions within the system.

groundwater extraction (Karkheh, Indus–Ganges); climate change (a pending concern in all case studies); and more insidiously, reductions of water quality to the point where water becomes virtually unusable by one or more production systems or sectors (Yellow, Indus–Ganges).

What makes a big river basin more or less resilient to water scarcity? Once a point of water scarcity is reached, the social system is forced to reorganize and adapt to the

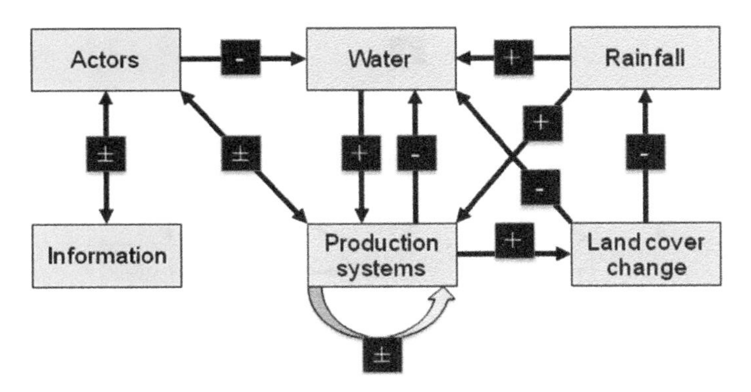

Figure 3. Generic system diagram. Derived from the model in Figure 2, indicating the strongest interactions (and whether increasing or decreasing) in the downstream component during a period when agriculture is expanding and rainfall is reliable In this instance, households and other actors are withdrawing water; production systems are using water; and land-cover change is further reducing the amount of water available. Provided that there is sufficient rainfall (that is, the positive influence of climate on water quantity is large enough), there will be sufficient water for the net effect of water on production systems to remain positive. The system can remain in its current regime (and keep growing) as long as the magnitudes of the interactions indicated by pluses and minuses complement one another sufficiently well for excess water to remain in the system. However, if the demands on water become excessive (whether by direct or indirect pathways), water scarcity will limit production and create a crisis in the social–ecological system. The ability of the system to navigate this crisis while keeping essential system elements intact (for example, particular social groups, human wellbeing, ecosystems, food production systems, and/or culture and ways of life) represents its resilience.

new regime that now dominates the system. The way in which this reorganization occurs, and the management trajectory that the system subsequently follows, can be expected to have a large influence on the overall resilience of the system. Resolution of management problems in many real-world situations is hindered by ineffective institutions, corruption and incompetence within key organizations. Substantial delays can occur between crossing the water quantity threshold and the implementation of an effective management response, particularly in systems where social networks and management institutions are weak and actors have little history of communicating with one another or coordinating their use of water at an appropriate scale.

In some of the case studies in the BFPS, such as the Yellow River, the point at which water scarcity becomes important has long since been passed. The persistence of desirable social and ecological elements in these systems depends on reaching a successful two-pronged solution that: (1) creates a socio-politically acceptable compromise under current conditions; and (2) in the longer term, addresses the slow variables of reductions in water quantity and increasing demands on resources.

In other large-basin SESs, such as the Andes, water scarcity looms as a possible future shock to the system. In cases where a threshold of water scarcity has not yet been reached, it should be possible to attempt to build resilience to water scarcity into the social system, through such activities as building stronger social networks, developing a better understanding of the limits on production systems, and working towards more effective integrated management over the entire basin.

From a resilience-oriented perspective, crossing the threshold into water scarcity can occur in a number of different ways. Although some of these pathways may ultimately be beneficial for the long-term resilience of the case study basin to climate change (for example, if crisis leads to better integration of different management actions), entering into water scarcity can easily push a large-basin SES into one or more problem situations from which escape can be difficult.

For example, a common management syndrome, termed "fixes that fail" (Senge 1990), occurs when a solution that is effective in the short term has unforeseen longer-term consequences. This is very similar to the typical complex systems dynamic in which changes in a slow variable (often one which is considered unimportant by decision makers) gradually undermine attempts to manage a faster variable (Carpenter and Turner 2000). A big river basin of the kind discussed here can become trapped in this dynamic through attempts to alleviate poverty. Poor households are often perceived as being trapped by circumstances that make it difficult to raise the starter capital that is needed to create the means to provide a steady income (Yunnus 1998). Being able to afford a vehicle with which to transport produce to market, for instance, can make a huge difference to a farmer's earning potential. One plausible solution to getting households out of the poverty trap is for actors in the system, such as governmental agencies and non-governmental organizations (NGOs), to provide arable land, technical support and training to would-be farmers. The logic of this solution appears sound: that financially poor households will then be able to both feed themselves and sell their surplus, gradually lifting themselves out of poverty and ultimately becoming self-sustaining. In practice such programmes are often accompanied by, or part of, resettlement schemes such as those in the Amazon Basin (Imbernon 1999).

If the human population is large, however, fostering additional farming activities without appropriate zoning and enforcement of good farming practices can result in widespread clearing of native vegetation and increasing demand for water (Figure 4). This in turn sets in place a series of slower environmental changes, such as reductions in infiltration

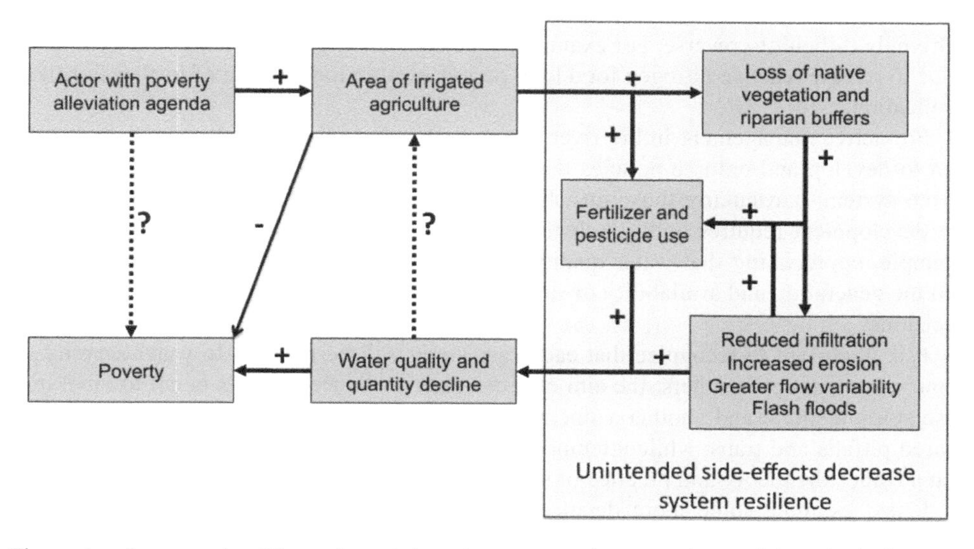

Figure 4. An example of how attempts to reduce poverty in a complex social–ecological system can have unintended side-effects. Gains in poverty alleviation achieved by increased food production may be offset by declines in human wellbeing resulting from worsening water quality and quantity. This figure also ignores the potential human population increase that may follow from increased food production. In the worst-case scenario, poorly conceived poverty reduction programmes may actually worsen the problem that they are attempting to solve.

rates, increased erosion of topsoil, and increases in pesticide and fertilizer use. These changes can gradually accumulate until a point is reached at which agricultural production becomes heavily dependent on external inputs. In the worst cases, water quality and quantity are reduced, water storage is difficult because of high sedimentation rates, landslides and floods become commoner, and soil fertility is considerably lower. The net outcome of naïve policies aimed at agricultural expansion can thus be to leave poor households worse off than before, while creating significant environmental problems that may take decades to fix.

At some point this kind of dynamic will also create difficulties for the agencies concerned, requiring a complete rethinking of current policies and an attempt to push the system on to a new trajectory. Whether or not a feasible solution can be achieved is path dependent, in the sense that it will depend heavily on the degree to which other options remain open within the system. If the number of people who have been allocated farmland in the basin is too high for the river basin to cope with, the entire river basin SES may enter a long-lasting poverty trap. Escaping this predicament requires altering the current regime, focusing policies on different goals, and possibly transforming the system by reducing the resilience of some attributes and enhancing the resilience of others. For example, introducing policies that focus on urban job creation may attract people into towns, reducing the pressure on soil and water resources. On the island of Puerto Rico, for instance, expansion of the pharmaceutical industry following tax reductions ultimately led to a forest cover increase from 4% of the island's area to over 40%, with corresponding improvements in water quality and quantity (Gonzalez 2001, Lugo and Helmer 2004).

Agricultural activities can create situations in which different biophysical regimes exist in at least three different areas: soils, hydrology and the atmosphere (Gordon *et al.* 2008). Gordon *et al.* (2008) review 10 different agriculturally linked regimes that can lead to alternate stable states in big river basins. Shifts between some of these regimes can be extremely difficult to reverse. For example, once soils are high in nutrients, they can continue to release reactive nitrogen for a long period, even in the absence of further fertilizer application.

Pro-active management in big river basins with substantial agricultural demands will aim to develop and enforce policies that recognize the longer-term and slower variables in the system, particularly those on which farming activities depend. This level of policy development requires both a holistic perspective on what constitutes the system (for example, appreciating that water quality and quantity are closely linked to ecosystems) and the generation and availability of sufficient high-quality information on which to base decisions.

It is important to recognize that each case study will be resilient to water scarcity in some ways and not in others; the aim of cross-case resilience analysis is not to label one case study resilient and another vulnerable, but rather to draw out and explore potential shared pitfalls and traps, while attempting to identify and test general guiding principles that advance the theory and practice of social-ecological research.

I next examine some more detailed examples from different case studies from the perspective of the preceding analysis.

Case study resilience

A first basic separation of the nine case studies can be made on the basis of water quantity. In two catchments, the Yellow River and the western half of the Indus–Ganges, the SES is well over the water scarcity threshold. The Yellow River in particular has run dry on several occasions, with a set of resulting problems. The Limpopo, the Karkheh and the Nile are relatively close to the water scarcity tipping point but are not necessarily there yet, except perhaps in drought years. The Niger, the Volta, the Mekong and the Andes are systems in which water scarcity is not currently considered a major issue, although it may of course become so in the future.

Although water is the first and most important identity criterion for a big river basin, other important elements (see Tables 2 and 3) typically include long-term residents of the catchment; commercial, large-scale players who contribute to national food security and economic growth; unique ecosystems, such as mangroves and wetlands; and a range of ecosystem services that are integral to the wellbeing of people living in the basin. If these elements are lost, the system can be said to have not been resilient.

Each basin has one or more "burning issues" that may eventually lead to the system entering crisis mode (if it is not there already) and ultimately to the loss of one or more important aspects of system identity. These issues are thus the areas in which basin resilience can be considered weakest. They are summarized for each of the eight case studies in Table 4.

As Table 4 shows, there is a wide range of complex dynamics in progress in each of the case study systems. The most fundamental common problem is probably the role of increasing demand for water from large-scale users, who include industry, irrigation farmers and cities. As demand increases, these systems must develop ways of reducing environmental impacts, allocating water equitably, and leaving some leeway in the system for periods of reduced flow. In fundamental terms, net population expansion must be slowed and water use efficiencies in industry and agriculture will need to be built in at an early stage. Once a large basin system has crossed the water scarcity threshold, with numerous water-demanding actors already in place, corrective action becomes increasingly difficult.

One of the most important concerns from a management perspective is that attempts to manipulate the system should not create further problems. As discussed previously, one of the greatest dangers is that naïve policies and unfettered, unregulated development may ultimately make worse the very problems (poverty, food production, slow economic growth) that they are trying to resolve. People have a tendency to see solutions through their own expert lenses. For example, development agencies often see the root problems in big river basins as institutional; scientists may see them as being driven by water flows, deteriorating soil quality, or other biophysical components of the system; and demographers may blame population growth. In many cases disciplinary experts assume that implementing "fixes" in their own area of expertise will solve the problems. Economists demand better pricing and incentive structures; development agencies want greater institutional accountability; and ecologists want restored riparian buffers and ecological flows. In each of these cases, remedial actions offer only partial solutions. While each may have some desirable outcomes, none on their own is adequate to solve the broader problem. Long-term sustainability in nearly every case will require the implementation of a set of carefully negotiated tradeoffs between different actors in the system, including not only human needs but also minimal flows that will keep relevant ecosystems functioning and diverse (see Zwarts *et al.* [2006] for example). Diversity in both the human and the ecological realm offers an important buffer against future change (Yachi and Loreau 1999, Norberg *et al.* 2008);

Table 4. Burning issues in eight big river basins.

Question	Burning issues in the basin
Limpopo, Southern Africa	Natural resource-based economic development plans have been scaled up in each of the four riparian states. Infrastructure development has lagged behind, and previous investments are not optimized. Limpopo Basin stakeholders include large groups of rural poor who rely on natural resource-based livelihoods for survival. Major threats include: (1) An already overcommitted river basin where new water resources development is being planned and implemented. This increases pressure on ecologically sensitive areas, such as wetland ecosystems; (2) Competing uses including extremely high-value commodities (platinum and others); large-scale commercial irrigated agriculture; poorly resourced smallholder agriculture; tourism—much of it in areas with unsecured land and water rights; (3) The likelihood that climate change will further reduce available water and increase already unpredictable water events (floods, drought, rainy seasons—onset, duration, accumulation, and so on); (4) Limited institutional function in water governance despite numerous structures and policies toward that end.
Karkeh, Iran	There is growing competition for water among different sectors in the basin, which is already closed. Increasing demand for water in industrial and domestic sectors will reduce water available for agriculture and the environment. At the same time, food demand is rising due to population growth. In addition, the current rate of ground water extraction in the basin is much more than the aquifer's capacity. As a result, the groundwater table is dropping by about one metre every year. Lastly, existence of the Ramsar "Hoor-al Azim" swamp is in danger, partly due to low inflow from Karkheh River to the wetland.
Nile, North and East Africa	The Nile Basin supplies water to about five million ha of irrigated agriculture. The present irrigation demand consumes about two-thirds of the Nile River flow. Potential future problems include expansion of irrigation, climate change and unilateral approaches to basin development and management. The irrigation demand is expected to double in the near future to feed the rapidly growing population in the basin. The Nile Basin is not managed in an integrated and sustainable manner; rather the riparian countries are independently developing and managing their own parts of the basin, creating a major challenge for efficient and sustainable utilization of the Nile Basin water resources.
Volta, West Africa	The basin contains some of the world's poorest countries, and most are getting poorer. Increases in food production are well behind demographic increase, mainly as a result of poor soils, lack of inputs and production means (oxen and tools), bad access to markets, poor infrastructure, lack of education and health care, and so on. Institutions are midway between traditional hierarchies and modern state laws and hence are frequently dysfunctional.
Niger, West Africa	The major threats in the Niger Basin revolve around environmental variability and institutional weaknesses. The population suffers from high ambient poverty levels (economy, health, education, infrastructure, and so on) and is affected by regular disasters (drought, flooding, and cholera outbreaks to name a few). The basin also suffers from insecurity issues (ethnic tensions in the North with the Touareg in Niger and Mali, and religious tensions in northern Nigeria and widespread insecurity in southern Nigeria), as well as from corruption and inefficient governance. Additional pressure is mounting due to extreme population growth, climate change, land appropriation, rising fuel and food prices, and uncontrolled development (land-use changes, pollution, dam development and so on). The lack of established and effective institutions also has a detrimental impact on integrated water resource management at the basin scale, as several dam developments are planned, which are likely to increase the strain on blue water resources and downstream (often transboundary) users. Agriculture is predominantly subsistence agriculture and rainfed, exposing it to variations in climate. Yields are low and currently rely on land extension (rather than yield increases) to feed a growing population.

Yellow River, China	The Yellow River basin is faced with absolute water scarcity and poor water quality and little improvement is expected for the future. The basin faces severe competition between irrigation water uses mid-stream and partially downstream and rapid urban-industrial development downstream. No additional irrigation expansion is planned, but the government policy of food-self sufficiency in key grains will continue to put pressure on the basin's water resources. The river has been running dry; no flow was recorded at the downstream station, for five to 226 days during 1972–99. After 1999, the Yellow River Conservancy Commission was given the task to ensure that some flow reaches the river mouth. Despite this, the small downstream wetland area remains threatened from canalization, oil development and rapid urbanization. Its area declined by 50% over the last 20 years. Other urgent issues include population growth, further urban–industrial development, climate change and high sediment deposition rates. Water allocation between regions is still based on an allocation plan issued in 1987, which has not been revised to reflect the change in water demand. Water quality in the river is very poor, due to a lack of treatment stations and limited enforcement of existing environmental regulations in the basin. The silt content of the water is very high. New policies of silt flushing have helped reduce the risk of downstream flooding, but are a further drain on water availability for irrigation.
Indus-Ganges Basins, India and Pakistan	These basins are facing two entirely different kind of crisis. The western part of the Ganges and the Indus are witnessing an over-exploitation of the groundwater resources and a continuous decline in the water table making the dependent agriculture hydrologically/financially unsustainable. This region is considered the global hotspot for groundwater over-use. The eastern part of the Ganges has a lot of surplus water but lacks human, financial and technical capital to make good use of the available water, keeping millions of the population in poverty and low productivity. Climate change is the major threat to the basins which will seriously alter the flow regime and intensity and frequency of the extreme events: floods and droughts and temperature changes. Related issues include low productivity in a large part of the basin; a high concentration of poverty, especially the eastern section; deterioration of public irrigation systems (canals) and unsustainability of ever-increasing groundwater irrigation; and the transboundary nature of the basin in the volatile South Asian context.
Mekong Basin, Asia	The key issues in the Mekong Basin stem from rising pressure on the natural resource base. The growing population needs hydropower and food. The population of the Lower Mekong Basin is expected to increase from the current 65 million to about 90 million by 2050, and the proportion of urban dwellers from about 20% to about 40% or about 36 million. Economic growth is around 4.5% per annum. Total food demand will increase at a rate greater than that of population alone, due to rising incomes and changing diet preferences consequent upon urbanization. A further threat to food supplies comes from climate change. Projections suggest that the current rate of growth in agricultural production will meet future demands; but capture fisheries will probably fail to meet rising demand, since production is unlikely to go up and impoundments may lead to declines. There is widespread poverty in the Lower Mekong region. The people in Cambodia and Lao PDR are among the poorest in the world, and in the northeastern part of Thailand and the provinces of Vietnam that are part of the basin, many people suffer from severe poverty. Poverty is closely related to access to water and cultivable land, as well as to fish.

(*Continued*)

Table 4. (Continued)

Question	Burning issues in the basin
	Although water availability is not considered a huge concern in the Mekong Basin (except in certain areas, such as northeast Thailand, during the dry season), alterations to flow regimes by impoundments and irrigation diversions are impacting river ecology, fish production, access to water, and food security. Changes in the natural flow regime may alter the environment of fisheries in the Tonle Sap and elsewhere. Altered low flows may lead to salinity intrusion into the Mekong Delta, thus altering the balance of rice and shrimp production, which in turn may affect food security and incomes.
Andes, South America	Areas of stress are usually associated with degradation of the natural environment and its impact on the quality and quantity of water supply; increasing demands for water and energy from increasingly large, urbanized populations; and the juxtaposition of water-demanding activities with others (such as mining, wastewater, oil and gas) which can spoil water for downstream uses. Climate change is an emerging threat, particularly to sensitive headwater ecosystems such as cloud forests and paramos. Areas relying on snowmelt may suffer. Access to water is also important in this system and may lead to conflict. Land and water legislation and policy are variable and sometimes poorly developed in Andean countries and this can seriously undermine the sustainability of common resources.

This table summarizes some of the problems that seem most likely to move individual river basins in this set of case studies into a new social–ecological regime. Note that for obvious reasons, this list largely excludes unexpected perturbations and surprises.

change imposes selective pressure on social–ecological systems, and it is more likely that important system elements will be able to persist into the future if a range of potential solutions is available. Similarly, it is important that managers in SESs think carefully through the implications of capital in all its forms (ecological, social, financial, human and physical) for resilience (Pretty 2008). Capital can play an important enabling function during times of change (Adger 2003, Deutsch *et al.* 2003, Brand 2009) and conversion of capital from one form to another, such as from ecological to financial, may alter system resilience to future perturbations.

Ecosystem services are fundamental to the wellbeing of human populations living in big river basins, and development often requires tradeoffs between different ecosystem services (Rodriguez *et al.* 2006, Mainuddin *et al.* 2009). I asked individual case study teams to identify key tradeoffs and compromises, as well as important actions that would be relevant in different basins for the sustainable management of ecosystem services. These responses, together with some of my own interpretations of individual basin reports, are summarized in Table 5.

As these responses indicate, there are a number of apparent needs that are integral to the long-term resilience of most of the case studies. Chief among these is the need to develop more effective ways of integrating water resource management such that it bridges the gaps between upstream and downstream users, and between different sectors, more effectively. Effective management will require effective enforcement of regulations, another major problem.

Table 5 suggests some possible pathways to greater resilience, but it does not offer a comprehensive route to resilience for any single river basin. Resilience to climate change at present appears to be higher in basins that are not yet water limited. Most big river basins are experiencing high population growth, however, and increasing demand for food (and hence, agricultural expansion). Building resilience to climate change in each of these systems will require proactive, forward-looking management and regulation. It is also important to note that resilience is not always desirable (dysfunctional management systems being a case in point). Resilience will need to be built in some system components and reduced in others; and in some case studies, such as the Yellow River, transformation of the SES (with the loss of some or all of its current "key" elements, for good or for bad) seems virtually inevitable. Olsson *et al.* (2004), based on their study of a successful transformation of a wetland SES (Kristianstad) in southern Sweden, suggest three key steps towards social transformation: preparing the system for change, seizing a window of opportunity, and building social–ecological resilience of the new desired state. Transformation in Kristianstad was facilitated by the creation of a new bridging organization which was able to bring together key actors.

Managers and water users in many cases are aware of many of the issues and possibilities listed in Table 5. Awareness of solutions, however, does not always translate into solutions; action occurs at the end of a political process in which different needs and values are considered and alternative solutions are negotiated (Cronon 2000). The development and implementation of more effective management strategies in most case study basins is currently being hindered by a wide range of political variables, including historical incompatibilities, power struggles, a lack of trust and overlapping (and sometimes contradicting) legislation. For example, in some of the BFP case studies, different water management districts refuse to share their water flow data to provide a comprehensive picture of basin-wide flows. There is also a paucity of relevant science-based information on which appropriate legislation can be based. Some of the major uncertainties and information gaps in the case study basins are summarized in Table 6.

Table 5. Tradeoffs and other compromises perceived by basin case study teams as needed for improving the sustainability of big river basins; and suggestions for necessary changes to enhance resilience in each basin.

Basin	Compromises or tradeoffs with relevance for sustainability	Pathways proposed to lead to greater resilience
Limpopo, Southern Africa	Economic vs. natural capital: sustainable management of ecosystem services, and particularly reductions in water use, will have to be paid for. Short-term vs. long-term benefits: rural households will have to be compensated for the loss of economic opportunity that they face when conserving the resource base. Tradeoffs between sectors: industry, particularly mining and agriculture, will have to balance their needs with those of rural communities.	Improve equity, reduce marginalization of rural poor by education and building local capacity and institutions. Introduce more effective policing and enforcement of existing regulations. Improve pricing and prioritization structures for smallholder water use. Develop more effective ways for different countries in the basin to interact and co-manage water resources, through strengthening LIMCOM. Planning being done now must be based on best available climate forecasts under the assumption that there will be less water, more major weather-related events, and increased temperatures.
Karkeh, Iran	Anthropogenic vs. ecosystem needs: ecosystems need reduced groundwater usage and incorporation of environmental flows into decision making. Upstream vs. downstream tradeoffs: downstream water laws need to be revised. High vs. low economic value activities: possibility to shift water away from (lower-value) food production and towards hydropower and urban uses. Specifically, difficult tradeoffs expected between hydropower and irrigation needs.	Extend trade networks to meet future food demand. Develop a larger set of non-agricultural ways of generating income. Need to improve water management and allocation processes. Improve water use efficiency of irrigated agriculture. Map water production to identify opportunities for increasing water quantity.
Nile, North and East Africa	Upstream vs. downstream tradeoffs: differences in water access, demand and management must be resolved. Water quantity available to humans vs. water for ecosystems (including wildlife and grazers) on floodplains in the Sudd area (Jonglei Canal debate). Consumptive vs. non-consumptive uses (for example, hydropower vs. irrigation).	Create enabling environment (basin-wide institution with law-enforcement capacity, water management policy, strategies and regulation). Improve productivity of farmland and agricultural water management. Reduce water waste/loss and increase availability Consider climate change and environmental impact assessment during project planning and design.

	Multivariate tradeoffs between different, potentially conflicting uses of water (for example, irrigation, fishing, hydropower, drinking, washing, transport, repository for waste).	Attract investment funds (economic capital). Resolve political differences between nations, particularly Egypt and other water users. Build management capabilities of agricultural and ranching communities.
Volta, West Africa	Few explicit tradeoffs at present because water volume not limiting, although rainfall is highly variable and soils are poor. Future tradeoffs between storage and irrigation and cropping vs. ranching systems seem likely.	Introduce better management for grasslands, particularly for livestock with access to water and use of crop residues (livestock perceived as only possible use of grasslands in the northern drier part of the basin). Develop reforestation programmes, reduce deforestation rate. Soil use currently unsustainable; introduce small-scale irrigation and better farming practices. Possibly regulate fisheries and limit expansion of rice production into small inland valleys.
Niger, West Africa	Impoundments to support hydropower and irrigation during low-flow periods vs. downstream river flow for herders, fishermen and rice cultivators. Livestock ranching of nomadic herds vs. agriculture (which restricts water access and cattle movements). Tradeoffs between different crops with different values and water demands (for example, rice vs. market gardens). Customary laws vs. governmental legal system (effectiveness of different systems for water management differs; plurality can create tenure insecurity and negatively impact development).	Improved agricultural productivity. Improved water productivity and/or more effective water management (ground water extraction, dam construction, and so on). Inclusion of marginalised sectors, such as fishers and nomadic pastoralists, in decision-making and planning processes. Conflict resolution between nomads and farmers. Resolution of land tenure issues stemming from legal pluralism (traditional vs. legal systems). Improved education and access to clean water.
Yellow River, China	As a closed basin, use of water by any sector almost inevitably entails a tradeoff with other sectors. Irrigated agriculture in middle and upper sections of basin vs. industrial, domestic and ecosystem demands downstream: Transboundary (province) water diversion vs. water demand, particularly in downstream areas within the basin;	Enforce water quality regulations to ensure that less industrial waste and domestic sewage enters the river. Eliminate fertilizer subsidies, which lead to nutrient runoff and poor water quality. Find new ways to compensate provinces for using less water, particularly upstream provinces where benefit per unit of water used is lower (however, given the reuse of flows within the basin, the potential for saving is also limited).

(*Continued*)

Table 5. (Continued)

Basin	Compromises or tradeoffs with relevance for sustainability	Pathways proposed to lead to greater resilience
	Waste removal vs. other water quality-related services (wastewater discharge high in upper and middle sections).	Enhanced cooperation between the ministries of agriculture (MOA) and water resources, as the MOA's policies tend to increase pressure on water resources without taking these into account in their decision-making processes.
	Fertiliser use for food production vs. downstream water quality. Flood control vs. water storage. Poverty alleviation (via irrigated agriculture) vs. ecosystem integrity.	Reduce agricultural water use, depending on how much water use efficiency has been increased.
	Decentralization of water rights vs. central control. Investing for rural labour expansion in agriculture vs. investing in non-agricultural job creation.	Develop a single rational allocation plan between regions, depending on whether the present allocation plan can be revised. Limit industrial development where there is insufficient water to meet its needs. Clarify nature of water rights and consider creating a more effective water market.
Indus-Ganges Basins, India and Pakistan	In the Indus, a closed basin, use of water by any sector almost inevitably entails a tradeoff with other sectors. Upstream vs. downstream uses. Tradeoffs between economic returns on different crops and on monoculture vs. diverse cropping systems. Reliance on formal vs. informal institutions (tradeoffs in flexibility, responsiveness, enforcement, and adherence to rules). Policies to encourage groundwater pumping (short-term gains) vs. groundwater recharge (long-term sustainability).	Improve food security for the region, especially the poor population. Develop new cropping patterns in response to climate change. Create sufficient attractive off-farm employment (involve farmers in other sectors). Manage extreme events better (improve contingency preparedness) and reduce vulnerability of the population. Provide energy security to the rural population through alternative fuels. Improve markets and infrastructure. Empower poor people, especially women.
Mekong, Asia	Development of hydropower dams vs. downstream water use and ecosystem function (including production of fish for capture).	Improve tenure and access rights of the poor to natural resources; give marginalised communities a voice. Maintain sufficiently good downstream water quality and quantity to support fisheries.

	Development of new irrigation system and increasing total crop production, again vs. downstream water use and ecosystem function (including production of fish), but also vs. other land uses. Values and livelihoods from upstream crops vs. other ecosystem services and values (for example, erosion control in steep uplands and livelihoods from forested / natural areas). Limit dry season irrigation vs. seawater intrusion in the delta (and consequent balance of rice and shrimp farming in the delta).	Improve current institutions (for example, strengthen Mekong River Commission) and develop a more holistic institutional perspective (that is, not only focused on growth and development). Improve local and regional governance; develop ways to prevent national political interests (for example, indebtedness to China for aid) from becoming too dominant in decision making.
Andes, South America	Upstream vs. downstream activities and impacts (for example, mining vs. domestic water use). Local vs. national benefits, increase investment in water treatment for urban centres, with local costs but national benefits. Economic benefits vs. ecological integrity (poor environmental record of agriculture in the region). Tradeoffs between competing land uses in steep areas (for example, agricultural production systems, major dam projects, inter-basin transfers, mining). Downstream irrigation vs. water supply to major cities.	Manage land use more carefully (for example, deforestation impacts on water and sediment inputs to hydropower dams; or agricultural impacts on water quality). Improve management of wastewater and effluents from cities and industry. Improve political stability, reduce rural violence, implement better governance. Develop ways of including currently marginalised groups in decision-making processes in their own regions. Introduction of compensation schemes (PES/PWS) to encourage better resource use.

This table combines responses to a questionnaire and my own interpretations of information presented in case study reports.

Table 6. Major information needs in eight big river basins, as identified by case study teams.

Question	Major information needs identified by case study teams
Limpopo, Southern Africa	Actual water use, by sector, per country is still difficult to pinpoint. Better data from Botswana. The changing, unpredictable political situation in Zimbabwe and potentially South Africa make generational planning and assessment difficult. Political priorities may change quickly, thereby realigning the economic landscape and placing pressure on natural resources. These dynamics are poorly understood. Likely impacts of climate change.
Karkeh, Iran	Better hydrological data, both for flows in basins and groundwater–surfacewater interaction. Understanding how to develop and implement groundwater policy. Better understanding of linkages between water and poverty in the basin. Relationships between production systems not typically included in basin water assessments (for example, ranching, forestry, fisheries) and water supply and demand. Likely climate change impacts on freshwater availability in basin. Impacts of Iranian government's decision to remove subsidies on energy and agricultural inputs.
Nile, North and East Africa	Little data available on water requirements of wetland ecosystems. In addition, contributions made to the Nile by the Sudd wetland system (one of the largest in the world), and possible impacts on the Sudd of the Jonglei Canal, are poorly understood. The relationships between biophysical variables and ecological indicators are poorly understood. Climatic and hydrologic monitoring data are weak; lack of basin-wide monitoring strategy; water quality information lacking; climate change projections uncertain.
Volta, West Africa	Quality/reliability of published data from state agencies is often not known (for example, agriculture production, livestock, cultivated area, fine-scale poverty data). Data gaps from countries that have only a small part of the basin. Large gaps in hydrological (river discharge) data. Similarly, groundwater resources poorly understood. Climate projections uncertain. Sustainability of fisheries on Lake Volta unclear.
Niger, West Africa	Population growth data. Impacts of climate change uncertain. Need to research to determine how to improve agricultural output in local conditions. Ways to improve uptake of agricultural productivity interventions. Ways to improve governance and integrated water resource management.

Yellow River, China	Need monitoring, information and awareness generation on water quality situation in the basin. Information exchange needs between ministries of agriculture and water to assess impacts of agricultural policies on water availability. Key uncertainties need to be resolved: cost of poor water quality to the public and the country; cost of continuing business as usual; cost of rigid water allocation planning (for example, province-level quota systems). Approaches to drought management (lack of policies given past focus on flood control). Impacts of climate change on both surface and groundwater supply. How much water use efficiency can be increased in agricultural sector, and likely impacts of adopting agricultural water-saving technologies on local and basin-level water use. Irrigated and rainfed area by crop is unknown. Crop evapotranspiration (ET) by month by county (or better resolution). Water requirements for downstream ecosystems (estimates from the various sources are very different).
Indus-Ganges Basins, India and Pakistan	Due to transboundary nature and low trust level among the riparian countries, water resources data for basin-level analysis are unavailable. A similar situation also exists for groundwater/flood data (and other resources). Impact of climate change upon Himalayan water resources is not well understood and there are conflicting scientific evidence. Poor knowledge of use by sector and revenues generated for water services. Impact of water resources development on local ecology, biodiversity, livelihoods, displacements and other costs.
Mekong, Asia	Impact of dams, new irrigation systems and upstream development on the lower part of the basin, on the capture fisheries and on the environment and ecology. Understanding relationships between food security and poverty reduction. Impact of climate change on flows, agricultural productivity, fisheries ecology, environment and sea level rise.
Andes, South America	Better integrated modelling and research that is capable of moving beyond single issues to looking at multiple issues and unforeseen consequences. Bridging science and policy to provide policy support. Better measurement and monitoring of some of the fundamentals (for example, rainfall and distribution of water treatment plants). Better information on the role of poverty in determining water access. Better measures of livelihoods and wellbeing, beyond purely economic measures.

Further details on these needs are available in individual chapters of *Water International* 35 (5) (2010).

Addressing these information needs would presumably make the case study basins more resilient, in the sense that the facts that are needed for informed decision making would be more readily available. It is, however, a mistake to assume that simply getting the science or even the number and nature of institutions right will solve the problems; the focus in most cases will need to be on process, and in particular on developing flexible, transparent, inclusive political approaches to dealing with and resolving water allocation issues.

One of the hardest questions to answer in the context of assessing the resilience of big river basins is that of whether pro-active managers are thinking about (and preparing for) the right future. Surprise is an integral element of complex systems and it is quite possible that new, unexpected threats to the integrity of big river basins will emerge from unexpected quarters (see, for example, the discussion in Gordon *et al.* 2008). For example, none of the BFP case studies has considered the possible implications for basin sustainability of such possible future trends as increased production of biofuels (de Fraiture *et al.* 2008), outsourcing of agricultural production from wealthier countries to Africa (Daviron and Gibbon 2002), or the potential for anthropogenic influences to create "unhealthy landscapes" that are susceptible to novel forms of water-borne or water-associated pathogens (Patz *et al.* 2004).

The future will not necessarily conform to the present. Building resilience to surprise is more a case of building adaptive capacity (for example, through developing monitoring schemes with feedback loops to management, setting in place appropriate decision-making processes, building capital stocks, maintaining diversity and redundancy, and ensuring that some "slack" remains in key components of the system) than of managing for a specific outcome or goal. Managing for resilience inevitably comes at a cost, and many complex systems will (whether intentionally or not) tend to reduce adaptive capacity by shifting towards greater efficiency during periods of relative stability when diversity, redundancy, and under-utilization are seen as wasteful (Holling and Meffe 1996). One of the central challenges for big river basin management is thus to put in place the structures (institutions, policies, expertise) that are needed to cope with times of change, and to retain them through periods of relative constancy when their direct relevance may be less obvious.

Concluding comments

This brief analysis of the resilience of big river basin SESs in this set of case studies has raised some important questions and recommendations in two realms: (1) conceptual and scientific understanding; and (2) practical recommendations for managing these systems towards greater resilience in desired characteristics.

In the conceptual and scientific realm, standard approaches to river basin management (including most development and water allocation programmes) have yet to fully grapple with the questions that are raised by scientific uncertainty and system complexity. All of the case studies face potentially large uncertainties in the estimation and prediction of a wide range of important variables, ranging from likely future rainfall and runoff through to more diffuse variables such as changes in water demand and the future role of technology in water-use efficiency. Uncertainty exists not only in the estimation of such variables, but also in our understanding of the nature of the relationships between variables. Lake eutrophication offers a good example of a well-studied, non-linear relationship in freshwater systems (Carpenter 2003), but a wide range of other relationships are likely to show significant non-linearities and complex behaviours, particularly if climate change or water extraction push key variables beyond their normal range of variation (Holling and Meffe

1996, Gordon *et al.* 2008). Understanding and incorporating such uncertainties into models and decision-making tools remains an important goal for research in many big river basins.

In the practical realm, a number of general and specific suggestions emerge from the preceding discussion. Complex systems theory highlights the value of maintaining diversity, redundancy, and capital within a given system as well as of fostering learning, innovation and adaptive capacity where possible (Holling and Meffe 1996, Norberg and Cumming 2008). The temptation to blame problems on a single cause or driver (such as population growth or institutional failure), and build expectations around working towards a single overarching solution or "panacea", must be avoided (Ostrom 2007). Big river basins are complex systems that are composed of multiple interacting variables, which change at different speeds and occupy different hierarchical levels. System manipulations, even of biophysical components of the system that are supposedly well understood, may have unintended consequences (Patten *et al.* 2001). Pro-active management will be most effective (and most capable of making necessary mistakes, and learning from them) if it occurs via politically acceptable and participatory processes (Adger and Jordan 2009), uses the best possible science, and takes a holistic perspective that seeks to incorporate social, ecological, and economic elements in problem definition and solution.

Acknowledgements

I am grateful to Simon Cook for inviting me to contribute this paper and to the nine contributing BFP project teams for their willingness to respond to my questions and provide me with relevant materials. Most of the text in Tables 2 to 6 was generated by the case study authors, although I may have subsequently introduced errors. I would particularly like to thank Amy Sullivan (Limpopo); Poolad Karimi (Karkheh); David Molden, Seleshi B. Awulachew and Solomon S. Demissie (Nile); Jacques Lemoalle (Volta); Andrew Ogilvie and Jean-Charles Clanet (Niger); Claudia Ringler, Jinxia Wang and Ximing Cai (Yellow); Bharat Sharma (Indus-Gangetic); Mac Kirby and Mohammed Mainuddin (Mekong) and Mark Mulligan (Andes). Simon Cook, Tassilo Tiemann and an anonymous reviewer provided useful comments on earlier versions of this paper.

Note

1. Ludwig (2001) defines "wicked" problems as those that include not only multiple disciplines, stakeholders and epistemologies, but also that cannot be separated from issues of values, equity and social justice.

References

Adger, W., 2003. Social capital, collective action, and adaptation to climate change. *Economic Geography*, 79 (4), 387–404.

Adger, W.N., 2006. Vulnerability. *Global Environmental Change*, 16 (3), 268–281.

Adger, W.N. and Jordan, A., 2009. Sustainability: exploring the processes and outcomes of governance. *In*: W.N. Adger, and A. Jordan, eds. *Governing sustainability*. Cambridge: Cambridge University Press, 3–31.

Agarwal, C., *et al.* 2002. *A review and assessment of land-use change models: dynamics of space, time, and human choice*. General Techical Report NE-297. Newtown Square, PA: US Department of Agriculture, Forest Service, Northeastern Research Station, 61.

Anderies, J.M., 2004. Minimal models and agroecological policy at the regional scale: an application to salinity problems in southeastern Australia. *Regional Environmental Change*, 5 (1), 1–17.

Anderies, J.M., Janssen, M.A., and Ostrom, E., 2004. A framework to analyze the robustness of social-ecological systems from an institutional perspective [online]. *Ecology and Society*, 9 (1), 18. Available from: www.ecologyandsociety.org/vol9/iss1/art18/ [Accessed 15 November 2010].

Brand, F., 2009. Critical natural capital revisited: ecological resilience and sustainable development. *Ecological Economics*, 68 (3), 605–612.

Carlson, J.M. and Doyle, J., 2002. Complexity and robustness. *Proceedings of the National Academy of Sciences of the United States of America*, 99 (Suppl. 1), 2538–2545.

Carpenter, S., 2003. *Regime shifts in lake ecosystems: patterns and variation*. Excellence in Ecology Series Number 15. Oldendorf/Luhe, Germany: Ecology Institute.

Carpenter, S.R. and Turner, M.G., 2000. Hares and tortoises: interactions of fast and slow variables in ecosystems. *Ecosystems*, 3 (6), 495–497.

Carpenter, S., *et al.*, 2001. From metaphor to measurement: resilience of what to what? *Ecosystems*, 4 (8), 765–781.

Checkland, P., 1981. *Systems thinking, systems practice*. New York: Wiley.

Checkland, P., 2009. *Systems thinking, systems practice (with 30-year retrospective)*. New York: Wiley.

Clark, J.S., *et al.*, 2001. Ecological forecasts: an emerging imperative. *Science*, 293 (5530), 657–660.

Clark, M.J., 2002. Dealing with uncertainty: adaptive approaches to sustainable river management. *Aquatic Conservation: Marine and Freshwater Ecosystems*, 12 (4), 347–363.

Cronon, W., 2000. Resisting monoliths and tabulae rasae. *Ecological Applications*, 10 (3), 673–675.

Cumming, G.S., 2011. *Spatial resilience in social-ecological systems*. Berlin: Springer.

Cumming, G.S., *et al.*, 2005. An exploratory framework for the empirical measurement of resilience. *Ecosystems*, 8 (8), 975–987.

Cumming, G.S. and Collier, J., 2005. Change and identity in complex systems [online]. *Ecology and Society*, 10, 29. Available from: http://www.ecologyandsociety.org/vol10/iss1/art29/ [Accessed 15 November 2010].

Cumming, G.S., Cumming, D.H.M., and Redman, C.L., 2006. Scale mismatches in social-ecological systems: causes, consequences, and solutions [online]. *Ecology and Society*, 11 (1), 14. Available from: http://www.ecologyandsociety.org/vo11/iss1/art14/ [Accessed 15 November 2010].

Daviron, B. and Gibbon, P., 2002. Global commodity chains and African export agriculture. *Journal of Agrarian Change*, 2 (2), 137–161.

de Fraiture, C., Giordano, M., and Liao, Y., 2008. Biofuels and implications for agricultural water use: blue impacts of green energy. *Water Policy*, 10 (Suppl. 1), 67–81.

Deutsch, L., Folke, C., and Skanberg, K., 2003. The critical natural capital of ecosystem performance as insurance for human well-being. *Ecological Economics*, 44 (2–3), 205–217.

Ernstson, H., *et al.*, in press. Scale-crossing brokers and network governance of urban ecosystem services: the case of Stockholm, Sweden. *Ecology and Society*, (in press).

Gonzalez, O.M.R., 2001. Assessing vegetation and land use changes in northeastern Puerto Rico: 1978–1995. *Caribbean Journal of Science*, 37 (1–2), 95–106.

Gordon, L.J., Peterson, G.D., and Bennett, E., 2008. Agricultural modifications of hydrological flows create ecological surprises. *Trends in Ecology and Evolution*, 23 (4), 211–219.

Hahn, T., *et al.*, 2006. Trust-building, knowledge generation and organizational innovations: the role of a bridging organization for adaptive comanagement of a wetland landscape around Kristianstad, Sweden. *Human Ecology*, 34 (4), 573–592.

Holling, C.S., 1973. Resilience and stability of ecological systems. *Annual Review of Ecology and Systematics*, 4, 1–23.

Holling, C.S., 2001. Understanding the complexity of economic, ecological, and social systems. *Ecosystems*, 4 (5), 390–405.

Holling, C.S. and Meffe, G.K., 1996. Command and control and the pathology of natural resource management. *Conservation Biology*, 10 (2), 328–337.

Imbernon, J., 1999. A comparison of the driving forces behind deforestation in the Peruvian and Brazilian Amazon. *Ambio*, 28 (6), 509–513.

Issaka, R.N., *et al.*, 1997. Geographical distribution of selected soil fertility parameters of inland valleys in West Africa. *Geoderma*, 75 (1–2), 99–116.

Levin, S.A. and Lubchenco, J., 2008. Resilience, robustness, and marine ecosystem-based management. *BioScience*, 58 (1), 1–6.

Ludwig, D., 2001. The era of management is over. *Ecosystems*, 4 (8), 758–764.

Lugo, A.E. and Helmer, E.H., 2004. Emerging forests on abandoned land: Puerto Rico's new forest. *Forest Ecology and Management*, 190 (2–3), 145–161.

Mainuddin, M., Kirby, M., and Chen, Y., 2009. *Fisheries productivity and its contribution to overall agricultural production in the Lower Mekong River Basin*. Colombo: Challenge Program on Water and Food.

Martin, R. and Sunley, P., 2006. Path dependence and regional economic evolution. *Journal of Economic Geography*, 6 (4), 395–497.

Norberg, J. and G.S. Cumming, eds, 2008. *Complexity theory for a sustainable future*. New York: Columbia University Press.

Norberg, J., *et al.*, 2008. Diversity and resilience of social-ecological systems. *In*: J. Norberg and G.S. Cumming, eds. *Complexity theory for a sustainable future*. New York: Columbia University Press, 46–79.

Olsson, P., *et al.*, 2007. Enhancing the fit through adaptive co-management: creating and maintaining bridging functions for matching scales in the Kristianstads Vattenrike biosphere reserve, Sweden [online]. *Ecology and Society*, 12 (1), 28. Available from: http://www.ecologyandsociety.org/vol12/iss21/art28/

Olsson, P., Folke, C., and Hahn, T., 2004. Social-ecological transformation for ecosystem management: the development of adaptive co-management of a wetland landscape in southern Sweden [online]. *Ecology and Society*, 9 (4), 2. Available from: http://www.ecologyandsociety.org/vol9/iss4/art2/ [Accessed 15 November 2010].

Ostrom, E., 2007. A diagnostic approach for going beyond panaceas. *Proceedings of the National Academy of Sciences of the United States of America*, 104 (39), 15181–15187.

Ostrom, E., 2009. A general framework for analyzing sustainability of social-ecological systems. *Science*, 325 (5939), 419–422.

Patten, D.T., *et al.*, 2001. A managed flood on the Colorado River: background, objectives, design, and implementation. *Ecological Applications*, 11 (3), 635–643.

Patz, J.A., *et al.*, 2004. Unhealthy landscapes: policy recommendations on land use change and infectious disease emergence. *Environmental Health Perspectives*, 112 (10), 1092–1098.

Polis, G.A., Anderson, W.B., and Holt, R.D., 1997. Toward an integration of landscape and food web ecology: the dynamics of spatially subsidized food webs. *Annual Review of Ecology and Systematics*, 28, 289–316.

Polis, G.A., Power, M.E., and Huxel, G.R., eds, 2004. *Food webs at the landscape level*. Chicago: University of Chicago Press.

Pretty, J., 2008. Agricultural sustainability: concepts, principles and evidence. *Philosophical Transactions of the Royal Society B-Biological Sciences*, 363 (1491), 447–465.

Rodriguez, J.P., *et al.*, 2006. Trade-offs across space, time, and ecosystem services. *Ecology and Society*, 11 (1), 28.

Rosegrant, M.W., *et al.*, 2000. Integrated economic–hydrologic water modeling at the basin scale: the Maipo river basin. *Agricultural Economics*, 24 (1), 33–46.

Scheffer, M., 2009. *Critical transitions in nature and society*. Princeton: Princeton University Press.

Scheffer, M. and Carpenter, S.R., 2003. Catastrophic regime shifts in ecosystems: linking theory to observation. *Trends in Ecology and Evolution*, 18 (12), 648–656.

Senge, P.M., 1990. *The fifth discipline: the art and practice of the learning organization. Appendix 2. Currency*. New York: Doubleday.

Stirling, A., 2003. Risk, uncertainty and precaution: some instrumental implications from the social sciences. In: M.L. Berkhout and I. Scoones, eds. *Negotiating environmental change: new perspectives from social science*. Cheltenham: Edward Elgar, 33–76.

Walker, B. and Salt, D., 2006. *Resilience thinking: sustaining ecosystems and people in a changing world*. Washington, DC: Island Press.

Waltner-Toews, D., Kay, J.J., and Lister, N.-M.E., 2008. The ecosystem approach: uncertainty and managing for sustainability. New York: Columbia University Press.

Waltner-Toews, D., *et al.*, 2003. Perspective changes everything: managing ecosystems from the inside out. Frontiers in Ecology and the Environment, 1 (1), 23–30.

WCED, 1987. *Our common future*. Report of the World Commission on Environment and Development (WCED). Oxford: University Press, 43–66.

Yachi, S. and Loreau, M., 1999. Biodiversity and ecosystem productivity in a fluctuating environment: the insurance hypothesis. *Proceedings of the National Academy of Sciences*, 96 (4), 1463–1468.

Yunnus, M., 1998. Alleviating poverty through technology. Science, 282 (5388), 409–410.

Zwarts, L., *et al.*, 2006. The economic and ecological effects of water management choices in the upper Niger River: development of decision support methods. *International Journal of Water Resources Development*, 22 (1), 135–156.

The nature and impact of climate change in the Challenge Program on Water and Food (CPWF) basins

Mark Mulligan, Myles Fisher, Bharat Sharma, Z.X. Xu, Claudia Ringler, Gil Mahé, Andy Jarvis, Julian Ramírez, Jean-Charles Clanet, Andrew Ogilvie and Mobin-ud-Din Ahmad

In this article the authors assess the potential impacts of projected climate change on water, livelihoods and food security in the Basin Focal Projet basins. The authors consider expected change within the context of recently observed climate variability in the basins to better understand the potential impact of expected change and the options available for adaptation. They use multi-global circulation model climate projections for the AR4 SRES A2a scenario, downscaled and extracted for each basin. They find significant differences in the impacts (both positive and negative impacts) of climate change, between and within basins, but also find large-scale uncertainty between climate models in the impact that is projected.

Climate variability in basins

While some of the Challenge Program on Water and Food (CPWF) Basin Focal Projects (BFPs) have carried out detailed hydrological or water resource modelling of the impacts of climate change projections, not all BFPs did so since this was not one of their main objectives. Moreover, BFPs used different scenarios, different global circulation models (GCMs), and different downscaling methods for the basins in which they undertook to model the impact of climate change. In this paper we use common datasets for historic and projected climates to compare historic climate trends and projected climate change scenarios within and between the basins. We review the conclusions of the BFPs with respect to climate variability and change within basins and we present results for the Indus-Ganges, Yellow River and Andes basins where different water resources or hydrological models (WEAP, SWAT and FIESTA[1] respectively) have been used to model the hydrological impacts of projected climate change.

Before we can understand climate change in basins, we must understand the climate norms and variability within them. The data in Table 1 were re-gridded to 1 km^2 spatial

Table 1. Climate variability for 10 basins 1901–96.

African basins	Limpopo	Niger	Nile	Volta
Mean precipitation (1901–96) mm/month (mm/yr)	44 (530)	82 (988)	51 (609)	79 (953)
Mean temperature (1901–96) °C	21	27	25	28
Interannual variability, temperature[1]	5.7	5.7	4.7	5.8
Interannual variability, precipitation[2]	2.8	2.1	1.9	2.1
Asian basins	**Karkeheh**	**Mekong**	**Yellow**	**Ganga**
Mean precipitation (1901–96) mm/month	34 (402)	137 (1647)	32 (384)	89 (1072)
Mean temperature (1901–96)	15.4	25.1	6.7	21
Interannual variability, temperature[1]	6.0	3.8	0.8	4.4
Interannual variability, precipitation[2]	4.4	1.5	14.5	4.0
South American basins	**Andes**	**São Francisco**		**Global (land)**
Mean precipitation (1901–96) mm/month	63 (751)	85 (1018)		60 (714)
Mean temperature (1901–96)	13.2	24		8.1
Interannual variability, temperature[1]	0.6	5.1		7.5
Interannual variability, precipitation[2]	8.9	2.1		2.4

Notes: [1] Standard deviation of annual mean temperature (1901–96); [2] Coefficient of variation of annual mean precipitation (1901–96).
Source: New *et al.* (2000).

resolution using nearest-neighbour interpolation, since data were not available that would support more sophisticated downscaling. The data were then cut to the basin boundaries, as defined by the CPWF[2], for consistency with other datasets that we used in hydrological modelling. We analysed monthly data from 1901 to 1996 to provide a measure of mean temperature and precipitation as well as their inter-annual variability. We calculated the mean values for global land areas using the same methods and data to place the basins in a global context. As might be expected, long-term temperature and precipitation vary considerably between basins. For example, the inter-annual variability for annual mean temperature is higher in absolute terms but lower relative to mean temperature for the African basins, whereas the opposite is often the case for Asian and Latin American basins.

There is significant between-basin variation in the long-term mean annual precipitation inputs to the African basins from 530 mm/yr (Limpopo) through 990 mm/yr (Niger). The most temporally variable African basin is the Limpopo with a mean basin inter-annual variability (coefficient of variation) of 2.8% at the basin scale. The inter-annual variation of individual precipitation stations may be much higher than the mean basin value but the mean basin value is more representative of the basin-wide impact of any variability and is thus of greater policy relevance. Mean annual temperature for the African basins varies from 28°C (Volta) to 21°C (Limpopo) with similar levels of inter-annual variability for temperature between them. For the Asian basins mean precipitation varies from 400 mm/yr (Karkheh) to 1650 mm/yr (Mekong). Inter-annual, basin-wide variability of precipitation is greatest by far for the Yellow River at 14.5%, and least for the Mekong at 1.5%. Inter-annual variability in temperature for these basins is greatest for the Karkheh (standard deviation [SD] = 6°C) and least for the Yellow River.

For the Latin American basins mean precipitation is greatest for the São Francisco (1020 mm/yr) but inter-annual, basin-wide variability is much greater for the Andes at 8.9%. Mean annual temperature is much greater for the São Francisco and inter-annual temperature variability is also greater (SD = 5.1°C). For context, mean long-term values for global land areas are 710 mm/yr (precipitation) and 8.1°C (temperature). The significant differences of temperature and precipitation between basins determine the context of their water and food productivity but the context also depends on seasonality and the spatial variability of climate within basins.

Figure 1 shows the mean historic (1901–96) seasonality of temperature for the basins. Clearly, some have strong seasonality of temperature. The hottest basins are the Niger, Volta, Mekong and Nile, which also have the lowest seasonal temperature range. The coldest basins are the Yellow River and Andes. Basins with the strongest temperature seasonality are the (high-latitude) Yellow River and Karkheh and those with the least seasonality are the Mekong and Nile, the Andes (which have significant presence in both hemispheres) and the São Francisco. The southern hemisphere basins, as expected, show a reversed timing of temperature seasonality compared with the northern basins.

Figure 2 shows the mean historic (1901–96) seasonality of precipitation for the basins. Precipitation seasonality is greatest for the monsoonal Mekong and Ganges basins but also for the Niger and Volta. The São Francisco shows a strong and opposite precipitation

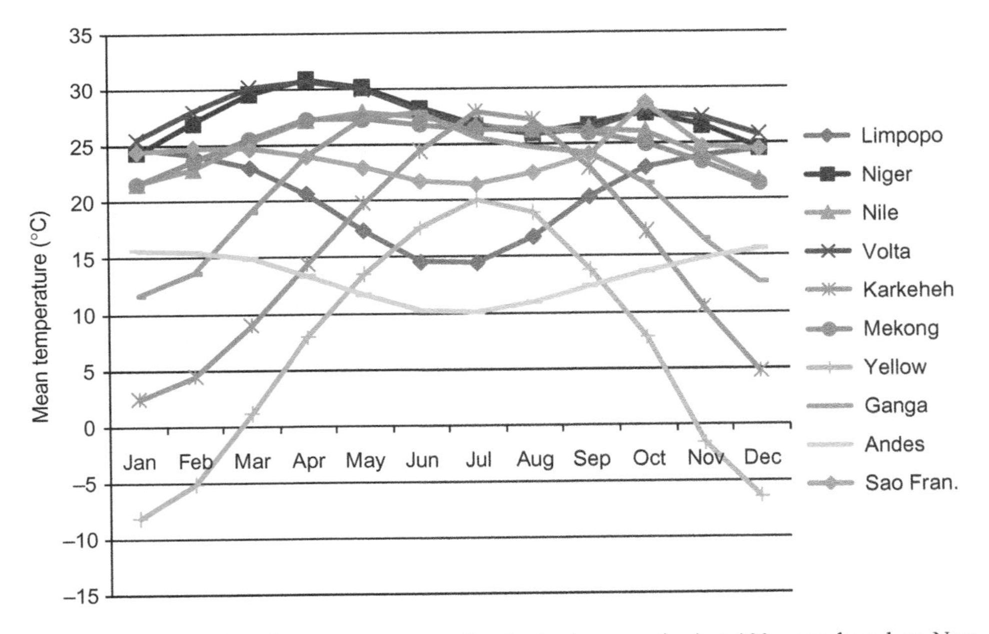

Figure 1. Seasonality of mean temperature for the basins over the last 100 years based on New *et al.* (2000).

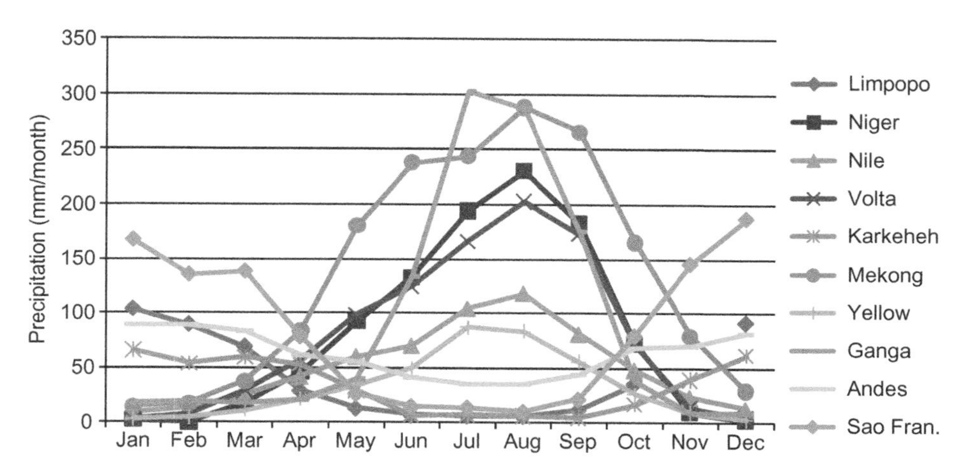

Figure 2. Seasonality of mean precipitation for the basins over the last 100 years based on New *et al.* (2000).

seasonality to the other basins. The Andes shows very little precipitation seasonality at the basin level, perhaps because the northern and southern hemisphere seasonalities cancel each other out at the cross-equatorial basin scale.

Clearly the highly seasonal basins, especially those that combine low precipitation and high temperature in the same season, are particularly vulnerable to constraints on water availability and reductions in water productivity during the dry season; even their annual water balance (that is, rainfall minus evapo-transpiration) may be substantial. Thus the Ganges, the Niger, and the Volta are seasonally very susceptible to crop failure. For the Nile, rainfall variability is one of the main factors that controls food insecurity in the region, although rainwater harvesting can mitigate in some areas (Nile Basin team 2010). For the Limpopo, seasonal variability makes farmers vulnerable to crop failure and livestock loss.

Farmers relying on rainfed agriculture are vulnerable to the variability and unpredictability of rainfall. Those with resources to invest in storage and water-saving technology are better able to withstand these shocks but few can do so year after year. Moreover, farmers see investing in storage and water saving as a risky venture and see low input (and thus low output) subsistence as a means of minimizing the risks associated with climate variability (Kato *et al.* 2009, Limpopo Basin team 2010). In the Volta, rainfall variability and associated lack of water security is one of the three main causes of poverty in Ghana. In the other Volta countries, poverty is greater in the drier north than the wetter south (Volta Basin team 2010). There is high variability between years and seasonally, which adversely impacts crop yields and inflows to the Akosombo/Kpong hydropower scheme (Volta Basin team 2010). For the Mekong, rainfall also varies considerably on an inter-annual basis: severe droughts in 2010 led to adverse impacts on both crops and fisheries (Mekong Basin team 2010)

Recent climate trends

We analysed the New *et al.* (2000) dataset to identify any major change in climate in the basins over the last century by comparing the mean temperature and precipitation for the period 1900–50 with 1950–96, at the basin scale (Table 2). There were small reductions in all of the African basins in mean annual precipitation between the first and second

Table 2. Observed climate change over the twentieth century, comparing means for 1900–50 with means for 1950–2000.

African basins	Limpopo	Niger	Nile	Volta
Precipitation, mm/month (mm/yr)	−1.2 (− 13.8)	−1.3 (− 15.9)	−0.4 (− 4.8)	−0.3 (− 3.4)
Temperature, °C	0.1	−0.9	−0.6	−0.4
Asian basins	**Ganga**	**Karkeheh**	**Mekong**	**Yellow**
Precipitation, mm/month (mm/yr)	−1.25 (− 15)	−0.04 (− 0.4)	1.22 (14.6)	0.53 (6.3)
Temperature, °C	−0.4	0.1	−0.2	−0.9
South American basins	**Andes**	**São Francisco**		**Global (land)**
Precipitation, mm/month (mm/yr)	−0.2 (− 2.8)	−2.3 (− 28.1)		1.0 (12)
Temperature, °C	−0.02	−0.1		−0.04

Source: New *et al.* (2000).

half of the twentieth century, with the strongest decrease at the basin scale (16 mm/yr) in the Niger, followed by the Limpopo (−14 mm/yr). These basin-scale means may differ from data for particular rainfall stations or for parts of the basins, but again are of greater relevance for policy than the results from single stations.

For the Asian basins, the there was little basin-wide change in precipitation in the Karkheh, while the both the Yellow River and the Mekong increased and the Ganges decreased. Temperature changes little in the Asian basins. In the South American basins, there was little change in precipitation in the Andes, but the São Francisco decreased by 28 mm/yr. This analysis indicates that global land temperatures changed little over the century, although the Intergovernmental Panel on Climate Change (IPCC) (2007) reports that warming over land and oceans over the period 1906–2005 was 0.74 ± 0.18°C. Much of the warming occurred in recent years and at high latitudes, however, rather than between the first and second halves of the twentieth century in the tropics. Globally over land, precipitation increased by 12 mm/yr. From physical principles one might expect higher sea-surface temperatures to increase precipitation but changes in *global* precipitation have not been detected so far (Lambert *et al.* 2005). However, observed latitudinal and regional redistribution of precipitation is consistent with the patterns of change that GCMs predict in response to greenhouse forcing (Zhang *et al.* 2007).

In field studies of river discharge, it is very difficult to separate the impact of changes in land use and irrigation extraction on flow from those caused by climatically driven, decreased precipitation or increased evapo-transpiration over time. Moreover, given the short periods and limited extents of most field-based studies and records of river discharge, it is a challenge to understand a basin's temporal variability and distinguish between normal variability and the effects of climate change from measured data of stream flow. Thus, few of the BFP basins have reported on the long-term variability of water balances or stream flows in relation to climate. Most have focused instead on the current water situation based on climatic means. There are some exceptions, however.

Masih *et al.* (2009, 2010) show that the stream flow in the Karkheh varies greatly within and between years. Peak and low flows can differ by as much as tenfold (at Jalogir, the mean flow in April is 386 m^3/sec compared with 41 m^3/sec in September). Masih *et al.* (2010)

also found significant long-term trends in stream flow for all stations, with December flows increasing and May flows decreasing. Declining low flows were more important in the upper parts of the basin, whereas higher winter flows and flooding were confined to the middle parts of the basin. The trends were attributed to changes in precipitation rather than to changes in water use or land use. Less rainfall in April and May gave lower flows while increased winter rain (particularly in March) coupled with warming increased flooding.

Based on data of temperature, precipitation and evaporation from 80 stations, the Yellow River has become warmer and drier from the 1960s to 2000 (Xu *et al.* 2007). Winter temperatures have increased significantly, and 65 of 77 stations show decreasing mean annual precipitation, especially in the autumn, summer and spring, with small increases in the winter. There is strong spatial variability, with summer rainfall increasing around Yinchuan but decreasing elsewhere. The long-term annual runoff in the Yellow River basin (YRB) is 58 km^3, but it has decreased annually since the 1950s (Ni *et al.* 2004, Xu *et al.* 2007) to an average of 43 km^3 in the 1990s, a decline of 25%. The decline is attributed to a combination of less rainfall and higher rates of water withdrawal for irrigation. The Yellow River ran dry, for 15 days, for the first time in 1972. From 1985 to 1997, it ran dry nearly every year with the dry period becoming progressively longer (133 days in 1966 and 226 days in 1997 [Xu *et al.* 2005]). Recently, droughts are significantly more frequent in the YRB and floods are significantly less frequent. For example, overbanking at Huayuankou occurred almost annually during the 1950s but less than every three years from 1986 to 2000 and not at all since 2000.

The Niger Basin is said to have been in drought since the 1970s (Conway and Mahé 2009, Conway *et al.* 2009), with the drought deeper in the west than in the east and south (Mahé and Paturel 2009). The 1980s were the driest decade of the twentieth century. After severe droughts in the 1970s and 1980s, rainfall increased after 1993 but levels are still low (Mahé *et al.* 2009, Niger Basin team 2010). The impacts of these droughts in the region differ geographically according to their intensity and the bioclimatic context and human pressure on the land.

Climate change in basins: impacts and vulnerabilities

Climate change projections for a region, including some of those carried out for the BFPs, often involve downscaling and applying the outcomes of a single GCM, based on one emission scenario. To put this in context, there is a wide range of emission scenarios, published by the IPCC in its *Special report on emissions scenarios* (SRES) (Nakicenovic *et al.* 2000). Typically, we use scenario A2a. In the A2 family of SRES scenarios, high-income but resource-poor regions shift toward advanced post-fossil technologies (renewables or nuclear), while low-income resource-rich regions generally rely on older fossil-fuel technologies. Human population reaches 15 billion by 2100 (Nakicenovic *et al.* 2000) and CO_2 emissions increase nearly linearly to 2100; 780 ppmv by 2050, more than double the 1850 level of 270 ppmv (Nakicenovic *et al.* 2000). A2a is thus a high-carbon scenario, but is realistic given recent history and is aptly called the "business as usual" scenario.

The IPCC based their fourth assessment report (AR4, IPCC 2007) on 21 GCMs, which can give significantly different projections for a particular climate datum for a given geographical site, and can produce significantly different agricultural impact (Semenov and Stratonovitch 2010). It is therefore important to use a multi-model approach to avoid errors in forecasting the impacts of climate change.

Ramírez and Jarvis (2008) used a delta method to downscale the monthly means of GCM projections to a 1-km grid, which we coupled with the 1km-gridded monthly means of the WorldClim (Hijmans *et al.* 2005) database, representing the 1950–2000 mean climate. We extracted downscaled GCM scenarios for the SRES A2a scenario for the CPWF basins, and calculated basin mean change between 1950–2000 (indicative of the "present") and 2040–69 (the so-called 2050 forecast) for annual mean temperature and precipitation. We present these as differences between the basin means of 17 GCMs[3] for the "present" and 2050. For each case we provide the basin maximum, minimum and mean for temperature and precipitation change respectively. We also include mean data for the appropriate continents. The continents are as defined by the HYDRO1k dataset (USGS 2010) and the basins according to the CPWF definition.

In examining the scenarios for change for each basin we focus on the change in mean temperature and precipitation but also on the maximum and minimum change projected in each as a measure of within-basin spatial variability. As a measure of variation between GCM predictions (that is, projection uncertainty), we use basin means of the GCMs and calculate the standard deviation for annual mean temperature and for annual total precipitation. We also give the corresponding basin minimum and maximum uncertainties between models, which are a measure of spatial variability in projection uncertainty. We now examine these multi-model scenarios, continent by continent.

African basins

Scenarios of change

For the African basins (Table 3) we see a mean projected warming of 2.7–2.9°C by the 2050s compared with 2.8°C for Africa as a whole. This warming is very significant compared with observed twentieth-century change in temperature. The mean 17-GCM standard deviation of annual mean temperature between is high at 1.6°C for Africa as a whole and is also significant for all of the basins, indicating a high degree of disagreement between model projections (that is, uncertainty) for these basins. Precipitation shows an increase for all African basins except the Limpopo, and Africa as a whole shows an increase in precipitation of 87 mm/yr over the period. The projected precipitation increase is greater for the Nile and Volta than the Niger. In all cases except the Limpopo the 17-GCM standard deviation in annual total precipitation projections is significant (>100 mm/yr), again indicating a high level of uncertainty in the magnitude of projected wetting. All basins show significant within-basin spatial variability for temperature (for example, T_{min} 2°C to T_{max}

Table 3. Change in temperature and precipitation in five African basins from 1950–2000 to 2050 under the A2a emissions scenario and the standard deviation between GCM projections for 17 GCMs for each basin.

Africa	Temperature						Precipitation					
				Standard deviation						Standard deviation		
Basin	Min	Max	Mean	Min	Max	Mean	Min	Max	Mean	Min	Max	Mean
	°C	*°C*	*°C*				*mm*	*mm*	*mm*			
Africa	1.3	3.6	2.8	0.28	2.8	1.6	−208	588	87	0.0	571	198
Limpopo	2.3	3.4	2.9	0.85	1.6	1.2	−174	50	−5	0.0	223	37.1
Niger	2.0	3.4	2.7	0.55	1.9	1.2	−40	252	66	1.0	354	124
Nile	1.9	3.7	2.8	0.80	2.4	1.3	−23	558	86	0.0	360	99.8
Volta	2.0	3.2	2.7	0.68	1.8	1.2	−17	251	80	1.2	347	144

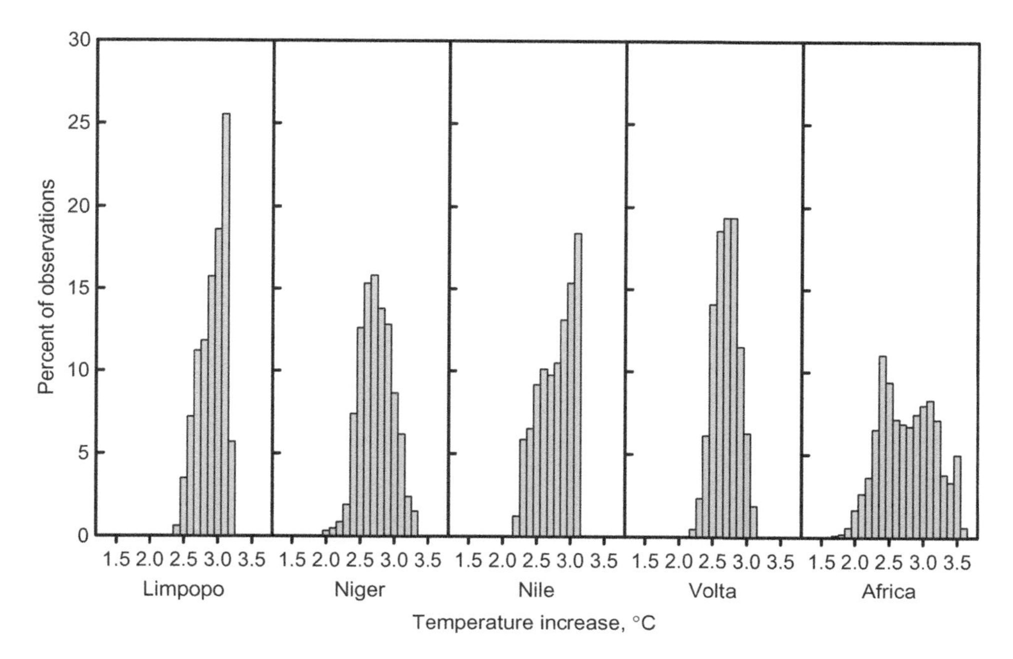

Figure 3. Frequency histograms for temperature change for African basins (mean of 17 GCMs for each basin).

3.4°C for the Niger Basin) and, in particular, precipitation. All basins have some areas with projected drying and others with projected wetting (for example, the Nile with P_{min} −23 mm/yr and P_{max} 558 mm/yr).

Figure 3 shows frequency distributions for temperature change for the African basins. These are drawn to the same horizontal scale with frequency expressed as a percent of observations (that is, percent of basin area). These frequency distributions serve to show the range and spatial variability in projected change within and between the basins. Within-basin variability in the projected temperature change is often greater than the differences between basins and this variability cannot be seen clearly in the tabulated basin means given above. We see that for the African continent as a whole the multi-model mean projected temperature change varies significantly, with a few areas showing change of around 1.5°C, most areas warming between 2.0°C and 3.0°C and some areas warming greater than 3.0°C. The Limpopo and Nile basins in particular have significant basin areas towards the high end of expected temperature changes.

The frequency distributions for precipitation (Figure 4) indicate a wide range of projected change for Africa with much of the continent showing little change but some areas showing increases of a few hundred mm/yr and a few areas showing small decreases in precipitation. The Niger Basin shows wetting of around 100 mm/yr throughout as does the Volta and the Nile, whereas the Limpopo shows significant drying throughout.

We can expect all African basins to be water-stressed by the increased evaporative demand resulting from the projected warming. Where this is offset by precipitation gains, which may result from the higher temperatures and thus higher evapo-transpiration (Zhao and Molders 2008), the overall impact may be minor. Basins such as the Limpopo that can expect combined warming and declining rainfall in an already-marginal environment, agricultural productivity will fall and poverty may increase significantly.

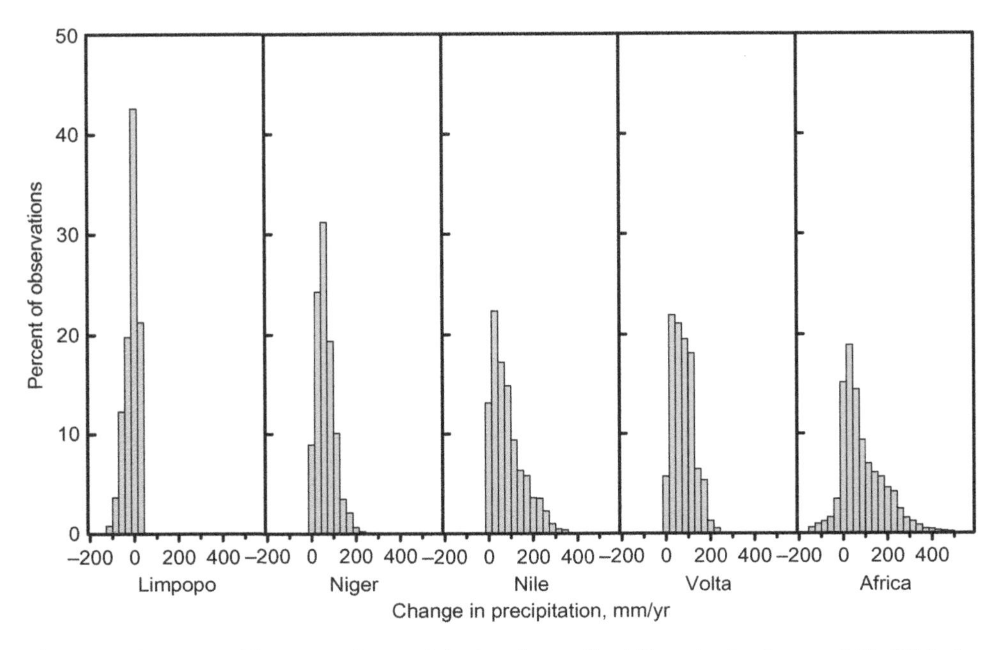

Figure 4. Frequency histograms for precipitation change for African basins (mean of 17 GCMs for each basin).

Asian basins

Scenarios of change

For the Asian basins (Table 4), mean warming varies from 2.0°C (Mekong) to 3.4°C (Karkheh) and the mean for Asia as a whole is 3.0°C. The standard deviation between GCMs for projected temperature is twice as high for Asia as for Africa and the GCMs show significant disagreement for the temperature for the Karkheh and Ganges in particular. Precipitation for Asia as a whole will increase 127 mm/yr with very significant variability between GCMs (Psd_{mean} of 2000 mm/yr), that is, significant uncertainty. Change in mean precipitation for all the Asian BFP basins increases from 270 mm/yr (for the

Table 4. Change in temperature and precipitation in five Asian basins from 1950–2000 to 2050 under the A2a emissions scenario and the standard deviation between GCM projections for 17 GCMs for each basin.

Asia	Temperature						Precipitation					
				Standard deviation						Standard deviation		
Basin	Min	Max	Mean	Min	Max	Mean	Min	Max	Mean	Min	Max	Mean
	°C	*°C*	*°C*				*mm*	*mm*	*mm*			
Asia	1.2	4	3.0	0.36	2.4	3.0	−1730	2073	127	65.3	11518	2005
Ganges	2.1	3.7	2.7	1.35	2.2	1.8	56.7	502	274	357	4624	1349
Mekong	1.6	2.3	2.0	0.85	1.7	1.4	−66.3	97.7	27.2	1064	3771	1736
Yellow	2.6	3.3	3.0	0.81	1.5	1.0	64.7	184.6	124	185	1168	570
Karkheh	3.1	3.5	3.4	1.53	1.8	1.7	−86.9	134	19	236	518	366

Ganges) to 20 mm/yr (Karkheh), with much lower increases in precipitation for the Karkheh and Mekong than the other basins. The precipitation projections for the Ganges and the Mekong in particular are subject to significant disagreement between GCMs (Psd_{mean} >1000 mm/yr for both). There is significant spatial variability of temperature within basins but they all show warming throughout. The Ganges and Yellow River both show wetting throughout, while the other basins show wetting in some areas and drying in others (P_{min} and P_{max} are opposite sign).

The frequency distributions of temperature change for the Asian basins are shown in Figure 5. For the entire continent warming is bi-modally distributed, reflecting differences between the continental and oceanic parts of the region. Overall warming is 1.5°C–4°C, with peaks at 2.5°C and 3.5°C. Warming within the Ganges varies from 2°C in much of the basin but up to 3.5°C in places. The Karkheh is much less variable with warming concentrated in the range 3.0–3.5°C. Warming is slightly less for the Yellow River, from around 2.5°C to 3.5°C and lowest for the Mekong at 1.5°C–2.3°C. The Asian basins are mainly projected to be wetter (Figure 6). The increase in precipitation in 75% of Asia is small, although some areas will increase up to 750mm/yr. The Ganges is projected to be 0–500 mm/yr wetter, while in the Karkheh there will be small, balanced increases and decreases. The Mekong will increase mainly 50 mm/yr and the Yellow 50–150 mm/yr.

Again, any increase in rainfall may be cancelled out by the increased annual evapotranspiration caused by warming resulting in little impact, especially in wet basins like the Mekong. The seasonality of changes in rainfall and temperature will be critical, especially for the highly seasonal (monsoonal) Asian basins. The impact will also be greater water-scarce basins (for example, Karkheh) or those with high population densities (Indus, Ganges and Yellow).

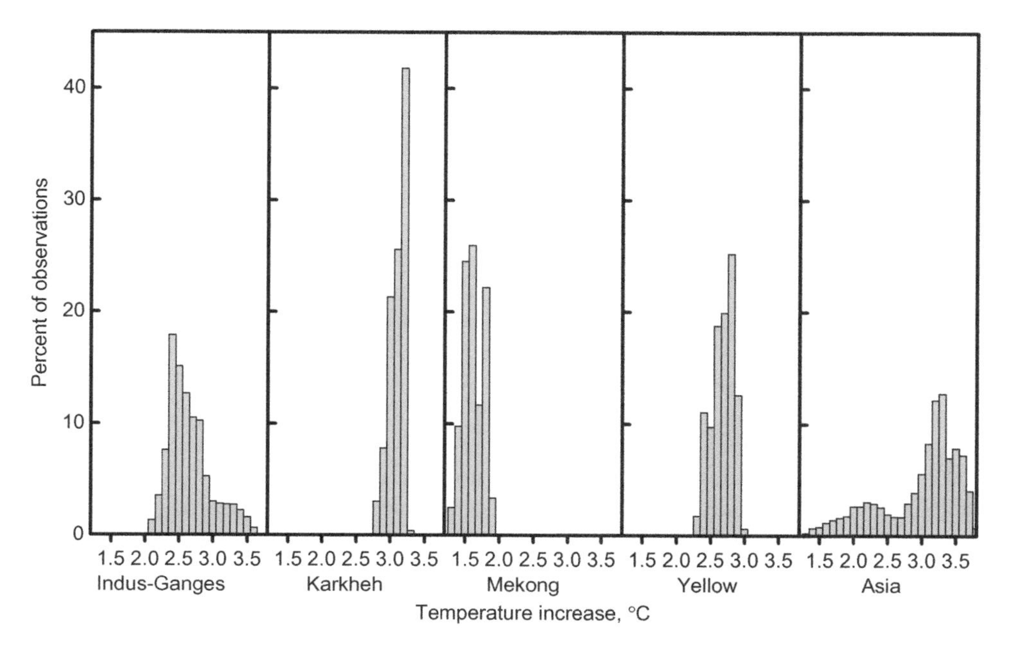

Figure 5. Frequency histograms for temperature change for Asian basins (mean of 17 GCMs for each basin).

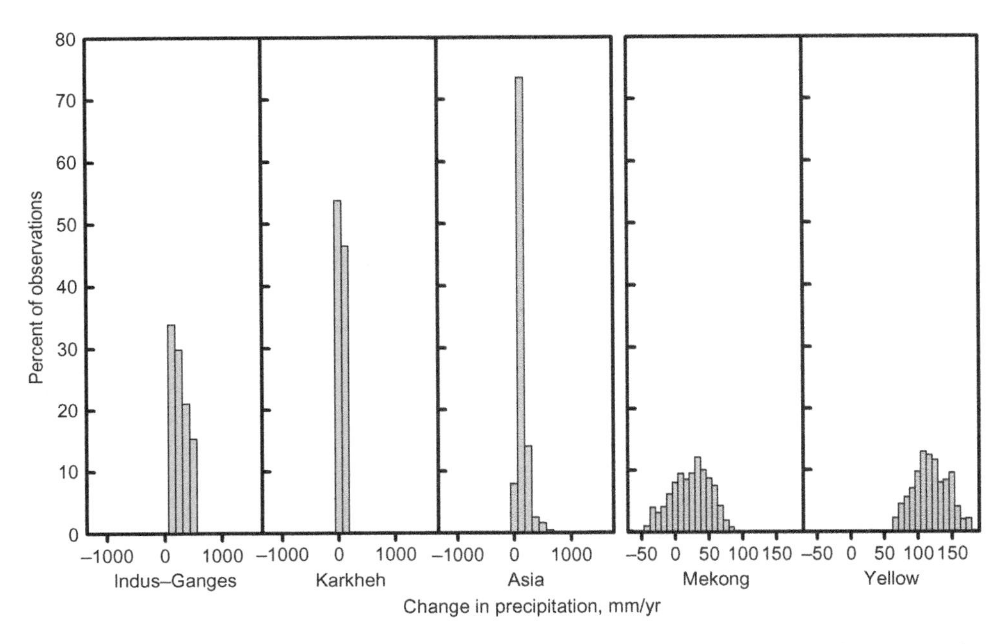

Figure 6. Frequency histograms for precipitation change for Asian basins (mean of 17 GCMs for each basin).

The Asian basins highlight the impacts of climate change relevant to understanding water, productivity and poverty in basins. The populous Yellow River basin has always flooded frequently – it is called the "river of sorrow" – with the downstream reaches raised some metres above the level of the surrounding floodplain by the continuous deposit of loess eroded from the northwestern loess plateau, whence the name Yellow River. For centuries, basin management has attempted to control flooding, to which the downstream reaches are particularly vulnerable. With increased demand for water upstream, both for irrigation and other uses, hydrological droughts, not floods, are now the main climate-related hazard. The northwestern loess plateau is especially drought-prone with droughts two years in three, nine of them severe, from 1950–74 (Yellow River project team 2009). During the past 20 years, there have also been several droughts in the upper and middle reaches of the basin, severe enough to cause losses in grain yield and shortages of drinking water. Note that it is likely that these problems result from increasing use of water rather than climate change.

The Yellow River project team used the HadCM3 GCM to assess the impact of climate change on stream flow in the headwaters of the YRB (Collins *et al.* 2001). It projects that maximum air temperatures in spring increase by +1.5°C, +3.0°C and +4.1°C for 2020, 2050, and 2080 respectively, although minimum air temperature increased less, while annual precipitation increased by 3.5%, 6.4% and 8.7%. The team used statistical downscaling (Dibike and Coulibaly 2005) to incorporate the output into the soil water assessment tool (SWAT) hydrological model (Sun and Cornish 2005), which is used extensively in hydrologic studies (TAMU 2010). While precipitation was projected to increase, annual flows in the headwater catchments declined by 89 m^3/s, 117 m^3/s and 152 m^3/s for 2020, 2050 and 2080 because higher temperatures increased evapo-transpiration (Xu *et al.* 2009). Compared with HadCM3, the multi-GCM mean we calculated shows a basin mean increase in precipitation of around 32%, which suggests that the decline in flow may not be as severe as Xu *et al.* forecast. We did not repeat the SWAT modelling however.

For the Indus and Ganges basins, Christensen *et al.* (2007) indicate that the Tibetan region could warm by 3.8°C by 2100, confirmed by Sharma *et al.* (2009). These predictions are significantly higher than our multi-model projection of 2.7°C for the entire Ganges Basin. The Christensen *et al.* (2007) temperature trend was used by the Indus-Ganges team to parameterize the WEAP (Yates *et al.* 2005) model in the Ganges and the Indus basins. WEAP models biophysical water balance, including glaciers. The team used scenarios of uniform temperature increase over a 20-year period of 1, 2 and 3°C. The last two scenarios are extreme given that our 50-year multi-GCM projections given above are only 2°C over most of the basin. Precipitation was not changed, although increases are projected.

The scenarios' effect on stream flows lessen from upstream to downstream as the contribution from upstream precipitation dominates the extra flows from increasing glacier melt and as higher temperatures increase evapo-transpiration in the plains downstream, which reduces stream flows. The extra flow from glacial meltwater under these scenarios is thus rather unimportant at downstream sites such as Farakka. Upstream, however, for example in the upper Ganges at Haridwar and Narora, or in mountainous sub-basins (for example, at the Tehri Dam) the extra flow is significant. Table 5 summarizes the WEAP output for the Indus and the Ganges, showing clearly that there are important changes upstream as a result of glacial melting, especially for the 3°C scenario. Glacial melting could not continue indefinitely and flows would decline once the glaciers were exhausted. Perhaps more important than total annual flows is the change in seasonality, as winter precipitation generates runoff instead of falling as snow and being stored as ice for release the following spring and summer. This may have implications for winter flooding as well as dry season flows, especially in the upper basin areas.

Changes in streamflow for the Indus are similar to the Ganges Basin but greater in magnitude due to the higher contribution of glacier and ice melt to the Indus flows. Even 1°C warming increases the flows at both Tarbela Dam and Sukkur Barrage by nearly 10% and even more with greater warming (Table 6), indicating increased flood risk, especially upstream.

Generally, additional water from glacier melt is not available when it is most needed during periods of low flow. The Upper Ganges Canal currently withdraws about 6 km^3/year at Haridwar, which is about the same as the expected extra flow as a consequence of the 3°C scenario. If this extra water is to be used, it would need to withdrawn, which means substantially increasing the irrigation infrastructure, or stored in existing or new dams. It is unlikely that the high-volume winter flows could be captured, but the lower-volume flows in

Table 5. WEAP-simulated average change in annual streamflow at locations of the Ganga basin with temperature increasing gradually over 20 years, as compared to the reference period (1982 to 2002).

Scenario	Average change in annual stream flow at					
	Tehri Dam	Haridwar	Narora	Tajewala	Delhi	Farakka
	km³/yr	*km³/yr*	*km³/yr*	*km³/yr*	*km³/yr*	*km³/yr*
+1°C over 20 yr	+0.6 (+ 8)[1]	+1.9 (+ 6)	+1.8 (+ 8)	+0.2 (+ 2%)	negligible	+4.2 (+ 1)
+2°C over 20 yr	+1.2 (+ 17)	+3.9 (+ 13)	+3.6 (+ 15)	+0.3 (+ 4%)	negligible	+8.8 (+ 3)
+3°C over 20 yr	+1.9 (+ 26)	+6.0 (+ 20)	+5.4 (+ 23)	+0.4 (+ 5%)	negligible	+13.4 (+ 4)

Note: [1] Values in parenthesis are the % change relative to the reference period (1982 to 2002).

Table 6. Simulated average change in annual streamflow in some locations of the Indus Basin with temperature increasing gradually over 20 years, as compared to the reference scenario (1982–2002).

Scenario	Average change in annual stream flow	
	Inflow to Tarbela Dam	Flow at Sukkur Barrage
	km^3/yr	*km^3/yr*
+1°C over 20 yr	+6.6 (+9)[1]	+8.1 (+10)
+2°C over 20 yr	+15.2 (+21)	+18.4 (+22)
+3°C over 20 yr	+22.6 (+31)	+28.5 (+34)

Note: [1]Values in parenthesis are the % variation relative to the reference period (1982 to 2002).

April and May could be captured, either by dams or be used for recharge of over-exploited aquifers in both the Indus and the Ganges. These engineering solutions could substitute for the current functionality of storage as ice of winter snow and its release as melt water in the dry season. Were annual rainfall to increase as projected by the multi-model scenario (and especially if this were to occur in the dry season), there may be little impact of this glacial melting on river flows and even benefits of climate change for water productivity.

South American basins

Scenarios of change

Projected warming for the South American basins is around 2.7°C for the Andes and 2.6°C for the São Francisco with the continent as a whole expected to warm by 2.6°C by 2050 (Table 7). Differences between GCMs in temperature projection for South American basins are lower (Tsd_{mean} = 0.6°C- 0.9°C) than for the Asian and African basins which increases our confidence in them. Precipitation is expected to increase by 175 mm/yr for South America as a whole, but by only 137 and 101 mm/yr respectively for the Andes and São Francisco basins. The disagreement between GCMs for precipitation is very high, especially for South America as a whole but also for the basins (Psd_{mean} of 882 and 991 mm/yr for Andes and São Francisco respectively) so the precipitation projections are highly uncertain. Spatial variability within the basins is significant with warming varying from 1.5°C to

Table 7. Change in temperature and precipitation in two South American basins from 1950–2000 to 2050 under the A2a emissions scenario and the standard deviation between GCM projections for 17 GCMs for each basin.

Basin	Temperature						Precipitation					
				Standard deviation						Standard deviation		
	Min	Max	Mean	Min	Max	Mean	Min	Max	Mean	Min	Max	Mean
	°C	*°C*	*°C*				*mm*	*mm*	*mm*			
South America	0.5	3.3	2.6	0.0	4.2	0.8	−140	1292	130	0.0	11510	2023
Andes	1.5	3.6	2.7	0.0	3.9	0.9	−95	999	137	0.0	7663	882
São Francisco	1.8	3.0	2.6	0.01	2.6	0.6	−107	254	101	406	1990	991

3.6°C for the Andes and 1.8°C to 3.0°C for the smaller São Francisco. Precipitation change for both the Andes and the São Francisco is also spatially variable with wetting in some parts of the basins and drying in others.

The frequency distribution for temperature change (Figure 7) within the Latin American basins indicates that multi-GCM mean temperature change varies from 1.0°C–3.0°C for South America as a whole (with most areas at the warm end of this range). Since the Andes Basin covers a wide latitudinal range, it shows warming from 2°C through to almost 3.5°C with most areas warming by 2.5°C–3.5°C. The São Francisco Basin, being more spatially restricted and largely confined to lowlands, shows changes from 2°C–3°C with most areas at the higher end of the range.

The frequency distribution for precipitation change (Figure 8) shows wetting for the continent as a whole, with most between 0 and 300 mm/yr but some as much as 600 mm/yr and some areas drying slightly. For the Andes there is a very spatially variable change with a positively skewed distribution of 0–500 mm/yr more precipitation but some areas drying slightly. The within-basin variability of precipitation change is less for the São Francisco, with most areas showing gains of precipitation of less than 200 mm/yr or losses of less than 100 mm/yr.

The Andes Basin team used the multi-GCM scenarios for the Andes to carry out an equilibrium run of the FIESTA hydrological model (Mulligan and Burke 2005). FIESTA is a grid-based, diurnal-in-monthly timestep, water balance model that is designed specifically for application up to continental scales and for tropical montane environments. It has a sophisticated treatment of spatial water balance based on wind-driven rainfall and fog inputs minus actual evapo-transpiration. Version 1 of FIESTA has no snow component, so climate change effects on glacial melting are not included in the analysis. However, since

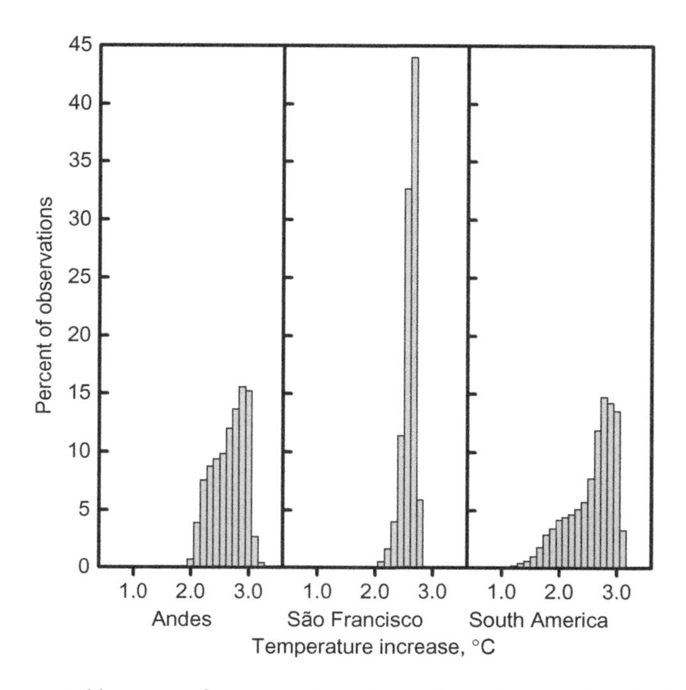

Figure 7. Frequency histograms for temperature change for Latin American basins (mean of 17 GCMs for each basin).

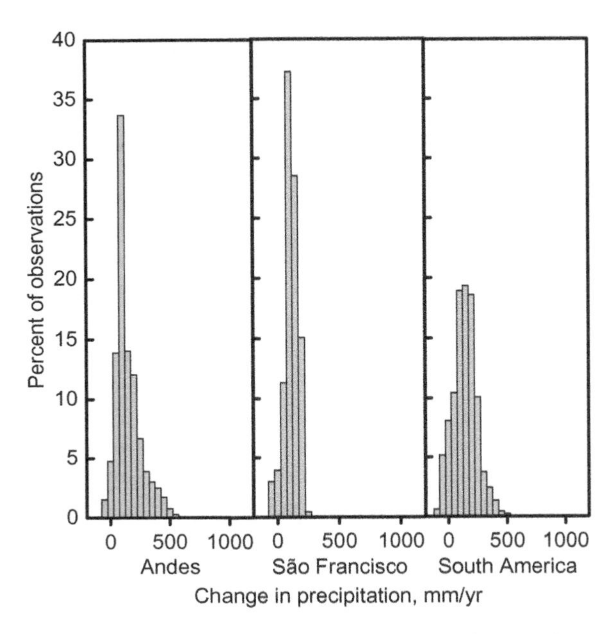

Figure 8. Frequency histograms for precipitation change for Latin American basins (mean of 17 GCMs for each basin).

glacial melting is not an extra input but rather a transient storage of existing inputs, it affects the seasonality of available water where there is snow and ice upstream, but not the amount of total available water, at least in the long term. Incorporating snowmelt for the Andes may thus have little impact on the water balance changes shown here. Figure 9 maps the mean temperature and precipitation change for the 17 GCMs for the Andes. Temperatures are expected to increase throughout the basin, but by much more in the high southern Andes and the northeastern Andes. Precipitation is also expected to increase throughout the basin by around 70–100 mm/yr but by 300–1000 mm/yr in parts of the Andean foothills in Colombia, Ecuador, Chile and Bolivia.

The actual magnitude of change is highly uncertain in some areas, however. Figure 10 shows the spatial distribution of standard deviation in mean annual temperature and annual total precipitation between the 17 GCMs. For temperature, the SD is within 1.0°C or less for most of the high Andes but for parts of the northern and eastern Andes is about 2.0°C or more. For precipitation, SD is less than 1000 mm/yr for most of the Andes but peaks to more than 8000 mm/yr in isolated spots where downscaled GCM results disagree on large-scale changes in precipitation. This is not surprising given GCMs' coarse resolution relative to the spatial variability of precipitation, and differential representation of the Andes topography and the complexity of precipitation-forming mechanisms in the region.

The seasonality of change (Figure 11) indicates a weak seasonality in precipitation change in the Andes (compared with the strong seasonality in precipitation change for the Amazon, not shown). Seasonality in temperature change is considerable however, with much greater warming in the central Andes from June–September (Figure 12), which is the dry season for this zone. This kind of seasonal change has the potential for high impact on water resources and agricultural productivity.

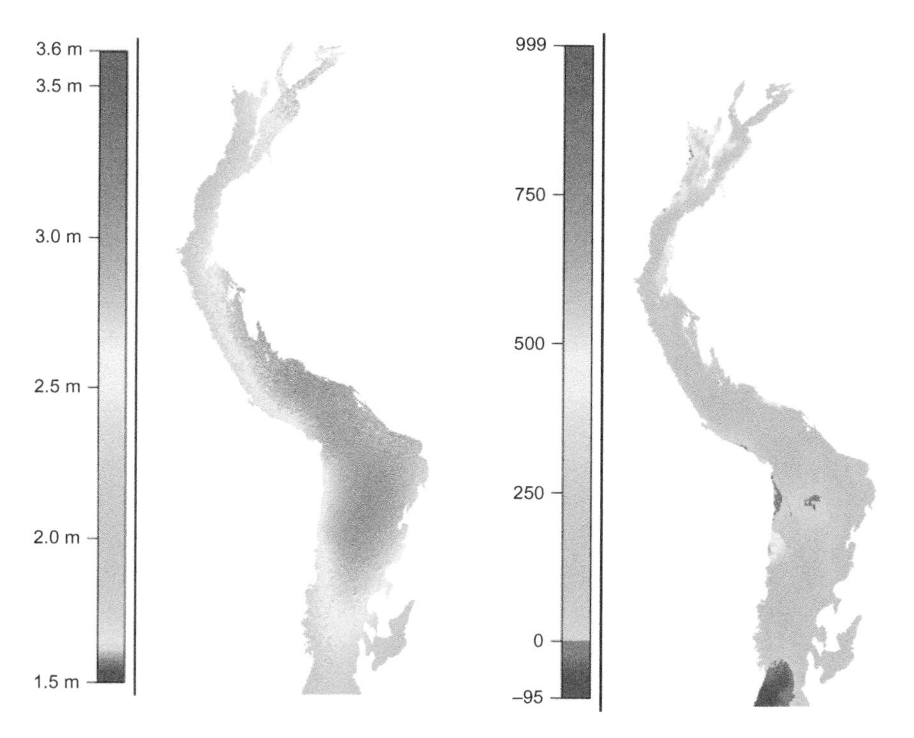

Figure 9. Temperature (°C, left) and precipitation (mm/yr, right) change from current to 2050s as the mean of 17 GCMs SRES A2a scenario.

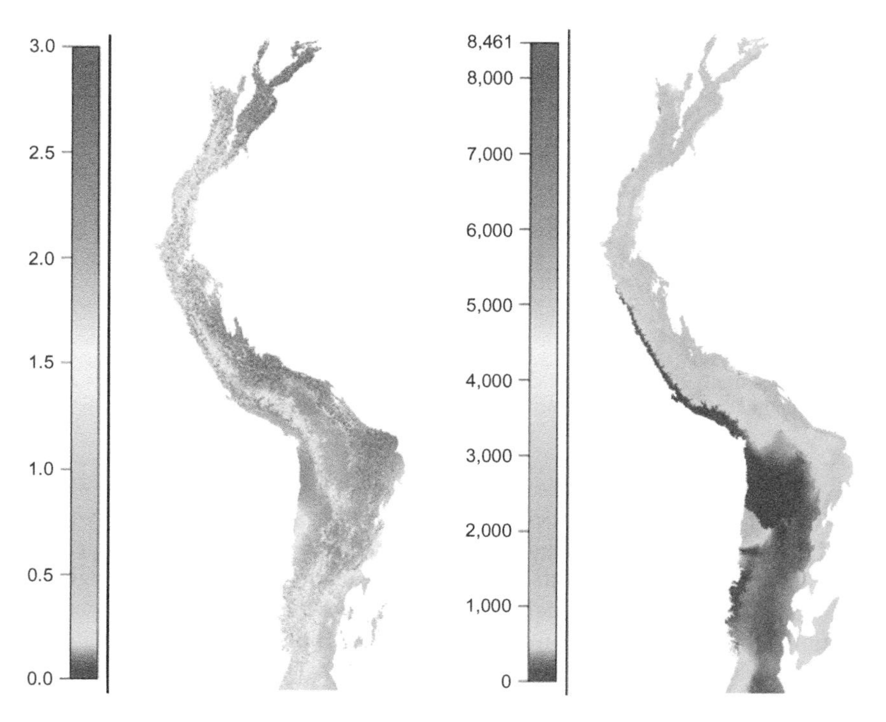

Figure 10. Standard deviation (uncertainty) of 17 GCM projections for temperature (°C) and precipitation (mm/yr) change in the Andes.

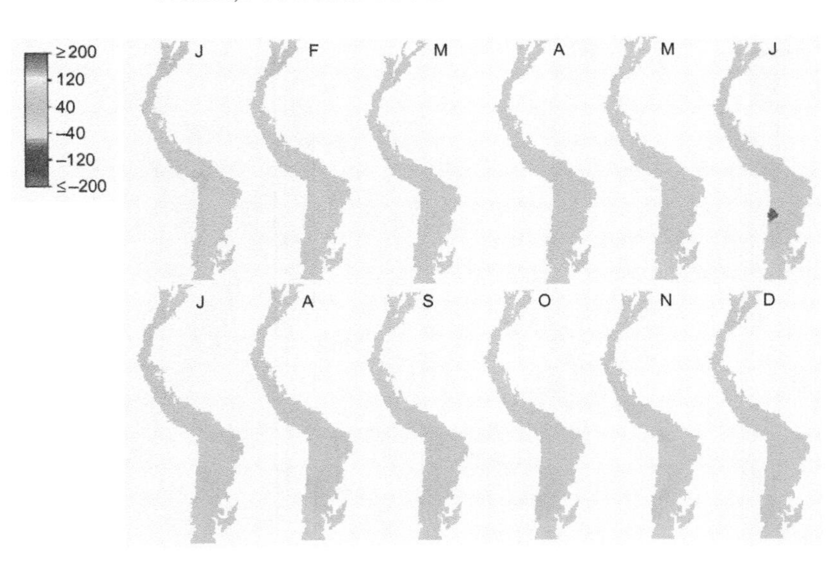

Figure 11. Seasonality of precipitation change expected to the 2050s for the Andes as the mean of 17 GCMs (January to December from top left to bottom right).

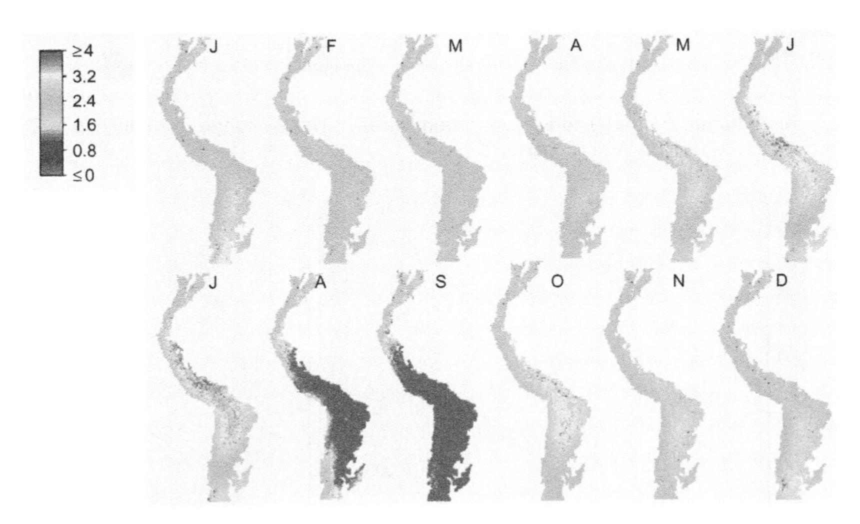

Figure 12. Seasonality of temperature change (°C) expected to the 2050s for the Andes as the mean of 17 GCMs (January to December from top left to bottom right).

The multi-model mean projection for the Andes is a basin-wide average warming of 3.6°C by the 2050s and a basin-wide increase of precipitation of 137 mm/yr by the 2050s. We used fully spatially distributed monthly data for this mean scenario as temperature and precipitation inputs to the FIESTA delivery model, assuming no change in vegetation cover. This scenario generates a mean increase in water balance (rainfall minus actual evapotranspiration) for the basin of 75 mm/yr compared with the baseline simulation, using WorldClim data to represent the climates for 1950–2000. This basin-mean value disguises significant within-basin variation with many areas showing small increases or decreases in water balance and a few areas showing much larger increases or decreases (Figure 13).

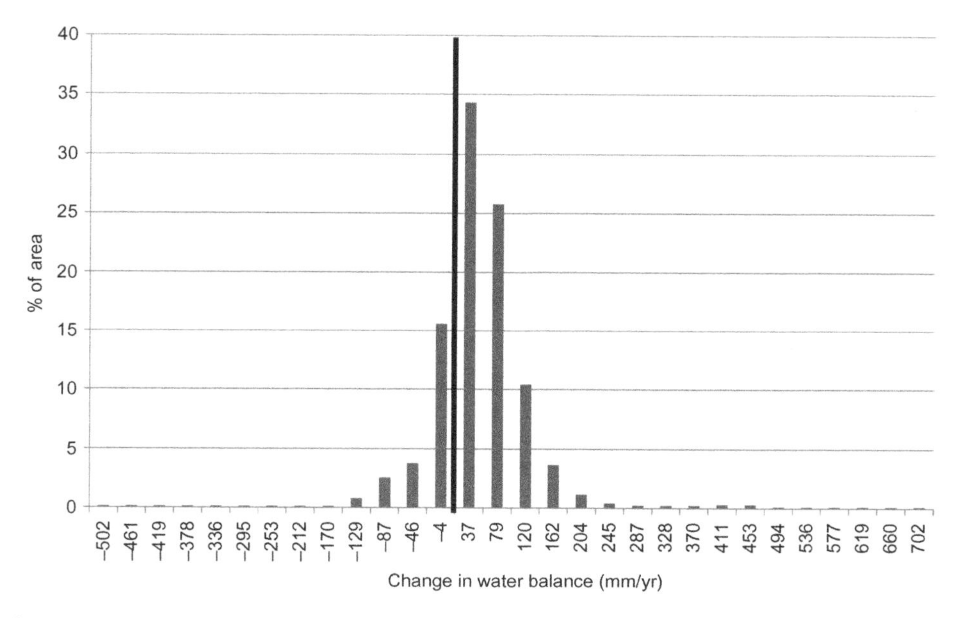

Figure 13. Impact of a mean climate change scenario for SRES AR4 A2a (mean of 17 GCMs) on water balance for the Andes.

Most of the Andes show an overall wetting (despite the increased temperature) with the impact on water balance mainly a function of the higher rainfall. Results may differ in places if humidity, wind speed and solar radiation are also affected by climate change and input to the model.

Likely costs, food security and livelihood impacts of climate change

The BFPs carried out little work on the costs, food security and livelihood impacts of climate change since this was not part of their remit. As part of some of the BFPs, hydrological models were driven with GCM scenarios to understand the hydrological impacts of climate change, but crop simulation models were not used to examine the impacts of climate change. Moreover we have to look beyond climate change to consider what its impact in a basin will be. We need also to consider the current levels of water availability and productivity, population and trends of population growth, and the technological and economic developments in a region, which might reduce dependence on agricultural livelihoods as well as the current sustainability of production. The BFPs do report some potential impacts of climate change for their basins and we report those by continent here, together with conclusions based on the outcomes of the analyses of cross-basin water availability and climate change.

African basins

The African basins show significant projected increases in temperature but, with the exception of the Limpopo, also show increases in precipitation. The projected change in precipitation is high compared with the historic inter-annual variability for African basins and relative long-term stability of twentieth-century rainfall. The projected wetting suggests a relatively minor impact, except that the African basins are have some of the lowest GDPs (Cai *et al.* 2011), the lowest current water balances, and the greatest levels of poverty. All of these make these basins vulnerable to the changing balance of increased evapo-transpiration (resulting from the projected warming) and increased rainfall. Some parts of the basins will be much more affected than others as the projected rainfall and temperature changes are variable within basins. Moreover some parts of the African continent will be much more or less affected than the BFP basins. Either way, full hydrological modelling for the basins, incorporating the relevant changes in all climate variables, and impacts on vegetation cover would be necessary to understand fully the implications of these changes.

The Volta Basin team (2010) report that agricultural production in Benin could decrease by 10–20 % by 2050 as a result of climate change. For Togo, warming combined with little rainfall change would produce only minor effects on agricultural and livestock production. For Ghana, rainfall is estimated to decrease, which, combined with further agricultural development, might considerably increase the demand for irrigation water. For Burkina Faso, warming would be coupled with slight wetting so that for the wettest parts of the country cotton production may increase. Results from climate change modelling for the Niger are uncertain and contradictory, however on average there is a trend towards warming, an increase in variability and extreme events, a later start to the rainy season, more dry spells, with more rain in central parts of West Africa and less in the far West (Niger Basin team 2010). For the Nile, Conway (2005) reviewed 10 studies on climate change, concluding that while temperatures are expected to increase, there is uncertainty about the direction and magnitude of future changes in rainfall. Elshamy *et al.* (2009), analysed rainfall and potential evapo-transpiration from 17 GCMs from the IPCC 4th assessment report, for the period 2081–98, to calculate Blue Nile runoff at the Ethiopia–Sudan border. All the GCMs showed warming of 2–5°C, and an increase of potential evapo-transpiration of 2–14%. However, the models disagreed on changes in precipitation, with values varying from -15% to +14% between GCMs with the ensemble mean showing no change. Assuming no or only modest changes in rainfall, increased evapo-transpiration may reduce the water balance of the upper Blue Nile Basin.

Overall, there is great uncertainty, with increased variability and unreliability of rainfall a common feature of the projections. Increased water productivity, drought mitigation of rainfed cultivation through proper soil and water management techniques, and the development of irrigation are clear requirements to confront these changes in subsistence agriculture (Volta Basin team 2010).

Asian basins

The projections for the Asian basins indicate significant warming but also significant increases in rainfall, which again are much greater than the historic inter-annual variability or long-term trend over the twentieth century. Some basins, such as the Mekong, are already very wet so this wetting may have little impact on agriculture. Other basins are vulnerable to change because of high populations (Indus and Ganges), low precipitation (Karkheh) or high population combined with intensive irrigated agriculture (Yellow River). If the wetting is spread evenly through the year or concentrated in the existing dry season, productivity may increase. Conversely, and more likely, if the increase is more wet-season rain, increased productivity is unlikely. Where the increases in precipitation are small they will likely be lost to higher evapo-transpiration. Continued industrialization and de-agriculturalization of the Indus, Ganges and Yellow River basins and reduced rate of population growth may reduce their considerable vulnerability to climate change.

For the Karkheh, climate change may reduce production of summer crops and fruit as well as ecosystem health, but is unlikely to have direct impact on livelihoods are due to government livelihood support and Iranian petroleum assets. Ardoin Bardin *et al.* (2009) used several GCMs to project the effect of climate change on the Niger. They concluded precipitation in the northwest of the basin would decrease for the next decades but would increase slightly in the southeast. Increased evapo-transpiration caused by higher temperatures and reduced precipitation will likely give decreased in runoff and river flows.

For the Yellow River basin, projected climate change would raise the annual water deficit by 4.21 km^3, and the river discharge would fall by 2.18 km^3. A key concern for the Yellow is not only fulfilling productive water demands, but also maintaining sufficient flow (15 km^3 in the rainy season) to transport the sediment burden downstream and thus avoid breaching the dykes. Much of the Yellow's infrastructure was designed for flood control, but climate change and droughts will bring new challenges (Yellow Basin team 2009).

The Mekong is expected to warm 1–3°C by 2050 (Mekong Basin team 2010), particularly January to May, and the greatest warming in the eastern highlands. The dry season is expected to lengthen and intensify, and the rainy season is expected to shorten and intensify, with much more rainfall in the wettest months particularly in parts of the Lao PDR. Overall, however, there will be only modest increases in annual rainfall in most parts of the Mekong. There may be more seasonal droughts, more floods, and saltwater intrusion and flooding in the delta caused by sea level rise (Hoanh *et al.* 2003). The longer dry season in many parts of the basin will reduce agricultural production in the face of increasing food demand and growing populations (Hoanh *et al.* 2003).

South American basins

Projections for the South American basins indicate warming and wetting at much greater rates than the historic level of inter-annual variability and the twentieth-century long-term trend. For the Andes the warming, and particularly the wetting, are highly spatially variable from minor drying in some parts to significant wetting in others. In the wetter parts of the northern Andes the wetting may reduce agricultural productivity, but in the dry southern Andes there may be important increases in productivity, food security and livelihoods. Agriculture will need to adapt to these novel climates. Given the spatial heterogeneity of the Andes, some climates, such as *paramos*, will disappear.

Agricultural mitigation and adaptation of water systems to climate change: practices, options and constraints.

Most studies of climate change focus on anthropogenic greenhouse-induced global climate change, its regional manifestation and resulting impacts. There are other forms of climate change, however, notably local and regional changes driven by the impact of surface albedo on the energy balance of the land surface rather than the energy balance of the atmosphere. Changes in albedo occur with large-scale changes in land use such as deforestation and irrigation. A number of studies have investigated the impact of deforestation for pasture in the Amazon and tropical Africa (Nobre *et al.* 1991, Eltahir and Bras 1996, Zheng and Eltahir 1997, Roy and Avissar 2002, Voldoire and Royer 2004, D'Almeida *et al.* 2007) and its likely impact on regional patterns of precipitation. Evapo-transpiration, aerodynamic roughness and thus convective uplift and cloud formation are less under pasture, or similar land uses, compared with forest. While the physical mechanism is clear, there is little observational evidence that land-use change in the Amazon has affected either cloud frequency or patterns of precipitation patterns in the region (Amazon Initiative Consortium, Ecosystems and Poverty [ESPA-AA] 2008). Extreme model scenarios of total forest conversion of the Amazon, however, do indicate the potential for significant regional climate change to occur as a result (Nobre *et al.* 1991, Roy and Avissar 2002, D'Almeida *et al.* 2007). Large-scale irrigation projects do result in "greening up" of nearby drylands by generating subtle, local near-surface temperature cooling (Sacks *et al.* 2008), but there is little evidence that irrigation causes significant or permanent large-scale changes in climate. The effects of changed surface energy balance may be important locally, but greenhouse-induced climate change is locally and globally more significant and, as we will see later, agriculture is an important contributor.

In addition to its contribution to the surface energy balance by changing surface albedo, agriculture is a major component of the global carbon dioxide, nitrous oxide and methane budgets. Agriculture is thus a factor in climate change induced by anthropogenic greenhouse gas (GHG) emissions. Cropland, pastures and plantations cover some 50% of the Earth's land surface, and agriculture contributes 10–12% of all anthropogenic GHG emissions (Smith *et al.* 2007), including about 60% of all nitrous oxide and 50% of the methane. Nitrous oxide is emitted through injudicious use of nitrogenous fertilizers. Ruminants are the most important global emitters of methane, while anaerobic decomposition in flooded rice is another. Carbon dioxide is emitted through all agricultural activities that consume fossil fuels including mechanized water pumping and redistribution, soil management, planting and harvesting crops and food processing. Of course, agriculture also sequesters atmospheric carbon dioxide during the process of photosynthesis but this sequestration is short-lived compared with the carbon stored in soil and woody biomass of most natural ecosystems, especially forests. Nevertheless agricultural emissions (*excluding* electricity and fuel use) are roughly in balance with sequestration (Smith *et al.* 2007).

Better land and water management can significantly reduce GHG emissions by agriculture and the gains are cost-competitive compared with equivalent gains from mitigation in the industrial, energy, transportation and forestry sectors (Smith *et al.* 2007). Mitigation options include: (a) improved land, waste and water management practices to increase efficiency; (b) improved livestock, feed and manure management; and (c) restoration of degraded lands. The option with the greatest potential is improved soil management to increase carbon accumulation in soils and also reduce carbon, methane and nitrous oxide emissions from soils. Substituting biofuels for fossil fuels in agriculture also has potential (Smith *et al.* 2007). These measures are often synergistic with the goals of poverty alleviation and sustainable development and have the co-benefits of increased efficiency, reduced costs and reduced environmental impacts, although there are sometimes tradeoffs with other environmental problems such as habitat loss.

The nature of mitigation options, co-benefits and tradeoffs vary with the geographic location and the time horizon under consideration and thus requires significant geographically focused research. Nevertheless, improved agricultural practices seem an easy way to achieve poverty, sustainability, water *and* climate change goals and should thus be a key focus of agricultural development and climate change policy. Each basin will need to adopt locally specific strategies of adaptation and mitigation in line with the context of the basin and the extent of climate change expected.

We summarize options for adaptation suggested by the basin teams. The Karkheh could intensify winter cropping and supplemental irrigation in rainfed areas (especially in the wetter upper Karkheh, where only one or two irrigations could boost agricultural production [Ahmad *et al.* 2009]). In the dry areas of the Limpopo, conservation farming and rainwater harvesting using small reservoirs and other water storage techniques, combined with value-chain and finance facilities could improve productivity (Limpopo Basin team 2010). In wetter areas, the focus needs to be on market linkages, value chains and increased competitiveness. The Niger Basin team (2010) report that current farmer strategies to reduce rainfall-related risks tend to prevent intensification. Solutions to reduce the risk of crop failure, such as rainwater harvesting and drought-tolerant crops, are necessary to allow farmers to invest in yield-boosting fertilizer and other inputs. Niger farmers also need access to training and to be linked to input and output markets and financial services as well as having secure access to land and water. Mitigation strategies including early warning systems and water and food storage options are required to help reduce the impact of extreme climatic events. All of these require good governance. For the Nile, adaptation strategies to cope with seasonal water shortages, such as seasonal livestock migration or feed import are important (Nile Basin team 2010).

For the Asian basins, the Mekong Basin team (2010) report various potential adaptations including: developing crop cultivars that can tolerate drought, salinity and acidity; improvement and development of water supply and drainage systems; constructing saline-control structures and early-warning systems for extreme events; increasing public awareness of and action on water quality and hygiene, and diversification of livelihoods away from agriculture. For the South American basins, the Andes Basin team (2009) report that improved land management in watersheds and the implementation of payments for watershed services (PWS) schemes can help adapt watersheds to the projected effects of climate change in the region. Given the extreme spatial variability of the Andes for a positive adaptation or intervention to be successfully replicated elsewhere, a careful analysis of, and adaptation to, local conditions is mandatory.

Implications for sustainable poverty alleviation

The impacts of climate change are clearly highly uncertain and spatially highly variable, both within and between basins. Climate change will occur in the context of an extremely variable biophysical, social, agricultural and economic landscape, and can only be understood within the context of this landscape. Figure 14 shows the population (density and change) and development (GDP and change) context for the CPWF basins (based on the analysis in Mulligan *et al.* [2011] which shows water resources variables). The climate change context (precipitation and temperature change) is also shown within the context of the basins' current climate characteristics (seasonality, inter-annual variability and change of precipitation as a percentage of the current water balance, that is, relative change). Each variable (except precipitation change) is plotted as a percentage of the highest value for all basins so the basin with a variable of 100% is the basin with the highest value of that variable. Precipitation change is expressed as a percentage of the highest minus the lowest value for all basins, thus separating positive changes (wetting) from negative changes (drying).

Of all basins, the Ganges has the highest GDP, population density and projected population growth, but also the highest precipitation seasonality. It has medium warming and significant wetting (though not the greatest in relation to current water balance), but the sheer population load of the basin and already strong dry season make this basin vulnerable to change. For the Ganges, even in the absence of climate change there will be serious limitations to water availability when the unsustainably extracted groundwater

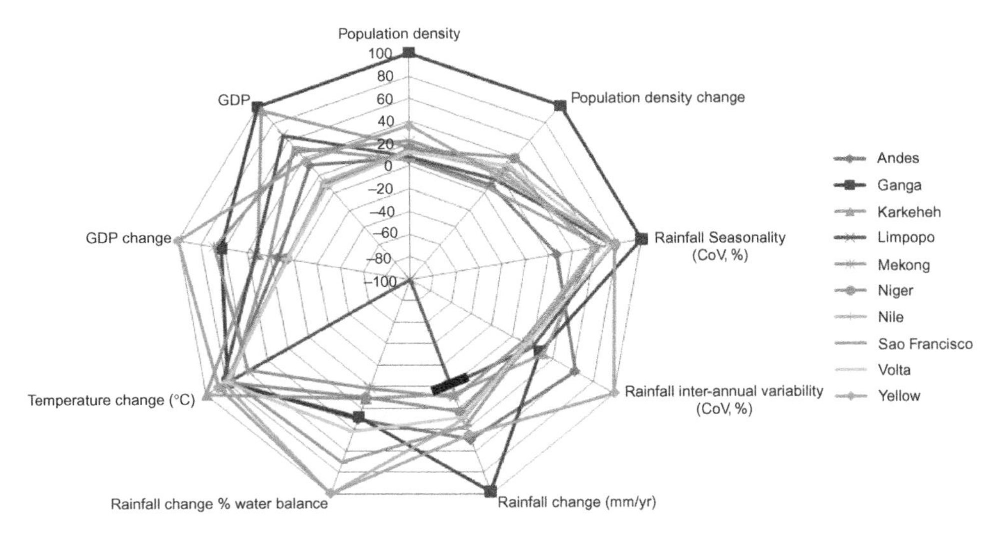

Figure 14. Trajectories of change and their context for the CPWF basins. The black bar indicates zero change for rainfall.

resource is depleted (Tiwari *et al.* 2009). The Yellow River has the greatest inter-annual precipitation variability and high precipitation seasonality with projected average wetting, but also warming that is especially important in relation to its current water balance. GDP is expected to grow significantly, however, but the population is unlikely to grow, which provides some resilience to climate change. Warming is expected to be greatest in the Karkheh with some wetting, which is small compared to its current water balance. GDP is high and is expected to grow. The basin is not under high population pressure, so its outlook is positive compared to some other basins. For the Mekong, a small increase in precipitation is projected, which is insignificant compared with its current water balance. GDP is expected to grow, facilitating adaptation and resilience.

The Andes will also be wetter although this is small in relation to its mean water balance. The Niger, Nile and Limpopo are all extremely seasonal with strong inter-annual precipitation variability. Precipitation will change little, but the change is a significant proportion of the current water balance for the Nile (wetting) and the Limpopo (drying) but not the Niger. GDP in these basins is low GDP and low growth; there are few options for adaptation or resilience to climate change, except for the Nile, which has significant potential to store water that is starting to be developed.

The patterns of warming, wetting and drying are highly spatially variable within each of the basins (given their considerable extent) such that while Figure 14 is representative for the basin as a whole, areas within each basin may be quite different, and indeed may look like another basin. The impacts of climate change on flows are cumulated and integrated downstream, so that stream flow reflects the basin mean, but local impacts on soil water, groundwater and (non-irrigated) crop production will be highly variable within the basins. Where basins are transnational, the relationship between these distributions and the distributions of population, urbanization and GDP will determine the local impacts on people and poverty.

It is important to note that we have focused on annual integrals of rainfall and temperature as a first step towards understanding the impacts of climate change impacts in basins. The full impact of climate change will depend upon changes to all climate variables, including relative humidity, cloud cover, solar radiation and wind speed. Moreover impact will be highly dependent on the seasonality of change, as we show for the Andes above, in relation to growing seasons and on the magnitude and frequency of extreme events including drought and floods, with their associated costs on livelihoods. These changes are not revealed by changes to monthly or annual precipitation and temperature. In general, however, the impact of glacier melt will be minor in comparison to the projected rainfall changes even for those basins draining from the Himalayas.

There are many impacts of climate change that will affect people, agriculture and livelihoods in ways we do not account for here. Rise in sea level will impact low-lying areas of basins through loss of agricultural land and intrusion of saline water. Climate variability and change, unsustainable land use and poor agricultural practices may increase land degradation, with impacts on agriculture beyond the direct effects on water resources we discuss here.

Shifts in climatic belts may generate changes in land suitability that are difficult to adapt to under the complex patterns of land tenure that dominate in many of these basins. This may result in the need for both people and activities to migrate to areas that become more favourable, locally, regionally, nationally or internationally. Where areas become less suitable for agriculture, diversification of livelihoods away from agriculture will be necessary, which together with urbanization, industrialization, and watershed management for the provision of ecosystem services are notable examples. Agriculture is not a universal option for improving livelihoods especially under conditions of increased climatic variability. Institutions will need to adapt to this fact, in particular by facilitating links to markets, and educational opportunities necessary for populations to engage in non-agricultural sectors.

Institutions will also need to plan to adapt water management to address variability and change in flows in the face of uncertain outcomes. Such adaptations will include hard options such as water storage, water harvesting, irrigation and other technologies applied sustainability and soft (policy) adaptations in land and water management. Where these adaptations can improve sustainability and eco-efficiency they will have benefits for long-term prosperity in the region and are an "easy win" mitigation strategy to climate change.

Conclusions, key uncertainties and research priorities

The answer to the question: "What will be the nature and impact of climate change in the BFP basins?" is clearly "We do not know". Climate projection is not an exact science. The 17 different GCMs produce 17 similar, but different, views on the climate future for each basin for the given A2a scenario. We emphasize that A2a is only one of many possible scenarios, which means we are only making informed guesses as to the kind of changes that will occur. The guesses follow the principles expected from the basic physics of climate and indicate a warming and wetting of the basins when a consensus view of the 17 GCMs is taken. The picture is often very different if only one or two GCMs are used, which we urge should be avoided where possible. The literature contains many impact studies derived from one GCM applied to simple hydrological models to estimate the impacts of temperature change, or temperature and precipitation change, but the results then depend on which GCM was used.

Many hydrological studies of the impact of climate change do not run hydrological models to examine the impacts of the chosen scenario but simply extract the rainfall and temperature predictions for the region of interest. There are few basin-scale studies of impact; most applications use regional or continental scenarios as indicators of basin-scale

Table 8. Climate change impacts and adaptations by basin.

Basin	What will climate change do to food security in my basin?	Which investments should I make to counteract potential adverse impacts to food security?
Andes	May improve food security in the south and central Andes.	Soil erosion management, water storage for the dry season, compensation for watershed management and benefit sharing.
Indus-Ganges	May increase or decrease productivity depending on impact of warming on evapo-transpiration.	Basin highly vulnerable because of unsustainable water source, dependence on agriculture and dense populations. Storage of winter flows in dams and managed groundwater recharge for supply of dry season irrigation.
Karkheh	May fall because of strong warming of already highly seasonal basin with high summer temperatures and low rainfall.	Industrialization and livelihood diversification. Water storage and irrigation where self-sufficiency is important.
Limpopo	May fall because of rainfall decline coupled with warming. Depends on seasonality of change.	Livelihood diversification, rainwater harvesting and storage.
Mekong	Likely little impact from warming and rainfall but may be affected by flooding and sea level rise.	Stormflow management, livelihood diversification, water quality and sanitation management.
Niger	Possible improvement because of wetting, depending on impact of warming in already very hot basin and also on seasonality of change.	Reduce impacts of rainfall shocks and seasonality through rainwater harvesting, water storage, groundwater exploitation and early warning systems.
Nile	Variable from north to south of basin. No change if increased evapo-transpiration is offset by rainfall gain.	Reduce impacts of rainfall shocks and seasonality through rainwater harvesting, seasonal livestock migration.
São Francisco	Little impact given abundance of water, unless reductions occur in dry season.	Dry season water storage, livelihood diversification.
Volta	Possible improvement because of significant wetting but depends on seasonality of rainfall change in relation to growing season and temperature change in relation to high temperature season.	Rainfall harvesting and storage for dry season irrigation.
Yellow	May improve because of higher temperatures increasing growing season in highly temperature-seasonal setting and because of rainfall gains but depends upon seasonality of change.	Livelihood diversification. Management of seasonal and inter-annual variability through storage, recharge and rainwater harvesting.

change rather than downscaling to the basin scale. Since changes in water balance will reflect the effects of warming on evapo-transpiration as well as effects of changing rainfall, it is important to use hydrological models that fully represent evapo-transpiration in relation to the resident vegetation. We then need to combine the multi-model climate scenarios with catchment models and crop simulation linked to poverty and livelihoods. In this way we begin to understand the impacts of climate change on water, food and poverty in basins. To model yield we must also account for CO_2 fertilization, which may serve to reduce the impact of projected climate change on yield (Lin *et al.* 2005).

No empirical work covers this entire spectrum, meaning that many of the impact studies to date represent only part of the system. We also need to analyse climate change in relation to the context of climate variability and the present adaptation of agricultural systems to this variability. Moreover, climate will not change in isolation. All of the basins will experience socio-economic, technological, political and environmental change in addition to climate change. These will impact on the basins' resilience and capacity for adaptation to climate change. Models of the impact of climate change are not complete unless they are seen within the context of the developmental and economic development that led to climate change in the first place, which is indeed apparent in the relevant SRES scenario. Overall, focusing on a single-GCM scenario analysis will almost certainly give it a biased view of change. Moreover, a simple examination of rainfall and precipitation change will fail to incorporate the climatic, land surface and socio-economic feedbacks in the climate–water–agriculture–poverty system, many of which will tend to confer resilience. As we move from simple scenario application to more sophisticated integrated modelling of the climate–water–agriculture–poverty system, we will tend to produce more robust and less extreme scenarios of change.

Given the uncertainty of the SRES scenarios, the difficulties in downscaling, and using the scenarios in impact studies, we should focus more on understanding the sensitivity of systems to climate change. In that way we could avoid the scenario-dependence as well as the GCM-dependence of our analyses. The Andes Basin team did such a sensitivity analysis (Andes Basin team 2009) as did Jones *et al.* (2006).

So, what will climate change do to food security in the BFP basins? Which investments should we make to counteract potential adverse impacts to food security? In Table 8 we summarize our best estimate of expected changes and possible adaptation strategies, by basin.

Table 8 shows that the effects of climate change in the relatively short term of 40 years to 2050 will not all be bad, the outcomes often depending on the seasonality of the change relative to growth seasons and the balance between temperature and rainfall changes. The basins that we identify as having little resilience, such as the Limpopo and the Niger will undoubtedly be worse off in terms of food security than they currently are. The populous basins (Indus-Ganges and the Yellow River) will also face problems of food security but their increasing economic strength may enable them to offset problems through trade. The final consideration is that change within many of the basins is highly spatially variable and this will require site-specific solutions with a view to upstream influences and downstream effects, rather than solutions at national or basin scales.

Acknowledgements

We acknowledge the Challenge Program on Water and Food for their funding of the individual basin focal projects. We also thank the CPWF BFP team and anonymous reviewers for their comments, which much improved this paper.

Notes

1. WEAP = water evaluation and assessment planning; SWAT = soil and water assessment tool; FIESTA = fog interception for the enhancement of streamflow in tropical areas.
2. Data were not available for analysis of the Indus Basin, so we show only historic, present and multi-GCM results for the Ganges.
3. Bjerknes Centre for Climate Research, Norway, BCM2.0 Model; Canadian Centre for Climate Modelling and Analysis, Coupled Global Climate Model 2; Canadian Centre for Climate Modelling and Analysis, CGCM3.1 Model, T47 resolution; Canadian Centre for Climate Modelling and Analysis, CGCM3.1 Model, T63 resolution; Centre National de Recherches Meteorologiques, Meteo France, France, CNRM-CM3; Commonwealth Scientific and Industrial Research Organization Atmospheric Research (CSIRO), Australia, Mk3.0 Model; National Oceanic and Atmospheric Administration (NOAA) Geophysical Fluid Dynamics Laboratory CM2.0 Model; NOAA Geophysical Fluid Dynamics Laboratory, CM2.1 Model; Goddard Institute for Space Studies, Atmosphere-Ocean model; Hadley Centre for Climate Prediction, Meteorological Office, UK, HadCM3 Model; Institut Pierre Simon Laplace/Laboratoire de Météorologie Dynamique/Laboratoire des Sciences du Climat et de l'Environnement (IPSL/LMD/LSCE), France, CM4 V1 Model; Center for Climate System Research/National Institute for Environmental Studies/Frontier Research Center for Global Change (CCSR/NIES/FRCGC), Japan, MIROC3.2, high resolution; CCSR/NIES/FRCGC, Japan, MIROC3.2, medium resolution; Meteorological Institute of the University of Bonn, ECHO-G Model; Max Planck Institute for Meteorology, Germany, ECHAM5/MPI OM; Meteorological Research Institute, Japan, CGCM2.3.2a; National Center for Atmospheric Research (NCAR), USA, PCM1.

References

Ahmad, M.D., *et al.*, 2009. Mapping basin-level water productivity using remote sensing and secondary data in Karkheh River basin, Iran. *Water International*, 34 (1), 119–133.

Amazon Initiative Consortium, Ecosystems and Poverty (ESPA-AA), 2008. *Challenges to managing ecosystems sustainably for poverty alleviation: securing wellbeing in the Andes/Amazon: situation analysis prepared for the ESPA programme* [online]. Belem, Brazil: Amazon Initiative Consortium. Available from: http://www.ecosystemsandpoverty.org/ [Accessed 16 December 2010].

Andes Basin team, 2009. *Report of the Andes Basin Focal Project* [online]. Colombo: CPWF. Available from: http://www.waterandfood.org [Accessed 19 November 2010].

Ardoin Bardin, S., *et al.*, 2009. Using general circulation model outputs to assess impacts of climate change on runoff for large hydrological catchments in West Africa. *Hydrological Sciences Journal–Journal des Sciences Hydrologiques*, 54 (1), 77–89.

Cai, X., *et al.*, 2011. Producing more food with less water in a changing world: assessment of water productivity in ten major river basins. *Water International*, 36 (1), 42–62.

Christensen, J.L., *et al.*, 2007. Regional climate projections. In: S. Solomon *et al.*, eds. *Climate change 2007: the scientific basis. Contribution of working group I to the fourth assessment report of the Intergovernmental Panel on Climate Change*. Cambridge: Cambridge University Press, 847–940.

Collins, M., Tett, S.F.B., and Cooper, C., 2001. The internal climate variability of HadCM3, a version of the Hadley Centre coupled model without flux adjustments. *Climate Dynamics*, 17 (1), 61–81.

Conway, D., 2005. From headwater tributaries to international river: observing and adapting to climate variability and change in the Nile basin. *Global Environmental Change*, 15 (1), 99–114.

Conway, D. and Mahé, G., 2009. River flow modelling in two large river basins with non-stationary behaviour: the Parana and the Niger. *Hydrological Processes*, 23 (22), 3186–3192.

Conway D., *et al.*, 2009. Precipitation and water resources variability in sub-saharan Africa during the twentieth century. *Journal of Hydrometeorology*, 10 (1), 41–59.

D'Almeida, C., *et al.*, 2007. The effects of deforestation on the hydrological cycle in Amazonia: a review on scale and resolution. *International Journal of Climatology*, 27 (5), 633–647.

Dibike, Y.B. and Coulibaly, P., 2005. Hydrologic impact of climate change in the saguenay watershed: comparison of downscaling methods and hydrologic models. *Journal of Hydrology*, 307 (1–4), 145–163.

Elshamy, M.E., Seierstad, I.A., and Sorteberg, A., 2009. Impacts of climate change on Blue Nile flows using bias-corrected GCM scenarios. *Hydrology and Earth System Sciences*, 13 (5), 551–565.

Eltahir, E.A.B. and Bras, R.L., 1996. Precipitation recycling. *Reviews of Geophysics*. 34 (3), 367–378.

Hijmans, R.J., *et al.*, 2005. Very high resolution interpolated climate surfaces for global land areas. *International Journal of Climatology*, 25 (15), 1965–1978.

Hoanh, C.T., *et al.*, 2003. *Water, climate, food and environment in the Mekong Basin in Southeast Asia – final report* [online]. International Water Management Institute, Mekong River Commission Secretariat, Institute of Environmental Studies. Available from: http://www.geo.vu.nl/~ivmadapt/downloads/Mekong_FinalReport.pdf [Accessed December 2006].

IPCC, 2007. *Climate change 2007. Working group I: the physical science basis* [online]. Cambridge: Cambridge University Press. Available from: http://www.ipcc.ch/publications_and_data/ar4/wg1/en/ch9s9-4.html#9-4-1 [Accessed November 2010].

Jones, R.N., *et al.*, 2006. Estimating the sensitivity of mean annual runoff to climate change using selected hydrological models. *Advances in Water Resources*, 29 (10), 1419–1429.

Kato, E., *et al.*, 2009. *Soil and water conservation technologies. A buffer against production risk in the face of climate change? Insights from the Nile Basin in Ethiopia*. International Food Policy Research Institute (IFPRI) Discussion Paper No. 871. Washington, DC: IFPRI.

Lambert, F.H., *et al.*, 2005. Attribution studies of observed land precipitation changes with nine coupled models. *Geophysics Research Letters*, 32, L18704, doi:10.1029/*2005GL023654*.

Limpopo Basin team, 2010. *Unreliable water, vulnerable populations and low water productivity: a role for institutions in the Limpopo Basin. Report of the Limpopo Basin focal project* [online]. Colombo: CPWF. Available from: http://www.waterandfood.org [Accessed 19 November 2010].

Lin, E.D., *et al.* 2005. Climate change impacts and crop yield and quality with CO_2 fertilization in China. *Philosophical Transactions of the Royal Society of London, Series B*, 360, 2149–2154.

Mahé, G. and Paturel, J.E., 2009. 1896–2006 Sahelian annual precipitation variability and runoff increase of Sahelian Rivers. *Comptes Rendus Geoscience*, 341 (7), 538–546.

Mahé, G., *et al.*, 2009. Water losses in the inner delta of the River Niger: water balance and flooded area. *Hydrological Processes*, 23 (22), 3157–3160.

Masih, I., *et al.*, 2009. Analysing streamflow variability and water allocation for sustainable management of water resources in the semi-arid Karkheh River basin, Iran. *Physics and Chemistry of the Earth*, 34 (4–5), 329–340.

Masih, I., *et al.*, 2010. Streamflow trends and climatic linkages in the Zagros mountains, Iran. Climatic Change. (Published online 16 January 2010). DOI 10.1007/s10584-009-9793–x.

Mekong Basin team, 2010. *Report of Mekong Basin Focal Project team*. Colombo: CPWF. Available from: http://www.waterandfood.org [Accessed 19 November 2010].

Mulligan, M. and Burke, S.M., 2005. *FIESTA: fog interception for the enhancement of streamflow in tropical areas* [online]. London: Kings College. Available from: http://www.ambiotek.com/fiesta [Accessed 19 November 2010].

Mulligan, M., *et al.*, 2011. Water availability and use across the CPWF basins. *Water International*, 36 (1), 17–41.

Nakicenovic, N., *et al.*, 2000. Special report on emissions scenarios: a special report of working group III of the Intergovernmental Panel on Climate Change [online]. Cambridge: Cambridge University Press. Available from: http://www.grida.no/climate/ipcc/emission/index.htm [Accessed 19 November 2010].

New, M., Hulme, M., and Jones, P., 2000. Representing twentieth-century space-time climate variability. Part I: development of 1901–96 monthly grids of terrestrial surface climate. *Journal of Climate*, 13 (13), 2217–2238.

Ni, J.R., *et al.*, 2004. On the variation of water resource structure in the Lower Yellow River. *Science in China Series E: Engineering and Materials Science*, 47 Supp. I, 127–141.

Niger Basin team, 2010. *Report of the Niger basin focal project* [online]. Colombo: CPWF. Available from: http://www.waterandfood.org [Accessed 19 November 2010].

Nobre, A.C., Sellers, P.J., and Shukla, J., 1991. Amazon deforestation and regional cimate change. *Journal of Climate*, 4 (10), 957–988.

Ramírez, J. and Jarvis, A., 2008. *High resolution statistically downscaled future climate surfaces* [online]. Cali, Colombia: International Centre for Tropical Agriculture, CIAT. Available from: http://gisweb.ciat.cgiar.org/GCMPage/docs/Downscaling-WP-01.pdf [Accessed 19 November 2010].

Roy, S.B. and Avissar, R., 2002. Impact of land use/land cover change on regional hydrometeorology in Amazonia. *Journal of Geophysical Research*, 107 (1), 1–11.

Sacks, W.J., *et al.*, 2008. Effects of global irrigation on the near-surface climate. *Climate Dynamics*, 33 (2–3), 159–175.

Semenov, M.A. and Stratonovitch, P., 2010. Use of multi-model ensembles from global climate models for assessment of climate change impacts. *Climate Research*, 41 (1), 1–14.

Sharma, E., *et al.*, 2009. Climate change impacts and vulnerability in the eastern Himalayas, Kathmandu, Nepal: International Centre for Integrated Mountain Development (ICIMOD).

Smith, P., *et al.*, 2007. Agriculture. *In:* B. Metz, *et al.*, eds. *Climate change 2007: mitigation. Contribution of working group iii to the fourth assessment report of the Intergovernmental Panel on Climate Change*. Cambridge: Cambridge University Press, 497–540.

Sun, H. and Cornish, P.S., 2005. Estimating shallow groundwater recharge in the headwaters of the Liverpool plains using SWAT. *Hydrological Processes*, 19 (3), 795–807.

TAMU, 2010. Applications of the SWAT Model. College Station, TX: Texas A&M University (TAMU) [online]. Available from: http://swatmodel.tamu.edu/applications [Accessed October 2010].

Tiwari, V.M., Wahr, J., and Swenson, S., 2009. Dwindling groundwater resources in northern India, from satellite gravity observations. *Geophysical Research Letters*, 36, L18401, doi:10.1029/2009GL039401.

USGS, 2010. *HYDRO1k elevation derivative database* [online]. Reston, VA: United States Geological Survey (USGS). Available from: http://eros.usgs.gov/#/Find_Data/Products_and_Data_Available/gtopo30/hydro [Accessed 23 November 2010].

Voldoire, A. and Royer, J.F., 2004. Tropical deforestation and climate variability. *Climate Dynamics*, 22 (8), 857–874.

Volta Basin team, 2010. *Report of Volta Basin Focal Project*. Colombo: CPWF. Available from: http://www.waterandfood.org [Accessed 19 November 2010].

Xu, Z.X., *et al.*, 2005. An overview on water resources in the Yellow River basin. *Water International*, 30 (2), 225–238.

Xu, Z.X., Li, J.Y., and Liu, C.M., 2007. Long-term trend analysis for major climate variables in the Yellow River basin. *Hydrological Processes*, 21 (14), 1935–1948.

Xu, Z.X., Zhao, F.F., and Li, J.Y., 2009. Response of streamflow to climate change in headwater catchment of the Yellow River basin. *Quaternary International*, 208 (1–2), 62–75.

Yates, D., *et al.*, 2005. WEAP21. A demand-, priority-, and preference-driven water-planning model: part 1, model characteristics. *Water International*, 30 (4), 487–500.

Yellow Basin team, 2009. *Yellow River basin: living with scarcity* [online]. Colombo: CPWF and Washington, DC: IFPRI. Available from: http://www.waterandfood.org [Accessed 19 November 2010].

Zhang, X., *et al.*, 2007. Detection of human influence on twentieth-century precipitation trends. *Nature London*, 448 (7152), 461–465.

Zhao, L. and Molders, N., 2008. Interaction of impacts of doubling CO_2 and changing regional land-cover on evaporation, precipitation, and runoff at global and regional scales. *International Journal of Climatology*, 28 (12), 1653–1679.

Zheng, X. and Eltahir, E.A.B., 1997. The response to deforestation and desertification in a model of West African monsoons. *Geophysical Research Letters*, 24 (2), 155–158.

Connections between poverty, water and agriculture: evidence from 10 river basins

Eric Kemp-Benedict, Simon Cook, Summer L. Allen, Steve Vosti,
Jacques Lemoalle, Mark Giordano, John Ward and David Kaczan

The authors analysed livelihood conditions in 10 river basins over three continents to identify generalizable links between water, agriculture and poverty. There were significant variations in hydrological conditions, livelihood strategies and institutions across basins, but also systematic patterns across levels of economic development. At all levels, access to water is influenced by local, regional or national institutions, while the importance of national versus local institutions and livelihood strategies vary with economic development. The cross-basin analysis suggests a framework for thinking about water–agriculture–poverty links that can inform future research and policy development.

Introduction

The global water and food crisis that led to the Consultative Group on International Agricultural Research (CGIAR) Challenge Program on Water and Food (CPWF), and the Basin Focal Projects (BFPs) within it, can be seen, from a global level, as the result of three colliding factors: an increased demand for food to meet an increase in global population; rising incomes that drive increasing demand for water for food production and other purposes; and a renewable, but finite and already over-committed water resource (Molden 2007). This view motivates pessimistic scenarios that feature closing river basins, the breakdown of environmental security, deprivation of water resources to some groups, and curtailed livelihood options. But these are not the only or inevitable consequences of change, and represent a selection only of the range of possible outcomes as water and food systems develop. At smaller scales, individual case studies have illustrated important

With valuable input from Ahmad Assadzadeh, James Kinyangi, Mac Kirby, Chayanis Krittasudthacheewa, Marco Maneta, Don Peden, Jorge Rubiano, Marcelo Torres, Mobin ud-Din Ahmad and Wesley Wallender.

local variations on, and departures from, the broader global themes and their underlying assumptions.

Case studies, however, while informative and valid within their context, shed little light on the full range of conditions that are observed in river basins around the world. There are situations in which systems adapt to enable continued development, or where people are impacted by important but less noteworthy trends, both of which are less likely to be studied and reported on than those that entail more dramatic and easily recognized problems. Moreover, case studies, in isolation, cannot explain generalizable conditions of global significance. The top-down, global approach and the bottom-up, case study approach can thus lead to two extreme views. At the global level, there is a clear picture of increasing stress on natural resources but unclear impact on communities and households, while detailed case studies present clear local accounts but offer little information about how representative they are of the range of situations that are likely to exist.

This paper provides an intermediate view between these two extremes by presenting evidence from the 10 river basins of the BFPs to describe more fully the link between water, agriculture and poverty. The basin papers in the September 2010 special issue of *Water International*[1] provide evidence about the links between water and poverty from a range of case studies, using a variety of analytical techniques. A specific focus of the research was to determine whether there is a link between agricultural water productivity and poverty. The questions addressed by the Basin Focal Projects were broader than this, however, and revealed an important variation between basins at different levels of development. Accordingly, we ask how agricultural water use may impact development, broadly conceived, within river basins.

Water, poverty and "water poverty"

There are at least two ways to think about water and poverty. First, we can ask, how do water-related constraints and opportunities contribute to poverty and its alleviation? Second, we ask, what are water-specific forms of deprivation? The first framing points to links between water and poverty, where "poverty" is conceived in broad terms. The second framing leads to the concept of "water poverty". An important conclusion from the BFP research is that the first approach is more analytically tractable than the second; moreover, it is arguably more relevant for policy. The dominant approach within the water field, however, has been the second, water poverty, approach. Accordingly, we review those ideas briefly here.

There are multiple definitions for "water poverty" (Sullivan 2002, Black and Hall 2004, Cook and Gichuki 2006). The influential Black and Hall (2004) definition is a functional poverty definition, in that it lists observable deprivations associated with water risks and constraints. It also includes an implicit institutional context, introduced by way of explicit categorical inequalities, that is, inequalities arising from socially recognized categories, such as ethnicity, religion or gender (Tilly 1998), specifically, those affecting slum dwellers, women and girls. Cook and Gikuchi (2006) illustrate the underlying causes of agriculturally based water poverty, highlighting the role of low water productivity in the dynamics of poverty. Their framework encompasses assets and livelihood strategies by discussing the importance of livestock, crops and water infrastructure to the poor. This more expansive view is captured well by the sustainable livelihoods framework (DFID 1999), which is discussed below. Sullivan (Sullivan 2002, Sullivan and Meigh 2003) takes a functional definition of water poverty and makes it operational by constructing a water poverty index, which is a hierarchical aggregate. The water poverty index is a weighted sum of component indicators that measure water resources, water use, access to water, water-management capacity, and ecosystem needs. The bottom of the hierarchy is a set of specific indicators that are aggregated to form the component indicators.

Background: poverty and livelihoods

For a term that has such wide currency, "poverty" is an elusive concept. In its *Handbook on poverty and inequality* the World Bank defines poverty as "a pronounced deprivation in well-being" (Haughton and Khandker 2009), but this is rather vague and does not immediately suggest paths to identify and alleviate poverty. In practice, the World Bank uses the now-dominant approach to measurement, a consumption or income-based poverty line: those below the line are considered to be poor, and those above the line are non-poor. While a poverty line operationally defines who is poor, it is not a definition of poverty in itself. Rather, it derives from an assumption that people would obtain what they need to live if they could, and if they do not, it is a symptom of their poverty. For this reason, as with the original poverty line (Orshansky 1965), many national poverty lines are based on the cost of a minimally nutritious basket of food, on the assumption that food is the most basic necessity and hence an inability to obtain food is a good indicator of overall deprivation.

Metrics tend to create their own reality as policy increasingly seeks to change the value of the metric rather than the underlying reality it is meant to represent (Scott 1998, Molle and Mollinga 2003). This is true also of poverty lines; over time the emerging defects of using them as guides to policy have been addressed by refining the concept (Haughton and Khandker 2009) and by exploring alternative approaches to measuring and defining poverty (Sen 1999, Carter and Barrett 2006). Here we adapt and extend the useful classification scheme of Carter and Barrett (2006), and we discuss the following poverty concepts: definitions based on static and dynamic financial flow; definitions based on static and dynamic assets; functional definitions; and definitions based on capability.

Measures of poverty based on financial flow

Definitions based on financial flow focus on income or expenditure flows. Static measures of financial flow assume that people have relatively stable incomes or expenditures, which largely remain below or above a poverty line. An indicator based on this concept can be calculated using standard household surveys without the need of panel data that track individuals or households over time. But it cannot distinguish between chronic poverty, where people remain poor for many years, and transitory poverty, in which a significant number of people move into and out of poverty (Carter and Barrett 2006). In contrast, dynamic measures of financial flow capture changes in income and expenditure for individuals over time. This requires panel data, which are becoming more widely available, but are still challenging to collect and less readily available than are cross-sectional "snapshots" over time. Within the family of metrics of financial flow, the best option is generally considered to be a dynamic consumption measure. Dynamic measures distinguish between chronic and acute poverty, while metrics of consumption are superior to income as measures of a household's ability to meet its needs. This is because while income can fluctuate rapidly and widely, through saving, consumption smoothes out these fluctuations (Carter and Barrett 2006).

Measures of poverty based on assets

The argument for measures of poverty based on consumption rather than income points to an important factor, which measures of financial flows miss. People and households accumulate assets when their incomes allow them to do so and make use of those assets to meet their needs in lean times. Sufficient assets also allow them to undertake new initiatives, such as expanding a farm, digging a well, or buying an animal. The advantage of measuring assets rather than consumption is that it can serve to distinguish between structural and random (stochastic) factors (Carter and Barrett 2006). Structural factors are reflected in a steady accumulation or drawdown of assets across many households, while stochastic factors are more acute and less broad-based. As with measures based on financial flows, dynamic measures of assets that make use of panel data are more revealing than static measures, but more challenging to acquire.

Functional poverty definitions

Neither indicators based on financial flows nor on assets are direct measures of the "pronounced deprivation of well-being" that characterizes poverty. An alternative approach is to adopt a functional definition of poverty that identifies specific forms of deprivation and measures them. Most definitions of water poverty (that is, water-specific deprivation) fall into this category by counting, for example, those without access to safe water, at risk of flooding, and other specific hazards and constraints (Sullivan 2002, Black and Hall 2004, Cook and Gichuki 2006). Functional measures are most appropriate for qualified types of poverty, such as water poverty, food poverty and health poverty. They become unwieldy when applied to poverty more broadly, because specific manifestations of poverty can vary widely from one situation to another.

Institutional poverty analysis

One of the most creative thinkers about poverty, inequality and development is the economist Amartya Sen. He has elaborated a capability-based view of poverty, in which poverty is a reflection of the "substantive freedoms [an individual] enjoys to lead the kind of life he or she has reason to value" (Sen 1999). This notion of poverty as freedom emphasizes the impact of the institutions within which individuals and households make their decisions and pursue their livelihoods. It recognizes, for example, that in many societies a woman has fewer chances to live the life that she values than a man with the same assets and for this reason she experiences a deprivation that is not captured either by flow- or asset-based measures. The major role that institutions play in determining livelihoods and poverty outcomes is an important component in the analysis of rural livelihoods.

Livelihoods

Conceptions of poverty have evolved in tandem with concepts of development, and in particular sustainable development, because poverty is expected to decrease with development. In Amartya Sen's framing, the link is explicit: development is the removal of "unfreedoms" that limit people's capabilities (Sen 1999). The asset and capabilities

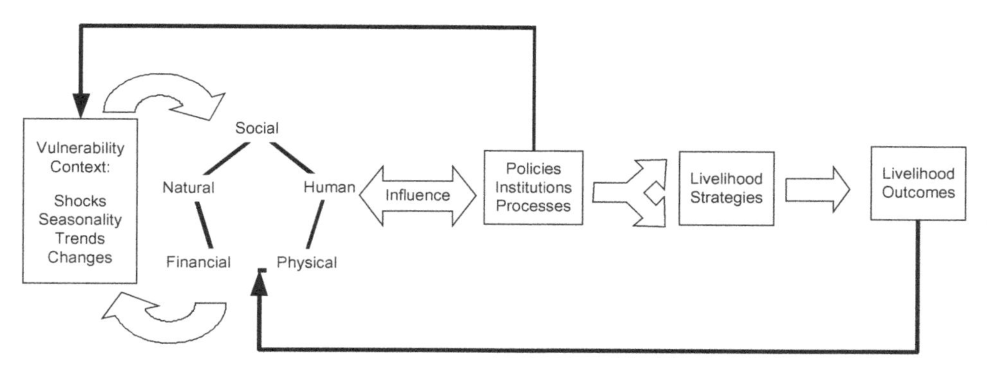

Figure 1. The DFID sustainable livelihoods framework.

approaches to poverty are merged in a view of livelihoods that grew out of dissatisfaction with the views of rural livelihoods prevalent in the 1990s and that are reflected in the UK Department for International Development's (DFID) sustainable livelihoods framework (Scoones 1998, Bebbington 1999, DFID 1999). In this framework (Figure 1), households deploy their financial, physical, human, social and natural assets using livelihood strategies to meet their livelihood goals. They do this within a vulnerability context, characterized by shocks, trends and cyclical changes, and moderated by the formal and informal institutions within which they operate.

The sustainable livelihoods framework is a usable way of thinking about development and poverty, including within the water resources context (Nicol 2000). It encompasses an asset-based approach to analysing livelihoods and embeds them within an institutional context. It also draws upon resilience concepts in its focus on fluctuations in the natural, economic and social environment (Baumgartner and Högger 2004).

Review of evidence from the basins

The basin papers describe basin-specific poverty analyses. They make clear that each of the basin teams of the BFPs followed a unique approach to understanding and analysing water-related poverty. Techniques ranged from scoping methods with low data requirements, to intensive data analysis with significant data requirements. Regardless of the amount of data involved, the general process used in the different basins included:

- choosing indicators of poverty and water poverty;
- identifying candidate causal or correlated variables;
- creating maps of variables and looking for patterns;
- carrying out statistical analysis and modelling, such as systems or hydrological models, Bayesian methods, and spatial statistical techniques, to explore relationships; and,
- using models for hotspot analysis, investigating causality, and scenarios.

We elaborate on these steps in the next section.

Methods

The motivation for carrying out a water and poverty analysis is to identify ways to reduce or eliminate poverty through appropriate interventions. Knowledge of where water-related poverty exists and why it is there informs the interventions. Therefore, the different BFP basins made use of either general poverty indicators or specific indicators of water and poverty. General measures of poverty included financial flow variables (such as the proportion of the population below an income or expenditure-based poverty line); asset inventories; and functional, outcome-based indicators (such as infant mortality, nutritional status, education, life expectancy, and child mortality and morbidity). Water-related indicators included exposure to hazards (for example, flood risk, drought prevalence, and water-borne or water-related disease), climate data (such as rainfall and remotely sensed normalised difference vegetation index, NDVI), and provision of water infrastructure (such as access to irrigation, access to safe water and sanitation, and water productivity). Some basins also created summary indicators. For example, the São Francisco project constructed a novel index of water availability, while the Mekong project constructed an aggregate index for water-related poverty.

With the chosen indicators, several of the basins mapped poverty, which revealed important large-scale patterns and suggested relationships. At its most basic, poverty mapping is simply the process of putting poverty indicators on a map and looking at them, which was done at an early stage in the Volta and the Mekong to orient the study. Such analyses can reveal compelling large-scale patterns; for example, the Volta and São Francisco basins, which run on a north–south axis, have a strong rainfall gradient, and poverty levels vary, more or less systematically, along that gradient. Similarly, the Yellow River, the Indus, and the Ganges have pronounced upstream–downstream poverty gradients. Complementing this "map and look" approach are semi-formal methods for aggregating poverty indicators into an overall poverty index (as in the Mekong), and formal methods, such as spatial statistical analysis (as in the Niger).

Most of the BFPs carried out non-spatial statistical analyses and modelling that explored the relationships between water and poverty variables. As these constitute the bulk of the poverty discussion within the basin-specific papers, they will not be repeated here. Rather, we focus on the outcome of the analyses, which is to reveal patterns of correlation between water-related explanatory variables and poverty variables.

The "development trajectory"

We take the current development status of the basin as an organizing principle for the framework we develop in this chapter, since it determines the prevailing economic conditions that people are in, whether a basin is dominated by agriculture, by urbanization and industrialization, or is in transition from one to the other (World Bank 2007). The locations of the 10 Basin Focal Project river basins on the development trajectory are shown schematically in Figure 2. The predominantly agricultural basins Limpopo, Niger, Nile, and Volta, are characterized by a high contribution of agriculture to gross domestic product (GDP) and high rural poverty. The basins lying within more heavily industrialized countries, the Andes system of basins and the São Francisco, both have a low contribution of agriculture to GDP and low rural poverty. The transitional basins, Ganges, Indus, Karkeh, Mekong and Yellow, are intermediate between these extremes. As basins move along the trajectory, pervasive poverty gives way to isolated pockets of poverty within communities left behind in the overall economic development.

Poverty outcomes in the BFPs were found to depend on where each basin is located on the development trajectory, suggesting that poverty in general is a more useful analytical concept than "water poverty", that is, water-related manifestations of poverty. Moreover, as explained in the Background section above, poverty is best understood within a framework that sees households and communities making use of assets, moderated by the institutions

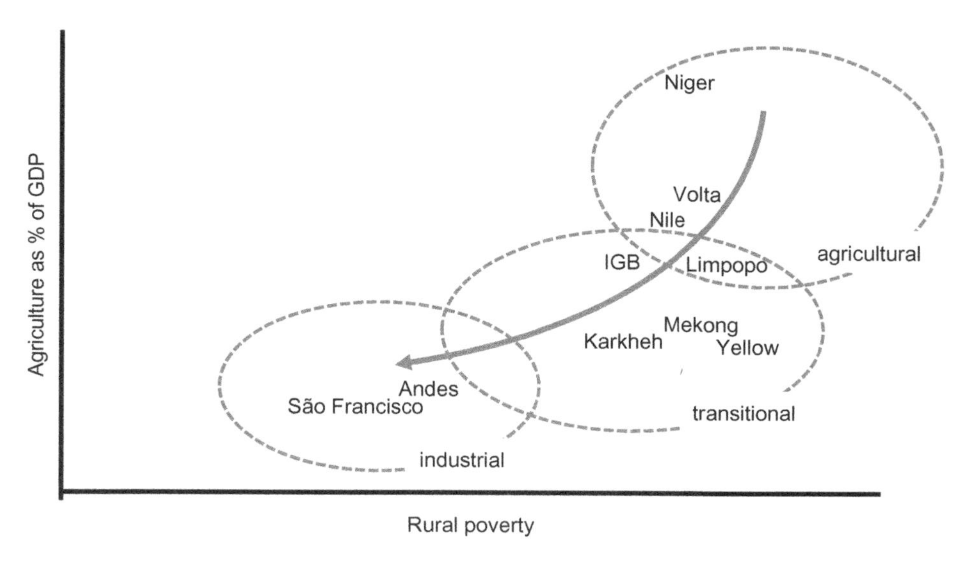

Figure 2. The basins along the development trajectory (World Bank 2007).

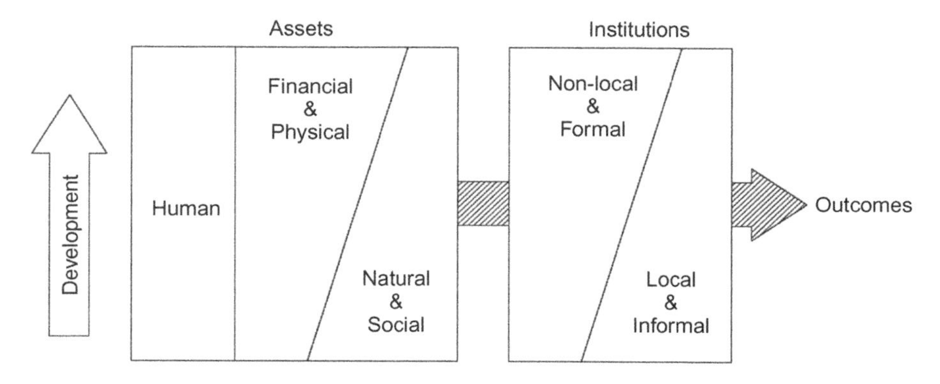

Figure 3. Assets and institutions along the development trajectory.

within which they operate, to achieve livelihood goals. Figure 3 summarizes results from the BFP basin studies. As communities, households, and basins move along the development trajectory in the course of national economic development, the mix of assets shifts from one in which natural and social capital are most important to one in which physical and financial capital play a larger role. At the same time, local and informal institutions decline in importance relative to formal institutions at the provincial, national and basin scale. At all levels of development, human capital is important. The changing role and form of livelihood assets and institutions with development. Figure 3 suggests some characteristic patterns in the 10 BFP basins.

Different aspects of water-related poverty play distinct roles at different levels of development. Table 1 summarizes conditions in basins according to their classification as agricultural, transition or industrial. Some caution is needed with this classification, as within any basin it is usually possible to find mixed classes. The specific, historically contingent, development path within a basin has a very strong influence on the conditions of the water and agricultural systems. It also influences the types of economic opportunities open to people and governments as they produce and consume, while the population and

Table 1. Basins at different development levels.

	Agricultural	Transitional	Industrial
Exemplar basins	Limpopo, Niger, Nile, Volta	Ganges, Indus, Karkheh, Limpopo (South Africa part), Mekong, Yellow	Andes, São Francisco
Role of agriculture in the national economy	Dominant. Agricultural development in many cases a key to broader economic development. Water productivity is very low in most places.	Agriculture a mainstay to rural livelihoods but competing with urban or industrial demands for water. Water productivity is extremely high in some areas.	Agriculture declining in importance as a source of livelihood for most of the population as alternative sources of income develop. Water productivity higher if measured by monetary value (i.e., farmers may grow low-yielding but high-value crops).
Poverty incidence: Indicators of wellbeing.	Widespread. High percentage, even if absolute numbers are low.	General, large numbers but lower percentage. Urban poverty increasing in importance.	Rural poor tend to be "left behind" general economic growth.
Physical infrastructure: road network, energy	Basic infrastructure is limited. A major constraint to agricultural development.	Pressure on pre-existing infrastructure. Substantial investment in infrastructure.	Continued investment.
Water resource development	Very little development of irrigation. Some hydropower. Less than 70% of the rural population has access to clean water supply/sanitation.	Extensive development of irrigation, in some cases to an unsustainable level. Hydropower or industrial users given high priority to meet demands of industrialization. Up to 80% with access to supply and sanitation.	Established. Further development of irrigation difficult due to increasing scarcity, while irrigation development not often targeted to the rural poor. Institutions developing to help share resources and benefits from water resource development.
Environmental security	Ecosystem services very important to specific groups (e.g. fishers and livestock herders) but these are generally informal and not valued in markets.	Major loss of ecosystem function. Ecosystem services not valued in markets. Fishers and smallholder livestock farmers declining. Aquaculture expanding.	Increasing attention to ecosystem function with emerging opportunities for trading of ecosystem services. Aquaculture increases in importance relative to capture fisheries. Livestock dominated by large-scale enterprises.

(Continued)

Table 1. (*Continued*)

	Agricultural	Transitional	Industrial
Vulnerability to water-related hazards	Very little protection. Major impact of health on livelihoods through sickness and disease. Livelihood systems rely on risk avoidance.	Moderate protection through engineering.	Engineering and institutional protections developing.
Development of markets and financial institutions	Semi-subsistence farming dominates, although most populations are linked to markets. Local informal institutions.	Active development of markets. Financial services not available to all or for all desired investments. Diminishing importance of local institutions.	Commodity and high-value crops dominate. Widely available financial services. Relatively large role for government institutions.

scale of economic activity within a basin strongly influences the pressures exerted on the natural environment.

Agricultural basins

The predominantly agricultural basins of the BFP basins, the Limpopo, Niger, Nile and Volta are all in Africa. Within these basins, poverty is widespread and heavily concentrated in rural areas. People are largely unprotected from hazards, even recurring, and therefore anticipated, hazards such as seasonal variations in rainfall and endemic water-related diseases.

Crop agriculture is predominantly rainfed, while livestock and fish make important contributions to household incomes and income diversification. Fish and livestock provide essential livelihoods to certain groups, such as pastoralists and freshwater fishers, who are facing increasing pressures on aquatic and land resources. Water productivity is typically very low, in part due to limited markets for outputs and inputs, and in part as a result of risk management strategies that seek to maintain a minimum guaranteed output at the expense of maximizing average output.

Households derive much of their own food from subsistence agriculture and, compared to transitional and industrialized basins, operate relatively independently from state organizations. State-provided infrastructure, such as roads and irrigation, and services, including education, are limited in scope. The dominance of local institutions in agricultural basins often means inconsistencies and conflicts between the plans of the state and their implementation on the ground. At the same time, local institutions ensure a minimal safety net through communal use of resources, although sharing output makes it hard for farmers to invest time and resources into improving their productivity, as the benefits are captured by everyone.

Transitional basins

The transitional basins, the Ganges, Indus, Karkeh, Mekong and Yellow, have developed substantial non-agricultural activities but agriculture remains a mainstay of rural life. These are "patchy" basins containing substantial areas that could be classified as either agricultural or industrial. These basins contain the largest populations of the BFP basins. The numbers of poor are very large, even though the proportion of poor to non-poor is substantially lower than in the agricultural basins. One of the characteristics of transitional basins is that rural development becomes a priority for governments, and in some of these basins, such as the Karkheh and the Ganges, we see considerable political pressure to stabilize the rural economy.

As illustrated in papers on the Yellow (Ringler *et al.* 2010) and Indus-Ganges (Sharma *et al.* 2010), irrigation is highly developed in the transitional basins, and has enabled the populations to expand to levels that now seem, in some parts of the basins, difficult to sustain. Agriculture provides a livelihood for many and in places is at or near to its potential maximum productivity. Partially as a consequence of major expansion of agriculture, ecosystem services have been impacted considerably. Fish and livestock have declined in overall importance, although they are dominant livelihoods for some of the poorest communities, and both livestock and fish continue to play a role in livelihood diversification. In the Mekong and, to a lesser degree, the Ganges Delta, fish remains a major source of livelihood support that is under increasing pressure as development massively increases the demand for hydropower and irrigation water. In the Indus and the Yellow basins, which are drier, conflicts over water use threaten continued development.

Industrialized basins

The Andes collection of basins and the São Francisco, both in Latin America are classified as industrialized. While neither of them is dominated by industrial production, they are within countries that have significant industrial production, and this affects the employment opportunities, level of infrastructure, and government services available to rural populations. In both of them, agriculture accounts for less than 10% of the annual increase in gross domestic product (GDP), although in Brazil, agriculture is actually increasing in importance as a result of strong growth of commercial agriculture amongst which there remain large pockets of poor small-scale farmers. Rural poverty persists in these areas, but it tends to be more localized, and is characterized as areas that have been "left behind" by the surrounding economic development. In the São Francisco, resource-poor smallholders do not generally benefit from the economic industrialization. They find it hard to gain entry into larger-scale farming and processing operations, and increasingly sophisticated agricultural markets. Moreover, they often do not have access to the resources to adapt to the major changes in the agricultural landscape.

While the poorer areas of these basins have better access to state-controlled services compared to agricultural and transitional basins, they are still marginalized in comparison to other parts of the basin. Access to water has greatly shaped agricultural development in the São Francisco Basin but concern over access to water in these basins is shared with concerns regarding access to education, markets and finance. Water-related hazards, such as flooding and drought, continue to be a problem, but institutions, financial assets, and infrastructure are sufficiently well developed that communities are able to recover from most events.

Results: a poverty and water framework

Earlier in this paper we argued that poverty is a multi-faceted phenomenon, and traced a history of thinking about poverty. In reviewing evidence from the basins we also identified the critical importance of a basin's stage of development to an analysis of water and poverty links. So that we can capture the various aspects of poverty revealed in the basin studies, we combine elements of functional, asset-based, and capability-based definitions of poverty to construct a poverty and water framework. We identify the following aspects of water-related poverty:

- *Scarcity:* where people are challenged to meet their livelihood goals as a result of water scarcity;
- *Access:* where people lack equitable access to water;
- *Low productivity:* where people acquire insufficient benefit from water use;
- *Chronic vulnerability:* where people are vulnerable to relatively predictable and repeated water-related hazards such as seasonal floods and droughts, or endemic disease; and
- *Acute vulnerability:* where people suffer an impaired ability to achieve livelihood goals as a consequence of large, irregular and episodic water-related hazards.

While there are dependencies between these aspects, for example, productivity and vulnerability are both dependent to some extent on scarcity and access, to an important degree they act independently. In particular, institutions mediate the link between scarcity and vulnerability and between scarcity and access, while high productivity can lessen vulnerability in water-scarce areas. Thus, the five aspects of water-related poverty are related to the institutional, variability, and asset components of the sustainable livelihoods framework (Figure 4). Deprivation as a result of water scarcity reflects a lack of natural assets; equitable access is determined largely by institutions; vulnerability to water-related hazards is largely (although not entirely) due to variability in the natural environment; low water productivity is affected by household and community assets, such as access to markets or knowledge; and loss of livelihood due to change is a consequence of variability in the external natural, economic, and social environment.

The poverty and water framework along the development trajectory

Of the different aspects of water-related poverty (Figure 4), inequitable access emerges at all levels of development. Local institutions, basin-scale institutions, geography and hydrology appear to determine whether development and poverty reduction will be broadly or narrowly based. In case studies carried out in northeast Thailand, which suffers from an extended dry season, poor groundwater quality, and floods in the rainy season, local norms favour a broad distribution of benefits from improved production. Perhaps for this reason, small-scale, local initiatives have performed better than large-scale, state-sponsored irrigation projects. In contrast, in the Niger Basin, diverse and fragmented local institutions lead to inconsistent implementation of large-scale projects. Benefits are shared inequitably, which explains the weak (or negative) relationship between water productivity and poverty that was highlighted in the Niger paper (Ogilvie *et al.* 2010). The effects of geography and hydrology can be seen in several basins: in the Andes, where water access aligns with the north–south rainfall gradient and vertical climatic gradients; in the

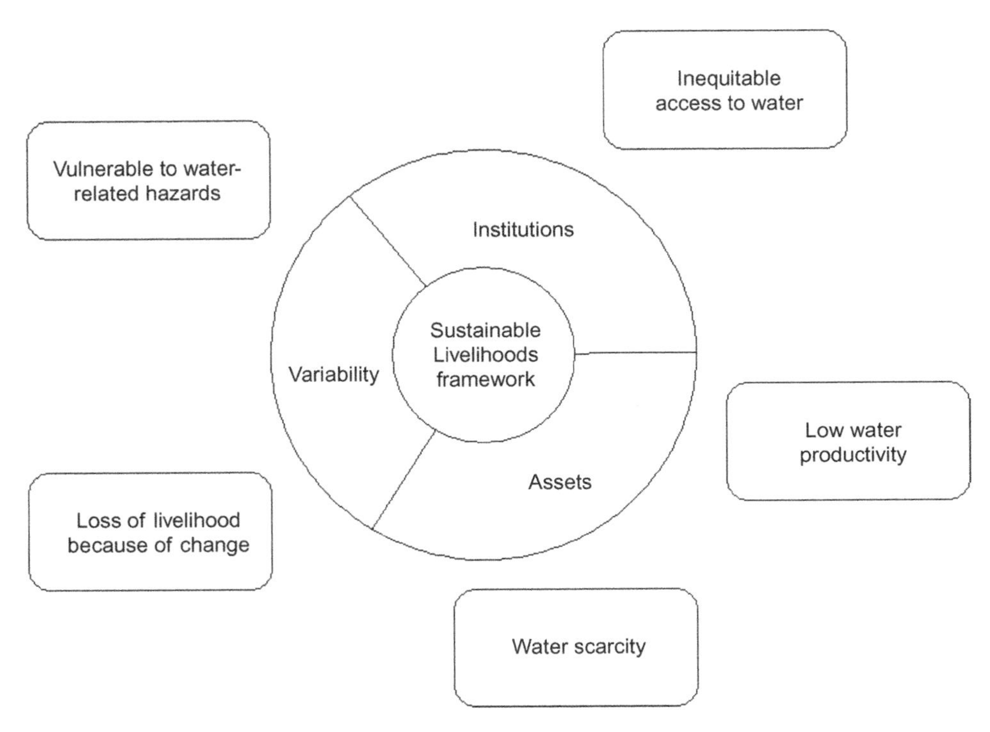

Figure 4. The poverty and water framework and its connection to the sustainable livelihoods framework.

Volta and São Francisco, where poverty follows the rainfall gradient; and in the distinct poverty trajectories of the upper and lower parts of the Ganges, Indus, Limpopo, Nile and Yellow.

Unlike access to water resources, other aspects of water-related poverty play different roles at different stages of the development trajectory. For agriculturally dominant basins, water scarcity is common, exacerbated by a lack of storage, and water productivity is an effective lever for development, if the benefits are broadly shared, and households suffer from chronic water-related hazards. As basins become more industrialized, water scarcity becomes less common or less severe, and water productivity becomes one of many inter-related factors that impact upon poverty levels. Households and communities are more vulnerable to acute water-related hazards, that is, hazards that happen rarely but have a large impact.

Water-related interventions along the development trajectory

As shown in Figure 2, agriculture plays a smaller role in the economies of basins that are closer to the industrial end of the development trajectory, and they have a lower incidence of rural poverty. Poverty reduction means, in practice, movement along the trajectory from the upper right of the figure towards the lower left. A consequence of this, as we argue below, is that water-related interventions are more or less effective depending on where a basin lies on the trajectory. These differences can be understood from the changing mix of livelihood assets shown in Figure 3.

Within agricultural basins, development of agriculture is often a pre-requisite to other forms of development. Until recently, standard agricultural development theory argued that rising agricultural productivity was essential to raising rural incomes, as it enabled rural populations to diversify into non-agricultural activities (Timmer 1998). Following recent extensive research into rural livelihoods, the current understanding is more nuanced (FAO 1998, World Bank 2007), but rising agricultural productivity has been identified as a key factor in the transition out of rural poverty in several countries (World Bank 2007). Local activities and innovation are essential, and a primary goal is to reduce barriers to effective and equitable institutions. These activities often require the development of infrastructure and services around rural populations. However, as at any stage of development, institutions are important, and these interventions may be ineffective if the benefits are captured by elites.

Irrigation may have substantial impacts, but only if other contributing factors are also improved, including markets and financial institutions, and if local institutions are supportive. As described in the papers on the agricultural basins (the Limpopo, Sullivan and Sibanda 2010; the Niger, Ogilvie *et al.* 2010; the Nile, Awulachew *et al.* 2010; and the Volta, Lemoalle and de Condappa 2010), there is very little irrigation at present, and only limited water is available to expand irrigation coverage. As smallholder production is dominated by rainfed agriculture, marginal improvements in rainfed agriculture, if they are widely shared, are likely to have a larger impact than irrigation expansion. Moreover, field-scale innovations can be carried at relatively low collective risk, and can support the development of human and social capital that make larger-scale improvements more successful.

In transitional basins (the Ganges and the Indus, Sharma *et al.* 2010, Karkeh, Ahmad and Giordano 2010, the Mekong, Kirby *et al.* 2010, and the Yellow, Ringler *et al.* 2010), access to water resources or to the benefits they generate are of greater importance to the poor than water scarcity or basic provision of infrastructure. In each of these basins, except the Mekong, the poorest areas are those without irrigation. At the same time, extensive irrigation has provided water to farmers at the cost of increasing pressure on scarce water resources. The Mekong is a wet basin, and large-scale irrigation dominates only in the delta; in other parts of the basin, farmers use small-scale irrigation systems. Consequently, investments in infrastructure and development of institutional capacity to manage water resources are needed, as with the agricultural basins, but under conditions of increasing pressure. Infrastructure and institutional capacity, in turn, can help to manage chronic hazards as substantial improvements are made in water supplies and sanitation, together with flood control. Given the large numbers of people in these basins, secure provision of basic services has a significant impact on wellbeing and national development goals. Within existing transitional basins, there is limited scope for further development of large-scale irrigation and there is already a high level of productivity in some irrigation areas (for example, in the Yellow and Ganges), suggesting that improvement of rainfed agriculture in the poorest parts of these basins may be overlooked as a source of change, while diversification through aquaculture and livestock can help to smooth variations in income.

Within industrialized basins, represented here by the Andes (Mulligan *et al.* 2010), and the São Francisco (Vosti *et al.* unpublished data), the opportunities for improvement in rural livelihoods arise less from improvements in the traditional agricultural sector than from salaried employment in the rapidly growing commercial sector, or from specialization within smallholder farming to capitalize on the development of new urban markets. In these basins, except in the poorest areas, which are pockets resembling agricultural or transitional basins, increasing water productivity is less a policy lever for poverty reduction than it is a strategy for the agricultural sector to meet its own goals. These goals themselves can help reduce poverty, via employment generation within and outside of agriculture. Water-related poverty persists, but strategies to reduce poverty, including water-related poverty, focus more on employment and market access than on water as such. In the São Francisco Basin, improved access to water may be necessary for reducing poverty in some parts of the basin, but will not be necessary in all areas, and is unlikely to be sufficient in any of them.

Conclusions

Poverty is a multi-dimensional phenomenon, and thinking about poverty has evolved over time as an appreciation of its complexities has grown. The links between water and poverty are also not simple and resist prescription. However, work in the BFPs revealed some common patterns and conclusions that can help to guide future policy and research. That work leads to the following conclusions concerning the nature of the relationship between water and poverty.

(1) From both an analytical and policy standpoint, it is more relevant to policy makers to understand the influence of water-related variables on general poverty and livelihood measures rather than to seek the meaning of indicators of "water poverty".
(2) There is no simple link between water scarcity and poverty because the nature of this relationship is strongly influenced by position along a "development trajectory". Although the development trajectory does not predict the character of water-poverty links, this condition is such a powerful factor that a first step in analysing the water–food–poverty links within a basin should be to determine where it lies along that trajectory.
(3) At any level of development, analysis of the links between water and poverty should take into account the livelihood strategies and institutional environment of the households at whom those interventions are targeted. The character of the relevant institutions and the mix of assets varies systematically with the households' and basin's development status.

Concerning interventions, we determined four different types of interventions from evidence within basins, each related to a different kind of livelihood capital.

First, interventions that seek to increase human capital are likely to be effective at any level of development, as long as they are matched to the needs and capacity of the community. Examples included improvements in human capital to support fisheries in the Volta; health and education in the upper Niger; education of farmers in the Indo-Gangetic basins in crop-specific practices; and education in the industrial Andean basins, since this was found to correlate strongly, and inversely, with poverty. Interventions such as the introduction of new management techniques, sharing knowledge about alternative crops, and individual and community capacity building can improve livelihoods and reduce poverty throughout the development trajectory.

Second, investments in natural capital are likely to be more effective at the agricultural stage of the development trajectory since people in these conditions rely most strongly on natural capital for their livelihoods. Nevertheless, realizing the benefits of investment in natural capital is also contingent on institutional support. Interventions such as rainwater harvesting, the development and support of water-user associations and other local water institutions, and techniques to improve green water use are likely to have a significant impact in agricultural basins. Analysis from the Niger, Nile and Volta emphasized the continued role of traditional institutions and the potential gains to rural livelihoods through improvements at the field scale.

Third, investments in water-related physical capital are likely to have a greater marginal impact on poverty at the agricultural and transitional levels of development, although individual improvements are unlikely to be successful without concurrent attention to surrounding infrastructure. Small reservoirs, small-scale multiple-use water systems, local road building, tube wells, small and large-scale irrigation, and similar interventions are more likely to reduce poverty levels where physical and financial infrastructure is not already well developed. While they are also important at the industrial level of development, in these situations they are best seen as strategic investments for regional development, rather than as mechanisms for poverty alleviation. Analysis from the Andean system of basins and the São Francisco, showed that poverty in these basins is strongly affected by national and regional institutions and by access to labour and agricultural markets, as well as to markets for non-agricultural goods produced in rural areas.

Fourth, at any level of development, the institutional context in which interventions are introduced is a strong influence on their success. The nature of dominant institutions varies as the basin passes through the agricultural, transitional and industrial stages of development. At the agricultural stage, the role of basin-wide institutions is less important to poverty reduction than are small-scale institutions. However, at the transitional and industrial stages, such large-scale institutions can be crucial for assisting those left in pockets of poverty as the basin experiences strong growth in population and economic activity. This was particularly apparent in the Indus, Ganges and Yellow River basins, where irrigation, which is more highly developed in some parts of the basin than others, is strongly correlated with lower levels of poverty. In the course of development the shift from local and informal institutions to non-local and formal ones can favour some groups and individuals at the expense of others or at the expense of the natural environment; as basins become more strongly industrialized, the economic capacity grows to invest in institutional processes to address any distortions.

Note

1. Andes (Mulligan *et al.* 2010), Indus-Ganges (Sharma *et al.* 2010), Karkheh (Ahmad and Giordano 2010), Limpopo (Sullivan and Sibanda 2010), Mekong (Kirby *et al.* 2010), Niger (Ogilvie *et al.* 2010), Nile (Awulachew *et al.* 2010), Volta (Lemoalle *et al.* 2010), and Yellow (Ringler *et al.* 2010). The report on the São Francisco is an internal BFP document, which will be published on the Internet in due course.

References

Ahmad, A. and Giordano, M., 2010. The Karkheh River basin: the food basket of Iran under pressure. *Water International*, 35 (5), 522–544.

Awulachew, A., *et al.*, 2010. The Nile basin: tapping the unmet agricultural potential of Nile waters. *Water International*, 35 (5), 623–654.

Baumgartner, R. and Högger, R., eds, 2004. *In search of sustainable livelihood systems: managing resources and change*. New Delhi: Sage Publications.

Bebbington, A., 1999. Capitals and capabilities: a framework for analyzing peasant viability, rural livelihoods and poverty. *World Development*, 27 (12), 2021–2044.

Black, M. and Hall, A., 2004. Pro-poor water governance. In: *Water and poverty: the themes*. Manila: Asian Development Bank, 11–20.

Carter, M.R. and Barrett, C.B., 2006. The economics of poverty traps and persistent poverty: an asset-based approach. *Journal of Development Studies*, 42 (2), 178–199.

Cook, S. and Gichuki, F., 2006. *Analyzing water poverty: water, agriculture and poverty in basins*. Working Paper Series. Colombo: CGIAR CPWF.

DFID, 1999. *Sustainable livelihoods guidance sheets*, London: Department for International Development (DFID).

FAO, 1998. *The state of food and agriculture 1998* [online]. Rome: Food and Agricultural Organization of the United Nations (FAO). Available from: http://www.fao.org/docrep/w9500e/w9500e00.htm [Accessed 11 November 2010].

Haughton, J. and Khandker, S.R., 2009. *Handbook on poverty and inequality*. Washington, DC: The World Bank.

Kirby, M., *et al.*, 2010. The Mekong: a diverse basin facing the tensions of development. *Water International*, 35 (5), 573–593.

Lemoalle, J. and de Condappa, D., 2010. Farming systems and food production in the Volta Basin. *Water International*, 35 (5), 655–680.

Molden, D., ed., 2007. *Water for food, water for life: comprehensive assessment of water management in agriculture*. London: Earthscan and Colombo: International Water Management Institute.

Molle, F. and Mollinga, P., 2003. Water poverty indicators: conceptual problems and poverty issues. *Water Policy*, 5 (5), 529–544.

Mulligan, M., *et al.*, 2010. The Andes basins: biophysical and developmental diversity in a climate of change. *Water International*, 35 (5), 472–492.

Nicol, A., 2000. *Adapting a sustainable livelihoods approach to water projects: implications for policy and practice* [online]. Working Paper 133. London: Overseas Development Institute (ODI). Available from: http://www.odi.org.uk/resources/download/1093.pdf [Accessed 11 November 2010].

Ogilvie, A., *et al.*, 2010. Water agriculture and poverty in the Niger River basin. *Water International*, 35 (5), 594–622.

Orshansky, M., 1965. Counting the poor: another look at the poverty profile. *Social Security Bulletin*, 28 (1), 3–29.

Ringler, C., *et al.*, 2010. Yellow River basin: living with scarcity. *Water International*, 35 (5), 681–701.

Scoones, I., 1998. *Sustainable rural livelihoods: a framework for analysis*. Working Paper 72. Brighton, UK: Institute of Development Studies.

Scott, J.C., 1998. *Seeing like a state: how certain schemes to improve the human condition have failed*. New Haven, CT: Yale University Press.

Sen, A., 1999. *Development as freedom*. New York: Anchor Books.

Sharma, B., *et al.*, 2010. The Indus and the Ganges: river basins under extreme pressure. *Water International*, 35 (5), 493–521.

Sullivan, A. and Sibanda, M.L., 2010. Vulnerable populations, unreliable water and low water productivity: a role for institutions in the Limpopo Basin. *Water International*, 35 (5), 545–572.

Sullivan, C., 2002. Calculating a water poverty index. *World Development*, 30 (7), 1195–1210.

Sullivan, C. and Meigh, J., 2003. Considering the water poverty index in the context of poverty alleviation. *Water Policy*, 5 (5/6), 513–528.

Tilly, C., 1998. *Durable inequality*. Berkeley: University of California Press.

Timmer, C.P., 1998. The agricultural transformation. *In*: C.K. Eicher and J.M. Staatz, eds. *International agricultural development*. Baltimore: Johns Hopkins University Press, 113–135.

World Bank, 2007. *World development report 2008: agriculture for development*. Washington, DC: The World Bank.

Institutions and organizations: the key to sustainable management of resources in river basins

Myles Fisher, Simon Cook, Tassilo Tiemann and James E. Nickum

Based on studies of water, poverty, and livelihoods in nine river basins reported in the first part of this book, we review the role of institutions and organizations. We attempt to identify their strengths and weaknesses and to generalise as to why they fail to address the basin-wide issues of water, poverty, and livelihoods. We attempt to show how a more comprehensive, integrated approach might change them to be more broadly relevant to basin-wide needs and to address the mismatch between development and the need to provide ecosystem services relevant to food, poverty, livelihoods, and sustainable ecosystems.

Introduction

Water, food, and livelihoods are linked with development in river basins. Their interactions determine how river basins sustain and fulfill human life. The basin case studies in this book illustrate the diverse links between water, food and development within nine river basins globally. These components are so closely linked that the traditional approach of focusing on one or the other obstructs the broader understanding that is necessary to explain constraints for the sustainable and effective establishment of livelihoods and their development, and formulate approaches to overcome them. The common thread across all the basin case studies is the central role of institutions and organizations in determining the outcomes of how access to and use of water influences livelihoods, not just of the rural poor, but people generally. Institutions are key components in determining the outcomes of development initiatives and how they impact poverty. We examine this aspect in more detail below.

To proceed, we must define institutions and organizations, which are often used synonymously. In this paper, we regard them as separate and distinct and take the Menard and Shirley (2008) definitions: "institutions are the written and unwritten rules, norms and constraints that humans devise to reduce uncertainty and control their environments ... . Organizational arrangements are the different modes of governance that agents implement to support production and exchange". In the text that follows, we consider 'institutions' and 'organizations' in line with these definitions. Institutions are rules and norms such as property and use rights, legal frameworks, official language, trust building mechanisms and social capital (especially between ethnicities as well as between administrative units), and regulatory culture. In contrast, organizations are bodies such as government departments, autonomous regulatory bodies, and non-government organizations. We acknowledge that institutions, as defined here, play a role in keeping poor people poor.

In our examination of the role of institutions and organizations in river basins, we use the concept of *ecosystem services* (ESs), which Daily (1997) defines as, "The conditions and processes by which natural ecosystems, and the species that make them up, sustain and fulfill human life". Building on this definition, we include in ESs all those goods and services that water provides within a river basin. This definition provides us with the tools to synthesize the broader understanding we seek.

A basin is composed of a mosaic of ecosystems, ranging from the built urban ecosystem through the various farmed and grazed lands that produce food, to pristine mountains, riverine wetlands and shorelines. Used in this way, ESs includes all the components of land and water use in a basin, not merely the maintenance of wetlands and similar ecosystems, which is the popular connotation that springs to mind in the context of 'ecosystem'.

We take the view that enabling the use of ESs controls how water impacts the mosaic of ecosystems in a basin, which in turn impacts people's livelihoods. ESs may be divided into *Provisioning services*:

- Supply of water for industrial and domestic use, and sanitation including regeneration processes such as detoxification/decomposition of wastes;
- Water and land for food production; and
- Water to generate hydropower;

and *Regulating services*:

- Drought mitigation and flood protection;
- Stabilization processes such as cycling and control of nutrient and gas fluxes and sediments; and
- Life-fulfilling options such as aesthetics, scientific, cultural or spiritual value as well as the provision of sufficient water to maintain those vulnerable ecosystems such as wetlands.

River basins can therefore be considered as systems that provide a range of ecosystem services upon which livelihoods of different groups depend. These groups appropriate, develop, share, and damage such services, using formal or informal rules (institutions). As they do so, they affect the livelihoods of others, and also affect the environment that provides ecosystem services. Governance and organizations determines the way in which such activities (institutions) interact, and are therefore key to the sustainable use of ecosystem services.

How institutions and organizations function in river basins

Frameworks to describe governance in basins

Research teams in all nine basins observed that national and often regional government organizations were typically divided into rigid compartments of water, agriculture, irrigation, and economic development. In contrast, activities at a community level often merged several of these compartments according to the resources or activities involved.

Historically, when we started the Basin Focal Projects (BFPs, see Fisher and Cook's Introduction of this book), we looked at a fairly narrow range of variables and activities involved in water and agriculture that we expected to be related to poverty. We quickly found that the reality within basins is that people use a selection of whatever resources are appropriate for their circumstances and are available to them. Water and agriculture are just two of many components that comprise this picture, so that we had to broaden the range of activities we measured to take account of what people within basins actually do.

We found that ultimately it is the environment that provides the resources that people use through ESs. Daily (1997, quoted above) elaborates the definition, "[ESs maintain] biodiversity in the production of ecosystem goods, such as seafood, forage, timber, biomass fuels, natural fibre, and many pharmaceuticals, industrial products, and their precursors". She adds that, "ecosystem services are the actual life-support functions, such as cleansing, recycling, and renewal ...". Daily does not include food production in this definition, which we would put as the most important ESs that water provides.

Some resources, such as soils and plant communities, are fixed in space but their characteristics vary over time, while others, such as water and products, flow freely between systems. Because of this, we allow our thinking to be constrained to these component processes, which might help us understand component processes of hydrology, agriculture, or economics. In doing so, we limit our ability to account for the interactions between them that control peoples' livelihoods in the real world. ESs are not used individually or by individuals, but collectively by communities of organizations. In this chapter, we therefore aim to reflect on the organizations that use ESs within river basins, without constraining ourselves to those that deal with water, or with crops, or with animals, or with industry.

Merrey and Cook (2011) review various frameworks used to describe governments in river basins and observe the disconnect that occurs between ministries that are responsible for agriculture and those that deal with water. They note that most research on governance and organizations in river basin has consisted of case studies or limited comparative analyses of specific processes, for example democratization, which is essentially about institutions.

Merrey and Cook (2011) report Svendsen *et al.*'s (2005) analysis of governance and institutions in basins based on a framework of "essential functions for river basin management" and their hypothesis that there is a minimum set of critical functions that must be fulfilled for effective river basin management. They outline Svendsen *et al.*'s (2005) two basic organizational patterns of basin governance: *centralized* ('unicentric'), where a single authority is responsible for river basin development and management, and *decentralized* ('polycentric'), where multiple organizations are involved.

Because of the political difficulty that nations face in ceding sovereignty to a supranational authority, unicentric basin governance is generally only feasible where basins are not shared among countries. Even basins wholly within the one country are rarely unicentric. Salman (2010) addresses some of these issues, particularly the role of downstream riparians in preempting the use by upstream riparians of the water that they contribute. He also shows the complexity for basin governance that is caused by the creation of the new state of South Sudan in the Nile Basin (Salman 2011).

Merrey and Cook (2011) identify that where upstream/downstream relations are critical, there is a need in transboundary basins for arrangements that emphasize the need to:

- Avoid conflict between organizations managing different activities that are juxtaposed within basins;
- Emphasize institutional support in the poorest areas of basins to build livelihoods and reduce vulnerability;
- Put in place mechanisms that share resources and benefits that recognize traditional functions while introducing more modern ones; and
- Recognize the emerging opportunity for sharing ESs.

A brief review of governance and organizational requirements in basins

Here we attempt to answer the questions, what is it that organizations need to enable them to improve the livelihoods of the poor? How do we define organizational success and failure? We summarize the main issues that are associated with organizational factors in basins, as reported by the authors of the basin case studies in this book (Table 1). We attempted to classify the issues in each basin on the basis of their success or failure. Some of them did not separate as either success or failure, so we added other categories where appropriate.

An important characteristic of organizations in the river basins in the case studies is the disconnect between the government agencies responsible for agriculture and food security, and those responsible for water management (Table 1). Agricultural water is often an orphan, with several agencies responsible but with no clear demarcation between them. At the political level, agriculture ministries are often more focused on the pricing of inputs and outputs, and broad policies, with, at best, a small unit devoted to water management at the local level. In contrast, ministries in charge of water affairs give highest priority to domestic water and sanitation services, and major storage infrastructure (with, again, irrigation often excluded or given low priority). In transboundary river basins, agricultural water is often given lower priority than controlling pollution or floods. Agricultural agencies may take an interest in promoting "green" water, management of rainwater at the local level for crop production, while in contrast water agencies typically ignore green water and focus on "blue", i.e., surface water. Then there is groundwater, which is often under the control of yet another agency, if it is controlled at all. Responsibilities are therefore fragmented at multiple levels of government. As a consequence, the systems involved are often so poorly understood that there can be no integrated approach to river basin management.

Table 1. Issues associated with institutions and organizations in the nine river basins included in this book and assessment of their success or failure. Some issues are not easily classified as successes or failures and we identify them separately. Where the citation is undated, it refers to the appropriate chapter in this book.

Basin	Institutional and organizational issues implicated in development processes	Authors
Andes	Failure Unequal access to land and water resources; Legacy of political violence; Discrepancy between urban rich and rural poor, 'traditional' and modern institutions; and Subversion of administering organizations and diversion of resources by populist politicians. Partial success Ecosystem services emerging as an important factor in rural development of upper catchments.	Mulligan *et al.*, *El País* (2010)
Indus-Ganges	Failure Fragmentation of water resource management in the Ganges, and low water productivity in the Indus; and Fragmentation of water and agricultural support. Encouraging trend Transboundary issues are high on the political agenda, including downstream impacts on Bangladesh; and Impacts of Punjab development on the Indus.	Sharma *et al.*
Karkheh	Modest success Strong rural development policies. Development issue Aspiration to food self-sufficiency may over-stress the water resource.	Ahmad and Giordano
Limpopo	Failure Legacy of past colonial influences on sharing of land and water resources – high levels of inequity between privileged minority and the vast majority; Complexity of institutional landscape at regional, national and local scales; and Weakness of institutions and organizations at the broader scale. Success In contrast with the failures, traditional management systems, which are institutions at the local scale, are robust.	Sullivan and Sibanda
Mekong	Failure Mismatch between 'strong' and 'weak' development trajectories; and No binding trans-boundary water management treaty. Potential failure Transboundary impacts of hydropower development on people whose livelihoods rely on the aquatic environment.	Kirby *et al.*
Niger	Failure Legal plurality between traditional and 'modern' institutions; and	Ogilvie *et al.*

Table 1. (*Continued*)

	Conflict between nomadic pastoralists and sedentary croppers Organizational constraint Strongly compartmentalized organizations and multi-factor constraints hinder development.	
Nile	Failure For most countries, while organizations formally have clear roles, in practice most are weak and ineffectively linked. Potential failure Transboundary tensions over sharing of water resources combined with examples of cooperative investments. Development issues Strongly contrasting interests of hydropower, irrigation, rainfed cropping, pastoralists and fishers; and Need for organizations to reduce vulnerability of the most sensitive ecosystems and communities.	Awulechu *et al.*, Haileslassie *et al.* (2009)
Volta	Failures Legalistic duality between traditional and 'modern' institutions such as land tenure; and Differential growth patterns of commercial and smallholder farmers. Modest success Transboundary management assisted by Volta Basin Authority.	Lemoalle and de Condappa
Yellow	On-going problems Closed basin. Successes Strong control of river flows through the Yellow River Conservancy Commission; Emergence of ecosystem trading as a potential solution; and Strong support for small-scale farmers.	Ringler *et al.*

As examples from the basin case studies in this book, "[A] critical lack of a cohesive, transparent social and [organizational] context conducive to agricultural investment" is identified by Ogilvie *et al.* as a key problem for agriculture in the Niger basin. In much of the Volta, "[F]ormal, national institutions [and organizations] coexist with the traditional structures of hierarchy. ... Decisions regarding land tenure and access to land and water, for example, are often made by the traditional authority, which reduces security of tenure, and can actively discourage investment in technologies that improve land and conserve resources ...". (Lemoalle and Condappa); and in the Nile, "[T]here are numerous ... [organizations] involved in the management of water and agriculture across the basin, [but] there is still a dire need for improved human and [organizational] capacity to implement programmes for the benefit of the rural poor" (Awulachew *et al.*).

The function of organizations can help identify the role for interventions

In considering the role of governance and organizations in enabling the benefit from ecosystem services (ESs) we see three broad types of function, which work at different scales.

(1) *Enabling gain from ESs*: Organizations enable the assembly of food production systems through the institutions of secure land and water rights, financial institutions to support markets and supply chains. Supply chains enable value adding. Energy production is enabled by secured rights, financing and distribution. This function operates at a national to global scale.

(2) *Enabling democratic distribution of ESs and the benefits from them*: Both institutions and organizations determine who is entitled to resources and the benefits that they provide. Stable institutions and organizations enable the development of markets and trading systems. A goal of water democracy is to establish transparent and informed systems to encourage equitable distribution. This function operates primarily at local to national scales.

(3) *Protecting essential or irreplaceable ESs*: Who protects environmental flows? Communities may preserve critical flood-control areas from agricultural or urban development. Others may conserve grazing land for drought emergencies. Increasing attention is devoted to the use of land for regulation of water flow and quality, sequestration of carbon and biodiversity conservation.

Because the requirements that have to be managed to make ESs work are closely-linked and operate at very different social scales, it is very difficult to set up proper governance, institutions, and organizations. In other words, the scales appropriate to natural processes do not match the scales of the social processes needed to support them. Moreover, administrative boundaries that delimit institutional authority rarely match natural boundaries, leading to anomalies.

Demand for these functions varies widely between basins (Table 2). They require different capabilities and inputs, for example, increased gain (Type 1) might be triggered by a technological change such as new germplasm, embedded in a new production system, whereas improved distribution (Type 2) depends almost entirely on social and political change. Valuation of regulating ESs requires clear scientific insight of these processes, many of which are difficult to assess. They may be protected by regulation such as prohibiting building on flood zones or promoting trading such as permitting sale of water rights.

The types of interventions therefore track closely the variation in the function of organizations. Agricultural research has traditionally focused on technology to improve productive capacity, for which organizations are needed to support technological development (e.g. plant breeding centres) and to transfer the technology such as extension services. Attention is now shifting towards a more systemic approach to include improvements in supply chains or market functions around production technology, which really move them towards Type 2 institutions. Interventions to improve the function of organizations that support sharing of resources or benefits include information and communication technologies. Interventions to support organizations protecting essential ESs include quantitative modeling to support informed dialogue and agreement over water.

Andes

Some gain is already in progress, but there are critical issues about sharing of natural resources, particularly realistic valuation of the ESs that support urban water supply, and protecting resources in the face of land use change, mining, which is often illegal and therefore uncontrolled, climate change, and urbanization.

Table 2. Demand for the three functions described above by basin. The summary is discussed in more detail in the following text. *** = high, ** = moderate, * = low, – = little or none.

	Andes	Indus-Ganges	Limpopo	Mekong	Niger	Nile	Volta	Yellow
Gain	–	**	***	*	**	***	***	–
Sharing	***	***	***	***	**	**	**	**
Protection	**	***	–	**	*	*	–	***

Indus-Ganges

Improved food productivity is essential but is not expected to come from the 'conventional' areas, where productivity is already at or near maximum, but new areas such as West Bengal, where productivity is currently very low. A critical institutional issue is political development to cope with unregulated water management and the hi-jacking of politics by vested interests.

Limpopo

The issue is to enable substantial gains in productivity, particularly in rainfed agriculture, while also enabling the institutional re-distribution of land and water rights from a pre-existing pattern.

Mekong

The major issue of the Mekong is to balance productivity gain from hydropower with the need to ensure institutional distributed rights to water by other users, and the protection of difficult-to-define ecosystem services.

Niger

The overriding need is for increased food security in rainfed agriculture, while also conserving institutional entitlements to land and water for a range of different communities (e.g. transhumant herders and sedentary croppers).

Nile

Improving rainfed agricultural productivity is an increasing focus in the Nile, while re-distribution of water rights between riparian countries is also a major interest.

Volta

The clear goal is to improve rainfed agricultural productivity through the institutional establishment of functional market-based systems, protection from water-related health hazards, and the development of secure land rights. At the same time, organizational distribution of water entitlements is managed at basin scale through the office of the VBA.

Yellow River

Agricultural productivity is at or near maximum attainable in parts and is moving towards value-adding, and non-farm income. Industry is increasing in importance and is attracting a greater share of water entitlements, with the introduction of the institution of trading systems. Additionally, the necessity of environmental flow was realized somewhat after the basin closure in the late 1990's and is now a prime objective of the YRCC.

Do governance and organizations adapt to meet the demands of their constituents?

From Table 1, it is clear that there are a number of different themes that emerge as important in any given context. A few themes are common across the basins:

- The need for transboundary river basin governance and organizational arrangements where upstream-downstream relations are critical, which is even important in the Yellow where provinces do not always accede to the dictates of the central government;
- Conflict between organizations managing different activities that are juxtaposed within the basin;
- The need for institutional and organizational support in the poorest basins to build livelihood systems and reduce vulnerability;
- The need for institutional resource and benefit-sharing mechanisms that recognize traditional functions while introducing more modern ones; and
- The emerging opportunity for communities to share ecosystem services equitably for the same or complementary uses.

We note that whilst it is conventional to think of organizations as ordered hierarchical entities within a system of governance, in many of the basins studied, no such clear-cut order exists at the community level. A competing concept of bricolage[1] has been suggested (Cleaver 2001) that describes a less structured reality. Whatever situation predominates on the ground, it is clear in virtually all cases that the activities of people transcend the bounds of systems that might be described as 'water', 'food' or 'development'. We therefore posit that both institutions and organizations proscribed as working exclusively in one or other of these categories is not only irrelevant, but actively discourages a more holistic approach to the problems that people in basins face.

Adaptation to development

Here we consider how institutions and organizations adapt as development takes place. The basic concept of development within river basins, as elsewhere, is that communities follow a development trajectory in which economic activity progressively expands, and with it, the demand for ESs. The pattern of use of ESs changes over time, with winners and losers. As demands for resources increase, this is reflected to some degree by the institutions and organizations that help moderate and direct behaviour. In many instances, problems of development relate less to the condition of natural resource endowments but to the way people manage, share, or otherwise alter them. Sometimes, usually because of lack of awareness of the consequences of their actions, people abuse the resources. Illegal mining in the Andes is a case in point (Mulligan *et al.* this book), although it is doubtful if the miners are entirely unaware of the pernicious consequences of their activities.

The basin reports earlier in this book describe three broad conditions of development:

- *Under-development*: in which poverty is correlated with low agricultural or economic activity. This condition occurs over much of Africa, but is also seen in parts of Asia and Latin America. Water and land scarcity is not an issue, although drought hazard may be. Economic water scarcity, as in the Limpopo, is another example;
- *Transitional development*: severe pressures during growth spurts in the absence of corresponding institutional and organizational developments. This condition occurs when the governments are unprepared for the pressures that development places on existing institutions and organizations. Illegal mining and downstream contamination in the Andes is an example.
- *Over-development*: in which the populations within basins and the activities that support them, agricultural, industrial and urban, have expanded to the point where river basins are stressed. This is clearly the case with the Yellow, and is a potential problem with the Karkheh and the upper parts of the Limpopo.

As societies develop, people use water and food systems in complementary ways that can be missed by an exclusive approach. An important example is the low water productivity of rainfed crops in the Niger, the Volta, by smallholders in the Limpopo, and in the upper riparian countries of the Nile. Consequently the broader perspective looks at a range of ESs provided by river basins on which people depend. As global population increases, ESs will come under increasing pressure and many commentators have observed situations in which river basin systems have been pushed to failure (e.g., Pierce 2007). The question is: Is breakdown inevitable? Can it be avoided by appropriate evolution in the institutions and organizations? We attempt to answer these questions with examples from the case-study chapters in the earlier part of this book.

Evolution of institutions and organizations

Institutions and organizations evolve as demand calls and resources permit. The most successful include the institution of 'water democracy', in which all legitimate actors are represented and given power in a transparent and informed process. Failure includes situations in which power inequalities are exploited, in which some groups, which includes the environment, are denied a legitimate voice. This is a problem in the South African part of the Limpopo where large (and efficient) commercial farms attempt to maintain their privileged position to the detriment of poor small holders elsewhere in the basin. In the Mekong, it is likely that development of hydropower dams will damage the livelihoods of those who depend on the Tonle Sap and other capture fisheries in the basin and threaten the fertility of the delta region in Vietnam due to reduced flow and invading sea water. The challenge is to include all stakeholders in the decision-making process and that benefits from development are shared equitably and to act according to the results of scientifically sound analyses of the potential gains and losses of interventions independent of the political gains an intervention, such as a big dam, might bring.

How development processes can lead to institutional and organizational failure

Imagine the condition that pre-dates economic development. Human population is zero. Under these conditions, stabilizing and regeneration of ESs dominate. As human populations grow, demand for food, water supply and sanitation increase. Land- and water-use changes modify processes, ultimately to the point at which such processes are compromised.

Industrial and agricultural development was promoted in the Yellow River until the end of the 1990s with scant regard for the ESs needs of the river downstream. The central government responded when the river failed to reach the sea for over 200 days in 1997, reallocating 4.8 km^3 from irrigation without direct compensation to the affected farmers for the loss of water, although subsidies were given to the provinces to carry out water-saving engineering projects, like canal lining. The Yellow River Conservancy Commission is charged with the overall management of the Yellow River, but intransigence of officials at the provincial and lower levels is often a barrier to effective management. Perverse incentives in which local officials, and the central government, seek to maximize yields rather than use water efficiently complicate the issue.

Egypt is by far the most developed country in the Nile basin. Together with Sudan, it invokes agreements from colonial times that did not consider upstream riparians, despite the upstream countries disavowing the agreements at or soon after independence (Salman 2010). The Nile Basin Initiative, created in 1999 as a mechanism to share the benefits of the Nile water, has had limited impact. In 2011, Ethiopia, Kenya, Rwanda, Tanzania, and Uganda signed a Comprehensive Framework Agreement, which Egypt and Sudan vigorously opposed. There are indications that Egypt may be prepared to soften its stance following the country's 2011 political changes and the creation of the new state of South Sudan.

In the Upper Limpopo in South Africa there is the juxtaposition of large, and efficient, commercial producers with subsistence smallholders in the former 'homelands'. The commercial producers have much greater economic and political influence than the subsistence smallholders so that the government finds it politically difficult to change the situation. Moreover, fundamental agrarian reform would likely be as economically damaging as it has been in neighboring Zimbabwe.

The green revolution of high yielding wheat varieties responsive to fertilizer brought about astonishing development in the upstream Ganges especially in Punjab. Punjabi farmers irrigate with pumped subsurface water for which the electricity is provided by the state free of cost. This perverse incentive has produced an unsustainable system with withdrawal from the aquifer far exceeding recharge. Without an institutional change to complete cost recovery, which farmers fiercely resist, the system is doomed to fail.

The Farakka Barrage, constructed by India 10 km upstream from the border with Bangladesh, diverts the Ganges from its original course through Bangladesh to the Hooghly River in India. India closes the barrage during the dry season so that no Ganges water reaches Bangladesh but opens it when the Ganges floods. In this case there is no transboundary organization and despite numerous attempts the issue is still unresolved.

In some basins, such as the Indus, the Nile, and the Yellow the basic river hydrology fails. As a consequence the river can no longer provide the downstream ESs necessary to maintain ecosystem integrity. The consequences include seawater penetration, loss of wetland habitat, accelerated siltation, and so on.

Conclusions: how institutions and organizations must adapt to support equitable development

The above examples provide some examples of factors that prevent organizations from doing what they should. We now attempt to look in more detail at what goes wrong and suggest some solutions.

What defines organizational failure?

It is easiest to define organizational failure by reference to an "ideal" system. Failure can then be determined as that difference between the ideal and the actual condition that it is reasonable to attribute to organizations. So what defines an ideal system?

An ideal river basin is where the basin sustainably provides maximum ESs to humanity. We emphasize that this is different from an undisturbed or pristine system, which is likely to be in a state of equilibrium. In contrast, providing for human needs disturbs the innate equilibrium of a system. The key is to devise systems that allow the ESs to satisfy human needs but allows them to continue to support other essential functions. The characteristics of an ideal system include:

- The sum of all ESs provides maximum utility;
- No harm at the local level: Negative local interactions are avoided (e.g. land productivity is not increased at the cost of aquatic ecosystem function);
- No harm (scale-independent). Negative temporal and geographic impacts are avoided. Long-term benefits are not sacrificed to acquire short-term benefits; nor are long-range benefits sacrificed to meet local needs;
- Within the constraint of 'no harm', exchange between sub-systems is not constrained.

These characteristics probably describe a Utopian system because we cannot provide an example of a basin that meets all of them. Nevertheless, we think it is useful conceptual framework that allows us to estimate quantitatively organizational and institutional success and failure.

Institutions and organizations hold the key

We view institutions and organizations as key to solving the water and food crisis caused by the increasing global population. Analysis of the detailed conditions within the nine basins reported in this book makes clear that the problem is not so much one of resource constraint (although it is in places), but one of resource use and distribution of rights and benfits. It is institutions and organizations that determine the benefit derived from use of ESs, the way the benefits and the resources are shared, and what penalties there are for abusing them.

The definition of success or failure of institutions and organizations

Hydrology, agriculture, and economics all provide a range of measures of success, none of which seem appropriate to determine the function of institutions and organizations as they influence those ESs that support livelihoods. Indeed, if we focus only on improving livelihoods, we are in danger of encouraging institutions and organizations to exploit ESs only to support human livelihoods without regard for the impact that might have on other components of the ecosystem. Unfortunately, that happens in many cases. Nevertheless, we need a description of success for which we propose a working definition: A successful institution or organization is one that obtains maximum collective gain from ESs while minimizing loss.

Identifying causes for success or failure

Although a detailed examination of whether any organization or institution is successful or not would undoubtedly identify a large number of contributing factors, we believe that they can be grouped into three broad causes:

- Ability of the organization or institution to adapt themselves as conditions change in response to development (e.g., as population demographics change, or as demands on resources become unsustainable, as in the overexploitation of subsurface water); and
- Congruence of the institution's or organization's activities with the biophysical conditions under which the target population obtains their livelihoods (e.g., the soils, landforms, rainfall distribution, probability of drought, probability of flooding, and so on).

Cork (2003) identifies a number of questions underpinning how institutions and organizations determine the use of ESs. Briefly they are:

- Who has rights to access the natural resources that provide the ESs? Are groups excluded? How are rights allocated? Can they evolve? These are questions about institutions, but are critical to the functioning of organizational arrangements.
- How are rights for use specified? What responsibilities are associated with the right to use?
- To what extent are rights connected spatially? Are rights in tune with the underlying dynamics of resource processes, stocks and flows?
- Can the institutions and organizations change to keep pace with changed demands?

The first two of Cork's (2003) questions address fundamental questions of equity that should be the cornerstone of any organization responsible for overseeing the distribution of ESs. There are many examples in the basin chapters where there are inequities ranging from the village level to internationally. The third question addresses sustainable use of ESs and concepts of integrated natural resources management. Once again there are examples in the closed basins (Yellow, Nile, and Upper Limpopo) and the overuse of subterranean water in both the Indus and the Ganges. The final question about institutional and organizational flexibility parallels our first and third bullet points above.

What goes wrong?

There are a number of common factors that we identify from an analysis of the nine basin chapters. We summarise them below.

- *Uncertainty*: the consequences of a particular action are not known, not quantified, and hence are not understood, which prevents informed agreement amongst parties.
- *Short-term*: long-term impacts are often of little interest to politicians. Those who recognize the need for a longer-term view (e.g. local communities) may be the least empowered.
- *Local interests only*: strong local concerns often override the need to minimize off-site impacts.
- *Organizational compartmentalization*: the relevant ministries (irrigation, agriculture, agriculture) do not talk together, and worse jealously guard their own turf.
- *Corruption*: lack of transparency or accountability leads to corruption, or at least the suspicion of corruption, which damages legitimacy.
- *Power inequalities*: obstruct stable, long-term agreement over shared resources. Examples are political violence in many of the Andean countries, and disenfranchisement of substantial groups in South Africa.
- *Lack of capacity*: in economies where agriculture is dominant, that is in the least developed countries, the level of investment in human as well as physical capital is insufficient to empower institutions and organizations to administer or regulate the use of ESs.

Final comments

We have attempted to identify the institutional and organizational strengths and weaknesses revealed in the nine basin studies in this book. Building on that analysis we have attempted to generalise on the factors that contribute to the success of institutions and organizations, or conversely what contributes to their failure. Unfortunately, weaknesses and failures appear to prevail. Of the factors that we identify, short-term views and local self-interest, in some places resulting in blatant corruption, are the most pervasive. Giordano and Wolf (2003) identify cases in which transboundary arrangements have been successful where nations have negotiated water treaties. These appear to depend on the political goodwill between the riparians involved. In contrast, the transboundary organizations reported here appear to have had little success, mainly because countries are unwilling to cede their sovereign power. In these cases it seems likely that they can only provide venues at which common interests can be discussed.

Institutional and organizational success occurs when they 'catch up' with demand. Often this process is obstructed by factors such as partial jurisdiction, uncertainty, local interests, power inequalities and lack of capacity.

Acknowledgements

We thank the CPWF for the opportunity to include this chapter in this book. We recognize that the Merrey and Cook (2011) paper, which focused on the 'bricolage' process, was the precursor of this chapter. We hope that we have fairly represented that paper in the body text, and we thank Dr Merrey for allowing us to refer to a pre-publication version.

Notes

1 "'Institutional bricolage' [is] a process by which people consciously and unconsciously draw on existing social and cultural arrangements to shape institutions in response to changing situations. The resulting institutions are a mix of 'modern' and 'traditional', 'formal' and 'informal'." (Cleaver 2001).

References

Cleaver, F., 2001. Institutional bricolage, conflict and cooperation in Usangu, Tanzania. IDS Bulletin, **32,** (4): 26–35.

Cork, S.J., 2003. The nature and value of ecosystem services in Australia. *In*: N. Allsopp *et al.*, eds. Rangelands in the new millennium. Proceedings of the VIIth International Rangelands Congress, 26 July–1 August, 2003, Durban, South Africa. Aitkenvale, Australia: International Rangelands Congress, Inc.

Daily, G.C., 1997. Introduction: what are ecosystem services? *In*: G.C. Dailey, ed. Nature's services: Societal dependence on natural ecosystems. pp.1–10.Washington DC: Island Press.

El País, Cali 2010. "El PIN maneja a la CVC y Acuavalle": Gobernador. 21 September, 2010.

Giordano, M. and Wolf, A., 2003. Sharing waters: Post-Rio International water management. Natural Resources Forum, **27**, (2), 163–171.

Haileslassie, A., Peden, D., Gebreselassie, S., Amede, T., Descheemaeker, K. 2009. Livestock water productivity in mixed crop–livestock farming systems of the Blue Nile basin: assessing variability and prospects for improvement. Agricultural Systems, **102**, (1–3), 33-40.

Menard, C. and Shirley, M.M. 2008. Introduction. *In*: C. Menard and M. M. Shirley, eds. Handbook of new institutional economics. p. 1. Berlin: Springer.

Merrey, D.J. and Cook, S., 2011. Progressing from research to improved livelihoods in developing country river basins: Fostering institutional creativity. Water Alternatives (in press).

Pierce, F., 2007. *When the rivers run dry*. London: Eden Project Books.

Salman, S. 2010. Downstream riparians can also harm upstream riparians. The concept of foreclosure of future uses. Water International, **35**, 350–364.

Salman, S., 2011. The new state of South Sudan and the hydro-politics of the Nile Basin. Water International, **36**, 154–166.

Svendsen, M., Wester, P. and Molle, F., 2005. Managing river basins: An institutional perspective. Chapter 1 *In*: Svendsen, M., ed., Irrigation and river basin management: Options for governance and institutions, pp. 1–18. Wallingford, UK: CABI and Colombo, Sri Lanka: IWMI.

Index

Page numbers in *Italics* represent tables.
Page numbers in **Bold** represent figures.

For Product Safety Concerns and Information please contact our EU representative GPSR@taylorandfrancis.com Taylor & Francis Verlag GmbH, Kaufingerstraße 24, 80331 München, Germany

Printed and bound by CPI Group (UK) Ltd, Croydon, CR0 4YY
01/05/2025
01858463-0001